WMU Studies in Maritime Affairs

Volume 7

More information about this series at http://www.springer.com/series/11556

Lawrence P. Hildebrand • Lawson W. Brigham •
Tafsir M. Johansson
Editors

Sustainable Shipping in a Changing Arctic

Editors
Lawrence P. Hildebrand
World Maritime University
Malmö, Sweden

Lawson W. Brigham
International Arctic Research Center
University of Alaska Fairbanks
Fairbanks, AK, USA

Tafsir M. Johansson
Word Maritime University
Malmö, Sweden

ISSN 2196-8772 ISSN 2196-8780 (electronic)
WMU Studies in Maritime Affairs
ISBN 978-3-319-78424-3 ISBN 978-3-319-78425-0 (eBook)
https://doi.org/10.1007/978-3-319-78425-0

Library of Congress Control Number: 2018944449

Printed on acid-free paper

This Springer imprint is published by the registered company Springer International Publishing AG part of Springer Nature.
The registered company address is: Gewerbestrasse 11, 6330 Cham, Switzerland

Foreword

As the President of the World Maritime University (WMU), it is my pleasure to introduce the seventh volume of the WMU Studies in Maritime Affairs book series published by Springer, titled *Sustainable Shipping in a Changing Arctic*. This series was launched in 2013 to encourage academics and practitioners from all areas of specialization across the field of maritime affairs to contribute to the expansion of knowledge through publications of the highest quality and market relevance. Previous books in the series include *Farthing on International Shipping* (2013), *Piracy at Sea* (2013), *Maritime Women: Global Leadership* (2015), *Shipping Operations Management* (2017), *Corporate Social Responsibility in the Maritime Industry* (2018), and *Trends and Challenges in Maritime Energy Management* (2018).

With this book series, the WMU aims to further develop capacity and expertise in maritime education and training; maritime law and policy; maritime safety and environmental administration; port management, shipping management, and logistics; maritime energy management; and ocean sustainability, governance, and management. The book series also serves as a platform for promoting and advancing the UN 2030 Agenda for Sustainable Development and the marine-related Sustainable Development Goals, particularly Goal 14 on oceans as well as the interconnected Goals 4 (quality education), 5 (gender equality), 7 (affordable and clean energy), 9 (industry, innovation, and infrastructure), 13 (climate action), and 17 (partnerships).

WMU is a postgraduate maritime university established in 1983 by the International Maritime Organization (IMO), a specialized agency of the United Nations. It aims to further enhance the objectives and goals of IMO and IMO member states around the world. The fundamental objective of the University is to provide the international maritime community, and in particular developing countries, with a centre for advanced maritime and ocean education, research, scholarship, and capacity building and an effective means for the sharing and transfer of technology from developed to developing maritime countries, with a view to promoting the achievement, globally, of the highest practicable standards in matters concerning maritime safety and security, efficiency of international shipping, the prevention and control of marine pollution, including air pollution from ships, and other marine and related ocean issues. We also facilitate the harmonization, uniform interpretations, and effective implementation of maritime Conventions and related instruments.

This book gives particular emphasis to one such recent instrument, the IMO Polar Code for ships operating in Arctic and Antarctic waters. The Polar Code came into force on 1 January 2017. It marks a historic milestone in the work of the IMO by specifically addressing the importance of protection of the polar environments and going above and beyond those of existing IMO Conventions such as MARPOL, SOLAS, and STCW.

WMU is also a platform for knowledge generation, exchange, and dissemination, through seminars, workshops, and major conferences. Indeed, the genesis of this book was a successful conference, co-convened by WMU, IMO, and the Arctic Council's Protection of the Arctic Marine Environment (PAME) Working Group in Malmö, Sweden, in August 2015, titled *Safe and Sustainable Shipping in a Changing Arctic Environment* (ShipArc2015). This conference brought together leading figures in Arctic climate change, polar shipping, maritime law, environmental protection, and Arctic marine policy.

Sustainable Shipping in a Changing Arctic comes as a response to a profoundly changing Arctic marine environment with expanding marine use. It aspires to become a one-stop read for all interested parties from both the maritime business sector and academia. The chapters are written by world renowned academics and practitioners, all experts in their subject area. The book covers broad areas that focus on safe and sustainable shipping in a changing Arctic, a highly relevant topic that requires integrative knowledge and technical expertise spanning various disciplines.

This edited volume addresses a fundamental gap in the contemporary literature on the maritime Arctic. It offers a vital reference guide for Arctic and non-Arctic states and those with an interest in the Arctic, including the regulatory community, governments, the shipping industry, natural resources industries, and nongovernmental organizations. This volume will also serve as a teaching supplement in academic and professional maritime programmes.

I invite you to read this book and I am confident that you will find it relevant and responsive to your needs to know more about the new maritime Arctic.

World Maritime University, Malmö
Sweden

Cleopatra Doumbia-Henry

Contents

Introduction to the New Maritime Arctic

Lawson W. Brigham and Lawrence P. Hildebrand

Contents

Abstract Fundamental changes continue to reshape the maritime Arctic. Globalization (the linkage of Arctic natural resources to global markets), profound climate change, regional and global geopolitics, and challenges to the Arctic's indigenous people are all drivers of a new era at the top of the world. The Arctic Council's Arctic Marine Shipping Assessment released in 2009 continues to be a key, policy framework of the Arctic states for protection of Arctic people and the marine environment. An International Maritime Organization (IMO) Polar Code ushered in on 1 January 2017 a new era of governance for commercial ships and passenger vessels sailing in polar waters. Current Arctic marine commercial traffic is dominated by destinational voyages related to natural resource development, particularly along Russia's Northern Sea Route. New Arctic marine operations and shipping are emerging, but significant challenges remain including: effective implementation and enforcement

The research and editing of this chapter and volume was supported by U.S. National Science Foundation grant award 1263678 to L. W. Brigham and the University of Alaska Fairbanks.

L. W. Brigham (✉)
International Arctic Research Center, University of Alaska Fairbanks, Fairbanks, AK, USA
e-mail: lwbrigham@alaska.edu

L. P. Hildebrand
Ocean Sustainability, Governance and Management, World Maritime University, Malmö, Sweden
e-mail: lh@wmu.se

L. P. Hildebrand et al. (eds.), *Sustainable Shipping in a Changing Arctic*, WMU Studies in Maritime Affairs 7, https://doi.org/10.1007/978-3-319-78425-0_1

of the IMO Polar Code; a huge gap in Arctic marine infrastructure (hydrography and charting, communications, emergency response capacity, and more); enhancing the monitoring and surveillance of Arctic waters; the challenge of developing a set of marine protected areas and additional Polar Code measures for the circumpolar region; and, the need for large public and private investments, as well as potential public-private partnerships in the Arctic. Cooperation among the Arctic states, the non-Arctic shipping states, and the global maritime enterprise will be critical to effective protection of Arctic people and the marine environment, and developing sustainable strategies for the region.

Keywords Arctic marine traffic · Infrastructure · Polar Code · Arctic Marine Shipping Assessment · Arctic Council · Marine safety · Environmental protection

1 Introduction

Fundamental changes have come to the maritime Arctic early in the twenty-first century. The Arctic Ocean's sea ice cover is undergoing a profound transformation in extent, thickness and character (witness the transition from multiyear ice to dominant first year ice). The Arctic is becoming more integrated with the global economy through development of its vast natural resources including not only oil and gas, but also a suite of hard minerals such as cooper, nickel, palladium, zinc and more. These economic developments require new marine transportation systems that can operate safely and effectively in ice-covered waters. While the Arctic states have cooperated closely within the Arctic Council since its establishment in 1996, outside geopolitics and the involvement of many non-Arctic states in Arctic affairs have created a far more complex situation than at any time after the Cold War. Also, the process of the delineation of the outer continental shelf in the Arctic Ocean is simultaneously underway as provided for in the United Convention on the Law of the Sea (Article 76 of UNCLOS). The voices of the Arctic indigenous peoples are also being heard more clearly in the Arctic states, the Arctic Council, and international bodies such as the United Nations and the International Maritime Organization (IMO). Within this complex of changes is a continuing global process to establish an integrated system of rules and measures to govern Arctic marine safety and environmental protection. One of the central challenges is to preserve the basic principles of freedom of navigation (established by UNCLOS and customary law) with the rights of the Arctic indigenous people and the overall sustainable development of the region.

An important issue to address is how 'Arctic shipping' should be defined in this volume and within the larger community of Arctic Ocean users, regulators, and stakeholders. A narrow approach would be to focus solely on trans-Arctic voyages, potential shipping routes (although seasonal) across the Arctic Ocean, that have been promoted in the global media. However, most voyages today and those in the future are considered destinational in that a ship enters the Arctic, perhaps delivers and loads cargoes, and then sails out without transiting between oceans. Most Arctic tourist cruises are also destinational in taking passengers to locations such as Svalbard, Greenland and the Canadian Arctic. A more holistic approach, used in the Arctic Council's Arctic Marine Shipping Assessment (AMSA), is to include all vessels (100 tons or more was chosen in AMSA) that operate in the Arctic and are discharging into Arctic waters and releasing emissions into the lower atmosphere. Recent approaches in the Arctic Council use the more inclusive term 'Arctic marine operations and shipping' which includes all sectors such as fishing and especially offshore development, which normally requires an armada of support vessels around drilling rigs (PAME 2013). The main issue is that approaches need not be constrained by the perceptions and possibilities of trans-Arctic navigation. Using a broad definition such as 'marine operations and shipping' includes all vessels and better represents the future levels of Arctic traffic that can be used to develop measures to protect Arctic people and the marine environment (Brigham 2017).

2 Arctic Marine Shipping Assessment: Policy Framework

The AMSA 2009 Report released by the Arctic Council after approval by the Arctic Ministers (in Tromso, Norway at the Arctic Council Ministerial meeting in April 2009) remains an influential, strategic document. AMSA should be viewed in three key perspectives:

- As a baseline assessment of Arctic marine operations across all sectors using the AMSA database collected from the Arctic states and as an historic snapshot of Arctic marine use early in the twenty-first century.
- As a strategic guide for the Arctic and non-Arctic states, the Arctic indigenous people, the global maritime industry, and a host of actors and stakeholders who have interests in the future of the maritime Arctic.
- As a policy document of the Arctic Council since the AMSA 2009 Report and its recommendations were approved in consensus by the eight Arctic Ministers after a lengthy negotiation process led by the Senior Arctic Officials and Arctic state representatives in the PAME Working Group.

The study report highlighted 96 findings that included such fundamental topics as: UNCLOS being the legal framework for Arctic ocean governance; the IMO being the appropriate international body for the Arctic and non-Arctic states to turn to for issues related to Arctic maritime safety, environmental protection, and security (the report noted all eight Arctic states are active and influential IMO members); as of the release of AMSA in 2009 there were no mandatory IMO safety and protection rules or standards for polar ships, only voluntary guidelines; despite the profound changes in sea ice in recent decades the Arctic Ocean remains ice-covered for much of the year, not ice-free; nearly all current Arctic marine traffic is destinational not trans-Arctic voyaging (the same is true for traffic in 2017); the most significant threat of ships to the Arctic marine environment is the release of oil from accidental or illegal discharge; the key drivers and uncertainties of future Arctic navigation are Arctic natural resource development (and trade) and the state of governance for ships operating in the Arctic Ocean; a large number of uncertainties influence the future of Arctic marine activities, most significant being global commodity pricing (from the AMSA scenarios creation effort); and, Arctic shipping can have both negative and positive impacts on the social, cultural and economic conditions in coastal communities (from the AMSA Town Hall meetings) (AMSA 2009). All of these findings have significant implications for new regulatory and governance requirements for an increased number of ships operating in polar waters.

The 17 AMSA recommendations negotiated by the Arctic states are included in three, inter-related themes: I. Enhancing Arctic Marine Safety; II. Protecting Arctic People; and, III. Building the Arctic Marine Infrastructure (see Appendix B). Addressing all three themes is critical to achieving enhanced marine safety and environmental protection throughout the Arctic Ocean. Although AMSA did not focus on investment and funding these recommendations, it noted all would require broad international cooperation and likely public-private partnerships. The Arctic marine safety recommendations involved the IMO, particularly a focus on mandatory rules and regulations for ships operating in polar waters and moving beyond voluntary rules. The second set of recommendations considered a range of protection strategies and greater engagement with coastal communities. An important recommendation called for surveys of indigenous marine use, critical information given integrated, multiple use management approaches in the future (one example is ecosystem-based management), and the designation of special Arctic marine areas. Other significant impacts addressed by the AMSA recommendations include: invasive species, oil spills, effects on marine mammal, and air emissions (AMSA 2009).

The third theme of AMSA recommendations focused on the critical importance and lack of marine infrastructure throughout most the Arctic marine environment (AMSA 2009). The Arctic lacks a host of infrastructure that is central to marine safety and environmental protection including: hydrographic data and adequate charting; environmental monitoring and forecasting (sea ice, weather and icebergs); SAR capacity; environmental response capacity; ship monitoring and tracking systems (situational awareness which is reviewed extensively in this volume); salvage; deep water ports and port facilities; aids to navigation; adequate communications; and, more. The Arctic states in AMSA recognized that each marine infrastructure need will require significant and committed long-term funding. Public-private partnerships could be established for joint funding of ports, ship monitoring and surveillance systems, and communications systems.

In summary AMSA remains for the Arctic states and Arctic Council a key, framework policy document. Status reports for implementation of the AMSA recommendations have been issued by PAME and the Council at Ministerial meetings in Nuuk (2011), Kiruna (2013), Iqaluit (2015) and Fairbanks (2017). AMSA is important to this volume as many of its recommendations speak to the sustainability of Arctic marine operations and shipping and the need for a holistic, integrated approach to marine safety and environmental protection.

3 Forces of Change

The most visible driver of Arctic change and the one that garners the most global attention is the rapid retreat of sea ice at the top of the world. Perhaps it is also the most misunderstood factor with regard to the possibility for increased shipping. Clearly there is greater marine access throughout much of the Arctic Ocean due to the relentless decrease in sea ice extent and thickness observed in all seasons. However, it is critical to note from a marine transportation perspective that the Arctic Ocean remains fully or partially ice-covered for much of the autumn, winter and spring and only in summer will there potentially be long periods of ice-free conditions. It is not that marine transportation cannot flourish in these ice-covered conditions, it is that ice class commercial ships (one of seven Polar Classes, see Appendix G) will likely be required for most Arctic marine operations, once the IMO Polar Code comes fully into force on 1 January 2018 for ships built prior to 2017 (on 1 January 2017 all newly built ships have already come under the Polar Code). Quantifying this new Arctic marine access from Global Climate Model sea ice simulations of the future, and determining the lengths of the ice navigation seasons (using Polar Class ship capabilities) are two of the current research challenges that can provide key strategic information for planning Arctic marine transportation systems.

Globalization of the Arctic—the linkages of Arctic natural resources to global markets—was identified in the AMSA scenarios creation effort as the primary driver of the need for marine transportation systems in the Arctic Ocean (AMSA 2009). Arctic natural resource developments driven by global commodities prices remain a paramount factor influencing Arctic marine operations and levels of vessel traffic. For example, this factor is driving increases in traffic along the Eurasian Arctic and Russia's Northern Sea Route especially to the new liquified natural gas (LNG) terminal at Sabetta on the Yamal Peninsula. Also, offshore hydrocarbon development in the Norwegian Arctic in the Barents Sea requires significant marine support activity to exploratory drilling. A second major factor indicated in the AMSA scenarios work is the governance of Arctic marine activity described as the degree of stability of international rules and standards for marine use within the Arctic Ocean and for the global oceans. Implied by governance is the need for a stable, effective operating system of legal and regulatory structures; UNLOS provides the over-arching legal framework for the Arctic Ocean and the IMO Polar Code is a new governance regime for commercial ships in polar waters. Recent treaties negotiated by the Arctic states addressing Arctic Oil Spill Preparedness and Response, and Arctic Search and Rescue (See Appendices C and D), add to the web of emerging Arctic Ocean governance required in the twenty-first century. Future issues such as the designation of marine protected areas, new emissions controls (including black carbon), the use of heavy fuel oil, and further measures for control of discharges and evasive species, will set a more sustainable path for Arctic marine use.

Increases in Arctic marine traffic are of great concern to the Arctic indigenous people especially those who live in coastal communities. The voices and rights of these indigenous people, who have lived in the Arctic for millennia and used Arctic waters as a critical part of their survival, are being heard more clearly by the Arctic states, the Arctic Council (where they have representatives and delegations as Permanent Participants), and international forums such as the UN and IMO. The Arctic states have the challenges of protecting their Arctic (indigenous) citizens during this period of historic changes in the region, and providing avenues of economic stability, while at the same time following strategies for sustainable development in their northern regions. In summary the 'new' maritime Arctic mandates that governments and industry foster greater communication and involvement of the Arctic indigenous people in decision-making to respond to their range of concerns and interests with regard to Arctic maritime affairs.

4 Current Arctic Marine Traffic and Key Routes

The map in Fig. 1 presents the general geography of the eight Arctic states and the key marine routes early in the twenty-first century. Notable are the two historic major waterways: the Northwest Passage (NWP), a set of routes through the straits of the Canadian Arctic Archipelago that link Baffin Bay and the Atlantic with Bering Strait and the Pacific; and, Russia's Northern Sea Route (NSR), a set of routes (defined in Russian Federation law) between Kara Gate at the southern tip of Zemlya to Bering Strait. All of the Russian routes are incorporated in the exclusive economic zone out to 200 nautical miles including the waters of the Barents Sea where there is significant Arctic marine traffic (but not part of the NSR). The map also indicates additional marine routes around both Greenlandic coasts, into Hudson Bay, around Alaska, to Svalbard and in the Barents Sea.

The NSR is the one waterway that is showing significant increases in destinational traffic mostly to the Yamal Peninsula where Russia is developing an LNG terminal at Sabetta (on the western side of the Ob Gulf). A second key port complex is Novy Port, an oil export terminal 190 miles south of Sabetta also on the western shore of the Ob Gulf. Year-round navigation to both ports can be achieved with icebreaker escort. A fleet of fifteen icebreaking LNG carriers, ships of 300-meter length and that can carry 170,000 m^3 of liquefied gas, will call at Sabetta and carry LNG out of the Russian Arctic to global markets. These LNG carriers can operate independently without icebreaker support in modest ice conditions, and sail year-round westbound on the NSR to Russian and European ports. During the summer navigation season (3–4 months) the same ships will sail eastbound along the NSR to Bering Strait and into the Pacific to Asian markets. The first ship of this class, named *Christophe de Margarie*, underwent successful ice trails in March 2017 and made an historic passage from Hammerfest, Norway to Korea carrying LNG in summer 2017. On 8 December 2017 the LNG facility at Sabetta was opened by President Putin and the *Christophe de Margarie* loaded its first LNG cargo there initiating a new connection to global markets via the NSR (Staalesen 2017).

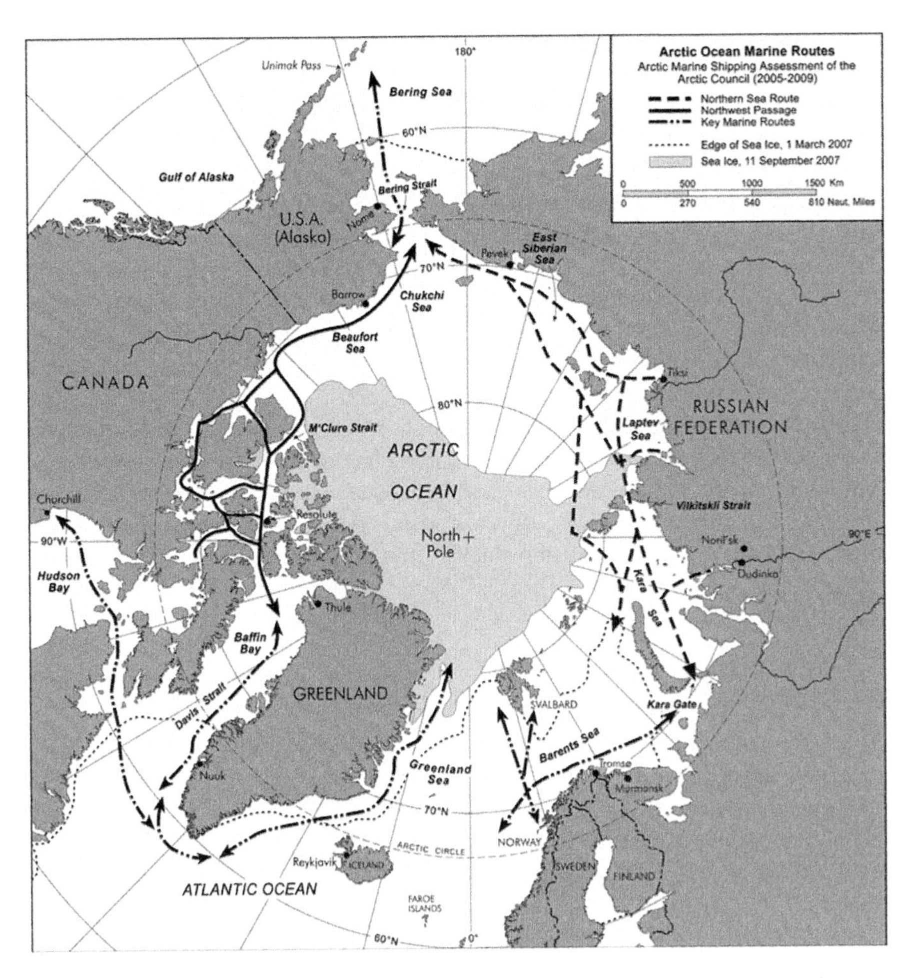

Fig. 1 Arctic Ocean marine routes. Source: L.W. Brigham, University of Alaska Fairbanks

It is also important to note that Arctic ship traffic has been maintained since 1979 from Dudinka, port city on the Yenisey River to Murmansk. This port links Norilsk, the industrial mining complex (world's largest producer of nickel and palladium) to the NSR and global markets. The marine transportation system consists of five, icebreaking carriers that do not normally require icebreaker escort. All of the aforementioned voyages are destinational. Full trans-Arctic voyages along the NSR have been less in number; during the 2011–2016 navigation season a modest 23 ships annually made a summer voyage across the full length of the NSR (Brigham 2016).

The NWP has experienced a modest growth of vessels making a full voyage from Atlantic to Pacific. However, only 290 vessels have made a complete voyage through the Northwest Passage (Pacific to Atlantic or vice versa) in its history from 1906 by the Norwegian Roald Amundsen in *Gjoa* to the end of 2017 (Headland

2017). During the past decade the vast majority of vessels making a full NWP passage have been private yachts and adventurers. During the 2017 navigation season 33 vessels made a complete NWP voyage; there were six notable voyages: an ice-capable commercial carrier, the *Atlanticborg*, carrying a cargo of aluminum from China to Quebec; the ice-strengthened cruise ship Bremen; the cruise ship *Crystal Serenity* with 2000 passengers; the Finnish icebreaker *Nordica*; the U.S. Coast Guard buoy tender *Maple*; and the Chinese icebreaking research vessel *Xue Long* (Headland 2017). However, this accounting of the (small) number of full NWP voyages does not reflect the majority of ship traffic in the Canadian Arctic. Most of the vessels in the region are sailing on destinational voyages primarily supporting Arctic communities and northern mines (for the export of bulk cargo) during the summer navigation season.

The Russian maritime Arctic and the offshore waters of Arctic Norway are the two regions which will likely witness increasing marine traffic in the decades ahead. The future of the North American maritime Arctic remains less certain and will plausibly experience increased traffic with rising global commodities prices. Arctic marine tourism may increase using smaller, polar expeditionary ships in summer.

5 New IMO Polar Code

The IMO International Code for Ships Operating in Polar Waters (Polar Code) is covered comprehensively in this volume. The importance and seminal nature of this new governance regime for polar waters cannot be over-stated. At its core the Code addresses marine safety and environmental protection issues for ships operating in cold, remote waters where maritime infrastructure is usually non-existent or very limited. However, the Polar Code is not a new IMO convention, but is a set of amendments to three IMO established instruments: the International Convention for the Safety of Life at Sea (SOLAS); the International Convention for the Prevention of Pollution from Ships (MARPOL); and, the International Convention on Standards of Training, Certification and Watchkeeping for Mariners (STCW). The central goal of the new Code is to create a uniform and nondiscriminatory set of rules and regulations for polar marine operators (Brigham 2017). The Code also includes a set of unified requirements seven Polar Class ships developed by the International Association of Classification Societies (see Appendix G for the Polar Class ship categories).

The Polar Code establishes binding or mandatory international standards for new and existing commercial carriers and passenger ships (500 tons or more) operating in Arctic and Antarctic waters. The Code covers a range of safety and protection issues: ship structural standards; required safety equipment; training and experience standards for the ship's officers and crew; and, environmental rules regarding oil, noxious liquids, sewage, and garbage. All of the maritime states, both flag and port states, have the challenge of implementing and enforcing the many elements of the Polar Code. The ship classification societies and marine insurance industry have key roles in evaluating the future risks of ships operating in polar waters and implementing these new uniform, international standards. The flag states will need to develop a process for issuance of the mandated Polar Certificate and foster the development of the ship-specific Polar Water Operational Manual which is now required for ships voyaging in polar waters.

The IMO Polar Code is only the beginning of a long process to further protect polar waters in an era of increasing polar marine operations. The IMO Polar Code is not comprehensive in that it does not in its initial version address such issues as black carbon, heavy fuel (in Arctic waters), ballast water discharges, an IMO emissions control area for the Arctic, and perhaps designation as a Particularly Sensitive Sea Area (PSSA). These issues will surely be addressed by IMO in the years ahead. During the 30th IMO Assembly in late 2017, a Polar Code 'Second Phase' was discussed which would address the issue of fishing vessels and smaller ships (under 500 tons and not covered by SOLAS) being included under the Code's requirements (IMO 2016).

6 Chapter Themes and Issues

This volume is focused on a broad set of challenges and issues related to sustainable shipping in a future Arctic, a region experiencing extraordinary change. The new IMO Polar Code provides a critical governance framework for polar operations of commercial ships and is comprehensively reviewed in the early chapters by IMO experts and academic scholars. Arctic ship monitoring and tracking is a fundamental element of infrastructure to provide effective enforcement of the Polar Code and other measures of safety and environmental protection. Tracking of Arctic ships by AIS is also a measure of prevention and enhanced safety as indicated in the chapters that provide a review of the latest uses of this technology to obtain a better understanding of real-time Arctic marine traffic patterns. Key chapters on Arctic Governance review the important legal implications of marine insurance and Arctic shipping, a look at the governance of biodiversity in the central Arctic Ocean, and how non-Arctic states view governance in their national Arctic strategies.

Effective measures of marine protection and emergency response capacities in the Arctic environment are critical requirements. Identifying potential marine protected areas (MPAs) and developing integrated strategies for Arctic oil spill response, especially in the remote central Arctic Ocean, are addressed in two comprehensive chapters. A review of the interactions of marine traffic and coastal communities of the Bering Strait region is a valuable chapter in that it presents the very real impacts increasing marine operations can have on local communities. Marine training (for example, ice navigation and emergency response) and capacity building are significant needs throughout the Arctic. Select requirements are covered in a set of chapters that includes a review of an industry research program on Arctic oil spill response. Sustainable Arctic business practices for offshore oil and gas in ice-covered waters are presented in a key review focusing on issues of resource allocation and operational management challenges. In summary, the chapters together represent a diversity of maritime challenges and issues and highlight the complexity of responses to greater use of Arctic waters and coastal environments.

7 Challenges and the Future

One of the interesting developments in Arctic affairs is that the response to increased marine operations and shipping in the Arctic has driven greater international cooperation among the Arctic states (in the Arctic Council) and within the IMO and other relevant bodies (Brigham 2011). The binding agreements of the Arctic states on Arctic search and rescue (2011) (Appendix D), and Arctic oil spill preparedness and response (2013) (Appendix C) indicate a strong willingness to reach consensus on practical maritime issues of near-term importance. The Arctic Council's AMSA represents a key policy framework and strategic guide that outlines the way forward by a unified group of Arctic states in protecting Arctic people and the marine environment. AMSA also showed the complexity of the drivers of future Arctic navigation in its set of plausible scenarios for the future (AMSA 2009). For the IMO, the marine insurance industry, and the global shipping enterprise, the Polar Code represents a new regulatory regime for polar waters and importantly, a set of uniform, non-discriminatory standards. However, the Polar Code presents a host of policy and practical challenges in its implementation as well as enforcement by the flag and port states. While the Polar Code is a seminal advance in governance of polar waters, the continued gap in maritime infrastructure (in hydrography and charting, aids to navigation, communications, salvage, port facilities, and more) hinders robust Arctic development (World Economic Forum 2014).

The current and future governance and regulatory instruments in the Arctic will require a continued close relationship between the Arctic Council and IMO, and consistent communication and involvement of the Arctic indigenous peoples. Information and data sharing among the Arctic states, indigenous groups, the maritime industry and all stakeholders must become the norm to achieve a greater understanding of the Arctic environment increasingly under profound change and stress. One of the key challenges will be for this diverse community of players to develop a more common understanding of what 'sustainable development' means in the context of increasing maritime use of the Arctic Ocean. The chapters in this volume will serve to highlight these challenges and portray issues in how sustainability can be reckoned with increasing use while embracing effective protection of Arctic peoples and the marine environment.

References

AMSA. (2009, April). *Arctic Marine Shipping Assessment (AMSA)*. Arctic Council, second printing.
Brigham, L. W. (2011). The challenges and security issues of Arctic Marine transport. In J. Kraska (Ed.), *Arctic security in an age of climate change*. Cambridge University Press.
Brigham, L. W. (2016). The realities and challenges of Russia's Northern Sea Route. *Fletcher Security Review*, *3*(1), (The Arctic Spotlight, Winter 2016 Issue).
Brigham, L. W. (2017). The changing maritime Arctic and new marine operations. In R. C. Beckman et al. (Eds.), *Governance of Arctic shipping* (p. 3). Brill.
Headland, R. K. (2017). *Transits to the end of the 2017 navigation season of the Northwest passage*. Scott Polar Research Institute, Cambridge University.
IMO. (2016). *International Code for Ships Operating in Polar Waters (Polar Code)*. Consolidated text of the Polar Code.
PAME. (2013, May 15). *Arctic Ocean review final phase 20112-2013 Report*. Arctic Council.
Staalesen, A. (2017, December 8). New era starts on the Northern Sea Route. *The Independent Barents Observer*.
World Economic Forum. (2014, January 22–25). *Demystifying the Arctic*. World Economic Forum Global Agenda Council on the Arctic. Davod-Klosters (Switzerland).

Part I
The Polar Code and Beyond

The International Code for Ships Operating in Polar Waters (Polar Code)

Heike Deggim

Contents

Abstract The International Code for Ships Operating in Polar Waters, better known by its short name “Polar Code”, was adopted by the International Maritime Organization (IMO) in 2014/2015. The Code became effective on 1 January 2017 upon entry into force of the associated amendments making it mandatory under both the International Convention for the Safety of Life at Sea (SOLAS) and the International Convention for the Prevention of Pollution from Ships (MARPOL). The Polar Code

The views expressed herein are those of the author and do not necessarily reflect the views of the International Maritime Organization.

H. Deggim (✉)
Marine Environment Division, International Maritime Organization (IMO), London, UK
e-mail: HDeggim@imo.org

L. P. Hildebrand et al. (eds.), *Sustainable Shipping in a Changing Arctic*, WMU Studies in Maritime Affairs 7, https://doi.org/10.1007/978-3-319-78425-0_2

marks a historic milestone in the Organization's work to protect ships and people aboard them, both seafarers and passengers, in the harsh and vulnerable environment of the waters surrounding the two poles, and at the same time protecting those environments. This chapter gives an overview of the requirements of the Code with regard to maritime safety and marine environment protection, also addressing its place in the existing global framework regulating international shipping. Associated training and certification requirements for officers and crew serving on ships operating in polar waters, as have been included in the International Convention on Standards of Training, Certification and Watchkeeping for Seafarers (STCW), are also described. The chapter finally examines what more can be done to ensure the safety of polar shipping, taking into account on-going discussions at IMO.

Keywords Polar Code · Regulatory framework · SOLAS · MARPOL · Heavy fuel oil

1 Introduction

Trends and forecasts indicate that polar shipping will grow in volume and diversity over the coming years. Commercial shipping and tourism are increasing. So-called eco-tourists are drawn by the breath-taking beauty of the polar landscapes, the chance to encounter some unique wildlife, and the sheer majesty of glaciers and icebergs. For ships carrying commercial cargo, northern sea routes offer the chance to considerably reduce journey distances between Europe and the Far East and thereby save on fuel, workforce and other operational costs. Also, the opportunities presented by the energy and mineral resources located in the areas around the poles are both impossible to ignore and another source of increased maritime traffic.

The challenges these developments bring need to be met without compromising either safety of life at sea or the sustainability of the polar environments. It cannot be denied that economic development and increasing commercial activity in the polar regions are controversial topics. There is an understandable and instinctive reaction, shared by many, against opening up two of the world's last remaining wilderness areas to exploitation. But the reality is that we cannot turn back a rising tide. The fact is that commercial activity and economic development in the polar areas are increasing, and increasing rapidly. The real issue is not whether this is a good thing; it is how to meet these challenges without compromising either safety of life at sea or the sustainability of the polar environments.

IMO's role is to ensure that the ships, and the people on them, which do operate in Arctic and Antarctic waters are safe and that their impact on the environment is minimal. The safety of ships operating in the harsh, remote and vulnerable polar areas and the protection of the sensitive environments around the two poles has always been a matter of concern for IMO and measures that specifically address shipping operations in those regions have been in place for several years.

However, with more and more ships navigating in polar waters, IMO has moved to address international concern about the protection of the polar environment and the safety of seafarers and passengers with the introduction of the mandatory Polar Code, for ships operating in Arctic and Antarctic waters. It entered into force on 1 January 2017 and it is the single most important initiative to establish an appropriate international regulatory framework for polar shipping. It is particularly important to keep in mind that the Polar Code requirements, which were specifically tailored for the polar environments, go above and beyond those of existing IMO conventions such as MARPOL and SOLAS. All the extensive safety and environmental regulations included in these and other IMO conventions are applicable globally and will still apply to shipping in polar waters. However, the Polar Code adds an additional layer on top, specifically for ships operating in these areas.

Operating ships in polar waters presents unique challenges. Poor weather conditions and the relative lack of good charts, communication systems and other navigational aids can pose serious problems. If accidents do occur, the remoteness of the areas makes rescue or clean-up operations difficult and costly. Extreme cold may reduce the effectiveness of numerous components of the ship, including deck machinery and emergency equipment. Ice can impose additional loads on the hull and propulsion system. To address these issues, the Polar Code sets out mandatory standards that cover the full range of design, construction, equipment, operational, training and environmental protection matters that apply to ships operating in the waters surrounding the two poles.

The Polar Code represents a major achievement in IMO's work to promote safe and sustainable shipping in all regions of the world, including the most challenging and difficult, and provides a strong regulatory framework aimed at minimizing the negative impact of shipping operations on the sensitive polar regions. The development and adoption of the Code has been achieved with the full participation, in the relevant IMO technical bodies, of not just the IMO Member States but also international organizations in consultative status, representing the shipping and shipbuilding industries, environmental interest groups, equipment manufacturers, seafarers' training providers and those which make up the maritime infrastructure, such as port and harbour authorities, pilots and hydrographers.

2 International Regulatory Framework for Shipping in Polar Waters

The United Nations Convention on the Law of the Sea (UNCLOS), which sets out the legal framework governing the rights and responsibilities of States in their use of ocean space, contains special provisions for ice-covered areas in Article 234. It confirms that "coastal States have the right to adopt and enforce non-discriminatory laws and regulations for the prevention, reduction and control of marine pollution from vessels in ice-covered areas within the limit of the exclusive economic zone".

IMO, as a specialized agency of the United Nations, is the global standard-setting authority for the safety, security and environmental performance of international shipping. Its main role is to create a regulatory framework for the shipping industry that is fair and effective, universally adopted and universally implemented.

Polar shipping always had a place in the work of IMO. The SOLAS Convention includes special requirements relating to shipping in polar waters in chapter V (Safety of navigation), concerning the collection of meteorological data, the Ice Patrol Service in the North Atlantic, ice information and danger messages. The 2008 Intact Stability Code, mandatory under SOLAS, contains a chapter dedicated to icing considerations. Under MARPOL Annexes I and V, the Antarctic is designated as a special area, prohibiting any discharge into the sea of oil or oily mixtures from any ship and principally the disposal into the sea of all plastics and other garbage.

While specially developed guidelines addressing international polar shipping had been in place since 2002, the IMO membership agreed in 2010 that the time had come to develop a legally binding instrument in order to provide a more comprehensive set of requirements to deal with the increased interests and traffic in the polar regions as well as the unique safety, operational, environmental and search and rescue concerns peculiar to these areas.

3 International Code for Ships Operating in Polar Waters (Polar Code)

3.1 Background

IMO's work to address the challenges posed by the increase in commercial shipping and tourism in polar waters goes back to the early 2000s. *Guidelines for ships operating in Arctic ice-covered waters* (MSC/Circ.1056) were first issued in 2002. IMO then received a request from the Antarctic Treaty Consultative Meeting (ATCM) to extend the Guidelines to also cover ships operating in the Antarctic. The need for this extension was particularly emphasized by a much-published accident happening in November 2007: the sinking of the cruise ship **MV Explorer** off King George Island, Antarctica, resulting in her crew and passengers drifting for 5 h in open-top lifeboats in sub-zero temperatures before being rescued, luckily with no casualties other than the ship herself. The outcome could easily have been very different.

Further work revising the Guidelines followed and in 2009 the IMO Assembly adopted the *Guidelines for ships operating in polar waters* (resolution A.1024(26)) which covered both Antarctic and Arctic waters. These non-mandatory Guidelines set out additional provisions, beyond existing requirements of the SOLAS and MARPOL Conventions, deemed necessary to ensure appropriate standards of maritime safety and marine pollution prevention for ships operating in polar waters.

Calls for the development of a mandatory Polar Code followed shortly after the adoption of the Guidelines and in 2010 IMO agreed to a proposal from several Member States to develop an internationally binding instrument specifically for polar shipping.

3.2 *Status and Structure of the Code*

The *International Code for ships operating in polar waters* (Polar Code) was adopted during the 94th session of IMO's Maritime Safety Committee (MSC 94) in November 2014 (Introduction and Parts I-A and II-B concerning safety measures) and the 68th session of the Marine Environment Protection Committee (MEPC 68) in May 2015 (Introduction and Parts II-A and II-B concerning pollution prevention measures), together with associated amendments to SOLAS and MARPOL to make the new Code mandatory under the two conventions. The Code became effective on 1 January 2017, upon entry into force of the aforementioned SOLAS and MARPOL amendments.

When adopting the Code, MSC and MEPC agreed that amendments to the Introduction of the Code, mandatory and applicable to both Parts, shall be adopted by both Committees in consultation with each other, whereas amendments to Parts I-A and I-B will be adopted by the MSC only and amendments to Parts II-A and II-B by the MEPC only. While parts I-A (Safety measures) and II-A (Pollution prevention measures) are mandatory under SOLAS and MARPOL, respectively, parts I-B (Additional guidance regarding the provisions of the Introduction and Part I-A) and II-B (Additional guidance regarding the provisions of the Introduction and Part II-A) are of a recommendatory nature. A consolidated text of the Code[1] has been prepared and will be maintained by the IMO Secretariat.

Each chapter in the Code principally set out goals, functional requirements and regulations. The chapters address general issues (definitions, survey and certification, etc.); the Polar Water Operational Manual (PWOM); ship structure; stability and subdivision; watertight and weather-tight integrity; machinery installations; fire safety/protection; life-saving appliances and arrangements; safety of navigation; communication; voyage planning; manning and training; prevention of pollution by oil; control of pollution by noxious liquid substances; prevention of pollution by sewage from ships; and prevention of pollution by discharge of garbage from ships. Appended to the Code are the Form of Certificate for ships operating in polar waters (Polar Ship Certificate) including the Record of Equipment and a Model table of contents for the PWOM.

---

[1]http://www.imo.org/en/MediaCentre/HotTopics/polar/Documents/POLAR%20CODE%20TEXT%20AS%20ADOPTED.pdf.

3.3 Objectives of the Code

The Polar Code supplements existing IMO instruments in order to enhance the safety of ships' operations and mitigate their impact on the people and the environment in the remote, vulnerable and potentially harsh polar waters. The goal of the Code is to provide for safe ship operation and the protection of the polar environment by addressing risks present in polar waters and not adequately addressed by other IMO instruments. Consequently, the Code takes a risk-based approach in determining the scope of regulations and adopts a holistic approach in reducing identified risks. It applies as a whole to both Arctic and Antarctic, taking into account the legal and geographical differences between the two areas.

The Code should ensure that ships operating in the Arctic and Antarctic regions comply with a globally agreed set of standards, which aim to ensure high levels of safety and environmental protection, both in the event of an incident and during routine operations.

3.4 General Requirements

General requirements of the Polar Code applicable to both Parts I and II are contained in the Introduction to the Code which contains the sections Goal; Definitions; Sources of hazards; and Structure of the Code.

Of particular importance, since requirements applicable to the categories differ, are the definitions of Category A, B and C ships (Introduction, paragraphs 2.1 to 2.3) as follows:

Category A ship means a ship designed for operation in polar waters in at least medium first-year ice, which may include old ice inclusions.
Category B ship means a ship not included in category A, designed for operation in polar waters in at least thin first-year ice, which may include old ice inclusions.
Category C ship means a ship designed to operate in open water or in ice conditions less severe than those included in categories A and B.

Ice strengthening is required, in accordance with the polar class assigned, for ships of categories A and B, but not for category C.

3.5 Maritime Safety Related Requirements

The safety measures in Part I-A of the Polar Code apply to new ships constructed on or after 1 January 2017. Ships constructed before 1 January 2017 will be required to meet the relevant requirements of the Code by their first intermediate or renewal survey, whichever occurs first, on or after 1 January 2018.

Part I-A consists of 12 chapters the contents of which are briefly described in the following:

Chapter 1—General

Definitions and requirements concerning survey and certification (every ship to which the Code applies shall carry a valid Polar Ship Certificate), performance standards (Polar Service Temperature (PST) shall be specified) and operational assessment (procedures or operational limitations are to be established).

Chapter 2—Polar Water Operational Manual (PWOM)

Requirements to provide the owner, operator, master and crew with sufficient information regarding the ship's operational capabilities and limitations in order to support their decision-making process.

Chapter 3—Ship structure

Requirements to provide that the material and scantlings of the structure retain their structural integrity based on global and local response due to environmental loads and conditions.

Chapter 4—Subdivision and stability

Requirements to ensure adequate subdivision and stability in both intact and damaged conditions.

Chapter 5—Watertight and weathertight integrity

Requirements to provide measures to maintain watertight and weathertight integrity.

Chapter 6—Machinery installations

Requirements to ensure that machinery installations are capable of delivering the required functionality necessary for safe operation of ships.

Chapter 7—Fire safety/protection

Requirements to ensure that fire safety systems and appliances are effective and operable, and that means of escape remain available so that persons on board can safely and swiftly escape to the lifeboat and liferaft embarkation deck under the expected environmental conditions.

Chapter 8—Life-saving appliances and arrangements

Requirements to provide for safe escape, evacuation and survival.

Chapter 9—Safety of navigation

Requirements to provide for safe navigation.

Chapter 10—Communication

Requirements to provide for effective communication for ships and survival craft during normal operation and in emergency situations.

Chapter 11—Voyage planning

Requirements to ensure that the Company, master and crew are provided with sufficient information to enable operations to be conducted with due consideration to safety of ship and persons on board and, as appropriate, environmental protection.

Chapter 12—Manning and training

Requirements to ensure that ships operating in polar waters are appropriately manned by adequately qualified, trained and experienced personnel.

Additional non-mandatory guidance is contained in Part I-B and concerns the determination of the Mean Daily Low Temperature (MDLT); limitations for operation in ice; the assessment required in Part I-A, section 1.5, for operational limitations and procedures to be included in the Polar Ship Certificate; performance standards; contents of the PWOM; navigation with icebreaker assistance; development of contingency plans; equivalent ice class; personal and group survival equipment; radars and charts; limitations of communication systems in high latitude; operation of multiple alerting and communication devices in the event of an incident; location and communication equipment to be carried by rescue boats and survival craft; and operations in areas with marine mammals or of cultural heritage and significance (Fig. 1).

3.6 Marine Environmental Related Requirements

The pollution prevention measures in Part II-A of the Polar Code are largely operational, relating mainly to discharge requirements, and apply to all ships, both new and existing, in line with the application requirements of MARPOL. While the Code contains requirements additional to those provided by MARPOL Annexes I, II, IV and V, it was felt that there was no need to introduce additional requirements with regard to Annex III (Regulations for the prevention of pollution by harmful substances carried by sea in packaged form) and Annex VI (Regulations for the prevention of air pollution from ships) which were considered to be sufficiently comprehensive to include polar shipping.

Part II-A consists of 5 chapters, the contents of which are briefly described in the following:

Chapter 1—Prevention of pollution by oil

Prohibits any discharge into the sea of oil or oily mixtures from any ship in Arctic waters (already prohibited in Antarctic waters by regulation 15.4 of MARPOL Annex I) and stipulates that all cargo tanks constructed and utilized to carry oil and all oil residue (sludge) tanks and oily bilge water holding tanks shall be separated from the outer shell by a distance of not less than 0.76 m.

Chapter 2—Control of pollution by noxious liquid substances in bulk

Prohibits any discharge into the sea of noxious liquid substances (NLS), or mixtures containing such substances, in Arctic waters (already prohibited in Antarctic waters by regulation 14.8.2 of MARPOL Annex II).

Chapter 3—Prevention of pollution by harmful substances carried by sea in packaged form

Intentionally left blank in the Code. Requirements of MARPOL Annex III apply.

Chapter 4—Prevention of pollution by sewage from ships

Prohibits discharges of sewage within polar waters except when performed in accordance with MARPOL Annex IV, subject to additional specific requirements as set out in the chapter.

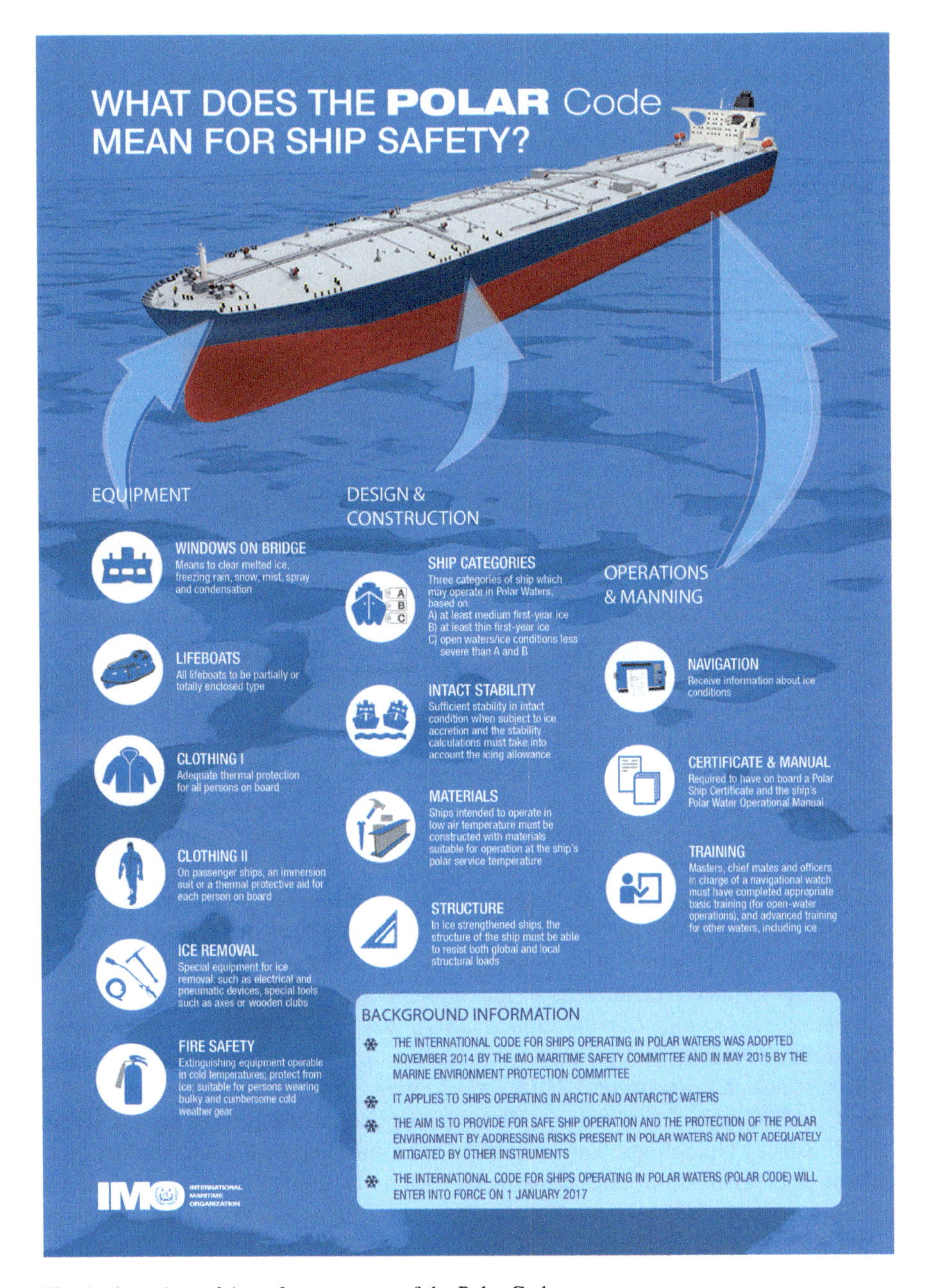

Fig. 1 Overview of the safety measures of the Polar Code

Chapter 5—Prevention of pollution by garbage from ships

Permits discharge of garbage into the sea in Arctic and Antarctic waters in accordance with regulations 4 and 6 of MARPOL Annex V, respectively, subject to additional specific requirements as set out in the chapter.

Additional non-mandatory guidance is contained in Part II-B concerning the requirements specified in chapters 1, 2 and 3; ballast water management; bio-fouling; and anti-fouling systems. In particular, ships operating in Arctic waters are encouraged to voluntarily apply regulation 43 of MARPOL Annex I (which prohibits the use or carriage of heavy fuel oils in the Antarctic area).

Figure 2 provides a brief overview of the pollution prevention measures of the Polar Code.

3.7 Related IMO Guidelines and Recommendations

To aid the development of the specific risk-based procedures (Part I-A, section 2.2) which need to be included in the PWOM required under the Code to be carried on board (Part I-A, paragraph 2.3.1), MSC 96 in May 2016 approved *Guidance on methodologies for assessing operational capabilities and limitations in ice* (MSC.1/Circ.1519). The Guidance is of an interim nature and intended to be reviewed four years after the entry into force of the Code, i.e. in 2021, based on experience gained with its application. It provides an example of an acceptable methodology for assessing limitations for ships operating in ice, the Polar Operational Limit Assessment Risk Indexing System (POLARIS), which has been developed incorporating experience and best practices from Canada's Arctic Ice Regime Shipping System, the Russian Ice Certificate supplemented by pilot ice assistance as prescribed in the Rules of Navigation for the Northern Sea Route; and other methodologies.

MEPC 70 in October 2016 and MSC 97 in November 2016 considered draft Polar Code-related amendments to the *Survey Guidelines under the Harmonized System of Survey and Certification (HSSC)*, 2015 (resolution A.1104(29)) and approved *Amendments to the Survey Guidelines under the Harmonized System of Survey and Certification, 2015, for ships operating in polar waters* (MSC-MEPC.5/Circ.12) which will be incorporated in the next update of the HSSC Survey Guidelines. The purpose of the Guidelines is the harmonization of survey and certification requirements of various IMO instruments, in particular the time periods for surveys so as to alleviate problems caused by survey dates, and intervals between surveys, which do not coincide.

MSC 97 also considered a request for clarifications on the application of the initial survey, maintenance surveys and certification, as required by paragraph 1.3 of chapter 1 of Part I-A of the Code and consequently approved *Unified interpretations of SOLAS regulation XIV/2.2 and paragraphs 1.3.2 and 1.3.6, Part I-A, of the Polar Code* (MSC.1/Circ.1562).

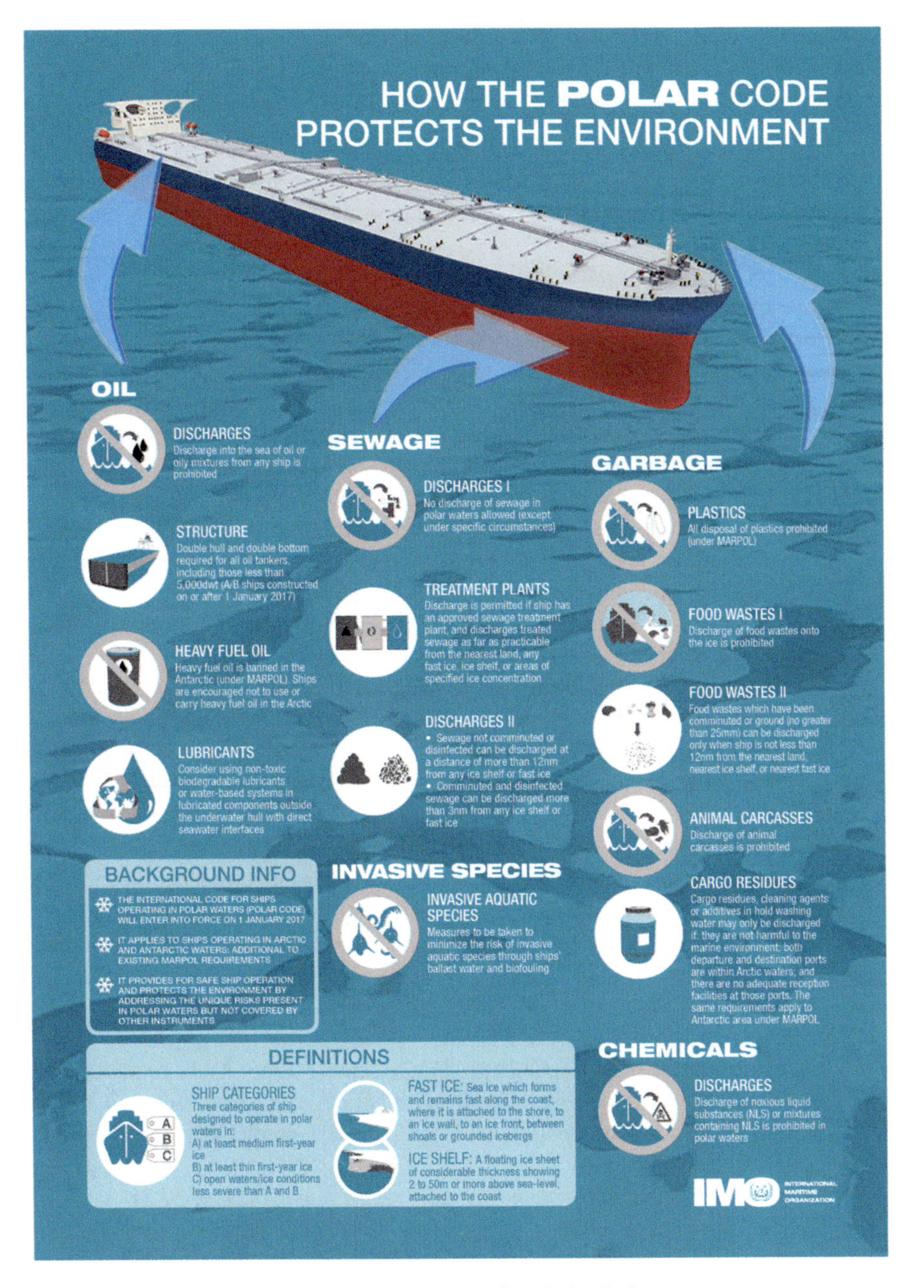

Fig. 2 Overview of pollution prevention measures of the Polar Code

Over the years, IMO also approved a number of other measures addressing directly or relating to polar shipping which include:

- *Pocket Guide for cold water survival*
  Regularly updated. Intended primarily for seafarers. Briefly examines the hazards of exposure to the cold that may endanger life and provides advice based on the latest medical and scientific opinion on how to prevent or minimize those dangers.
- *Enhanced contingency planning guidance for passenger ships operating in areas remote from SAR facilities*
  Approved in 2006. Requires that contingency plans for passenger ships for operating in areas considered to be remote from search and rescue (SAR) facilities should be prepared and that SAR co-operation planning arrangements should be enhanced for ships operating in areas remote from SAR facilities; and that the risks of remote area operation should be assessed and planned for.
- *Guidelines on voyage planning for passenger ships operating in remote areas*
  Adopted in 2007 in response to the growing popularity of touristic ocean travel and the desire for exotic destinations, which led to increasing numbers of passenger ships operating in remote areas. Detailed voyage and passage plans should include: safe areas and no-go areas; surveyed marine corridors, if available; and contingency plans for emergencies in the event of limited support being available for assistance in areas remote from SAR facilities; and additionally for ships operating in polar waters: conditions when it is not safe to enter areas containing ice or icebergs because of darkness, swell, fog and pressure ice; safe distance to icebergs; and presence of ice and icebergs and safe speed in such areas.
- *Guidance document for minimizing the risk of ship strikes with cetaceans*
  Approved in 2009. Provides guidance to Member Governments for reducing and minimizing the risk of ship strikes of cetaceans and sets out important general principles and possible actions that may be taken to reduce such risk. Encourages reporting of strikes to the global ship strikes database of the International Whaling Commission (IWC).[2]
- *Five new Arctic NAVAREAs/METAREAs*
  On "Full Operational Capability" (FOC) since 1 June 2011. Expanded the World-Wide Navigational Warning Service into Arctic waters, caused by the combination of increased business activity with less predictable, more extreme weather in the Arctic area.
- *Mandatory ship reporting system "In the Barents Area (Barents SRS)"*
  Adopted in 2012. Entered into force on 1 June 2013, requiring the following categories of ships to report to either Vardø VTS centre or Murmansk VTS centre: all ships of 5000 gross tonnage and above; where the tow exceeds 200 m; and any

[2]http://www.iwcoffice.co.uk/sci_com/shipstrikes.htm or by e-mailing the IWC Secretariat at: shipstrikes@iwcoffice.org.

ship not under command, restricted in their ability to manoeuvre or having defective navigational aids.

- *Guidelines for the reduction of underwater noise from commercial shipping to address adverse impacts on marine life*

  Approved in 2014. Provide guidance on reduction of underwater noise from the perspective of ship design (propellers, hull design, onboard machinery, technologies for noise reduction) and operation and maintenance (propeller cleaning, effective hull coatings, selection of ship speed, re-routeing and operational decisions to reduce adverse impacts on marine life).
- *Guide to oil spill response in ice and snow conditions*

  Approved in 2016. Identifies and describes those aspects of planning and operations directly associated with a response to a marine oil spill in ice and snow conditions anywhere in the world and assists managers and decision makers in recognizing and addressing key issues and potential response options on the strategic planning level.

Additionally, a large number of agreements, guidelines and recommendations for shipping in the Antarctic and Arctic have been developed by the Antarctic Treaty Consultative Meeting (ATCM)[3] and the Arctic Council.[4]

3.8 Further Work Related to the Polar Code

Once experience has been gained with the application of the Polar Code, it is anticipated that a second phase of work relating to the Code will commence, aimed at extending its application to non-convention ships operating in polar waters—this could include ships of a size below the application (tonnage) limits of IMO conventions, fishing vessels[5] and pleasure yachts. In this regard, MSC 97 in November 2016, having considered documents informing it of concerns regarding non-SOLAS ships operating in polar waters which do not fall under the requirements of the Code, reiterated its agreement that work related to the second phase, for non-SOLAS ships, should not begin until experience had been gained with the application of the Code to SOLAS ships.

In the meantime, following proposals from Member States, MSC 97 also instructed its Sub-Committees on Ship Systems and Equipment (SSE) and Navigation, Communications and Search and Rescue (NCSR) to review test and performance requirements for equipment on board ships certified to operate in polar waters, including, but not limited to, life-saving appliances, fire extinguishing media and navigation and radio-communication equipment.

---

[3]http://www.ats.aq.

[4]http://www.arctic-council.org.

[5]No internationally binding instrument regulating the safety of fishing vessels is currently in force.

NCSR 4 in March 2017, noting the general support for the reconsideration of performance standards for navigation and communication equipment in support of the implementation of the Polar Code, established a Correspondence Group and instructed it to develop a work plan listing all performance and test standards and requirements in need of revision in this respect; include the evaluation of specific additional conditions when approving navigation and communication equipment to be used for navigation in polar waters; consider interim solutions to address important matters at short notice; and consider alternative ways to address the work, such as the development of a separate consolidated performance standard, development of add-ons to existing performance standards, or a resolution. The group will report to NSCR 5, scheduled for February 2018.

SSE 4 in March 2017, having agreed to a plan that foresees the work to address additional testing and performance standards related to life-saving and fire-protection appliances and arrangements on board ships operating in polar waters to be completed at SSE 6 in 2019, also established a correspondence group to progress this work between meetings. The group was instructed to consider the evaluation of specific conditions when approving life-saving and fire-protection equipment for use in polar waters and develop relevant performance standards or add-ons to existing standards and will report to SSE 5, scheduled for March 2018.

4 Associated SOLAS and MARPOL Amendments Making the Polar Code Mandatory

The Polar Code is mandatory under both the SOLAS and MARPOL Conventions. Currently, SOLAS has 163 Contracting Governments and MARPOL 155 Parties and between them the two conventions cover 99.14% of the world's merchant shipping tonnage. The Polar Code applies to all ships operating in polar waters which have to comply with the requirements of SOLAS and MARPOL, typically ships over 500 gross tonnage and passenger ships carrying more than 12 passengers, operating internationally. This means that the Code applies, generally speaking, to all cargo ships, oil tankers, bulk carriers, container ships and cruise ships.

The two conventions have historically developed in a very different way and in a very different timeframe, with the earliest predecessor of SOLAS stemming from the year 1914 whereas the need for environmental shipping regulations only emerged much later, in the 1950s, with the increasing awareness of environmental issues in society. Consequently, the legal vehicles used to make the Code mandatory under the two conventions differ considerably.

SOLAS is generally regarded as the most important of all international treaties concerning the safety of merchant ships. It currently comprises 14 chapters, addressing a wide variety of ship-specific issues, from construction and stability over safety of navigation and radio communication to the carriage of cargoes and maritime security measures, to name just a few.

The Polar Code, i.e. its Introduction and Part I-A, was made mandatory under SOLAS by incorporating a new chapter XIV entitled "Safety measures for ships operating in polar waters" in the Convention, which was adopted on 21 November 2014 by means of resolution MSC.386(94) and entered into force on 1 January 2017.[6] The Code itself, which had been adopted by means of resolution MSC.385 (94), became effective on the date the new SOLAS chapter entered into force. Chapter XIV contains definitions; application provisions; requirements for ships to which the chapter applies; and provisions for alternative design and arrangements.

MARPOL is the main international convention covering prevention of pollution of the marine environment by ships from operational or accidental causes. The Convention consists of 6 annexes, each of which regulates a particular group of pollutants. The structure of the Convention necessitated separate amendments to each of its Annexes, as appropriate. The Code was therefore made mandatory by way of amendments to Annexes I, II, IV and V, while Annexes III and VI were considered to be sufficiently comprehensive to also cover polar shipping, with no need for additional requirements. Chapter 3 of Part II-A of the Code was left blank intentionally in order to keep the numbering of the chapters in this Part of the Code aligned with the numbering of the MARPOL Annexes.

The Polar Code, i.e. its Introduction and Part II-A, was made mandatory under MARPOL by amendments to Annexes I, II, IV and V which were adopted on 21 November 2014 by means of resolution MEPC.265(68) and entered into force on 1 January 2017. The Code itself, which had been adopted by means of resolution MEPC.264(68), became effective on the date the new MARPOL amendments entered into force. The amendments are structured along the lines of the corresponding SOLAS amendments and contain provisions for exemptions, waivers and exceptions; definitions; and special requirements in line with the subject of the respective MARPOL Annexes.

The geographical area definitions for the purposes of the Code are identical under SOLAS and MARPOL. "Polar waters" are defined as Arctic waters and/or the Antarctic area, with the latter being defined as the sea area south of latitude 60°S. "Arctic waters" are defined by a rather complicated description of geographical positions between Cap Kanin Nos (Russian Federation) and a point west of Greenland, completed by the circle formed at latitude 60°N and relevant coast lines. Figures 3 and 4 illustrate the area definitions which were developed specifically to define the application areas of the Code; other definitions are in use globally for different purposes.

[6] The time periods between adoption and entry-into-force are normally defined in the articles of conventions and differ between instruments. In accordance with the articles, the Committees may also choose different time periods at the time of adoption.

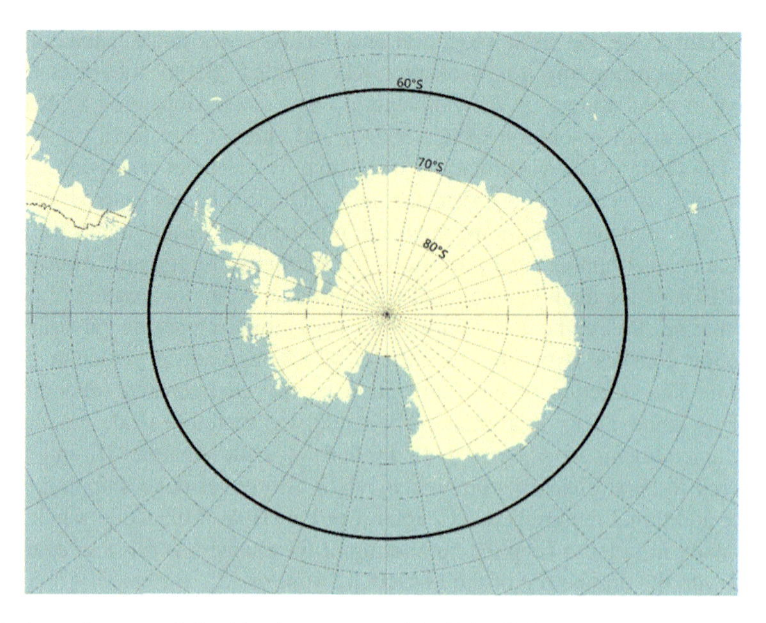

Fig. 3 Antarctic area as defined for the purposes of the Polar Code

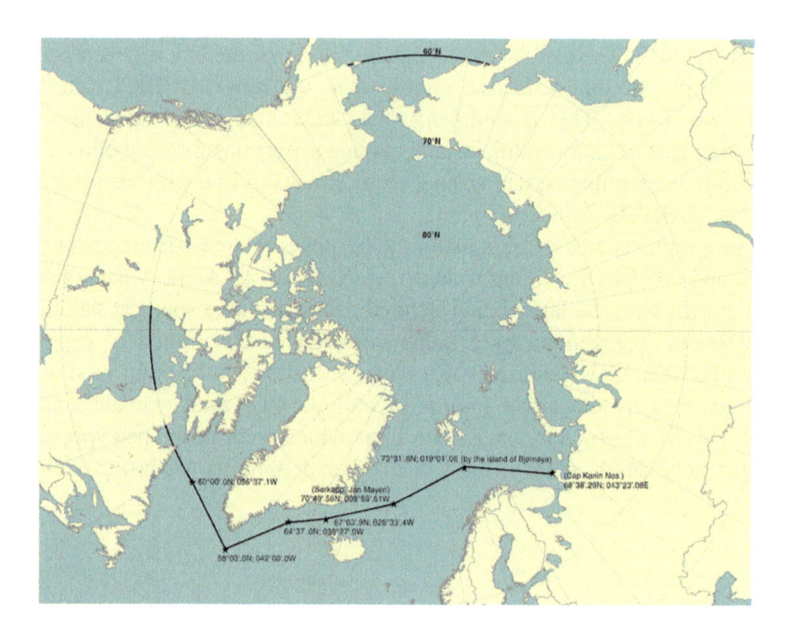

Fig. 4 Arctic waters as defined for the purposes of the Polar Code

5 Training and Certification Requirements for Officers and Crews on Ships Operating in Polar Areas

Mandatory minimum requirements for the training and qualifications of masters and deck officers on ships operating in polar waters were adopted by MSC 97 in November 2016 and are expected to become mandatory under the International Convention on Standards of Training, Certification and Watchkeeping for Seafarers (STCW) and its related STCW Code from 1 July 2018.

The requirements were developed by the Sub-Committee on Human Element, Training and Watchkeeping (HTW) and finalized at HTW 3 in February 2016. They include a new Regulation V/4 (Mandatory minimum requirements for the training and qualifications of masters and deck officers on ships operating in polar waters) for inclusion in the STCW Convention; as well as associated amendments to the STCW Code, including its Chapter V (Special training requirements for personnel on certain types of ships) concerning passenger management and a new section A-V/4 (Mandatory minimum requirements for the training and qualifications of masters and deck officers on ships operating in polar waters).

The new regulation V/4 of the STCW Convention requires that Masters, chief mates and officers in charge of a navigational watch on ships operating in polar waters shall hold a certificate in basic training for ships operating in polar waters and meet the standard of competence specified in section A-V/4, paragraph 1 of the STCW Code. In addition, Masters and chief mates shall also hold a certificate in advanced training for ships operating in polar waters, have at least two months of approved seagoing service in the deck department, at management level or while performing watchkeeping duties in an operational level, within polar waters or other equivalent approved seagoing service; and meet the standard of competence specified in section A-V/4, paragraph 2 of the STCW Code.

Section A-V/4 of the STCW Code sets out the detailed requirements for the required basic and advanced training in Tables A-V/4-1 (Specification of minimum standard of competence in basic training for ships operating in polar waters) and A-V/4-2 (Specification of minimum standard of competence in advanced training for ships operating in polar waters).

To assist Member States in complying with the new training requirements, HTW 4 in February 2017 validated two IMO model training courses: *Basic training for ships operating in polar waters and Advanced training for ships operating in polar waters,* which are expected to be published by the end of 2017. The purpose of the IMO model course programme is to assist and provide guidance to maritime academies and training institutes developing course programmes and syllabuses for seafarers seeking STCW certification.

6 Other Polar Shipping Related Issues

6.1 Objectives of the Code

MEPC's Sub-Committee on Pollution Prevention and Response (PPR) is currently considering the impact on the Arctic of emissions of Black Carbon from international shipping. The discussions are controversial and complicated due to widely differing views of IMO Member States on the matter.

However, PPR 2 in January 2015 agreed, and MEPC approved, a working definition for Black Carbon for international shipping, which is widely supported by the scientific community.[7]

Work continued at PPR 3 in February 2016 with the development of a draft measurement reporting protocol for Black Carbon, providing recommendations for the voluntary collection of Black Carbon data, including parameters for multiple measurement instrument technologies and a broad cross-section of current engine technologies, fuel types, and engine operating conditions. Member States were invited to use the reporting protocol and submit data to PPR 4, to facilitate its further refining.

PPR 4 in January 2017 considered a number of submitted results of Black Carbon measurements and again invited Member Governments and international organizations to use the protocol and submit further data and information derived from its application to its next session, in 2018. The objective is the identification of the most appropriate measurement method(s), focusing on fuel oils with a maximum sulphur content of 0.50% m/m, in light of the decision of MEPC 70 to confirm 1 January 2020 as the effective date of implementation of the global sulphur cap for ships' fuel oil. In between the meetings, a correspondence group is working on the finalization of the reporting protocol for voluntary measurement studies to collect Black Carbon data, based on experience to date of using the protocol, to improve usability and to address technical issues identified at PPR 4.

PPR 4 also agreed on a timeline for the completion of the work on Black Carbon emissions in the Arctic, envisaging that following finalization of the reporting protocol and identification of the most appropriate method for measurement of Black Carbon at PPR 5 (2018), appropriate control measures to reduce the impact of Black Carbon emissions from international shipping would be considered at PPR 6 (2019).

6.2 Use and Carriage of Heavy Fuel Oil (HFO) in the Arctic

MEPC 60 in March 2010 adopted a new regulation 43 (Special requirements for the use or carriage of oils in the Antarctic area) of MARPOL Annex I which prohibits

[7]Based on Bond et al. (2013).

the carriage in bulk as cargo, or carriage and use as fuel, of HFO in the Antarctic. The regulation entered into force on 1 August 2011. Furthermore, in 2014, amendments to regulation 43 were adopted, banning the use of HFO as ballast. These amendments entered into force on 1 March 2016.

There is currently no similar prohibition in place for the Arctic. The Polar Code only contains a recommendation in non-mandatory Part II-B which encourages ships to also apply regulation 43 when operating in Arctic waters.

Taking into account the potentially serious consequences of an HFO spill in the Arctic, this matter has since been raised by various Member States and organizations at MEPC meetings. However, for IMO to consider the introduction of any binding measures in this regard, a formal proposal from a Member State is necessary. Despite the frequent discussions and the serious campaigning by environmental organizations, to date no such proposal has been submitted for the consideration by MEPC.

In this connection it should be noted that MEPC 70 in October 2016, in a landmark decision for both the environment and human health, confirmed 1 January 2020 as the implementation date for a significant reduction in the sulphur content of the fuel oil used by ships. The decision to implement a global sulphur cap of 0.50% m/m in 2020 represents a significant cut to the 3.5% m/m global limit currently in place and is expected to substantially reduce the use of HFO globally.

6.3 Other Issues

There are a number of other issues related to shipping in polar waters which would merit careful consideration. These are matters that could be considered for inclusion in the Code directly or for principal regulation through the parent conventions. Any proposals to work on such matters would have to be considered and agreed by the IMO membership at large, also taking into account the agreement of the Committees that experience should be gained with the implementation of the Code before embarking on any amendments.

Matters for possible future consideration could include, but are not limited to, the following (listed in no particular order):

- establishment of an emission control area (ECA) with more stringent requirements for fuel oil used, for the Arctic in general or a specified area to demonstrate the need to prevent, reduce, and control emissions of NO_X or SO_X and particulate matter, or all three types of emissions from ships;
- extending the more stringent requirements that apply to the Antarctic as a Special Area under MARPOL Annexes I and V to the Arctic. Special Areas are defined under MARPOL as sea areas where, for recognized technical reasons in relation to their oceanographical and ecological conditions and to the particular character of their sea traffic, the adoption of special mandatory methods for the prevention of pollution from ships by oil or garbage is required. Under MARPOL, Special Areas are provided with a higher level of protection than other sea areas;

- further development and strengthening of the maritime infrastructure, in particular concerning the availability of port reception facilities;
- strengthening of search and rescue facilities under the Global Maritime Distress and Safety System (GMDSS) in polar areas;
- establishment of a comprehensive network of icebreaker support;
- addressing the unsatisfactory status of nautical charting for polar areas (according to information from the International Hydrographic Office (IHO), the chart coverage for Arctic and Antarctic areas at an appropriate scale is generally inadequate for coastal navigation and where charts do exist, their usefulness is limited because of the lack of any reliable depth or hazard information);
- ship incineration, including banning of it in ecologically sensitive areas or introducing a specified distance requirement from the ice-face and/or land;
- discharge of sewage through approved sewage treatment plants;
- control of discharge of grey water, i.e., the wastewater from galleys, showers, laundries, as well as food pulp, which could potentially cause harm to the environment due to concentrations of nutrients and other oxygen-demanding materials;
- measures to reduce underwater ship noise to minimize disturbance to marine life; and
- reduction and additional restrictions on ballast water discharges due to the great potential for major ecological impacts from species introduced via ballast water as ice cover recedes and seawater warms in polar areas.

7 Conclusion

There can be no doubt that the environmentally sensitive regions around the poles, two of the last remaining wilderness areas on earth, need to be protected and preserved. While guidelines to regulate international shipping in the Arctic and Antarctic have been in place for many years, with more and more ships navigating in polar waters IMO has moved to address international concern about the protection of the polar environment and the safety of seafarers and passengers on ships operating in these areas through the adoption of an internationally binding instrument, the mandatory Polar Code, which became effective on 1 January 2017 under the SOLAS and MARPOL Conventions. Its requirements, which were specifically tailored for the polar environments, go above and beyond those of existing IMO conventions which are applicable globally, including Arctic and Antarctic waters. The adoption of the Code is a major achievement in IMO's work to promote safe and sustainable shipping in all regions of the world, including the most challenging and difficult.

But this is not the end of the road. The Polar Code is a living instrument and will be under continuous review following experience gained with its implementation. Such reviews may result in amendments to its existing regulations and/or the guidance in its non-mandatory parts; but may also look at the introduction of completely new requirements, taking into account ongoing work at IMO as described in sections 6.1 and 6.2, as well as possible issues for further consideration as listed in section 6.3. The two IMO Committees involved in the development of the Code, MSC and MEPC, have already agreed that work on extending the applicability of the Code to non-convention ships, including fishing vessels and pleasure yachts, should commence in the not too distant future.

Acknowledgements I would like to acknowledge the support of my colleagues Lee Adamson and Tianbing Huang in the preparation of this chapter.

References

Bond, T. C., et al. (2013). Bounding the role of Black Carbon in the climate system: A scientific assessment. *Journal of Geophysical Research: Atmosphere, 118*, 5380–5552. https://doi.org/10.1002/JGRD.50171.

IMO MSC-MEPC.5/Circ.12. (2016). *Amendments to the Survey Guidelines under the Harmonized System of Survey and Certification, 2015, for ships operating in polar waters.* London: International Maritime Organization.

IMO PPR 3/22/Add.1. (2016). *Guide to oil spill response in ice and snow conditions, 2016.* London: International Maritime Organization.

IMO MSC/Circ.1056. (2002). *Guidelines for ships operating in Arctic ice-covered waters, 2002.* London: International Maritime Organization.

IMO Resolution A.1024(26). (2009). *Guidelines for ships operating in polar waters, 2009.* London: International Maritime Organization.

IMO MSC.1/Circ.1519. (2016). *Guidance on methodologies for assessing operational capabilities and limitations in ice, 2016.* London: International Maritime Organization.

International Convention on Standards of Training, Certification and Watchkeeping for Seafarers (STCW), 1978.

IMO Resolution A. 1104(29). (2015). *Survey Guidelines under the Harmonized System of Survey and Certification (HSSC), 2015.* London: International Maritime Organization.

United Nations Convention on the Law of the Sea (UNCLOS), 1982.

Arctic Maritime Safety: The Human Element Seen from the Captain's Table

Johanna Salokannel, Harri Ruoslahti, and Juha Knuuttila

Contents

Abstract The maritime industry is safety critical, where the element of uncertainty is present especially when entering high-risk shipping areas like the Arctic. The element of uncertainty increases, as the working environment gets more unpredictable and systems more complex. Unpredictability and complexity is making it difficult to define comprehensively and in advance which exact courses of action one should take when facing challenging ad hoc situations while navigating the Arctic. The human element is a vital part of successful and safe shipping in the Arctic. Recent resilience engineering and safety studies see the human element and their ability to adapt and adjust their performance to emerging situational needs, possible shortages in work descriptions and resources as a key to successful operations. High performance of the crew strongly contributes to the high performance of the ship where the captain plays a key role. This chapter addresses the safety issues in a

Juha Knuuttila was deceased at the time of publication.

J. Salokannel (✉)
Novia University of Applied Sciences, Aboa Mare Maritime Academy, Turku, Finland
e-mail: johanna.salokannel@novia.fi

H. Ruoslahti
Laurea University of Applied Sciences, Espoo, Finland
e-mail: harri.ruoslahti@laurea.fi

L. P. Hildebrand et al. (eds.), *Sustainable Shipping in a Changing Arctic*, WMU Studies in Maritime Affairs 7, https://doi.org/10.1007/978-3-319-78425-0_3

more holistic way including uncertainty and unpredictability as a part of safety management in the Arctic shipping.

Keywords Maritime safety · Human element · Shipping · Arctic shipping · Safety management

1 Introduction

The global climate change and melting sea ice has opened, at least for part of the year, new routes for shipping in the Arctic. The attractiveness of the new routes is in cutting the distance between the Pacific and Atlantic oceans enabling ships to save a considerable amount of time between ports, but navigating in the Arctic is not without risk. As a business case that saves time and resources on every journey makes the route attractive. But risks, if realized, might end up being more costly in increasing insurance premiums, damage to the ship, environmental pollution and putting crews at risk. However, regardless of the risks, the Arctic routes are an opportunity and shipping in the Arctic will most likely increase in the future.

In the maritime accident investigation, human error has been counted to be the cause in around 80–90% cases. Crews are still needed to sail ships and deal with daily challenges in high risk areas making the human element of a great interest from a safety management point of view.

This chapter focuses on the human element in the Arctic, where not all risks can be predicted and requirements for safety are high. Detailed descriptions of which action to take in occurring ad hoc situations are difficult, if not impossible to make. This means also that safety needs to be ensured with other means than just compliance to rules and/or best working practices. The objective of this chapter is to study how to ensure safety of navigation in the Arctic in situations where the exact risks are not known or when faced with safety critical situations that require rapid reaction and responding to when there are no sufficient instructions or experience to rely on.

The Arctic shipping routes are still unfamiliar and even there is information available, there is not necessarily that much experience amongst seafarers, if the routes open up for greater traffic. With time there will be experience based learning and shared information and lessons learnt. Before this happens, the safety of Arctic shipping needs to be created also with seafarers who most likely face unexpected and unpredictable situations. Compared to other high-risk areas like those with for example high traffic density or pirates, the exposure to the risks are usually much shorter compared to the Arctic where the journey can last around 2 weeks depending on the speed. Due to the remoteness of the area and lack of infrastructure, if something happens, help can be very far away.

2 Risks of Arctic Shipping

Working at sea, the seafarers are faced with various risk factors on a daily basis from harsh storms at sea to loading in ports. On the Arctic routes, risks are even higher. The extreme climate of the Arctic with its low temperatures, extraordinary light conditions and sudden storms as well as magnetic phenomena make the area quite distinctive. Navigating in the Arctic can pose challenges due to, for example magnetic compasses becoming unreliable at such high latitudes. GPS and GALILEO have reduced coverage, radio, satellite and communication signals are less reliable in such remote areas. Navigational charts and navigational information can be inaccurate and limited in number (Jensen 2007; Carpenter and Wyman 2014; Wright and Zimmerman 2015).

The International Maritime Organization (IMO), responsible for creating regulatory framework for international shipping industry, adopted the Polar Code (the International Code for Ships Operating in Polar Waters) in November 2014 and related amendments to the International Convention for the Safety of Life at Sea (SOLAS). This, however, does not solely ensure the safety of life, property and the environment while sailing in the Arctic. Restricted visibility due to fog and darkness, harsh weather, cold and violent storms still put serious demands on the crew. If a ship encounters difficulties, help, like Search and Rescue, repair and salvage services, can be very far away. In addition, if a crewmember is injured or becomes seriously ill, hospitalization poses challenges due to the remoteness of the Arctic routes.

As an example, the MV Clipper Adventurer cruise ship ran aground on 27th of August 2010 in the Coronation Gulf, Canada. Canadian Coast Guard dispatched on 28th the nearest icebreaker, Amundsen, to assist the ship that was 500 km away. The icebreaker was estimated to arrive at the scene at 09:00 next morning on 29th. Luckily, weather conditions were good, which made the ship stay safely stuck on a rock until help arrived (Stewart and Dawson 2011).

Not so fortunate, T/S Maksim Gorkiy, sailing from Iceland to Spitsbergen, collided with ice floe in heavy fog close to midnight between 19th and 20th of June, 1989. She was damaged as the ice ripped holes the hull, one of 10 meter long and some smaller ones to the bow. The ship started to sink. The first distress call was sent shortly after the accident. Even though a Norwegian Coast Guard rescue vessel, Senja, arrived at the scene within a few hours at 4:15 am, around 1.000 people had to abandon ship into lifeboats near freezing temperatures while 120 crewmembers stayed onboard fighting to keep the ship afloat (The New York Times; Marchenko 2015).

Floating ice poses also challenges for navigation. Small icebergs like growlers and bergy bits are difficult to detect with satellites and radar especially during rough weather as they are mainly submerged. Ice formation on deck and hatch covers can create problems for ship stability and deck equipment, which needs to be removed regularly. Entering an icy ship deck in darkness and harsh weather places the crewmembers at risk. Harsh conditions can also make the crew members more fatigue and affect daily work. Extreme cold can cause problems to the engine, fuel transfer and pumps needed for firefighting, which could freeze from excess water inside. Lacking or limited external facilities to repair breakdowns pose challenges and therefore many kinds of spare parts needs to be carried aboard. Whatever the situation or combination of the above mentioned and more, the crew is required to handle it. If help is far away, any small incident might escalate into bigger problems, therefore reaction time is of high importance. Any salvage operation in harsh, cold and dark weather will not be easy to complete. Therefore, having a qualified, well trained, and experienced crew becomes more important than ever (Carpenter and Wyman 2014).

The Polar Code recognizes that in the safe operation of a ship in the Arctic waters attention needs to be paid to the human element regarding their skills and knowledge. Therefore, all ships operating in polar ice-covered waters should carry at least one Ice Navigator. IMO defines the Ice Navigator as: "*any individual who, in addition to being qualified under the STCW Convention, is specially trained and otherwise qualified to direct the movement of a ship in the ice-covered waters*" (IMO 2010).

3 The Human Element and Human Error

The safety management of shipping has been focusing on unwanted outcomes, by investigating past accidents and predicting future risks and their probabilities. Naturally, it is important to understand what has gone wrong, and what could go wrong in the future, in order to create safety guards to prevent these from happening again, or to protect against their outcomes. However, the increase of automation and digitalization in our socio-technical systems has created processes and interconnections that are starting to be intractable making it both difficult to describe and predict all possible scenarios that might go wrong. (Hollnagel et al. 2011) Considering for example a ship that has interconnections from automated doors to navigational instruments and possible engine room systems that are connected to a remote service. The equipment can be installed at different times and new software integrated to old one. The complexity of these systems is making it difficult to know exactly to which all areas and how one failure affects. The complexity is also making it harder to detect failures.

When future risks are not entirely known, the element of uncertainty will enter the picture and managing it becomes of interest. The new Arctic shipping routes represent a high risk area with the element of uncertainty present, as all possible scenarios of what could go wrong cannot, at least not yet, be comprehensively predicted. Grote (2014 p. 71) writes, "*Uncertainty is at the heart of risk*". She argues that on top of acknowledging existing uncertainties one needs to understand that uncertainty cannot always be reduced completely. Therefore, uncertainty becomes a strategic question for a company dealing with risk management. With the understanding that uncertainty cannot be reduced entirely, maintaining a level of uncertainty, managing it and occasionally even increasing it, should be included in the decision-making of risk management in order to improve safety and pursue opportunity.

Aven and Krohn (2013) discuss probability and risk. They point out that probability is just one way to describe uncertainty, and that understanding risk should not be limited only to probabilities as it is too narrow view. They point out that when predicting the probability of a certain risk, a hazard or unwanted event, the probabilities can be the same for two different cases, but what is emphasized is the level of knowledge and data available regarding the phenomena. The "unthinkable" or the "unlikely" can be ignored due to assessments based on assumptions or beliefs, that these kinds of phenomena are not likely to happen. They argue, that a broader risk perspective is needed that go beyond the probabilities and avoiding only the undesired events. The perspective should also include how to improve performance with desired outcomes.

The safety of Arctic shipping, therefore, should not only rely on predicting risks and their probabilities, as the list will most likely be incomplete. Neither will investigating past accidents tell the whole truth about how to improve safety. To improve performance in order to improve safety requires understanding of what kind of performance to improve. Therefore, more research on what kind of performance leads to success, thus to better safety, is needed. To study shipping companies that have operated in the Arctic waters successfully could give maritime safety development very valuable input. Research on what makes them successful in the Arctic conditions, how their ships and crew manage critical situations and where their success originates from, surely will enlighten the safety development as much as studying past accidents and incidents.

Besnard and Hollnagel (2014) explain this idea quite well while arguing about some myths about safety and criticize the concept of human error. They give an example where a system is considered to be safe with a very low probability of failure of e.g. where 99,999 times out of 100,000, everything will go well. Then there is one unacceptable performance, an accident. The "human error" is considered to be the cause of that one accident, but attention is not paid to the 99,999 times where the same course of action has been a success. The human error has been seen to be the "cause" of the one unwanted event, but at the same time also humans are the source of the 99,999 times of success.

This principle applies to the Safety-II principles by Hollnagel (2014) and Resilience Engineering (Hollnagel et al. 2011), where the focus on improving safety should be on actions that go right as well as understanding what rarely goes wrong. Performance variability and the normal functioning of a system should be studied also in order to understand why the same behavior that usually goes right occasionally makes things go wrong.

Hollnagel (2009, 2014) argues also, that in the traditional way of looking at the human element and human error, humans are seen as the fallible component of a system like machines, where they either work as stated in work procedures or fail to follow them. The principle behind this is that written procedures are seen to be correct and the function of a system is predictable. Now, usually those who execute the work described in work process descriptions are often different from those who design the system, create and describe work procedures. The more complex the working environment becomes, the harder it will be to describe the work procedures and anticipate all conditions. Hence, the execution of work tasks exactly as described in the processes and work descriptions cannot always be done in all circumstances.

The decisions humans take in order to accomplish current work tasks, whether during normal operation or emergent disturbances, can often be based on limited resources like time, information, tools at hand etc. The decisions are made when if more time would have resulted in more information gained, lack of spare parts or tools at hand calling for improvisation or lack of manpower possibly relying on less qualified persons to do the work. The decisions and courses of action are done based on some level of uncertainty, hence the adjustments done in imperfect circumstances to complete work tasks are inevitably approximate. Reasons why performance mostly is a success is much the same as why performance at times may fail. People in general do not choose failure. Success to complete a task, due to incomplete work procedures and uncertainty, needs performance adaptability and variability. Therefore, this cannot be prevented in order to eliminate failures and hence, managing safety cannot only be by constraining daily work and decision making (Hollnagel 2009, 2014).

Best practices and written work procedures are important and the above does not exclude them. However, especially in high risk areas and when the element of uncertainty is present, merely following best practices and following rules, regulations and procedures does not always ensure safety.

Taking uncertainty into risk management and overall safety management of a ship means firstly that managing uncertainty will be part of strategic and operational decision making. In addition, the operation of a ship and its navigation cannot always be broken down into work procedures written in detail. Therefore, safety cannot be managed only by reducing uncertainty through standardization of work, routines, automation and stability. High level of routine, standardization and formalization requires that evolving events are predictable, that systems can be controlled and are tractable (Grote 2014).

4 The Element of Uncertainty

As uncertainty cannot be completely reduced from Arctic navigation, it should be included in the safety management and seen both as a positive and negative issue. Grote (2014) introduces a general framework to manage uncertainty where uncertainty is reduced, maintained or increased. In the traditional risk management, the objective is to reduce uncertainty, to stabilize, standardize and automate. Here control is centralized. Maintaining uncertainty again has the objective to be flexible and resilient towards uncertainty. It is understood that uncertainty cannot be completely reduced in complex environments. Leonhardt et al. (2009 p. 2) define Resilience Engineering: '*Resilience is the intrinsic ability of a system to adjust its functioning prior to, during, or following changes and disturbances, so that it can sustain required operations under both expected and unexpected conditions.*' The objective of maintaining uncertainty is that the system is tolerant to disturbances and can recover from them. In these kinds of environments, control needs to be decentralized, for example by empowering people.

Increasing uncertainty becomes relevant for example in the case of innovations and new ways of working, when better ways are sought and new ideas encouraged. When this happens, existing routines need to be left behind. Therefore, when innovation is needed in high-risk settings, uncertainty has to be increased, at least temporarily. Hence, stability and control will naturally be reduced. Also, when questioning authority or courses of action, uncertainty will be momentarily increased as questioning raises doubts about the current situation, and as new ways and possibilities are sought after. Often these doubts are also the reason why people may stay silent. The positive side of increasing uncertainty in decision making is seeing new angles and solutions to problems. This can lead to more successful end results or even prevent accidents from happening. However, when uncertainty is increased in critical situations by e.g. questioning decision making, it is important to know how to reduce uncertainty and regain control of the situation (Grote 2014).

High levels of routine, standardization and formalization are needed to create stability, predictability and control. This again reduces the need for ad hoc operations when courses of action are well enough known and described beforehand in work processes (for example the checklists ensuring that the important tasks have been included). In order to manage situations where uncertainty needs to be increased, control has to be decentralized to self-organizing units and performance is not controlled but shaped and directed. Also to increase learning, ad hoc situations should be favored. In these situations, humans are faced with uncertainty and new occurring needs to which they have to react (Grote 2014).

Doz and Kosonen (2008) write about strategic agility and leading with values. In traditional hierarchical, bureaucratic organizations, people were led by compliance to rules and regulations. However, when quick decisions and the agility to react fast are needed, the traditional way is a hindrance. A company should be lead collectively with normative, internalized operational framework and shared values. This way, quick decisions can be made when they need to be made in line with the overall principles, policies and values of the company.

The maritime domain is quite regulated and therefore cannot totally be run like an agile, constantly changing company. However, lessons can be learned on how to succeed with uncertainty and unpredictability by benchmarking the best companies operating in fast changing and turbulent environments. Further studies could include how these people are lead in areas with flexible, value-based rules and how the efforts of employees are directed towards the same goal. In addition, despite of the importance of the master as the leader of the ship, research in this area is very limited (Theotokas et al. 2014). Martínez-Córcoles et al. (2012) state that hardly any literature can be found regarding team leadership in safety performance settings. These are also issues that could benefit a more holistic view of safety management in the Arctic that cannot be based only on compliance to rules and working practices due to the unpredictability and uncertainties.

Standards and procedures are important and the maritime domain is quite regulated, therefore a balance between stability and flexibility in high-risk areas is a strategic risk management question of the company. Naturally it is important to follow the rules regulating the maritime industry: Safety of Life At Sea (SOLAS), The International Safety Management Code (ISM Code) and the Polar Code etc. However, IMO also states that there are areas that require continuous improvement and a culture of self-regulating.

IMO (IMO Safety Culture) defines safety culture as an organization that "*gives appropriate priority to safety and realises that safety has to be managed like other areas of business.*" IMO states also, that "*culture is more than merely avoiding accidents or even reducing the number of accidents, although these are likely to be the most apparent measures of success. In terms of shipboard operations, it is to do the right thing at the right time in response to normal and emergency situations*".

According to IMO, safety culture is to take root in the professionalism of seafarers, in their attitudes and performance; and highlights key activities "*to recognise that accidents are preventable through following correct procedures and established best practices, constantly thinking safety and seeking continuous improvement.*" The objective of safety management work should also be to "*inspire seafarers towards firm and effective self-regulation and to encourage personal ownership of established best practice*" (IMO Safety Culture). Clearly, there is a need for compliance to rules and regulations, but also when seeking continuous improvement, sometimes the old ways of working need to be left behind and new ways are introduced.

Balancing between compliance to rules and following best practices, and the flexibility to seek continuous improvement is important. Flexibility responds to uncertainty and stability answers to the need for control (Grote 2014). When entering areas like the Arctic where changing demands and unforeseen situations occur, the crew needs to have some room to operate, more decentralized authority to make decisions and adapt their behavior when facing ad hoc situations and to learn to gain experience. This requires flexibility and resilience to tolerate performance variability and disturbances. At the same time, when flexibility exists, individuals who are required to follow flexible rules should be well trained, educated and possess the right attitude of good Arctic seamanship. This needs to be supported by the organizational culture to ensure, that those who are performing under flexible rules can take advantage of the needed flexibility and not get confused, disorientated or violate the rules. The organization needs to build a culture, which is also in balance with control and accountability (Grote 2014).

5 Human Resources

Progoulaki and Theotokas (2009) state that human resources are considered to be very crucial to shipping companies in creating a competitive advantage. The high performance of a ship is the result of high performance of the crew. They emphasize the fact that high performance is a result of successful performance of the whole crew working as a team and not just the performance of individuals.

Theotokas et al. (2014) strongly emphasize the role of the ship master as the leader of a ship. Seafarers are living and working in a restrained space being long time away from home, continuously exposed to sea originated risks. The ship master as the leader of the ship and its crew is the key person for successful operations and hence leading also the safety of a ship which has been highlighted also by Martínez-Córcoles et al. (2012).

Liu et al. (2015) argue that team agility and rapid reaction is important to efficiently respond to the turbulent, competitive, and ever changing needs of the business environment. Also from a business perspective, Doz and Kosonen (2008) emphasize strategic agility as an answer to constant change, uncertainty and unpredictability. Successful, agile companies learn to operate in turbulent environments and under constant change where the achieved status is never taken for granted but must be constantly worked for. One aspect of agility is the collective commitment to goals, where the success of operations is the success of the whole company and not just the success of individuals.

6 Multicultural Shipping

Shipping has a global labor market, which also leads to multiculturalism onboard. To succeed in creating a high performance crew, the company should recruit high quality employees from the global labor market. They should be lead and motivated and the company should make an effort to ensure that they will stay with the company. Frequent turnover of crew can lead to loss of important human resources and tacit knowledge (Progoulaki and Theotokas 2009; Theotokas et al. 2014).

If crew turnover is high, rules and formalized working procedures are needed to ensure that the job gets done as required when one seafarer is changed for another. Formalization of work and work roles, written and enforced rules and procedures, high levels of routine can affect both flexibility and the social interactions of the crew. Social isolation and discrimination can occur onboard when seafarers are not trained to handle multiculturalism. Social and intercultural confrontations influence team cohesion negatively, affecting the performance of a crew as a team. To operate successfully, the crew, the people who live and work together, should have both the necessary official certificates and the personal ability to work as a member of the ship's team. When the crew consists of different nationalities with different cultural backgrounds and experiences, the role of the ship master as the leader of the ship becomes very important. The hard and soft skills of a master, the abilities to lead, motivate, inspire and empower the crew are vital for safe, effective and efficient operation of a ship (Progoulaki and Theotokas 2009; Theotokas et al. 2014).

Bergheim et al. (2014) studied the relationship between psychological capital (PsyCap), job satisfaction and safety perceptions in the maritime industry. PsyCap consists of four dimensions. Firstly one's belief to successfully execute and accomplish tasks. Secondly optimism, the tendency to have positive attitude towards the future events. Thirdly hope and a tendency to persistently pursue goals and change paths if needed to succeed. Fourthly resiliency, the ability to positively cope, tolerate and bounce back when faced with problems and challenges.

Their results indicated that PsyCap in the maritime industry, is positively correlated with safety climate when both personal and situational factors were relevant regarding workplace safety. They also argued, that safety climate perceptions could be more than just reflections of formal education and training in the job. It could reflect the individual motivational state of seafarers, which "*could be subject to training and leadership processes*" (2014 p. 31). They also argue, that PsyCap represents a new perspective for leadership and safety management to improve safety. The cultural backgrounds of the crew should also be taken into consideration, as different factors that influence safety climate could be dependent on the culture.

7 Continuous Improvement of Safety and Crew Involvement

Getting crewmembers to participate in safety related issues and activities are important. Safety participation of the crew, a proactive behavior towards safety, makes it possible to identify and detect non-conformities in processes, practices and the entire system. This is essential for continuous improvement and developing a good safety culture defined by IMO (IMO Safety Culture). Safety compliance is following rules and regulations, wearing personal protection equipment and performing activities needed to ensure workplace safety. Safety participation again is more of voluntary nature where the crewmembers for example voluntarily take initiative in safety tasks and safety improvement work. Personal motivation to participate in safety activities and safety knowledge are significant indicators of safety participation. It can be enhanced and significantly influenced by empowering leadership style (Martínez-Córcoles et al. 2012).

Murphy (2014) emphasizes the importance of leadership also. He is reflecting the principles from military world into the modern, complex business world with the elements of uncertainty and unpredictability present. Murphy (2014) emphasizes leadership, because it affects all aspects of a high-performing team and ensures their success as a team. Effective leaders do not only order, but also listen actively to ideas of their team members. This enables creative thinking and problem solving from new directions.

Unpredictability in the Arctic shipping routes can require at times creative thinking, fast responding and performance adjustments to occurring ad hoc situational needs. Team coherence, proactive behavior towards safety and continuous improvement can all be influenced by leadership. In these situations, also communication is highlighted, especially in safety critical situations.

8 Communication

Mazaheri et al. (2015) state in their case study on accident and incident reports on grounding, that appropriate communication and cooperation in studied incident cases stopped the situation from becoming serious. When inappropriate communication is present at the ship's bridge, information flow is interrupted. This will increase the likelihood of errors. They also point out that there is a strong link between inappropriate communication and personal factors in the incident reports in general, showing that the personalities of the crew affect safety through inappropriate communication. They also highlight proper interaction between the crewmembers.

Chauvin et al. (2013) highlights the same, where they state that most collisions are due to decision errors. Inter-ship communication problems and bridge resource management deficiencies are closely linked to collisions in restricted waters while having a pilot onboard. In cases of collision with another vessel while having a pilot onboard, 43 cases were linked to breakdowns in communication on the bridge, between the vessels or in the teamwork on the bridge.

Appropriate communication and speaking up, as in expressing one's mind or concerns aloud, are important for safety, as they open up new perspectives for decision-making and action. The master of the ship is in a key role to create an atmosphere and culture on the ship, where crewmembers feel free to express their minds, speak up and feel that their contribution is valued and appreciated. They also need psychological safety, where team members do not fear punishment or embarrassment when they question certain courses of action or come forward with new ideas for improvement. In time-critical situations, speaking up is emphasized. As pointed out earlier, appropriate communication could have prevented many accidents and near miss situations. However, speaking up should be done in a constructive, non-threatening way. Crewmembers should be encouraged to speak up, but at the same time understand that it is a two-way street. It is also important to be able to receive feedback and adequately react when spoken to (Grote 2014).

Grote (2014) points out too, that appropriate communication between the team and speaking up needs a general culture of trust, psychological safety and systematic training. These are organizational actions to support the teams and especially team leaders to create such a culture and routines that enable appropriate behavior to speak up and adequate reacting when spoken to.

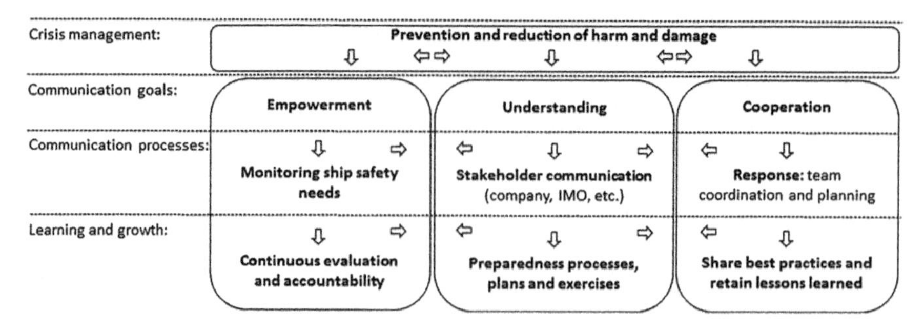

Fig. 1 Onboard crisis management communication framework (based on Palttala and Vos (2012): Strategy map for crisis communication supporting crisis management by public organisations)

Palttala and Vos (2012) have a Strategy map for crisis communication supporting crisis management, which can also be used to understand communication in the framework of continuous improvement and safety participation.

In Fig. 1, communication goals are divided into three: empowerment, understanding, and cooperation. On the communication process level every crewmember is empowered to actively participate in the monitoring of the ship's safety needs, while at the same time understanding the framework of company guidelines and formal regulations. Their successful cooperation is demonstrated in efficient and cohesive responses to changes in the environment. Team agility and rapid reaction, for example, are important to efficiently respond to the changing needs in the ship's environment. On a learning and growth level continuous evaluation, preparedness, and best practices promote accountability and retention of lessons learned.

As an example of this could be team work at the bridge with increased look out for growlers and submerged icebergs that are hard to detect. A multicultural crew with different cultural backgrounds, language barriers and difficulties in interpersonal cooperation can create challenges for efficient communication. Training and leadership can help to overcome these challenges.

9 Conclusion and Discussion

The Arctic is an environment where uncertainty and unpredictability are present. Hence, not all can be described in best practices to be followed neither can all risks be reduced, at least not yet. The human element is still needed to get the job done in all circumstances from normal operation to handling incidents and surviving accidents. IMO states that, safety culture should take root in the professionalism of seafarers, where competency, training and attitudes are important.

For the safe, sustainable and profitable shipping in the Arctic, attention needs to be paid to human resources. Hiring high performing seafarers from the global labor market to create a high performance team, and working for team unity and trust will enhance team performance and constant improvement. Strengthening the belief in safety-oriented actions and empowering every crewmember towards safety participation and constant improvement are important in creating a self-regulating culture. The crew also needs to believe that what they say and do matters and is valued.

Trust and psychological safety are important in promoting speaking up and appropriate communication within a crew team. In the modern multicultural environment of shipping, the hard and soft skills of a master, the abilities to lead, motivate, inspire and empower the crew are vital for safe, effective and efficient operation of a ship. As the seafarers are not usually trained to handle multiculturalism onboard, it is left for the captain of the ship to create such an environment, which promotes appropriate communication and teamwork.

In order to work with uncertainty, a high performing self-regulating crew needs a level of flexibility to cope with occurring ad hoc situations, to question current ways of working, and to make suggestions for continuous improvement. Making flexible rules is a strategic risk and safety management question of the company, because also compliance with official rules and regulations as well as best practices are still needed to ensure safety of Arctic shipping.

It is essential to ensure that the skills and knowledge of the crew develops and that the accumulated knowledge is kept within the ship and the company. This will enable the ship as an organization to learn from experience and improve constantly its performance and safety of Arctic shipping.

To lead and manage the safety of a ship is leadership and management of the people living and working in the ship. The execution of safety measures lie within the seafarers and their masters working at sea. They are the ones who react to and manage situations as they occur. They use their skills and knowledge to adapt to shortcomings in processes, work descriptions, equipment, and tools. They are the ones who face the sea-originated risks on a regular basis and fight for their survival in case of accidents. This chapter argues that from the captain's point of view, excellent seafarers, their competence, skills, collective attitudes and good Arctic seamanship are the key to a safer and more sustainable Arctic shipping.

References

Aven, T., & Krohn, B. S. (2013). A new perspective on how to understand, assess and manage risk and the unforeseen. *Reliability Engineering and System Safety, 121*(2014), 1–10.

Bergheim, K., Nielsen, M. B., Mearns, K., & Eid, J. (2014). The relationship between psychological capital, job satisfaction, and safety perceptions in the maritime industry. *Safety Science, 74* (2015), 27–36.

Besnard, D., & Hollnagel, E. (2014). I want to believe: Some myths about the management of industrial safety. *Cognition, Technology and Work, 16*(1), 13–23. hal-00720270.

Carpenter, G., & Wyman, O. (2014). *Arctic shipping: Navigating the risks and opportunities*. MARSH Risk Management Research. Retrieved May 25, 2015, from http://www.safety4sea.com/images/media/pdf/Arctic_Shipping_Lanes_MRMR_August_2014_US.pdf

Chauvin, C., Lardjane, S., Morel, G., Clostermann, J.-P., & Langard, B. (2013). Human and organisational factors in maritime accidents: Analysis of collisions at sea using the HFACS. *Accident Analysis and Prevention, 59*(2013), 26–37.

Doz, Y., & Kosonen, M. (2008). *Fast strategy – How strategic agility will help you stay ahead of the game*. Wharton School Publishing.

Grote, G. (2014). Promoting safety by increasing uncertainty – Implications for risk management. *Safety Science, 71*(2015), 71–79.

Hollnagel, E. (2009). *ETTO Principle: Efficiency – Thoroughness Trade-Off: Why things that go right sometimes go wrong*. Surrey: Ashgate Publishing Limited.

Hollnagel, E. (2014). *Safety-I and Safety-II: The past and future of safety management*. Surrey: Ashgate Publishing Limited.

Hollnagel, E., Pariès, J., & Woods, D. D. (2011). *Resilience engineering in practise: A guidebook*. Surrey: Ashgate Publishing Limited.

International Maritime Organization (IMO). (2010). Guidelines for ships operating in polar waters. *Resolution A, 1024*(26).

International Maritime Organization (IMO) Safety Culture. Retrieved June 1, 2015, from http://www.imo.org/en/OurWork/HumanElement/SafetyCulture/Pages/Default.aspx

Jensen, Ø. (2007). *The IMO guidelines for ships operating in arctic ice-covered waters. From voluntary to mandatory tool for navigation safety and environmental protection?* (FNI Report 2/2007). The Fridtjof Nansen Institute.

Leonhardt, J., Macchi, L., Hollnagel, E., & Kirwan, B. (2009). *A white paper on resilience engineering for ATM*. European Organisation for the Safety of Air Navigation (EUROCONTROL).

Liu, M.-L., Liu, N.-T., Ding, C. G., & Lin, C.-P. (2015). Exploring team performance in high-tech industries: Future trends of building up teamwork. *Technological Forecasting and Social Change, 91*(2015), 295–310.

Marchenko, N. (2015). *Arctic safety issues*. Retrieved July 10, 2015, from http://umanitoba.ca/faculties/management/ti/media/docs/N_Marchenko.pdf

Martínez-Córcoles, M., Schöbel, M., Gracia, F. J., Tomás, I., & Peiró, J. M. (2012). Linking empowering leadership to safety participation in nuclear plants: A structural equation model. *Journal of Safety Research, 43*(2012), 215–221.

Mazaheri, A., Montewka, J., Nisula, J., & Kujala, P. (2015). Usability of accident and incident reports for evidence-based risk modeling – A case study on ship grounding reports. *Safety Science, 76*(2015), 202–214.

Murphy, J. D. (2014). *Courage to execute: What elite U.S. military units can teach business about leadership and team performance*. Hoboken, NJ: John Wiley & Sons, Incorporated.

Palttala, P., & Vos, M. (2012). Quality indicators for crisis communication to support emergency management by public authorities. *Journal of Contingencies and Crisis Management, 20*(1), 39–51.

Progoulaki, M., & Theotokas, I. (2009). Human resource management and competitive advantage: An application of resource-based view in the shipping industry. *Marine Policy, 34*(2010), 575–582.

Stewart, E. J., & Dawson, J. (2011). A matter of good fortune? The grounding of the clipper adventurer in the northwest passage, Arctic Canada. *Arctic, 64*(2). Retrieved July 8, 2015, from http://pubs.aina.ucalgary.ca/arctic/Arctic64-2-263.pdf

The New York Times, All Safe in Soviet Ship Drama. Retrieved July 8, 2015, from http://www.nytimes.com/1989/06/21/world/all-safe-in-soviet-ship-drama.html

Theotokas, I., Lagoudis, I. N., & Kotsiopoulos, N. (2014). Leaders profiling of ocean going ship masters. *The Asian Journal of Shipping and Logistics, 30*(3), 321–343.

Wright, G. R., & Zimmerman, C. M. (2015). *Vector data extraction from forward-looking sonar imagery for hydrographic survey and hazard to navigation detection*. Oceans 2015, Washington, DC.

Beyond the Polar Code: IMO Measures for Assuring Safe and Environmentally Sound Arctic Navigation

J. Ashley Roach

Contents

Abstract The Polar Code has entered into force and the new polar seafarer requirements are expected to enter into force in July 2018. In the meantime the IMO is working on additional issues pertinent to operations in polar areas, such as risk assessment, additional performance and test standards, gathering data on non-SOLAS ships operating in polar waters, and amendments to the survey guidelines. There are additional measures that IMO might consider to strengthen safety

---

Captain J. Ashley Roach, JAGC, USN (retired), Office of the Legal Adviser, U.S. Department of State (retired).

J. Ashley Roach (✉)
Centre for International Law, National University of Singapore, Singapore

L. P. Hildebrand et al. (eds.), *Sustainable Shipping in a Changing Arctic*, WMU Studies in Maritime Affairs 7, https://doi.org/10.1007/978-3-319-78425-0_4

and environmental protections in the Arctic, including ships' routeing and reporting, VTS, port State control, MARPOL special areas, PSSAs, emission control areas, marine protected areas, ballast water and anti-foulants.

Keywords Polar Code · Port State control · ISPS Code · MARPOL special areas · Emission control areas · Marine protected areas · Ballast water control

1 Introduction

Although the Polar Code entered into force on 1 January 2017, and the new polar seafarer requirements to the STCW and Code are expected to enter into force 1 July 2018, the IMO is already considering new items. This chapter describes the IMO's ongoing work on Polar Code issues and discusses the various potential new measures available through the IMO that might affect Arctic shipping.

2 Current Work on the Polar Code

2.1 Risk Assessment

One matter being considered by the IMO is risk assessment, i.e. judging whether, when and where a particular class of ship may safely operate in the intended voyage in polar waters. This is of particular importance to the ability to obtain maritime insurance for voyages in polar waters (Kingston 2015).

Paragraph I-A/1.37 of the Polar Code provides that the Polar Ship Certificate 'shall reference a methodology to assess operational capabilities and limitations in ice to the satisfaction of the Administration, taking into account the guidelines developed by the Organization.' At MSC 94, the Committee agreed on the establishment of a correspondence group to prepare draft guidance on a methodology for determining limitations for operation in ice for structural risk assessment, and to exchange information on experience with operations in ice to validate guidance for operations in ice (International Maritime Organization 2014a). The correspondence group reported to MSC 95 that it made some progress on the development of the guidance but recommended the group be reestablished by MSC 95 (International Maritime Organization (Maritime Safety Committee) 2015a). Taking into account the need to make progress in light of the adoption of the Polar Code by MSC 94 and MEPC 68, MSC 95 decided to re-establish the Correspondence Group on the Development of guidance on a methodology for determining limitations for operation in ice, and associated draft MSC circular, for structural risk assessment and inclusion in the ship's documents, based on the documents previously submitted. MSC 96 approved the report in general and MSC.1/Circ.1519 on Guidance on methodologies for assessing operational capabilities and limitations in ice

(International Maritime Organization 2015b, 2016a, b, d). With regard to the future review of the Guidance, which could include discussion on the treatment of brash ice, the Committee agreed that this should be undertaken by the Ship Design and Construction (SDC) Sub-Committee, without a need for a new output, under the existing output 5.2.1.15 (Consequential work related to the new Code for ships operating in polar waters), in due course (International Maritime Organization 2016c).

2.2 *Additional Performance and Test Standards*

At MSC 95 the Committee also instructed the Sub-committee on Ship Systems and Equipment (SSE 3), in light of the adoption of the Polar Code, to consider whether additional performance or test standards for fire safety/protection and life-saving appliances and arrangements in relation to the Polar Code are necessary, and advise MSC 96 on the best way to proceed (International Maritime Organization 2015b, 2016e, g).

In a submission to SSE 3, it was recommended that the Sub-Committee advise MSC 97 that the International Life-Saving Appliance (LSA) Code should be further reviewed to identify and develop necessary amendments, with a view to meeting the additional demands that the Polar Code put on life-saving appliances and arrangements. The submission emphasized that any amendments would be additional performance and/or test criteria for the equipment and systems on board ships to which a Polar Ship Certificate is issued. For equipment and systems used on ships operating outside polar waters, the test regimes would remain unchanged (International Maritime Organization 2015c).

Following discussion, the Sub-Committee endorsed the view that additional performance and test standards for the equipment and systems on board ships operating in polar waters should be developed. In this connection, the Sub-Committee invited MSC 97 to endorse this decision and take action as appropriate. The Sub-Committee also invited interested Member Governments and international organizations to submit comments and proposals pertaining to the scope of work, type of equipment, etc. for consideration at MSC 97 (International Maritime Organization 2015d). At MSC 97 the Committee considered two documents proposing the Committee instruct SSE and NCSR Sub-Committees review, adapt and/or develop the necessary requirements (International Maritime Organization 2017a, b). The Committee noted this work was necessary to support implementation of the Polar Code, instructed SSE to review the LSA Code and relevant IMO resolutions to adapt current testing and performance standards to the Polar Code provisions or develop additional requirements if necessary, and develop guidance on extinguishing media at polar service temperature. The Committee instructed the Sub-Committee on Navigation, Communications and Search and Rescue (NCSR) to consider current communication requirements in SOLAS and the need for any amendments, and to consider the need for a new performance standard for GNSS compasses (International Maritime Organization 2016h).[1]

2.3 Extending the Polar Code to Non-SOLAS Ships

The Committee had previously agreed to consider extending the Polar Code to non-SOLAS ships after the Polar Code was adopted (International Maritime Organization 2011, 2012a, b).

MSC 95 noted two documents regarding incidents in polar waters involving non-SOLAS vessels (International Maritime Organization 2015e, f) and encouraged Member States and international organisations to submit information on incidents in polar waters to assist in assessing the potential scope of the Polar Code to non-Convention vessels operating in polar waters, for consideration at MSC 96 (International Maritime Organization 2015g). MSC 96 noted three submissions and invited addition information be provided to MSC 97 (International Maritime Organization 2016f).[2] MSC 97 considered two papers and noted their information would support the next phase of the Polar Code once it commences (International Maritime Organization 2017d).[3]

2.4 Polar Code-Related Amendments to the Survey Guidelines

MSC 97, with the concurrence of MEPC 70, approved MSC-MEPC.5/Circ.11 on *Amendments to the Survey Guidelines under the Harmonized System of Surveys and Certification, 2015 for Ships Operating in Polar Waters*. The Committee had considered a submission by IACS seeking clarification of paragraph 1.3 of chapter I-A of the Polar Code that was not specific as to which statutory certificate SOLAS regulation XIV/2/2 applied to passenger and cargo ships (International Maritime Organization 2016i). In approving the circular the Committee recognized that the

[1]In that regard Germany submitted a paper to NSCR 4, *Development of amendments to performance standards for navigation and communication equipment used in polar waters in support of the implementation of the Polar Code*, NCSR 4/28 (20 January 2017).

[2]Information was submitted to MSC 96 by New Zealand in MSC 96/24 (14 December 2015), by Iceland in MSC 96/24/3 (8 March 2016), and by FOEI and others in MSC 96/24/7 (8 March 2016).

[3]Two papers providing addition information were submitted to MSC 97, *International Code for Ships Operating in Polar Waters (Polar Code)*, MSC 97/21/8/Rev 1 (21 November 2016) (Chile) and *Non-SOLAS vessel operations in polar waters in preparation for work on phase 2 of the Polar Code*, MSC 97/21/10 (16 September 2016) (FOEI, WWW and the Pacific Environment). MSC 97/22, paras 21.9–21.10. These papers were submitted to SDC 4 as SDC 4/13 (9 December 2016) (FOEI, WWW and the Pacific Environment) and SDC 4/13/1 (8 December 2016) (New Zealand).

amendments should be incorporated in the draft Assembly resolution on the Survey Guidelines under the HSSC to be developed by the Sub-committee on Implementation of IMO Instruments (III 4) and considered by Assembly 30 (in November 2017) for adoption (International Maritime Organization 2017c).

As experience is gained in the years following its entry into force in 2017 and implementation, it can be expected that modification and improvements to the Polar Code will occur. However, it must not be forgotten that the Polar Code supplements and does not replace the many existing IMO conventions applicable to international shipping worldwide.

3 Potential New Measures

MEPC 67 stressed that any future amendments to the Polar Code to introduce additional or new environment-related requirements requires approval by the Committee as a new output in accordance with the Committee's Guidelines, MSC-MEPC.1/Circ.4/Rev.2, 8 June 2012, as amended (International Maritime Organization 2014b). Thus, unless a proposal fits under a continuous agenda item, it must first get approval from MSC/MEPC for a new work item/unplanned output, as described below. There are two relevant continuous agenda items:

- NCSR: Routeing measures and mandatory ship reporting systems; and
- MEPC: Identification and protection of Special Areas and PSSAs.

Discussed first are measures under the cognizance of the Maritime Safety Committee.

3.1 Routeing and Reporting Systems; Vessel Traffic Services

The Sub-committee on Navigation, Communications and Search and Rescue (NCSR), which reports to MSC, has several continuing agenda items, one of which is 'Routeing measures and mandatory ship reporting systems'. The Arctic 8 and Arctic Council Observer States are all party to the International Convention for the Safety of Life at Sea, 1974, as amended (SOLAS). Chapter V of the regulations annexed to SOLAS provides for the establishment of ships' routeing systems and ship reporting systems, which can be made mandatory if the IMO approves them (Regulations V/10 and V/11). SOLAS regulation V/12 provides for the establishment by parties of vessel traffic services where the volume of traffic or the degree of risk justified such services. These regulations are discussed next.

Ships' routeing systems are regulated by SOLAS Regulation V/10, which provides:

1. Ships' routeing systems contribute to safety of life at sea, safety and efficiency of navigation and/or protection of the marine environment. Ships' routeing systems are recommended for use by, and <u>may be made mandatory</u> for, all ships, certain categories of ships or ships carrying certain cargoes, when adopted

and implemented in accordance with the guidelines and criteria developed by the Organization.[4,5]

2. The Organization is recognized as the only international body for developing guidelines, criteria and regulations on an international level for ships' routeing systems. Contracting Governments shall refer proposals for the adoption of ships' routeing systems to the Organization. The Organization will collate and disseminate to Contracting Governments all relevant information with regard to any adopted ships' routeing systems.
3. The initiation of action for establishing a ships' routeing system is the responsibility of the Government or Governments concerned. In developing such systems for adoption by the Organization, the guidelines and criteria developed by the Organization shall be taken into account.
4. Ships' routeing systems should be submitted to the Organization for adoption. However, a Government or Governments implementing ships' routeing systems not intended to be submitted to the Organization for adoption or which have not been adopted by the Organization are encouraged to take into account, wherever possible, the guidelines and criteria developed by the Organization (see Footnotes 4 and 5).
5. Where two or more Governments have a common interest in a particular area, they should formulate joint proposals for the delineation and use of a routeing system therein on the basis of an agreement between them. Upon receipt of such proposal and before proceeding with consideration of it for adoption, the Organization shall ensure details of the proposal are disseminated to the Governments which have a common interest in the area, including countries in the vicinity of the proposed ships' routeing system.
6. Contracting Governments shall adhere to the measures adopted by the Organization concerning ships' routeing. They shall promulgate all information necessary for the safe and effective use of adopted ships' routeing systems. A Government or Governments concerned may monitor traffic in those systems. Contracting Governments shall do everything in their power to secure the appropriate use of ships' routeing systems adopted by the Organization.
7. A ship shall use a mandatory ships' routeing system adopted by the Organization as required for its category or cargo carried and in accordance with the relevant provisions in force unless there are compelling reasons not to use a particular ships' routeing system. Any such reason shall be recorded in the ships' log.

[4]Refer to the General Provisions on Ships' Routeing adopted by the Organization by resolution A.572(14)), as amended.

[5]The IMO Publication *Ships' Routeing* includes General provisions on ships' routeing, first adopted by IMO in 1973, and subsequently amended over the years. The provisions are aimed at standardising the design, development, charted presentation and use of routeing measures adopted by IMO. For additional information on ships' routeing, http://www.imo.org/OurWork/Safety/Navigation/Pages/ShipsRouteing.aspx.

8. Mandatory ships' routeing systems shall be reviewed by the Contracting Government or Governments concerned in accordance with the guidelines and criteria developed by the Organization.
9. All adopted ships' routeing systems and actions taken to enforce compliance with those systems shall be consistent with international law, including the relevant provisions of the 1982 United Nations Convention on the Law of the Sea.
10. Nothing in this regulation nor its associated guidelines and criteria shall prejudice the rights and duties of Governments under international law or the legal regimes of straits used for international navigation and archipelagic sea lanes.

Ship reporting systems are regulated by SOLAS Regulation V/11, which provides in part:

1. Ship reporting systems contribute to safety of life at sea, safety and efficiency of navigation and/or protection of the marine environment. A ship reporting system, when adopted and implemented in accordance with the guidelines and criteria developed by the Organization pursuant to this regulation, shall be used by all ships, or certain categories of ships or ships carrying certain cargoes in accordance with the provisions of each system so adopted.

2. The Organization is recognized as the only international body for developing guidelines, criteria and regulations on an international level for ship reporting systems. Contracting Government shall refer proposals for the adoption of ship reporting systems to the Organization. The Organization will collate and disseminate to Contracting Governments all relevant information with regard to any adopted ship reporting system.

⋮

8. All adopted ship reporting systems and actions taken to enforce compliance with those systems shall be consistent with international law, including the relevant provisions of the United Nations Convention on the Law of the Sea.

9. Nothing in this regulation or its associated guidelines and criteria shall prejudice the rights and duties of Governments under international law or the legal regimes of straits used for international navigation and archipelagic sea lanes.

The basic regulations on vessel traffic services are contained in SOLAS Regulation V/12, which reads in part:

1. Vessel traffic services (VTS) contribute to safety of life at sea, safety and efficiency of navigation and protection of the marine environment, adjacent shore areas, work sites and offshore installations from possible adverse effects of maritime traffic.

⋮

3. Contracting Governments planning and implementing VTS shall, wherever possible, follow the guidelines developed by the Organization.[6] The use of VTS may only be made mandatory in sea areas within the territorial seas of a coastal State.

[6]Refer to the Guidelines on Vessel Traffic Services adopted by the Organization by resolution A.857(20).

⋮

5. Nothing in this regulation or the guidelines adopted by the Organization shall prejudice the rights and duties of Governments under international law or the legal regimes of straits used for international navigation and archipelagic sea lanes.

IMO guidance for ship routeing and reporting systems is contained in various IMO resolutions.[7] Guidance for vessel traffic services is contained in a manual prepared by the International Association of Marine Aids to Navigation and Lighthouse Authorities (IALA) (IALA 2009, 2012).[8] Any new routeing system will have to take into account the seasonal presence of Arctic sea ice and how ships may navigate outside any routes to enhance safety due to the present sea ice conditions.

3.1.1 Mandatory Ship Reporting Systems in Straits

Mandatory ship reporting systems for straits used for international navigation have been approved by the IMO for the Straits of Malacca and Singapore (resolution MSC.73(69)), Torres Strait (resolution MSC.161(78)), Great Belt (resolution MSC.230(82)), Strait of Gibraltar (resolution MSC.300(87)), and The Sound (resolution MSC.314(88)).[9] In addition, pursuant to a Russian and Norwegian proposal, in 2012 the MSC adopted a new mandatory ship reporting system 'In the Barents Area (BARENTS SRS)' by Resolution MSC.348(91), 28 November 2012, effective 1 June 2013 (International Maritime Organization 2012c).[10] Reports are to be made to either Vardø VTS center or Murmansk VTS center. The reporting area is between 66°N and 72°N along the northern coast of Norway.

One might expect that proposals on routeing and reporting measures for the Bering Strait may well be introduced at future meetings of NCSR. Russia may

---

[7]The General Provisions on Ships' Routeing, *ibid*, reflect IMO, *General Principles for Ship Reporting Systems and Ship Reporting Requirements, including Guidelines for Reporting Incidents involving Dangerous Goods, Harmful Substances and/or Marine Pollutants*, Resolution A.851 (20) (27 November 1997) http://www.imo.org/blast/blastData.asp?doc_id=9884&filename=A%20851%2820%29.pdf; IMO, *Guidance Note on the Preparation of Proposals on Ships' Routeing Systems and Ship Reporting Systems for Submission to the Sub-Committee on Safety of Navigation*, MSC.1/Circ.1060 (6 January 2003); and IMO, *Amendment to the Guidance Note on the Preparation of Proposals on Ships' Routeing Systems and Ship Reporting Systems for Submission to the Sub-Committee on Safety of Navigation*, MSC.1/Circ.1060/Add.1 (26 May 2006) online: https://docs.imo.org/Shared/Download.aspx?did=37577. NCSR was tasked by MSC 95 to consider revisions of the *Guidelines and criteria for ship reporting systems* (MSC.43(64) as amended by MSC.111(73)), with a target completion year of 2017. IMO, *Report of the MSC on its 95th Session*, MSC 95/22 (19 June 2015), para. 19.12.3.

[8]For additional information on VTS, see http://www.imo.org/OurWork/Safety/Navigation/Pages/VesselTrafficServices.aspx.

[9]USCG, 'IMO MSC Resolutions', http://www.navcen.uscg.gov/?pageName=mscResolutions.

[10]A list of MSC resolutions may be found online: http://www.imo.org/KnowledgeCentre/IndexofIMOResolutions/Maritime-Safety-Committee-%28MSC%29/Pages/default.aspx.

also do the same for the Northern Sea Route, rather than rely solely on its unilateral measures.

3.2 *Port State Control*

Port State Control is the inspection of foreign flag ships in national ports to verify the condition of the ship and its equipment comply with the requirements of international regulations and that the ship is manned and operated in compliance with these rules. The basis in international law for port state control lies in provisions of the Law of the Sea Convention[11] and IMO treaties.[12] Guidance on the conduct of port state control is contained IMO Assembly resolution A.1052(27) (2011). In addition there are nine regional agreements on port state control to coordinate the inspections to focus on substandard ships and avoid multiple inspections.[13] It can be expected that the IMO Assembly resolution will in due course be revised to include guidance regarding the Polar Code.

3.3 *ISPS Code*

Following September 11, the IMO adopted special measures to enhance maritime security, as amendments to SOLAS (chapter XI-2) and the International Ship and Port Facility Security (ISPS) Code.[14] These are applicable to commercial ships that could be expected to traverse the Arctic Ocean, and will be applicable to ports on the rim.[15]

[11]United Nations Convention on the Law of the Sea, adopted 10 December 1982, 1833 UNTS 397 (entered into force 16 November 1994) (LOS Convention), articles 94(6) and 219.

[12]International Convention for the Safety of Life at Sea, adopted 1 November 1974, 1184 UNTS 277 (entered into force 25 May 1980) (SOLAS Convention), regulations I/19, IX/6.2, XI-1/4, XI-2/9; International Convention on Load Lines, 640 UNTS 133, London 5 April 1966 (entered into force 21 July 1968), article 21; International Convention for the Prevention of Pollution from Ships, adopted 2 November 1973, 1340 UNTS 61 (entered into force 2 October 1983) (MARPOL 73/78), Annex I articles 5 & 6 and regulation 11, Annex II regulation 16.9, Annex III regulation 8, Annex IV regulation 13, Annex V regulation 8, Annex VI regulation 10; International Convention on Standards of Training, Certification and Watchkeeping for Seafarers, with Annex (STCW), London 7 July 1978 (entered into force 28 April 1984), 1361 UNTS 75, article X and regulation I/4; International Convention on the Tonnage Measurements of Ships, 1969, with annexes, London 23 June 1969 (entered into force 18 July 1982), 1291 UNTS 3, article 121; International Convention on the Control of Harmful Anti-Fouling Systems on Ships, 2001, London 5 Oct. 2001 (entered into force 17 Sept. 2008), UNTS, IMO doc. AFS/CONF/26, article 11.

[13]See IMO, 'Port State Control' online: http://www.imo.org/en/OurWork/MSAS/Pages/PortStateControl.aspx.

[14]SOLAS Convention, *supra* note 11, Chapter XI-2 (Special Measures To Enhance Maritime Security) and the International Ship and Port Facility Security Code, adopted 12 Dec. 2002 by the Conference of Contracting Governments to the International Convention for the Safety of Life at Sea, 1974.

[15]*Ibid,* Regulation XI-2/2.

3.4 *Special Protection for Arctic Ocean Areas under the Cognizance of MEPC (Det Norske Veritas (DNV) 2014a)*

MEPC has as one of its continuing agenda items 'Identification and protection of Special Areas and PSSAs.'

3.4.1 MARPOL Special Areas

In Annexes I (Prevention of pollution by oil), II (Control of pollution by noxious liquid substances) and V (Prevention of pollution by garbage from ships), MARPOL defines certain sea areas as 'special areas' in which, for technical reasons relating to their oceanographical and ecological condition and to their sea traffic, the adoption of special mandatory methods for the prevention of sea pollution is required. Under the Convention, these special areas are provided with a higher level of protection than other areas of the sea.[16]

Annex I to MARPOL 73/78 contains regulations for the prevention of pollution by oil. The Annex provides for the establishment of special sea areas where for recognised technical reasons in relation to its oceanographic and ecological condition and to the particular character of its traffic, the adoption of special mandatory methods for the prevention of sea pollution by oil is required.[17] Guidelines on designating MARPOL Special Areas are contained in resolution A.1087(28), *2013 Guidelines for the Designation of Special Areas under MARPOL* (21 February 2014).

In respect of the Arctic Ocean, Part II-A of the Polar Code prohibits any discharge into the sea by oil or oily mixtures from any ship (regulation 1.1.1), and any discharge into the sea of noxious liquid substances or mixtures containing such substances (regulation 2.1.1). With regard to sewage and garbage from ships in Arctic waters, any such discharges are prohibited, except when in accordance with MARPOL Annexes IV and V and the requirements of Regulations 4.2.1–4.2.3 and 5.2.1 thereof.

As these prohibitions are more stringent than the normal restrictions in these MARPOL Annexes, it can be said that the Polar Code discharge restrictions in effect make the Arctic Ocean MARPOL Special Areas without saying so, particularly with the paucity of port waste reception facilities in the region.

---

[16]For a table listing all MARPOL Special Areas approved by IMO, see: http://www.imo.org/OurWork/Environment/SpecialAreasUnderMARPOL/Pages/Default.aspx. A prerequisite for the establishment of a MARPOL Special Area is the availability of adequate port waste reception facilities.

[17]MARPOL 73/78, Annex I, Regulation I/1.11. Resolution A.1087(28). 'Guidelines for the Designation of Special Areas under MARPOL' (2013), is not yet available at http://www.imo.org/KnowledgeCentre/IndexofIMOResolutions/Pages/Assembly-%28A%29.aspx.

In respect of the Antarctic area, any discharge into the sea of oil or oily mixtures, or noxious liquid substances or mixtures containing such substances, from any ship is already prohibited.[18] The discharge of garbage into several special areas, including Antarctica, is also prohibited.[19]

A prohibition on the use of heavy fuel oil (HFO) in the Southern Ocean was adopted by MEPC 60, effective 1 August 2011 (International Maritime Organization 2010), and amended by MEPC 67, effective 1 March 2016 (International Maritime Organization 2014c, d). A similar prohibition for the Arctic Ocean was considered at MEPC 70 and is being supported by the Arctic cruise industry (Clean Arctic Alliance 2016; International Maritime Organization 2016k; MarEx 2016).

3.4.2 Particularly Sensitive Sea Areas (PSSAs)

A PSSA is an area that needs special protection through action by IMO because of its significance for recognised ecological, socio-economic or scientific reasons, and which may be vulnerable to damage by international maritime activities.

Guidelines on designating a PSSA are contained in IMO resolution A.982 (24) *Revised guidelines for the identification and designation of PSSAs*, as amended (International Maritime Organization 2005, 2014e).[20] These guidelines include several criteria to allow areas to be designated as a PSSA if they fulfil a number of criteria, including: ecological criteria, such as unique or rare ecosystem, diversity of the ecosystem or vulnerability to degradation by natural events or human activities; social, cultural and economic criteria, such as significance of the area for recreation or tourism; and scientific and educational criteria, such as biological research or historical value.

An application for PSSA designation should contain a proposal for an associated protective measure or measures (APMs) aimed at preventing, reducing or eliminating the threat or identified vulnerability. Associated protective measures for PSSAs are limited to actions that are to be, or have been, approved and adopted by IMO.

When an area is approved as a PSSA, specific measures can be used to control the maritime activities in that area, such as routeing measures; strict application of MARPOL discharge and equipment requirements for ships, such as oil tankers; and installation of Vessel Traffic Services (VTS). Another routeing measure that can be used in a PSSA is an area to be avoided (i.e., an area within defined limits in which either navigation is particularly hazardous or it is exceptionally important to

[18] *Ibid*, Annex I, Regulation I/15.4 and Annex II, Regulation II/13.8.2.

[19] *Ibid*, Annex V, Regulation V/5.

[20] The guidelines update International Maritime Organization (2001). Consequential amendments to resolution A.982(24) proposed by MEPC 67/10 (22 July 2014) were deferred to MEPC 68, IMO, *Report of the MEPC on its 67th Session*, MEPC 67/20 (31 October 2014), para 10.1. MEPC 68 adopted resolution MEPC.267(68) on Amendments to resolution A.982(24) as set out in Annex 13 to MEPC 68/21/Add.1. International Maritime Organization (2015i).

avoid casualties and which should be avoided by all ships, or by certain classes of ships).

The guidelines provide advice to IMO Member Governments in the formulation and submission of applications for the designation of PSSAs to ensure that in the process, all interests—those of the coastal State, flag States, and the environmental and shipping communities—are thoroughly considered on the basis of relevant scientific, technical, economic, and environmental information regarding the area at risk of damage from international shipping activities.[21]

An approved PSSA is charted (International Maritime Organization 2002).[22] This serves to warn the mariner of the need for careful navigation.

Reporting of any subsequent developments or requirements for review are minimal at best. At MEPC 70 the Committee considered a proposal by the Russian Federation to introduce requirements to evaluate regularly the status and effectiveness of Special Areas and PSSAs (International Maritime Organization 2016j).

The Committee noted that with regard to the evaluation of existing PSSAs, in particular the effectiveness of APMs, MEPC 65 (May 2011) had requested Member Governments to submit such evaluations in accordance with paragraph 8.4 of the Revised PSSA Guidelines or to bring any concerns with the APMs to the IMO's attention so that any necessary adjustments may be made (International Maritime Organization 2013). It was noted that to date no specific evaluations have been received.

With regard to Special Areas, the Committee noted that there are no requirements to evaluate the effectiveness of such areas once they have been designated, although such an evaluation procedure could be incorporated in the *2013 Guidelines for the designation of Special Areas under MARPOL* (resolution A.1087(28)).

After debating the proposal, the Committee noted the Russian proposals and reminded Member Governments of their requirement to bring any concerns and proposals for additional measures or modifications to any APMs or PSSAs to the attention of the IMO, particularly if the levels of the threat from shipping have changed, so that any necessary adjustments made be made. Finally the Committee invited interested Member Governments wishing to amend the 2013 Guidelines to submit proposals for a new output to a future session, in accordance with the Committee's Guidelines (International Maritime Organization 2016k).

A study for the Protection of the Arctic Marine Environment Arctic Council Working Group (PAME) found support for a PSSA for the high seas area of the Arctic Ocean, with APMs of VTS, ship reporting system and ATBA, and establishment of one or more '[c]ore sea ice areas', but not a MARPOL Special Area (DNV Report 2014b).

---

[21]The International Chamber of Shipping has publicly stated its desire and willingness to be consulted by States contemplating PSSA submissions. For a list of IMO approved PSSAs, http://www.imo.org/OurWork/Environment/PSSAs/Ps/Default.aspx.

[22]See also IHO Chart Specification B-437.6; MPA Singapore, 'Symbols, Abbreviations, Terms and S-57 Objected used on Singaporean Nautical and Electronic Navigational Charts,' N22, at 37 (2011) http://www.mpa.gov.sg/web/portal/home/publications/chart-symbols-and-abbreviations.

3.4.3 Emission Control Areas (ECAs)

MARPOL Annex VI (Regulations for the Prevention of Air Pollution from Ships (1997)) establishes certain sulphur oxide (SOx) ECAs with more stringent controls on sulphur emissions. Annex VI was revised in 2008 to make the requirements more stringent, including particulate matter, and updated the regulations on the establishment of ECAs.[23] This may assist in dealing with the melting caused by the deposit of black carbon on the ice (Arctic Council 2001; International Maritime Organization 2015j, k).

3.4.3.1 Black Carbon

The IMO has been considering the impact on the Arctic of emissions of black carbon from international shipping. At MEPC 68, May 2015, the Committee agreed on a definition of Black Carbon for international shipping and noted the need for black carbon measurement studies so as to gain experience with the application of the definition and measurement methods. The Committee agreed on the need for a protocol for any voluntary measurement studies to collect data to identify the most appropriate measurement method(s) of black carbon emissions from international shipping. Finally the Committee noted that it was not possible at this stage to consider possible control measures to reduce the impact on the Arctic of emissions of black carbon from international shipping (International Maritime Organization 2015h; Fathom 2015).[24]

A study done for the World Wildlife Fund of Canada suggested that switching to liquid natural gas for Arctic shipping could greatly reduce the risks associated with the use of heavy (diesel) fuel oil, the major source of black carbon (Vard Marine Inc. 2015).

Several environmental groups provided information to MEPC 69 on the hazards posed by the use of HFO in the Arctic (International Maritime Organization 2016l). Differing views were expressed on this paper at MEPC 69. The Committee noted the paper and invited further proposals for a new output to address this matter to a future

---

[23]Amendments to MARPOL Annex VI adopted during the October 2008 session of the Marine Environment Protection Committee (by resolution MEPC.176(58)) included a revised regulation 14 on Sulphur Oxides (SOx) and Particulate Matter that permits the establishment of Emission Control Areas. Appendix III provides the criteria and procedures for designation of Emission Control Areas. These amendments entered into force 1 March 2010. The text of the revised Annex VI, IMO, MEPC 58/23/Add.1 Annex 13, online: http://www.imo.org/blast/blastData.asp?doc_id=10407&filename=176%2858%29.pdf. A list of ECAs may be found at http://www.imo.org/OurWork/Environment/SpecialAreasUnderMARPOL/Pages/Default.aspx.

[24]PPR 3 (February 2016) developed a draft protocol for any voluntary measurement studies to collect data. IMO, *Report to MEPC*, PPR 3/22, para. 8.10 and PPR 3/WP.4 Annex 1 (14 March 2016), and invited submission of data derived from its application to PPR 4. PPR 3/22, para 22.2.6. MEPC 70 noted these developments. MEPC 70/18, at 21 para 5.4. For further information on black carbon, see Fathom (2015).

session (International Maritime Organization 2016m).[25] MEPC 70, having noted the concerns expressed in four submissions to MEPC 70 regarding the protection of Arctic indigenous food security from the effects of shipping and the discussion and currently ongoing work on the use of HFO by ships operating in Arctic waters, invited Member States and other stakeholders to submit relevant information to future sessions, noting that further substantive work on these issues would require a new output (International Maritime Organization 2016n).

3.4.3.2 Arctic Council

Separately the Arctic Council has been working on black carbon. The 2013 Kiruna Declaration provided that the Ministers decided 'to establish a Task Force to develop arrangements on actions to achieve enhanced black carbon and methane emission reductions in the Arctic, and report at the next Ministerial meeting in 2015.'

The Task Force for Action on Black Carbon and Methane (TFBCM) was co-chaired by Canada and Sweden and included representatives from all Arctic States and most Permanent Participants. Arctic Council observers (including China, Germany, Japan, Republic of Korea, the European Union, the United Kingdom and the United Nations Environmental Program) also participated in various meetings of the Task Force, and relevant experts provided guidance, as required. The Task Force's outcome builds on previous technical work undertaken in the Arctic Council by an earlier Task Force on Short Lived Climate Forcers, the Arctic Monitoring Assessment Programme (AMAP), and the Arctic Contaminants Action Program (ACAP).

The Task Force, during the course of its six meetings, successfully delivered on its mandate and developed an Arctic Council Framework for Action on Enhanced Black Carbon and Methane emissions reductions (Arctic Council 2015a). As short-lived climate pollutants disproportionately impact the Arctic, their reduction will lead to benefits for the climate with important co-benefits for human health and air quality in the Arctic. This Framework represents a high level commitment of Arctic States to take mitigation action, but is not legally binding. It is an action-oriented document and includes work at the national, regional and global levels to reduce emissions of black carbon and methane.

The Framework lays out a common vision with enhanced, ambitious, national and collective action to accelerate the decline in overall black carbon emissions and to significantly reduce overall methane emissions. The work of the Task Force also resulted in the creation of an Expert Group with specific terms of reference to support progress on the implementation of the Framework and to continuously drive

[25] Submissions to MEPC 70 were made by FOEI and others, MEPC 70/17/4 (22 July 2016), Russian comments thereon, MEPC 70/17/9 (19 August 2016), by Canada and the United States, MEPC 70/17/11 (2 September 2016), and by FOEI, WWF and Pacific Environment on Arctic indigenous food security and shipping, MEPC 70/17/10 (19 August 2016).

ambition. It includes a further commitment to provide black carbon inventories starting in 2015 and provides guidance to report on national actions; to establish an aggregate summary of black carbon and methane emissions; and to adopt an ambitious, aspirational and quantitative collective goal on black carbon, and to consider additional goals, by the next Arctic Council Ministerial meeting in May 2017. Recognising that black carbon and methane emitted beyond the borders of Arctic States have a substantial impact on the Arctic, the Framework notes that Arctic States look forward to Arctic Council Observer states taking similar action.

The Framework also acknowledges that reducing anthropogenic carbon dioxide emissions remains the most important challenge to address global and Arctic climate change. Arctic States view the Framework as supporting and complementing the goals of the United Nations Framework Convention on Climate Change (UNFCCC).

Russia considers the expert group to be a working organ of the Arctic Council and to be a part of the Arctic Council structure. The participating states submit any national reports related to black carbon and methane emissions on their own initiative and within the framework of their participation in the work of the expert group. These reports are voluntary exchanges of information in accordance with international law and the national legislation of the respective participating state. In this context, 'high level political commitments' mean general guidelines for the further cooperation between the states on the issue of the regulation of the black carbon and methane emissions. The document of the Arctic Council 'Enhanced Black Carbon and Methane Reductions: An Arctic Council Framework for Action' will be implemented by the Russian Federation in the context of this understanding (Arctic Council 2015a).

At the 2015 Arctic Council Ministerial, the Ministers decided to implement the Framework for Action on Enhanced Black Carbon and Methane Emissions Reductions and established an expert group to report on progress to the SAO (Arctic Council 2015b).

3.4.4 Marine Protected Areas (MPAs)

MPAs are not adopted by the IMO. However, MPAs like any protected area, are regions in which human activity has been placed under some restrictions by other organizations in the interest of protecting the natural environment, its surrounding waters and the occupant ecosystems, and any cultural or historical resources that may require preservation or management. MPAs' boundaries will include some area of ocean, even if it is only a small fraction of the total area of the territory.[26]

With regard to MPAs in the Arctic, the Biodiversity Committee (BDC) of the OSPAR Convention for the Protection of the Marine Environment of the North-East

[26]For more information on MPAs, see online: http://ocean.nationalgeographic.com/ocean/take-action/marine-protected-areas/; and online: http://www.marine-conservation.org/what-we-do/program-areas/mpas/.

Atlantic is considering a proposal by Greenpeace and the World Wildlife Fund for a high seas MPA in that portion of the Arctic Ocean included in OSPAR Region 1. At its meeting 2–6 March 2015 the BDC agreed for a small drafting group should update the scientific evidence in the proposal before the OSPAR Commission would consider it (OSPAR Commission 2015). At its meeting 29 February–4 March 2016, the BDC endorsed the proposal as being scientifically robust and agreed to forward the draft proforma to OSPAR HOD for their consideration on the next steps, including with respect to the Arctic Council (OSPAR Commission 2016a). At the Commission's meeting 20–24 June 2016 in Spain, OSPAR agreed to invite OSPAR Contracting Parties, who are also Arctic Council States, to continue to provide information on further developments at the Arctic Council as appropriate (OSPAR Commission 2016b).

At its 2015 Ministerial, the Arctic Council Ministers also approved the Framework for a Pan-Arctic Network of Marine Protected Areas (Arctic Council 2015c),[27] and decided to continue work to develop such a network, based on the best available knowledge and science in order to strengthen marine ecosystem resilience, taking into account the cultural and sustainable use of marine resources (Arctic Council 2015d).

3.5 Ballast Water Controls

Consideration could be given to implementation of the ballast water rules under the 2004 IMO International Convention for the Control and Management of Ships' Ballast Water and Sediments.[28] The Convention enters into force 12 months after ratification by 30 States, representing 35% of world merchant shipping tonnage, on 8 September 2017.[29]

The adoption of the last set of Guidelines for the uniform implementation of the Convention and the approval and certification of modern ballast water treatment technologies have removed the last barriers to the ratification of the instrument and a significant number of countries have indicated their intention to accede to this Convention in the near future.[30]

Questions remain whether the technologies have been satisfactorily tested in polar waters.

[27]The framework is intended only for areas under the national jurisdiction of the Arctic States, and not for areas beyond national jurisdiction (page 5).

[28]The text of the Ballast Water Convention, and other information on the convention, online: http://www.imo.org/OurWork/Environment/SpecialAreasUnderMARPOL/Pages/Default.aspx.

[29]For the current ratification status, online: http://www.imo.org/About/Conventions/StatusOfConventions/Pages/Default.aspx.

[30]For further information on ballast water control, online: http://www.imo.org/OurWork/Environment/BallastWaterManagement/Pages/Default.aspx.

3.6 *Anti-Fouling*

The International Convention on the Control of Harmful Anti-fouling Systems on Ships, which was adopted on 5 October 2001, prohibits the use of harmful organotins in anti-fouling paints used on ships' hulls and establishes a mechanism to prevent the potential future use of other harmful substances in anti-fouling systems. The Convention entered into force on 17 September 2008. As of February 2017, there are 75 parties, including the Arctic 5 (Canada, Denmark, Norway, the Russian Federation and the United States), representing 93.71% of the world shipping tonnage.

Under the terms of the Convention, Parties to the Convention are required to prohibit and/or restrict the use of harmful anti-fouling systems on ships flying their flag, as well as ships not entitled to fly their flag but which operate under their authority and all ships that enter a port, shipyard or offshore terminal of a Party.[31]

An information paper was submitted to MEPC 69 about the risks that ballast water discharge and hull fouling pose to the Arctic, and also about possible mechanisms to control those risks (International Maritime Organization 2016o), which the Committee noted (International Maritime Organization 2016p).

4 Conclusions

First and foremost it must be reemphasized that the Polar Code and STCW amendments are not, and were never, intended to replace existing requirements for ships operating in polar waters. They are designed to supplement the existing requirements to account for the very different operating conditions in those waters that had not previously been dealt with.

Second, the Polar Code and STCW amendments are just the first version, and should be expected to be revised as experience with them occurs in the years after their entry into force in 2017 and 2018.

Improvements are already being considered, some in greater detail than others, although none are ready at the time of writing for adoption by the IMO.

There is a commitment to extend the application of the Polar Code to non-SOLAS and fishing vessels, and data is being collected in preparation therefor.

Further work is ongoing in various sub-committees to address issues such as risk assessment, additional performance and test standards, and amendments to the survey guidelines.

There are a number of areas where there may be new measures in the Arctic, such as routeing and reporting systems, vessel traffic services, and updating the port State control guidelines.

[31]For additional information on the Anti-Fouling Convention, online: http://www.imo.org/OurWork/Environment/Anti-foulingSystems/Pages/Default.aspx.

Other areas to enhance the protection of the Arctic marine environment could include MARPOL special areas, particularly sensitive sea areas, emission control areas, work on controlling or reducing the emissions of black carbon and GHG, and marine protected areas, as well as Arctic-specific controls on the use of heavy fuel oil, ballast water, and anti-foulants.

In addition, there are other efforts to protect and preserve the Arctic Ocean and its indigenous peoples, outside the purview of the IMO (and not discussed in this chapter), that include implementation of the agreements on search and rescue; oil pollution, prevention and response; and the new science cooperation agreement, developed by the eight Arctic states. In addition, multilateral negotiations are well underway to regulate the potential for commercial fishing in the high seas area of the Arctic Ocean.

References

Arctic Council. (2001). Arctic Monitoring and Assessment Program Working Group, *The Impact of Black Carbon on Arctic Climate*, AMAP Technical Report No 4 (2001). http://www.amap.no/documents/doc/the-impact-of-black-carbon-on-arctic climate/746

Arctic Council. (2015a). *Iqaluit SAO Report to Ministers*. https://oaarchive.arctic-council.org/bitstream/handle/11374/494/ACMMCA09_Iqaluit_2015_Iqaluit_SAO_Report_to_Ministers_formatted_v.pdf.pdf?sequence=1&isAllowed=y

Arctic Council. (2015b). *Iqaluit Ministerial Declaration*, 24 April 2015, para. 24. https://oaarchive.arctic-council.org/handle/11374/662

Arctic Council. (2015c). Protection of the Arctic Marine Environment Working Group, *Framework for a Pan-Arctic Network of Marine Protected Areas* (April 2015). https://oaarchive.arctic-council.org/handle/11374/417

Arctic Council. (2015d). *Iqaluit Declaration*, para. 40. https://oaarchive.arctic-council.org/bitstream/handle/11374/662/EDOCS-3431-v1-ACMMCA09_Iqaluit_2015_Iqaluit_Declaration_original_scanned_signed_version.PDF?sequence=7&isAllowed=y

Clean Arctic Alliance. (2016). Press Release, *Clean Arctic Alliance Response to Cruise Operators' support of Arctic Heavy Fuel Oil Ban* (17 November 2016). http://arcticjournal.com/press-releases/2706/clean-arctic-alliance-response-cruise-operators-support-arctic-heavy-fuel-oil

Det Norske Veritas (DNV). (2014). *Report on Specially Designated Marine Areas in the Arctic High Seas*, Report No/DNV Reg No: 2013-1442/17JMT1D-26 (11 March 2014) Part I. http://www.pame.is/index.php/projects/arctic-marine-shipping/specially-designated-marine-areas-in-the-arctic (finding it difficult to find support for a MARPOL Special Area for the high seas of the Arctic Ocean)

DNV Report. (2014), pp. 56–60.

Fathom. (2015). *Fathom Spotlight: 8 Things You Need to Know about Black Carbon and the Shipping Industry* (3 August 2015), *Ship & Bunker*. http://shipandbunker.com/news/features/fathom-spotlight/767141-fathom-spotlight-8-things-you-need-to-know-about-black-carbon-and-the-shipping-industry

International Association of Marine Aids to Navigation and Lighthouse Authorities. (2009). *Establishment of a Vessel Traffic Service beyond territorial seas*, IALA Guideline No. 1071 (9 December 2009).

International Association of Marine Aids to Navigation and Lighthouse Authorities. (2012). *Vessel Traffic Services Manual* (5th ed.). IALA, 2012. http://www.pmo.ir/pso_content/media/files/2013/1/22176.pdf

International Maritime Organization (Assembly). (2001). *Guidelines for the Designation of Special Areas under MARPOL 73/78 and Guidelines for the Identification and Designation of Particularly Sensitive Sea Areas*, Resolution A.927(22) (29 November 2001).
International Maritime Organization (Sub-Committee on Safety of Navigation). (2002). *Special Areas and Particularly Sensitive Sea Areas,* NAV 48/INF.2 (4 April 2002) (IHO).
International Maritime Organization (Assembly). (2005). *Revised Guidelines for the Identification and Designation of PSSAs*, Resolution A.982(24) (1 December 2005). http://www.imo.org/blast/blastData.asp?doc_id=11277&filename=A%20982%2824%29.pdf
International Maritime Organization (Marine Environment Protection Committee). (2010). *Amendments to the Annex of MARPOL,* Resolution MEPC.189(60) (26 March 2010), MEPC 60/22, Annex 10, adding chapter 9 to MARPOL Annex I.
International Maritime Organization (Sub-Committee on Ship Design and Equipment). (2011). *DE Report to the MSC,* DE 55/22 (15 April 2011), p. 23, para. 12.7.1.
International Maritime Organization (Sub-Committee on Ship Design and Equipment). (2012a). *DE Report to the MSC,* DE 56/25 (28 February 2012), p. 22, para. 10.7.
International Maritime Organization. (2012b). *Report of the Maritime Safety Committee on its 91st Session,* MSC 91/22 (17 December 2012), p. 35, para. 8.5.
International Maritime Organization. (2012c). *Annexes to the Report of the Maritime Safety Committee on its 91st Session,* MSC 91/22/Add.1 (17 December 2012), Annex 27.
International Maritime Organization. (2013). *Report of the Marine Environmental Protection Committee on its 65th Session,* MEPC 65/22 (24 May 2013), para. 9.7.
International Maritime Organization. (2014a). *Report of the Maritime Safety Committee on its 94th Session,* MSC 94/21 (26 November 2014), para. 3.62.
International Maritime Organization. (2014b). *Report of the Marine Environment Protection Committee on its 67th Session,* MEPC 67/20 (31 October 2014), p. 46 para. 9.21.
International Maritime Organization. (2014c). *Draft Report of the Marine Environment Protection Committee on its 67th Session,* MEPC 67/WP.1 (16 October 2014), paras. 7.26–7.27 adding the carriage of HFO as ballast to the prohibition.
International Maritime Organization (Marine Environment Protection Committee). (2014d). *Amendments to MARPOL Annex I (Amendments to regulation 43),* MEPC 67/7 (23 June 2014), Annex adding the carriage of HFO as ballast to the prohibition (Secretariat).
International Maritime Organization. (2014e). *Report of the Marine Environment Protection Committee on its 67th Session,* MEPC 67/20 (31 October 2014), para. 10.1.
International Maritime Organization (Maritime Safety Committee). (2015a). *Report of the Correspondence Group on the Development of guidance on a methodology for determining limitations for operation in ice,* MSC 95/3/7 (6 March 2015) (Norway).
International Maritime Organization. (2015b). *Report of the Maritime Safety Committee on its 95th Session,* MSC 95/22 (19 June 2015), p. 22, para. 3.91.
International Maritime Organization (Sub-Committee on Ship Systems and Equipment). (2015c). *Additional performance and/or test standards in support of the implementation of the Polar Code,* SSE 3/15/4 (15 December 2015) (Argentina, the Marshall Islands, New Zealand, Norway and Vanuatu).
International Maritime Organization. (2015d). *Report of the Sub-Committee on Ship Systems and Equipment on its 3rd session,* SSE 3/16, p. 50, paras. 15.15–15.16.
International Maritime Organization (Maritime Safety Committee). (2015e). *Request for data on incidents within polar waters,* MSC 95/21/3 (24 March 2015) (Iceland, New Zealand and South Africa).
International Maritime Organization (Maritime Safety Committee). (2015f). *Request for data on incidents within polar waters,* MSC 95/21/11 (14 April 2015) (Friends of the Earth International (FOEI) and Pacific Environment).
International Maritime Organization. (2015g). *Report of the Maritime Safety Committee on its 95th Session,* MSC 95/22, p. 85, para. 21.24.

International Maritime Organization. (2015h). *Report of the Marine Environment Protection Committee on its 68th Session*, MEPC 68/21 (29 May 2015), pp. 19–21, paras. 3.24–3.30.
International Maritime Organization. (2015i). *Report of the Marine Environment Protection Committee on its 68th Session*, MEPC 68/21 (29 May 2015), p. 48, para. 10.3.
International Maritime Organization (Sub-Committee on Pollution Prevention and Response). (2015j). *Report of the Working Group on Air Pollution from Ships*, PPR 2/WP.5 (22 January 2015), paras. 35–47.
International Maritime Organization (Sub-Committee on Pollution Prevention and Response). (2015k). *Report to the MEPC*, PPR 2/21 (16 February 2015), paras. 8.1–8.7.
International Maritime Organization (Maritime Safety Committee). (2016a). *Report of the Correspondence Group on the Development of guidance on a methodology for determining limitations for operation in ice*, MSC 96/3/4 (9 February 2016) (Norway).
International Maritime Organization. (2016b). *Report of the Maritime Safety Committee on its 96th Session*, MSC 96/25 (31 May 2016), paras. 3.74–3.78.
International Maritime Organization. (2016c). *Report of the Maritime Safety Committee on its 96th Session*, MSC 96/25, pp. 7–18, paras. 3.75–3.78.
International Maritime Organization. (2016d). *Report of the Maritime Safety Committee on its 96th Session*, MSC 96/25, p. 23, para. 3.9.
International Maritime Organization. (2016e). *Report of the Maritime Safety Committee on its 96th Session*, MSC 96/25, p. 23, para. 3.93.
International Maritime Organization. (2016f). *Report of the Maritime Safety Committee on its 96th Session*, MSC 96/25, paras. 24.1–24.3.
International Maritime Organization. (2016g). *Report of the Sub-Committee on Ship Systems and Equipment on its 3rd session*, SSE 3/16 (24 March 2016), paras. 15.15–15.16, 16.2.12.
International Maritime Organization. (2016h). *Report of the Marine Safety Committee on its 97th session*, MSC 97/22, paras. 8.27–8.32, 21.7–21.8.
International Maritime Organization (Marine Safety Committee). (2016i). *Clarification on the requirements related to the initial and maintenance surveys required by the Polar Code*, MSC 97/16/2 (16 September 2016) (IACS).
International Maritime Organization (Marine Environment Protection Committee). (2016j). *The need to evaluate the status of effectiveness of Special Areas and Particularly Sensitive Sea Areas*, MEPC 70/8/1 (15 August 2016) (Russia).
International Maritime Organization. (2016k). *Report of the Marine Environment Protection Committee on its 70th Session*, MEPC 70/18 (11 November 2016), paras. 8.6–8.10.
International Maritime Organization (Marine Environment Protection Committee). (2016l). *Heavy fuel oil use by vessels in Arctic waters*, MEPC 69/20/1 (12 February 2016) (FOEI, WWF, Pacific Environment and CSC).
International Maritime Organization. (2016m). *Report of the Marine Environment Protection Committee on its 69th Session*, MEPC 69/21 (13 May 2016), paras. 20.3–20.4.
International Maritime Organization. (2016n). *Report of the Marine Environment Protection Committee on its 70th Session,* MEPC 70/18, para. 17.20.
International Maritime Organization (Marine Environment Protection Committee). (2016o). *Ship-mediated bioinvasions in the Arctic: Pathways and control strategies*, MEPC 69/INF.17 Annex (9 February 2016) (FOEI).
International Maritime Organization. (2016p). *Report of the Marine Environment Protection Committee on its 69th Session*, MEPC 69/21, para. 4.24.2.
International Maritime Organization. (2017a). *Additional performance and/or test standards in support of the implementation of the Polar Code*, MSC 97/21/3 (18 August 2016) (Argentina, the Marshall Islands, New Zealand, Norway and Vanuatu).
International Maritime Organization. (2017b). *Comments on document MSC 97/21/3* MSC 97/21/12 (30 September 2016) (ICS and CLIA).
International Maritime Organization. (2017c). *Report of the Maritime Safety Committee on its 97th session,* MSC 97/22, paras. 9.22–9.23, 16.6–16.7.

International Maritime Organization. (2017d). *Report of the Maritime Safety Committee on its 97th session,* MSC 97/22, paras. 21.9–21.10.

Kingston, M. (2015). *Implementing the Polar Code; Education about requirements and fostering best practice in operational safety to make it work: Insurance Industry Contributions*, presentation at ShipArc2015, 26 August 2015. http://commons.wmu.se/shiparc/2015/allpresentations/31

MarEx. (2016). *Arctic Expedition Cruise Operators Support HFO Ban* (16 November 2016). http://maritime-executive.com/article/arctic-expedition-cruise-operators-support-hfo-ban

OSPAR Commission (The Convention for the Protection of the Marine Environment of the North-East Atlantic). (2015). *Summary Record of BDC 2015*, BDC 15/10/1-E, pp. 21–24, paras. 5.22–5.32. http://www.ospar.org/meetings/archive/biodiversity-committee-13

OSPAR Commission (The Convention for the Protection of the Marine Environment of the North-East Atlantic). (2016a). *Summary Record of BDC 2016,* BDC 16/9/1-E, pp. 18–19, paras. 5.8–5.16. https://www.ospar.org/meetings/archive/biodiversity-committee-631

OSPAR Commission (The Convention for the Protection of the Marine Environment of the North-East Atlantic). (2016b). *Summary Record of Meeting of the OSPAR Commission (OSPAR) Tenerife: 20–24 June 2016*, OSPAR 16/20/1-E, p. 23, para. 6.28. https://www.ospar.org/meetings/archive/ospar-commission-650

Vard Marine Inc. (2015). *Fuel Alternatives for Arctic Shipping* rev. 1 (20 April 2015). http://awsassets.wwf.ca/downloads/vard_313_000_01_fuel_alternatives_letter_final.pdf

Part II
Arctic Ship Monitoring/Tracking

Arctic Environment Preservation Through Grounding Avoidance

R. Glenn Wright and Michael Baldauf

Contents

R. Glenn Wright (✉)
GMATEK, Inc., Annapolis, MD, USA

World Maritime University, Malmö, Sweden
e-mail: glenn@gmatek.com

M. Baldauf
Hochschule Wismar, Institute of Innovative Ship Simulation and Maritime Systems, Rostock-Warnemuende, Germany

World Maritime University, MaRiSa Research Group, Malmö, Sweden
e-mail: mbf@wmu.se

L. P. Hildebrand et al. (eds.), *Sustainable Shipping in a Changing Arctic*, WMU Studies in Maritime Affairs 7, https://doi.org/10.1007/978-3-319-78425-0_5

Abstract Research results are described that explore technological innovation to reduce ship groundings and collisions by significantly increasing watchstander situational awareness to environmental conditions below the waterline. This is especially relevant to ship navigation in the Arctic requiring transit through shallow, draft-constrained coastal and archipelago waters that are relatively uncharted, lack aids to navigation, without adequate search and rescue facilities, and teaming with surface and underwater hazards to navigation. Such conditions and events create excessive risk to life and property through grounding and greatly expose the environment and wildlife to pollution damage through oil and chemical spills. Results of research accomplished to date are provided and strategies developed to enhance ship owner and operator diligence in better preparing for Arctic transits. Recommendations for future work in related capacities are also provided for enhancing the Polar Code, International Maritime Organization (IMO) carriage requirements and the Convention on Standards of Training, Certification, and Watchkeeping (STCW).

Keywords Grounding · Arctic navigation · Underwater sensing · Geo-referencing · GNSS · Spoofing · Forward-looking sonar

1 Introduction

According to statistics of accident investigation reviews issued from various sources ship groundings are the leading cause of large oil spills, yet human error remains one of the main reasons for their occurrence. The underpinnings of groundings contributed to or even caused by direct human error can in many cases be traced to the watchstander simply being unaware that grounding was about to occur. This illustrates an extraordinary lack of situational awareness regarding the specific underwater terrain in which the vessel was operating and a loss of orientation toward ongoing events (Endsley 1995; Lundberg 2015). Such occurrences may be due to the fact that the only direct knowledge available to the watchstander regarding the physical environment below the waterline being a two or three-digit number provided by an echo sounder that represents depth immediately beneath the keel—and nowhere else. Deduction of grounding implications based upon such scanty sensory inputs is problematic at best, and is easily overlooked if the watchstander is distracted by another task. Modern Integrated Navigation Systems according to IMO Performance Standards provide alarm facilities that can be configured to warn of crossing a safety depth contour that may assist in avoiding groundings (IMO MSC.252(83) 2007). However, these alarms do not work properly where few if any soundings exist within the electronic navigation chart (ENC) data, and in areas exhibiting rapid changes in depths as warning is likely to come far too late to avoid rising bottom terrain.

Research is presently underway leading to the invention and application of new technology that can aid in reducing vessel groundings. One focus specifically chosen for this research is the Arctic, where new inroads are being made to explore this pristine frontier and make it accessible to pioneers ready to exploit claims to natural resources and to open new trade routes. The geographical and ecological aspects of the Arctic as represented by the Northern Sea Route (NSR) and the Northwest Passage (NWP) are described to illustrate the unique characteristics of these routes that make them particularly vulnerable to grounding events, and why the consequences of groundings in these areas are magnified. Also discussed are the vulnerabilities of technology used routinely in worldwide ship navigation that are accentuated in the Arctic environment, considering both the limitations of equipment as well as susceptibilities to human factor deficiencies and man-made interference and disruption. The potential to contribute towards building new infrastructure in areas where the establishment of traditional infrastructure is difficult or impossible is identified. Accomplishments achieved to date are also cited that illustrate the potential to approach the goal of increasing watchstander situational awareness of the underwater environment along the route of transit, the problems encountered and associated limitations.

2 New Technology to Navigate the Global Ocean

The navigation capabilities of modern vessels plying the world's oceans include ENC representing the latest in multibeam full coverage hydrographic surveys contained within and displayed to watchstanders using Electronic Chart Display Information Systems (ECDIS). Vessel positioning is accomplished to accuracies of a few meters using Global Navigation Satellite Systems (GNSS) integrated with a growing number of sophisticated sensors and tracked by other vessels and land-based operators, service and support organizations using terrestrial and space-based Automatic Identification System (AIS) receivers. Physical aids to navigation (AtoN) are supplemented with AIS-based AtoN capable of being placed where physical aids cannot.

Unfortunately, little of this is true in the Arctic. Navigation charts are in many cases blank or inaccurate. Hydrographic surveys rarely exist and, if they do, are likely to be decade's old and performed using obsolete technology. Satellite positioning systems and communications can be unreliable. Physical AtoN cannot be placed throughout much of the Arctic due to ice movement and AIS-based AtoN lack infrastructure required for their use.

It is quite possible for mariners today to be in much the same situation as the explorers of the fifteenth and sixteenth centuries when aiming their bows towards the unknown. However, research is currently underway to help reduce this problem, the timing of which is especially significant in that the Arctic is experiencing increasing vessel activity resulting from the development of natural resources as evidenced by the 22 July 2015 granting by the U.S. Bureau of Safety and Environmental Enforcement of final approval for drilling by Shell Oil at the Burger Prospect in the Chukchi Sea off Alaska (Shell 2015). Transit shipping across the Arctic is also increasing to take advantage of shorter routes between Europe and Asia. The transit of 30 vessels through the NWP occurred in 2012 as compared to 2–7 vessels completing this transit each year from 1990 to 2006 (NTDENR 2015; Macfarlane 2012). The first large cruise ship, *Crystal Serenity* (68,000 grt, draft of 24.6 ft. (7.5 m.) with capacity of 1080 passengers and 635 crew) made this transit in 2016 and is taking reservations for their 2017 cruise (Crystal 2017). NSR traffic saw 46 vessels transiting in 2012 increasing to 53 in 2014 (NSRIO 2012, 2014). The potential exists between 2040 and 2059 for non ice-classed vessels to navigate previously inaccessible areas of the Arctic without icebreakers (Sullivan 2013). This is made possible in part due to diminishing sea ice coverage opening shipping routes that have previously been impassible. These routes include the NSR across the top of Russia and Norway, and the NWP along Denmark (via Greenland) and through Canada and along the United States that span thousands of nautical miles across shallow coastal seas and archipelagos where the potential risk for grounding is high. Accentuating these risks is the harsh climate that can cause the breakdown of machinery vital to safe vessel operation. The lack of adequate hydrographic survey and accurate nautical charts providing sparse depth information along much of this route can also render modern ECDIS equipment ineffective and unreliable.

3 The Arctic Routes

A series of different sailing lanes, the NSR from Kara Gate to the Bering Strait covers some 2200–2900 nautical miles depending upon route and ice conditions. Likewise, the NWP covers a distance of about 1300 nautical miles from Baffin Island to Banks Island plus another 1000 nautical miles from Banks Island to the Bering Strait. These routes are illustrated in Fig. 1.

The NSR consists of shallow waters along the length of the coastline from the Norwegian-Russian border in the west (in the Barents Sea) to the Bering Strait where average depths of the East Siberian and Chukchi seas are 58 and 88 m respectively, the Laptev Sea where 66% of its area along the coast is in depths of 100 m or less, the Kara Sea with an average depth of 90 m and the Barents Sea with 10–100 m depths along the coast in the southeastern region sloping to depths of 200–300 m to the northwest (AMSA 2009). Depth limitations at various straits along the route result in an overall controlling depth of 12.5 m (Carmel 2013).

The NWP is comprised of the marine routes between the Atlantic and Pacific oceans along the top of North America that span the straits and sounds of the Canadian Arctic archipelago and along the northern slope of Alaska in the United States (AMTW 2004). This archipelago is comprised of approximately 36,000 islands with variable depths, especially as the continental landmasses and islands are approached, providing a highly complex geography for vessel navigation (AMSA 2009a). The overall controlling depth is 10 m (Carmel 2013).

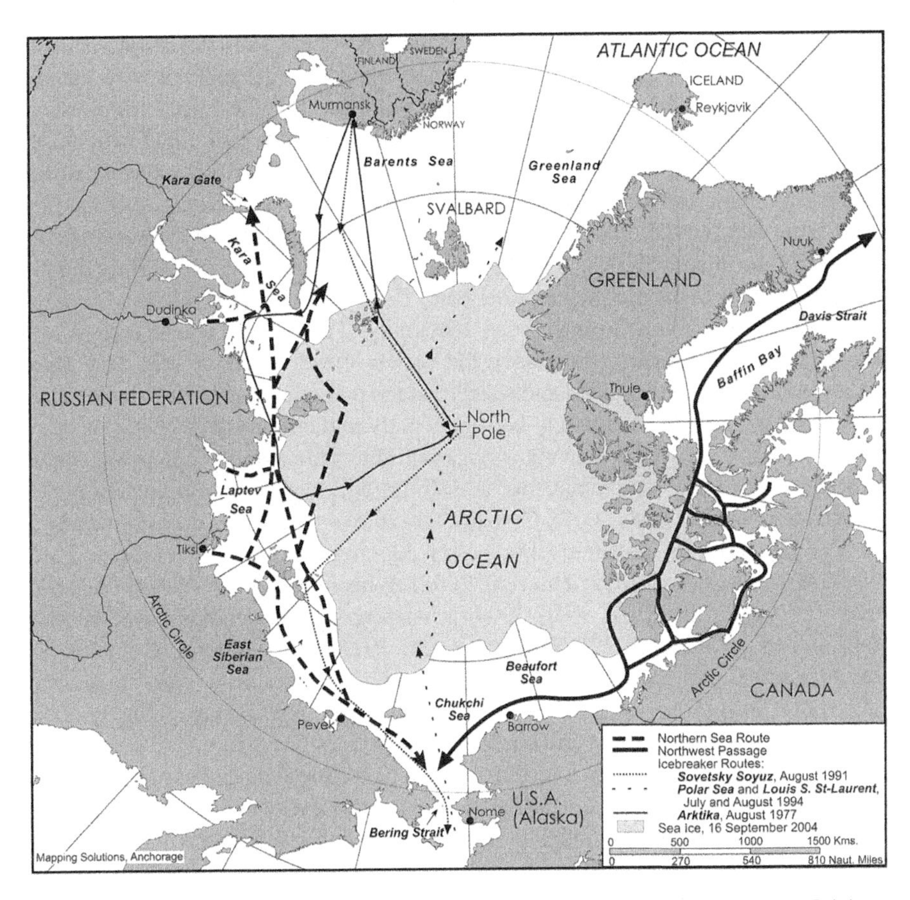

Fig. 1 Arctic Routes: Northern Sea Route (NSR) and Northwest Passage (NWP). Source: Brigham, L. and B. Ellis (Editors), 2004. Arctic Marine Transport Workshop, 28–30 Sept. 2004, Scott Polar Research Institute, Cambridge University. Mapping Solutions, Anchorage. Used by permission

4 Increasing Arctic Casualties and Groundings

The increase in shipping activity in the Arctic engenders corresponding increases in groundings that comprise a significant percentage of shipping casualties. Data depicting such a trend indicates shipping casualties in Arctic waters increased to 45 per year between 2009 and 2013 from only 7 during 2002–2007. Damage to machinery caused a third of these incidents, higher than the average elsewhere, reflecting the harsher operating environment (Allianz 2014a). When viewed in terms of total losses, the Russian Arctic/Bering Sea and the Canadian Arctic/Alaska experienced 46 (3.1%) of the 1462 total losses worldwide in the 10 year period between 2002 and 2011; increasing to 6 (5.1%) of 117 total losses worldwide in 2012 and 4 (4.3%) of 94 total losses worldwide in 2013 (Allianz 2014b). In 2014 there are 55 shipping casualties in Arctic waters as compared to just 3 a decade ago (Allianz 2015). The potential environmental impact of groundings may include containers lost overboard from vessels aground but otherwise intact that wash up in the Arctic. An incident occurred in 2004 when a container tank filled with plastic polymers was lost from a vessel enroute to Korea from South Africa 4 months earlier that washed up close to a remote Russian community in the Commander Islands in the Bering Sea (UNDP 2009). Attempts to move the container led to spilling about 15 tons of the chemicals. Local residents were poisoned and the community did not understand how to handle such an emergency (Eason 2015). Ståle Hansen, President and CEO of the Skuld Property and Indemnity (P&I) Club, recently cited cost estimates of $500 million to remove the wreck of the container ship M/V *Rena* that grounded in New Zealand and stated that the potential cost of such incidents in the Arctic would almost certainly be significantly higher (Hansen 2015).

Groundings involve a moving navigating ship either under command, under power, or not under command that is drifting, striking the sea bottom, shore or underwater wrecks (MSC 2008). Casualty data shows groundings are consistently at or near the top of the most frequent types of shipping accidents (TSB Canada 2013; NMD 2011; Butt et al. 2012; Zhu et al. 2002). Where accidents resulted in oil spills over 700 tons in the years 1970–2009, grounding was the leading cause at 36%, followed by collisions at 29% (Nicolas-Kopec 2012). Although not all groundings result in a total loss, the consequences of groundings can result in a serious accident in terms of cost to the safety of crew, vessel and cargo and damage to the environment due to physical impact and chemical spills.

The causes of groundings are many and include loss of machinery (e.g., engine, rudder, propeller and/or anchor); poor hydrographic surveys and nautical charts that fail to accurately depict bottom contours and identify hazards to navigation; human factors such as fatigue, errors in properly operating and interpreting ship sensors, indicators and alarms, and poor voyage planning. A recent report assessing probabilities for groundings based on accident reports indicates inadequate communication and cooperation on the bridge is the most significant contributing factor in a grounding accident (Mazaheri 2013). In 2008 poor hydrographic surveys were cited by the Helsinki Commission as a contributing factor to an increase in groundings in the Baltic, particularly in the shallow waters around Denmark (Eason 2010). These are just two of several factors that are of greater significance in the Arctic environment for ships that are poorly prepared and ill-equipped in terms of physical machinery, navigation equipment as well as crew training and experience that contribute towards the increase in accidents.

5 Challenges of Arctic Navigation

There are many operational, logistical and technical challenges to ship navigation in the Arctic. Those challenges having to do with expanding watchstander environmental situational awareness center on navigation charts, aids to navigation and sensors that provide direct insight into the environmental factors relevant to transit. In the Arctic, infrastructure is minimal and knowledge of the underwater environment is imprecise at best and wholly without survey in many areas. The risks and consequences of grounding dramatically increase due to potentially long rescue times and a lack of assets for salvage and environmental remediation in the event of casualty. These are key concerns identified by the Arctic Council for which action is being taken by member states and partners (AMSA 2009, 2015). This risk extends to both civilian and Government operations. Adm. Paul Zukunft, Commandant of the U.S. Coast Guard, when speaking about icebreaker capacity recently stated that the United States currently has no self-rescue capability in the higher latitudes (Zukunft 2015).

5.1 Navigation Charts

According to the U.S. National Oceanographic and Atmospheric Administration (NOAA) charting data in much of the Arctic is inadequate or nonexistent (ANCP 2013). The *U.S. Coast Pilot* states that much of the Bering Sea area is "only partially surveyed, and the charts must not be relied upon too closely, especially near shore. The currents are much influenced by the winds and are difficult to predict; dead reckoning is uncertain, and safety depends upon constant vigilance" (NOAA 16006 2010). NOAA has estimated that 4.7% of the U.S. maritime Arctic is surveyed to modern hydrographic standards (HSRP 2016). It has expanded the deployment of survey ships to improve navigational safety in the Arctic, conducting surveys of the ocean floor to measure water depths and search for navigational dangers and increasing Arctic charting activities in anticipation of growing vessel traffic in the region. Hydrographic projects for Alaska in 2015 cover an area of 2800 square nautical miles, plus the 12,000 linear nautical miles for a potential Arctic shipping route from Unimak Island to the Chukchi Sea (MAREX 2015). The NOAA Arctic Charting Plan providing information about existing, recently added, and proposed new ENC coverage in U.S. Arctic waters was recently released for public comment (ANCP 2015).

Most of Canada's Arctic waters have not been surveyed to modern standards except for Lancaster Sound, Barrow Strait, the Beaufort Sea, Amundsen Gulf, and the approaches to settlements and some mining sites. Spot soundings through the ice or reconnaissance track soundings are the only survey data available in much of the Arctic. In the Beaufort Sea, a route through the area with a large number of pingos has been surveyed in greater detail (NGIA 180 2014). It was reported the Canadian Hydrographic Service says that 10% of the Canadian Arctic has been surveyed to modern standards (AMSA 2009b).

Russian Arctic survey coverage along the NSR, illustrated in Fig. 2, appears more comprehensive than the NWP. The Russian Federation's Hydrographic Service reported to the Arctic Council since 2011 the Northern Sea Route's least-studied areas are being surveyed over a 31,000 km^2 area of the sea floor. Survey results have

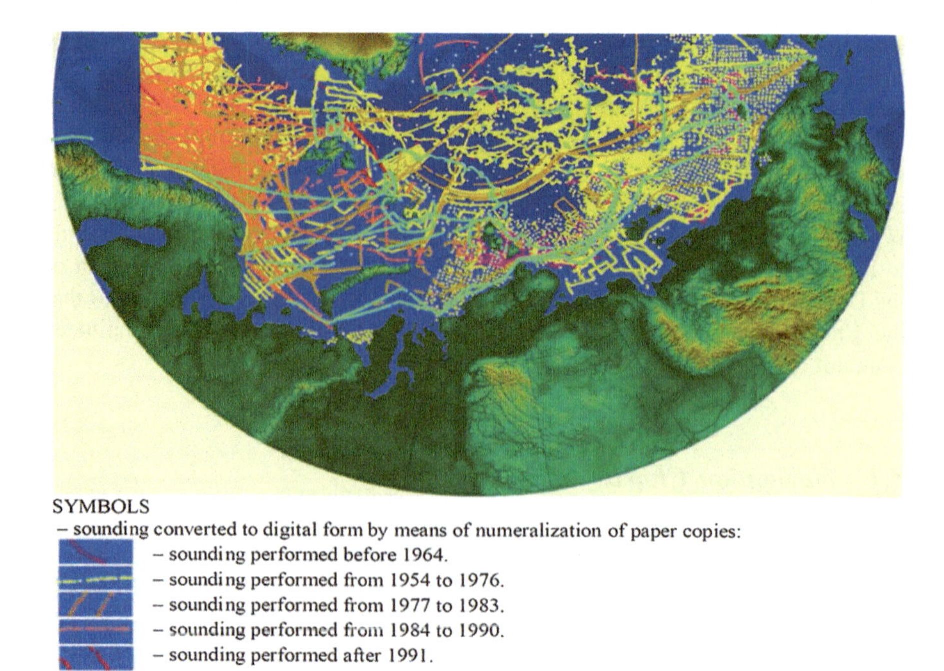

Fig. 2 Survey coverage of the Arctic based on the systematic sounding data (ARHC 2011)

been digitized and developed into electronic navigation charts. Sixty-eight marine electronic navigation charts were issued based on the survey results. In 2012 two hundred charts were developed (AMSA 2013).

Overall, of a total of 7,018,392 linear nautical miles (LNM) in the Arctic across the waters of Canada, Denmark (Greenland), Norway and the United States; 4,989,368 LNM (71.1%) of the depths remain unassessed by hydrographic survey (Hains 2014). Navigation charts depicting these areas are often blank and/or contain soundings that should not be relied upon. Of further concern are navigation charts for the Arctic that have been created according to different datums used to define the heights, depths and locations of hydrographic features. This requires extreme diligence to ensure accurate positioning when transiting areas using a chart with a datum different from that of the Global Navigation Satellite System (GNSS) in use, and when transiting areas from one chart to another where the charts were compiled using different datums.

5.2 *Electronic Chart Display and Information System (ECDIS)*

ENC include a Category of Zone of Confidence (CATZOC) rating that indicates the level to which the data meets minimum criteria in terms of position and depth accuracy (IHO 57). Except for ENC that contain the most recent modern surveys most Arctic charts do not have an assigned CATZOC value. This condition has direct implications on voyage planning and monitoring activities in terms of deriving keel depth and beam clearance safety margins when planning route(s) of passage. Safety contour and depth alarms required to be triggered by ECDIS would not be available in areas exhibiting a lack of soundings. Where soundings do exist on charts without an appropriate CATZOC rating, reliance upon them to set alarms would be foolhardy as the soundings should be assumed to be unreliable. In addition, incorrect setting of own-ship safety depth to a value greater than the current setting may provide an ECDIS display that omits any obstructions and shallows that may be present in the ENC.

5.3 *Aids to Navigation (AtoN)*

The inability to place navigational aids and buoys in constantly changing ice conditions produces increased risk in vessel navigation activity in the Arctic and sub-Arctic regions. This has resulted in there being few, if any physical aids to navigation existing along the coast and outside of heavily trafficked areas nearby and within ports. Buoy placement and maintenance activity in sensitive ecological areas is also problematic, and underdeveloped regions lack assets and infrastructure needed to provide navigation services. The absence of physical aids to navigation also eliminates the potential to use radar navigation to identify physical aids during times of poor visibility.

This shortcoming may be alleviated somewhat in the near term through the use of AIS-based AtoN at some locations in the Arctic. Deployment of this technology is currently being accomplished across the rest of the United States to supplement physical AtoN in areas where additional aids are beneficial to navigation. Their use in the Arctic is limited to areas where infrastructure exists to install and maintain these aids, and by the line of sight to their deployment locations due to very-high frequency (VHF) radio transmission characteristics.

5.4 *Communications*

The ability to reliably communicate in the Arctic region will become much more important over time as vessel activity increases. One factor that affects reliable communications is the large distances involved that favor the use of satellite and high frequency (HF) communications over shorter range VHF and ultra-high frequency (UHF) communications that is generally limited to the line of site. The negative influence of atmospheric effects on HF and satellite communications can make these methods unreliable during magnetic storms initiated by solar flare events. Heavy precipitation and antenna icing can also interfere with satellite communications. VHF and UHF communications is useful for communications between vessels, for tactical control and for emergency communications with mariners operating in the local area. Further research in the Arctic is needed to evaluate system coverage limitations and create improved signal modeling techniques.

There is also a lack of reliable navigation safety information to help mariners identify, assess, and mitigate risks the in Arctic region due to minimal Maritime Safety Information infrastructure in the region. The U.S. Coast Guard currently does not own any of its own electronic Maritime Safety Information (eMSI) infrastructure in Alaska, and is examining the use of existing AIS infrastructure to demonstrate the delivery of critical eMSI information to local mariners and ultimately improve navigational safety (CGRDC 2014). Similar efforts are also underway elsewhere including the European MonaLisa and its follow up projects Mona Lisa 2.0 and See Traffic Management Validation that among others, aim to integrate environmental sensitivity data and dynamic route planning within Maritime Spatial Plans, and the Accessibility for Shipping, Efficiency Advantages and Sustainability (ACCSEAS) project for improving maritime access to the NSR by developing and implementing eNavigation regional services and to prototype novel marine navigation and communication concepts (ACCSEAS 2015).

5.5 *Environmental Sensing*

Several technologies such as GNSS, radar and sonar fulfill mandatory carriage requirements on-board modern vessels worldwide, provide vessel positioning information and offer navigation assistance by directly sensing the immediate physical environment. However, each of these technologies exhibit specific limitations as a result of their use in the Arctic. For example, special challenges to GNSS exist at high latitudes due to low elevation with satellite inclination (55° for GPS). Limited services like Wide Area Augmentation System (WAAS) and dynamics associated with the auroral region and polar cap also make it difficult to predict ionospheric corrections and integrity (Kvam and Jeannot 2013).

The use of radar as a means to fix vessel position is also limited in the Arctic. Absent the presence of short range AtoN, radar piloting that relies on these aids while in inland, harbor and approach, and coastal navigation is not possible. Using radar to determine bearing and distance to known objects and land features is possible where islands are present at sea and mountains extend along the shoreline. However, much of the Arctic landscape is regular and featureless, sloping gradually from the shoreline with low and grassy inland terrain making the detection of natural features difficult at best. Since much of the area is uninhabited, there is also a lack of manmade objects from which bearings may be taken.

Sonar technology may be used to identify bottom features, gradients and depth contours useful to navigation. The echo sounder provides instantaneous depth information directly below the keel that can reassure the mariner there is sufficient water at the present time to avoid grounding. However, there is no assurance that the depth immediately forward of the bow is sufficient for continued navigation. Little navigational assistance is provided through the comparison of live echo sounder readings against navigation charts containing no or few soundings.

6 Enhanced Environmental Situational Awareness

The basis for an innovative means of vessel navigation in the world's uncharted frontier involves original research in the progression and integration of three seemingly unrelated technologies: 3-dimensional forward looking navigation sonar, Virtual (non-AIS) aids to navigation, and georeferencing of position to physical terrain features of the bottom. The approach described is eminently suitable for Arctic service in terms of usefulness, ease of deployment and low cost for installation and maintenance.

6.1 Navigation Sonar

Research has been performed by the authors on the use of 3-dimensional forward looking sonar as a means for improving safety of navigation. The premise of navigation sonar is based upon the assertion that own ship sensors should be adequate to detect soundings and bottom configuration both at its current position as well as forward of the vessel in the path of transit sufficient to ensure safe navigation. Such capability is especially appropriate where soundings and bottom configuration are inaccurate or not available due to poor or lack of accurate hydrographic survey. There is no IMO carriage requirement for navigation sonar at present.

The detection of bottom features, objects and soundings by determining range, azimuth and elevation information uses methods that can generally be described as variations on transmitting a steerable sonar signal ahead along the path of the vessel or by transmitting a single ping from which snapshots of the environment are obtained. Through this process a mosaic of the bottom topography and specific targets is created as the vessel proceeds on its course. Navigation sonar is capable of detecting and displaying the underwater environment looking ahead as much as 1000 m forward of the bow at speeds up to 25 knots. In shallower waters its range can extend from 8 to 20 times the depth ahead, depending on bottom and target conditions (FLS-1000, NOAS). It is most effective when the bottom topography slopes upwards, and when targets are large and consist of hard rock and/or coral that provide good acoustic signatures.

Enhancements to navigation sonar capabilities can be made to survey the sea bottom as well as to detect hazards attached to the bottom and floating in the water column to aid in Arctic navigation, where soundings on charts rarely exist and the vast majority of hazards to navigation have yet to be discovered. The utility of this technology as a means to avoid such hazards has been explored in a simulation of the M/V Costa Concordia disaster (Wright and Baldauf 2014a). Its use as a means to perform hydrographic survey has also been discussed in terms of International Hydrographic Office (IHO) standards (Wright and Zimmerman 2015; Wright and Baldauf 2016a). The potential exists to accomplish complete survey coverage with navigation sonar for the transit route with horizontal and vertical accuracies within IHO standards using this approach. Such data can supplement national hydrographic organizations' efforts in the collection of survey data in remote parts of the world and in areas lacking recent survey.

6.2 *Virtual Aids to Navigation*

Virtual AtoN based upon AIS technology are rapidly being deployed on a worldwide basis as a supplement to physical AtoN to increase hazard visibility to mariners. While this is a valuable technology, the use of AIS-based AtoN is severely restricted in the Arctic due to lack of existing infrastructure to provide power and communications for health monitoring, access for maintenance and VHF radio range limitations. The authors have shown it is possible to create (non-AIS) Virtual AtoN, defined by the International Association of Lighthouse Authorities (IALA) as something that "does not physically exist but is a digital information object promulgated by an authorized service provider that can be presented on navigational systems" (IALA O-143 2010). Methods used for the creation of Virtual AtoN have been described in a real-time, shipboard implementation of this technology (Wright and Baldauf 2014b, 2016b).

Virtual AtoN technology represents a major step beyond the capabilities of existing AtoN, although critical limitations exist that are inherent in their design and implementation. Qualifying as short range AtoN, Virtual AtoN appear neither visually nor directly on radar. AIS-based virtual AtoN appear on an AIS radar overlay and also on ENC/ECDIS displays. However, verification techniques to ensure AIS and ENC/ECDIS positions coincide have yet to be developed. Virtual AtoN may appear only on ENC/ECDIS displays. Such implementations can result in a potential "single point of failure" scenario that may cause false conclusions and possible system failures that may go undetected. A comprehensive Virtual AtoN verification approach to help overcome this deficiency and ensure virtual AtoN (AIS and non-AIS) are watching properly after deployment is possible using georeferencing techniques.

6.3 Georeferencing

Georeferenced navigation using terrain features and manmade object recognition has been in use for many years as a means of cruise missile and other unmanned aerial vehicle (UAV) navigation across the landscape. This has been made possible through highly detailed millimeter-resolution radar and laser-surveys of the land areas from aircraft and satellites, whereas the ocean depths are presently surveyed to a maximum resolution of about 5 km (Copely 2014).

A novel implementation of georeferencing has been examined based upon the extraction of features from both multibeam and navigation sonar data represented in the ENC, and correlation between vessel position indications and physical environmental features to verify that Virtual (and physical and AIS) AtoN are watching properly. Comparison between ENC soundings and echo sounder depths along the path of transit can be performed even if precise positioning information normally acquired using GNSS, AIS and other sources are unavailable due to a variety of manmade and natural events. The techniques used to accomplish this have been discussed based upon the results of a series of experiments conducted at sea (Wright and Baldauf 2015, 2016a). Environmental features are identified in terms of bottom depth soundings for the ENC and echo sounder as well as the difference between these two sources, differences in bottom slope and differences in the rate of change of bottom slope resulting in new capabilities to automatically detect discrepancies in either bottom conditions or GNSS positioning that may require additional caution.

7 Grounding Avoidance Strategies

An overall strategy to reduce the potential for grounding in the Arctic must originate during vessel preparation and outfitting for Arctic operations, continue through voyage planning, be implemented throughout the entire voyage to monitor its progress, and result in products that preserve the information and knowledge acquired during transit useful for crew debriefing and other purposes. Such products should also support the subsequent application of the information and knowledge gained to benefit future mariners who transit the same areas. Strategies must be developed in accordance with a Safety Management System that acknowledges the special circumstances of Arctic navigation and includes provisioning appropriate for correct vessel outfitting and crew training. Specific concepts that may be useful in developing strategies to avoid groundings are illustrated using the incidents cited. Applications and advantages that navigation sonar and Virtual AtoN may provide in such circumstances are also discussed.

Official reports commonly include accounts of groundings that have occurred due to a seeming lack of environmental situational awareness regarding the proximity of the keel of a vessel with the sea bottom. In several cases the watchstander was not even aware that grounding had occurred even several minutes after it happened.

Descriptions of incidents extracted from the Nautical Institute's Mariners' Alerting and Reporting Scheme (MARS) include:

- Chart corrections while navigating contribute to grounding (Report 201528)
- ECDIS unassisted grounding (201505)
- Self-induced fatigue contributes to grounding (201451)
- Fog bound grounding under pilotage (201420)
- Vessel strikes island (201356)
- Unpublished draft restrictions lead to grounding (201335)
- Improper bridge procedures and ECDIS use caused grounding (201257)
- Grounding in channel (200657)
- Cross checking positions lead to grounding (200524)
- Near grounding due to permanent highlighting of charts (200109)

These are a few of many examples of groundings in non-Arctic areas; many of which are highly travelled, well charted and represent the best of circumstances in terms of information and sensor availability. The Arctic is generally less than optimal in these terms. Groundings and their consequences can be considerably more serious in the Arctic where harbors of refuge are few, search and rescue facilities are scarce or non-existent, and salvage capabilities may take weeks or months to arrive on-scene due to weather and sea conditions. Several detailed accounts of groundings follow that will be used to illustrate grounding avoidance strategies based upon the results of research accomplished to date.

7.1 M/V Clipper Adventurer

On 27 August 2010 the cruise chip *Clipper Adventurer* ran aground in Coronation Gulf, Nunavut in the Northwest Passage on a shoal discovered in 2007, published in Canadian Notices to Shipping but not officially charted until June 2012. The Transportation Safety Board of Canada report concludes that the vessel ran aground after the bridge team chose to navigate a route on an inadequately surveyed single line of soundings, as shown in Fig. 3 (TSB Canada 2012a). This occurred in an area where the depths of the waters were virtually unknown.

Analysis M/V *Clipper Adventurer* was proceeding on autopilot at a speed of 13.9 knots when the vessel ran aground on hard rock shelf (TSB Canada 2012b). Bottom depth directly before the shelf, illustrated in Fig. 3, exceeded 100 m. Using navigation sonar with a 1000 m range, at a speed of 13.9 knots the vessel would have had approximately 128 s or 2.1 min advance warning before the grounding would have occurred based upon the following equation:

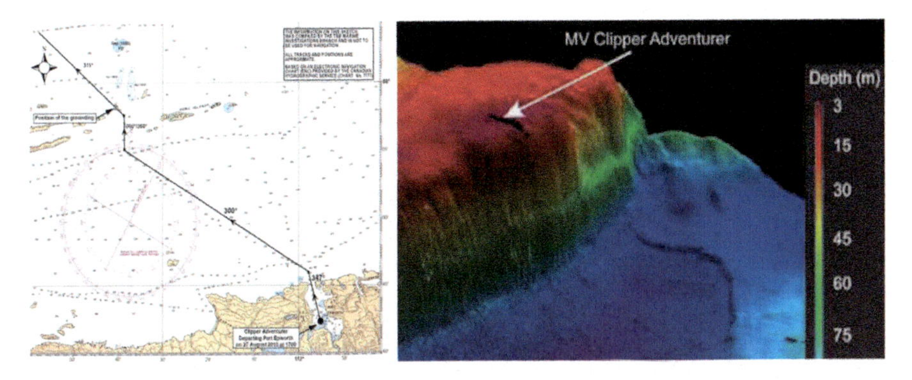

Fig. 3 Navigation Chart and Bottom Topography: *Clipper Adventurer* Grounding [59]

$$Advance\ Warning\,(\,sec\,) = \frac{MDR(\mathrm{m}) - SRR(\,\mathrm{sec}\,) - APT(\,\mathrm{sec}\,) - WRT(\,\mathrm{sec}\,)}{Vs\left(\frac{\mathrm{m}}{\mathrm{sec}}\right)} \quad (1)$$

where logical arguments may be made to establish values for each of these parameters:

Speed of Vessel (*Vs*) = 13.9 knots
Screen Refresh Rate (*SRR*) = 2 s
Maximum Detection Range (*MDR*) = 1000 m
Alarm Processing Time (*APT*) = 4 s
Watchstander Response Time (*WRT*) = 5 s
1 m/s = 1.9438 knots

The Marine Investigation Report identifies that navigation sonar was indeed installed on *Clipper Adventurer*, however it was not operational and in an unserviceable condition. This unit had a maximum detection range of 500 m and, if operational and the vessel was operating at 6 knots, the report concluded the crew would have had approximately 2 min advance warning of the shelf (TSB Canada 2012c). In either case there may not have sufficient sea room to halt all forward motion of the vessel. However a reduction in speed and course change may have altered the final circumstances such that the grounding may have been avoided or the severity of the grounding may have been lessened.

Virtual AtoN placed to mark the perimeter of the shelf would have been visible on ECDIS, providing warning adequate for the crew to take action to alter course and speed to avert grounding.

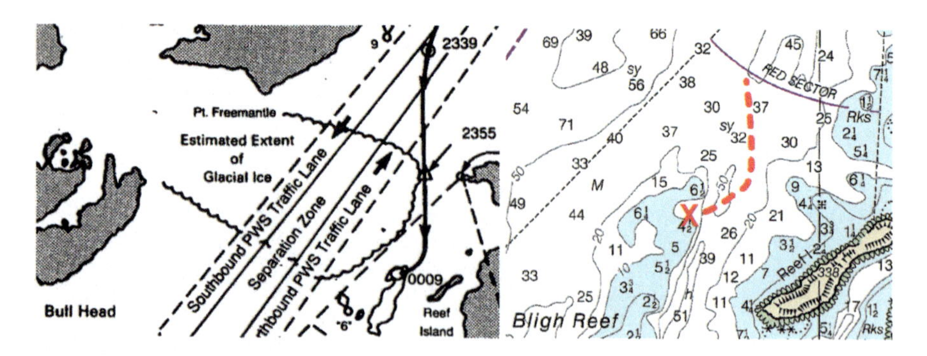

Fig. 4 *Exxon Valdez*—Glacial Ice Intrusion into Traffic Lanes; Bligh Reef Navigation Chart. Source: NTSB (1990b); Source: NOAA Chart #16708 (in fathoms)

7.2 M/V Exxon Valdez

On 24 March 1989 the U.S. Tankship *Exxon Valdez* loaded with 1,263,000 barrels of crude oil ran aground on Bligh Reef while departing the traffic separation scheme (TSS) to avoid ice. Approximately 258,000 barrels of oil spilled resulting in damage to the vessel and cargo estimated at $25 million and a cost for environmental cleanup in 1989 at $1.85 billion (NTSB 1990a). There was no indication in the report to contradict that the area was well surveyed and accurately depicted on nautical charts. The probable cause of the grounding was listed as failure to properly maneuver the vessel because of fatigue and workload along with several other factors.

Analysis *Exxon Valdez* was proceeding at night on autopilot at a speed of 16 knots when the vessel ran onto Bligh Reef NTSB 1990b. The depth of the hard bottom before the reef, illustrated in Fig. 4, averaged 30 fathoms sloping to 50 fathoms before rising to 4 fathoms at the reef. Neither navigation sonar nor ECDIS existed at the time of this grounding. A vessel transiting this same course today using navigation sonar with a 1000 m range at a speed of 16 knots would have had several minutes advance warning that they were in trouble before the grounding. First significant indication (other than the depth was half what is should have been in the inbound lane of the TSS) would have occurred around 1000 m out as the bow was directed towards Reef Island, where a solid wall to the water's edge directly in the path of the vessel would have been displayed. The second and final indication would have occurred after the turn away from Reef Island towards Bligh Reef, where another solid wall to the water's edge directly in the path of the vessel would have been displayed. Conceivably, navigation sonar might have been useful to navigate the channel between Reef Island and Bligh Reef once they found themselves in this situation.

Virtual AtoN placed to mark the perimeter of Bligh Reef, Reef Island and/or the channel between the two would have been visible on ECDIS, providing warning adequate for the crew to take action to alter course and speed to avert grounding.

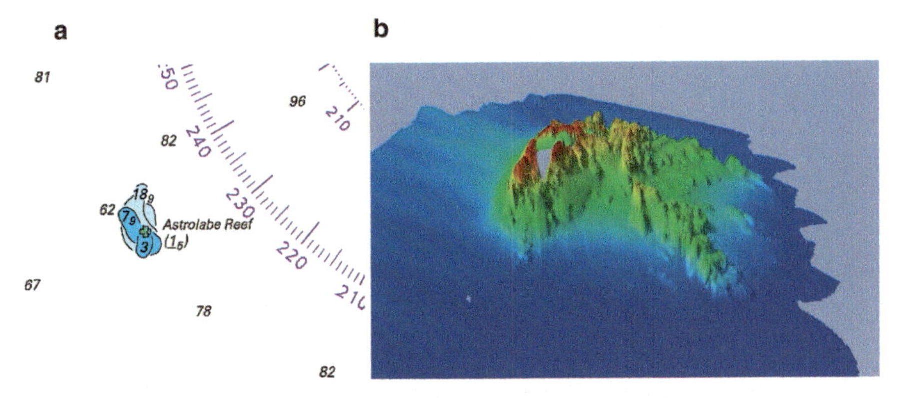

Fig. 5 Navigation chart and bottom topography: *Rena* Grounding. (**a**) Source: New Zealand Chart #541; (**b**) Data from WASSP 160F multibeam system. www.oceanDTM.com (used by permission)

7.3 M/V Rena

On 4 October 2011 the container ship *Rena* bound for the New Zealand port of Tauranga ran aground on Astrolabe Reef (illustrated in Fig. 5) at a speed of 17 knots. The master authorized the watchkeepers to deviate from the planned course lines on the chart to shorten the distance and expedite their arrival. Instead of passing two nautical miles north of Astrolabe Reef the second mate reduced the distance to one nautical mile to save time. He then made a series of small course adjustments towards Astrolabe Reef to make the shortcut. As a consequence *Rena* made a ground track directly for Astrolabe Reef. About 200 tons of heavy fuel oil were spilled in the accident, and a substantial number of cargo containers were lost overboard (TAIC 2014a).

Analysis *Rena* was proceeding at night on autopilot at a speed of 17 knots when the vessel ran onto Astrolabe Reef (TAIC 2014b). The depth of the bottom before the reef, illustrated in Fig. 5, averaged 80 fathoms rising to 4 fathoms at the reef. Were *Rena* equipped with navigation sonar with a 1000 m range at a speed of 17 knots the vessel would have had approximately 1.7 min advance warning before the grounding. The approaches to the reef would have been evident before then.

Although paper charts were being used, Virtual AtoN placed to mark the perimeter of Astrolabe Reef would have been visible on ECDIS, providing warning adequate for the crew to take action to alter course and speed to avert grounding.

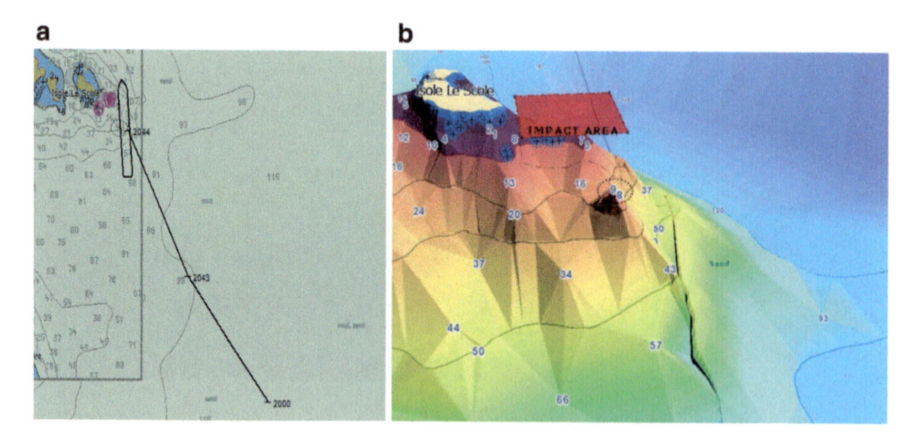

Fig. 6 Navigation chart and bottom topography: *Costa Concordia* Grounding. (**a**) Source: MCIB (2012b); (**b**) Source: http://blog.maxsea.com/ (used by permission)

7.4 M/V Costa Concordia

On 13 January 2012 the passenger ship *Costa Concordia* with 4229 persons on board while in navigation in the Tyrrhenian Sea off the coast of Italy collided with Scole Rocks adjacent to Isle de Giglio, as illustrated in Fig. 6. The ship had recently departed the port of Civitavecchia and was enroute to Savona, Italy. The hull was breached, thirty-two souls are dead or missing and the vessel was lost (MCIB 2012a). Human error is cited as a primary cause for this accident (Schröder-Hinrichs et al. 2012). Specifically, a chart of inadequate scale was used to superficially plan and execute a maneuver at night that was unsuitable in terms of distance from the coast and adjacent rocks and outcrops as well as depth below the keel. This grounding ultimately became the most costly maritime salvage event in history.

Analysis *Costa Concordia* was proceeding at night at a speed of 16 knots when the vessel ran onto Scole Rocks. The position marked 2000 on the chart to the left lies approximately 1100 m from Scole Rocks with depth in excess of 100 m with a mud bottom. The 100 m bottom contour is approximately 500 m further along the course, with upslope bottom rising to the 10 m bottom contour at the maximum range of the navigation sonar off the starboard bow. The bottom is also transitioning from mud to rock, with a resultant increase in acoustic reflectivity of the bottom material. Clear indications of the approaches to Scole Rocks would have appeared off the starboard bow on the navigation sonar display. Even with the switched off ECDIS alarms, such an indication potentially should have initially alerted the Master to the existence of a problem since the appearance of Scole Rocks would have been expected off the port bow based upon the passage planning. Furthermore, the depth contour displayed on the navigation sonar would clearly indicate insufficient depth in an unexpected location. Approximately 1.8 min advance warning could have provided of the pending grounding showing the approaches to Scole Rocks culminating with a wall extending up to the water's edge. Up until 150 m prior to grounding, an escape path to clear water would have been evident off the starboard bow. Baldauf and Wright (2014) provides a concept for triggering anti-grounding warnings taking into account simulation-based prediction of the ship's maneuvering behavior according to the prevailing circumstances of the concrete situation.

Virtual AtoN placed to mark the approaches to Scole Rocks would have been visible on ECDIS, providing warning adequate for the crew to take action to alter course and speed to avert grounding.

7.5 *MSV Fennica*

On 3 July 2015 the MSV *Fennica*, while enroute between Dutch Harbor Alaska and the Shell Oil drilling field in the Chukchi Sea, was holed when it traveled near a previously uncharted rocky shoal (Dlouhy 2015). The nautical chart of the inadequately surveyed area indicated several meters clearance to the bottom existed along the route, and tides were favorable. However, a deeper route of transit was available nearby that could have easily been used as an alternative to the shallower route actually taken.

Analysis At the time of this writing the official investigation into the grounding had just begun. The speed of the vessel at the time it was holed is unknown. The depth of the bottom, illustrated in Fig. 7, averaged 7–10 fathoms then around 15 fathoms until rising to 4 fathoms at the previously undetected shoal. Were MSV *Fennica* equipped with navigation sonar with a 1000 m range its useful range would have been restricted due to the shallow waters of the area to between 8 and 20 times the

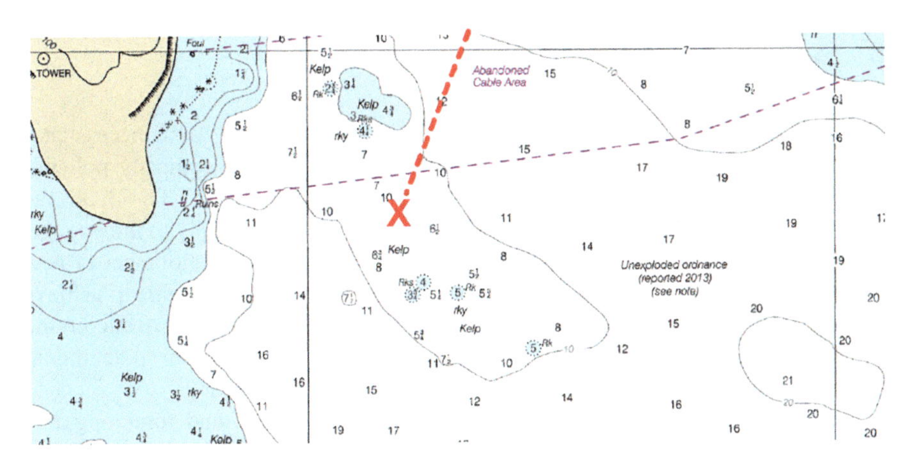

Fig. 7 Navigation Chart: Location where MSV *Fennica* was Holed. Source: NOAA Chart #16530 (Soundings in fathoms)

depth, or approximately 150–350 m range. At a speed of 10 knots the vessel would have had approximately 20–60 s advance warning before the vessel was holed, possibly providing sufficient time to lessen the severity of the incident. The approaches to the shoal would have been evident before then.

Virtual AtoN placed to mark the perimeter of shoal would have been visible on ECDIS, providing warning adequate for the crew to take action to alter course and speed to avert the incident.

7.6 *M/V* Petrozavodsk

On 11 May 2009 the 1250 gt. refrigerated cargo ship *Petrozavodsk* ran aground in heavy fog on the southern tip of Bjornoya (Bear Island) in the Arctic between Norway and Spitzbergen. Satellite tracking shows the ship held a steady course straight towards the shore and ran aground at a speed of 10 knots. The ship was declared a total loss. There were no injuries but up to 60 tons of fuel and other pollutants were spilled into the sea in an area with major sea bird populations. Many dead and injured seabirds were found along the coastline. The master and mate were reported to have high blood alcohol levels resulting in the vessel entering a protected area and running aground. Both were charged, convicted and sent to prison in Tromso, Norway (EMSA 2009).

Analysis This grounding is included to illustrate there is no fool-proof technological solution to prevent any situation where the master and mate are intent on criminally-negligent behavior that imperils the lives and safety of the crew and vessel.

8 Safety Management System

The unique risks and limitations associated with navigation in the Arctic necessitates a vessel safety management system that fulfills all appropriate company policies, port state regulations and IMO carriage requirements; as well as the COLREGS, the STCW convention, the Polar Code and other requirements that apply to the voyage. However, this is still not enough. The results of eNavigation, development of new carriage requirements and implementation of IMO initiatives take years to achieve fruition. A risk-adverse strategy must be followed towards equipping Arctic-bound vessels with navigation sonar in a manner that will increase their potential to successfully complete their voyage in this risk-prone environment.

The Arctic is a frontier that requires initiative, innovation and forethought to anticipate circumstances and events that are not likely to occur at lower latitudes and in areas more frequently traveled. This concept applies specifically to technology that may assist mariners today in reducing risk to navigation, yet no carriage requirements exist for their installation or use on vessels. One such example of progress in this area is the use of an Inertial Navigation System as a means to supplement GNSS positioning where atmospheric and other phenomena may interfere with reception. Many vessels in the Arctic and elsewhere are presently equipped with INS as a result of prudence and proper planning despite the absence of any specific requirement to do so. In many parts of the world eLORAN capabilities also exist and vessels are equipped with receivers as a means to supplement GNSS positioning. The authors propose that navigation sonar should be amongst this equipment on the bridge. The adoption of truly Virtual AtoN and their verification through the use of georeferencing techniques that would be immune to interference and spoofing attacks would also be a logical extension of this progress.

9 Conclusions

The technologies described provide new opportunities to increase maritime safety in the poorly-charted frontiers of the Arctic, the benefits of which would be applicable worldwide. Navigation sonar technology is not yet sufficiently tested or integrated on a widespread basis directly into existing navigation systems in its present form. Merely adding yet another sensor and display system to an already complex bridge environment without adequate engineering, planning and training is likely to make matters worse rather than improving them, and may result in increased risk to navigation similar to that which occurred upon the initial introduction of radar that lead to the collision between the M/Vs *Andrea Doria* and *Stockholm* (Halpern 2015). Adequate training for watchstanders in the use of navigation sonar by manufacturers should be mandated in anticipation of future enhancements to the STCW convention. Further, the Polar Code should be amended to stipulate the requirement for a second echo sounder is fulfilled using navigation sonar. Its use to acquire high resolution 3-dimensional hydrographic data to supplement traditional hydrographic survey assets as an independent third party data source should be actively considered by the International Hydrographic Organization and its member states.

Virtual aids to navigation that do not require physical infrastructure can be easily implemented and are best deployed to remote locations where infrastructure is scarce or nonexistent, and the cost and logistics of installation and maintenance is prohibitive. However, such systems can only be placed at locations where adequate hydrographic survey has performed and thorough knowledge of the seabed is available, and many remote areas in the Arctic are poorly surveyed if they are even surveyed at all. The same can be said for navigation and verification of position through georeferencing to seabed topography.

The cost of surveying the Arctic is very high, and such a crowd sourcing initiative using vessels of opportunity may help to expand survey coverage with little additional cost. However, it is also abundantly clear that payback in terms of cost avoidance of even one grounding requiring rescue or resulting in oil and/or chemical spillage in the Arctic would far outweigh any investment necessary to equip vessels and navigation systems with this navigation sonar equipment, and Virtual AtoN and georeferencing technologies. This will enable masters and crews alike to become familiar with operating and using the described technology as well as enjoy the benefits and enhanced safety provided by its capabilities.

Disclaimer The opinions, conclusions and recommendations within this chapter are solely those of the authors and do not represent any official position or endorsement of the United States Coast Guard, the National Oceanographic and Atmospheric Administration or any Government or non-governmental organization or entity.

References

ACCSEAS. (2015). ACCSEAS Final Report: Review of ACCSEAS Solutions through tests and demonstrations, 2015, IALA doc-id ENAV17-10.4.5.

Allianz. (2014a). Allianz global corporate and specialty. *Safety and Shipping Review*, 3.

Allianz. (2014b). ibid. 12.

Allianz. (2015). Allianz global corporate & specialty. *Safety and Shipping Review*, 3.

AMSA. (2009). Arctic Marine Shipping Assessment (AMSA) 2009 Report. *Arctic Council*, *5*, 23.

AMSA. (2009a). ibid. 20.

AMSA. (2009b). ibid. 127.

AMSA. (2013, May). Arctic Marine Shipping Assessment (AMSA). Status on Implementation of the AMSA 2009 Report Recommendations. *Arctic Council*, 22.

AMSA. (2015, April). Status on implementation of the AMSA 2009 report recommendations. *Arctic Council*, 16–19.

AMTW. (2004, September, 28–30). *Arctic Marine Transport Workshop*. U.S. Arctic Research Commission, 2.

ANCP. (2013, February 15). *Arctic Nautical Charting Plan, NOAA Office of Coast Survey*. Marine Chart Division, 7.

ANCP. (2015, June 5). U.S. Arctic Nautical Charting Plan: A Plan to Support Sustainable Marine Transportation in Alaska and the Arctic (*Draft for Public Comment*): NOAA Office of Coast Survey Marine Chart Division.

ARHC. (2011). National Report of Hydrographic Service of the Russian Federation Navy. In *2nd Arctic Regional Hydrographic Commission Meeting*, Copenhagen, 28–29 September 2011; ARHC2-02B. 8.

Baldauf, M., & Wright, G. (2014). New sensor technology integration for safe and efficient e-Navigation. In German Institute of Navigation (Ed.), *ISIS 2014 – International Symposium Information on Ships* (pp. 190–200). ISSN 2191-8392.

Butt, N., Johnson, D., Pike, K., Pryce-Roberts, N., & Vigar, N. (2012). *15 years of shipping accidents: A review for WWF* (p. 24). Southampton Solent University.

Carmel, S. (2013, July). The cold, hard realities of arctic shipping. *Proceedings Magazine, U.S. Naval Institute*, 139/7/1,325.

CGRDC. (2014). USCG Research and Development Center: Arctic Technology Evaluation 2014: August 8–30, 2014.

Copely, J. (2014, October 9). Just how little do we know about the Ocean floor? *Scientific American*. Retrieved July 19, 2015, from http://www.scientificamerican.com/article/just-how-little-do-we-know-about-the-ocean-floor

Crystal. (2017). Retrieved January 18, 2017, from www.crystalcruiseline.com/serenity/

Dlouhy, J. (2015, July 10). Damaged Arctic icebreaker's route questioned. *Fuel Fix*. Retrieved July 23, 2015, from http://fuelfix.com/blog/2015/07/10/damaged-arctic-icebreakers-route-questioned/

Eason, C. (2010, June 22). Baltic Sea nations urged to act as groundings rise. *Lloyds List*. Retrieved July 23, 2015, from http://www.lloydslist.com/ll/sector/regulation/article171780.ece

Eason, C. (2015, November 7). Finding the risks in Arctic shipping. *Lloyds List*. Retrieved July 22, 2015, from http://www.lloydslist.com/ll/sector/ship-operations/article451662.ece

EMSA. (2009). Maritime accident review 2009. *European Maritime Safety Agency, 16,* 29.

Endsley, M. R. (1995). Toward a theory of situation awareness in dynamic systems. *Human Factors, 37*(1), 32–64.

FLS-1000. FarSounder 1000 Navigation Sonar. Retrieved July 24, 2015, from http://www.farsounder.com/files/f31566_3d-sonar-brochure_3.0.pdf

Hains, D. (2014). Status of Arctic hydrography and nautical charting. In *Arctic Regional Hydrographic Commission (ARHC) PAME II-2014 Whitehorse Yukon Territory, Canada*, September 16, 2014, slide 16.D.

Halpern, S. (2015). *An objective forensic analysis of the collision between Stockholm and Andrea Doria*. Retrieved July 29, 2015, from http://www.titanicology.com/AndreaDoria/Stockholm-Andrea_Doria_Collision_Analysis.pdf

Hansen, S. (2015, July 9). *Unpredictable arctic leaves energy underwriters in the dark, Insurance Day*. editorial. http://www.skuld.com/about/skuld/press-cuttings/unpredictable-arctic-leaves-energy-underwriters-in-the-dark/24

HSRP. (2016, August, 30–September, 1). *Charting the U.S. Maritime Arctic*. Cleveland, OH: NOAA. Hydrographic Services Review Panel.

IALA O-143. (2010, March 5). IALA Recommendation on Virtual Aids to Navigation, O-143.

IHO 57. IHO Transfer Standard for Digital Hydrographic Data, Edition 3.1 - November 2000 Special Publication No. 57, Published by the International Hydrographic Bureau Monaco.

IMO MSC.252(83). (2007). *Revised performance standards for integrated navigation systems (INS)*. MSC.252(83). London: International Maritime Organization.

Kvam, E., & Jeannot, M. (2013, September 16–20). The Arctic testbed – providing GNSS services in the Arctic Region. In *Proceedings of the 26th International Technical Meeting of the ION Satellite Division, ION GNSS+ 2013* (pp. 890–901). Nashville, TN.

Lundberg, J. (2015). Situation awareness systems, states and processes: A holistic framework. *Theoretical Issues in Ergonomics Science*. https://doi.org/10.1080/1463922X.2015.1008601

Macfarlane, J. (2012). *A list of the full transits of the Canadian Northwest Passage 1903 to 2006*. Revised 2012. http://expedition-sailing-vessel.com/past-projects/10-past-projects/21-2000-the-northwest-passage

MAREX. (2015, June 9). NOAA to Boost Arctic Navigational Safety. *Maritime Executive*. Retrieved July 24, 2015, from http://maritime-executive.com/article/noaa-to-boost-arctic-navigational-safety

Mazaheri, A. (2013, November 20). *MMIMC contributors to groundings contributors to a grounding accident: What does evidence tell*. http://www.merikotka.fi/mimic/images/stories/MazaheriContributorsGroundings.pdf

MCIB. (2012a, January 13). *Marine Casualties Investigative Body (MCIB), Cruise Ship Costa Concordia, Marine casualty*. Report on the safety technical investigation (pp. 3, 13, 14). Ministry of Infrastructures and Transports (Italy).

MCIB. (2012b). ibid, 53.

MSC. (2008, December 18). *Casualty-Related Matters'*. Reports on Marine Casualties and Incidents, Ref. T1/12.01 International Maritime Organization. MSC-MEPC.Circ.3. Annex 2.

NGIA 180. (2014). *Pub. 180, Sailing Directions, Arctic Ocean* (11th ed., p. 68). National Geospatial-Intelligence Agency.

Nicolas-Kopec, A. (2012, December). *Transportation of HNS at Sea, Safemed II Project*. Malta: International Tanker Owners Pollution Federation.

NMD. (2011). *Marine casualties 2000–2010* (p. 6). Norwegian Maritime Directorate, Strategic Safety Department.

NOAA 16006. (2010). Bering Sea: Chart 16006, *U.S. Coast Pilot, Alaska: Cape Spencer to Beaufort Sea*. Washington, DC: NOAA, Chapter 8, paragraph 3.

NOAS. Sonardyne Navigation and Obstacle Avoidance Sonar. Performance Summary. www.sonardyne.com/product/navigation-obstacle-avoidance-sonar-system/

NSRIO. (2012). *NSR Transit 2012 (for 20.11.2012)*. NSR Information Office. http://www.arctic-lio.com/docs/nsr/transits/Transits_2012.pdf

NSRIO. (2014). *List of NSR transit voyages in 2014 navigational season*. NSR Information Office. http://www.arctic-lio.com/docs/nsr/transits/Transits 2014.pdf

NTDENR. (2015). Trends in shipping in the Northwest Passage and the Beaufort Sea (7.3). *State of the Environment Report*. Northwest Territories Environment and Natural Resources. Updated 29 May 2015. http://www.enr.gov.nt.ca/state-environment/73-trends-shipping-northwest-passage-and-beaufort-sea.

NTSB. (1990a). Grounding of the U.S. Tankship Exxon Valdez on Bligh Reef, Prince William Sound near Valdez, Alaska, March 24, 1989; National Transportation Safety Board Report NTSB/MAR-90/04, July 31, 1990. v.

NTSB. (1990b). ibid. Figure 2. 5.

Schröder-Hinrichs, J.-U., Hollnagel, E., & Baldauf, M., (2012, October). From Titanic to Costa Concordia - A century of lessons not learned. *WMU Journal of Maritime Affairs, 11*(2), 151–167.

Shell. (2015, July 22). *Shell Alaska Receives Final Permits to Drill*. Retrieved July 25, 2015, from http://www.shell.us/aboutshell/projects-locations/alaska/events-news/shell-alaska-receives-final-permits-to-drill.html

Sullivan, M. (2013, March 4). Global warming will open unexpected new shipping routes in Arctic, UCLA researchers find. *UCLA Newsroom*. http://newsroom.ucla.edu/releases/new-unexpected-shipping-route-243485

TAIC. (2014a). *Inquiry 11-204: Container ship MV Rena grounding, on Astrolabe Reef, 5 October 2011*. New Zealand: Transport Accident Investigation Commission.

TAIC. (2014b). ibid. 13.

TSB Canada. (2012a, April 26). Transportation Safety Board (TSB) Canada Report, Grounding of Passenger Vessel *Clipper Adventurer* 27 August 2010, Marine Investigation Report M10H0006, 29, 25.

TSB Canada. (2012b). ibid. 8.

TSB Canada. (2012c). ibid. 20.

TSB Canada. (2013). Transport Safety Board of Canada. *Statistics Summary Marine Occurrences*, 3.

UNDP. (2009). *UNDP Project Document*. 22.

Wright, R., & Baldauf, M. (2014a, June). Collaborative navigation through the establishment and distribution of electronic aids to navigation in real time. In *Joint Navigation Conference*. Orlando, FL: Institute of Navigation (ION).

Wright, R. G., & Baldauf, M. (2014b). Enhanced situational awareness through Multisensor Integration. In *INSLC Conference*, 15–17 September 2014. Buzzards Bay, MA: Massachusetts Maritime Academy.

Wright, R. G., & Baldauf, M. (2015). Physical characteristics of virtual aids to navigation; Activities in navigation. In A. Weintrit (Ed.), *Marine navigation and safety of sea transportation* (pp. 61–68). London: CRC Press.

Wright, R. G., & Baldauf, M. (2016a). Hydrographic survey in remote regions: Using vessels of opportunity equipped with 3-dimensional forward-looking sonar. *Journal of Marine Geodesy*. https://doi.org/10.1080/01490419.2016.1245266

Wright, R. G., & Baldauf, M. (2016b). Virtual electronic aids to navigation for remote and ecologically sensitive regions. *Journal of Navigation*, 1–17. https://doi.org/10.1017/S0373463316000527

Wright, R. G., & Zimmerman, C. M. (2015, October, 19–22). Vector data extraction from forward-looking sonar imagery for hydrographic survey and hazard to navigation detection. In *IEEE/MTS Oceans Conference*. Washington, DC.

Zhu, L., James, P., & Zhang, S. (2002). Statistics and damage assessment of ship grounding. *Marine Structures, 15*, 515–530.

Zukunft, A. P. (2015, July, 14). Coast Guard and the Arctic. In *6th Symposium on the Impacts of an Ice-Diminishing Arctic on Naval and Maritime Operations*. Washington, DC; reported from Sound Off, Workboat Magazine. Retrieved July 22, 2015, from http://www.workboat.com/sound-off/coast-guard-and-the-arctic?utm_source= Informz&utm_medium=Email&utm_campaign=eNewsletter#sthash.qAMXvcTR.dpuf

From Sensing to Sense-Making: Assessing and Visualizing Ship Operational Limitations in the Canadian Arctic Using Open-Access Ice Data

Mark A. Stoddard, Laurent Etienne, Ronald Pelot, Melanie Fournier, and Leah Beveridge

Contents

Abstract Vessels planning a passage in the Canadian Arctic face many risks, most notably from ice, extreme weather, remoteness, and uncharted or poorly charted bathymetry. For ship operators who view the Arctic as a relatively untouched area of opportunity, the desire to operate vessels in the Arctic brings new challenges and risks. This study introduces the Polar Operational Limitations Assessment Risk Indexing System (POLARIS) and demonstrates its use for assessing ship operational limitations using open-access historical ice information. The analysis of ship operational limitations in ice was aided by the construction of POLARIS scenario risk maps which were clearly demonstrated as a useful tool to support the strategic appraisal of ice conditions. Lastly, several use cases are provided to demonstrate how POLARIS and historical ice information can be used to support the strategic appraisal of ship operational limitations in ice.

M. A. Stoddard (✉) · R. Pelot · M. Fournier · L. Beveridge
Department of Industrial Engineering, Dalhousie University, Halifax, NS, Canada
e-mail: mark.stoddard@dal.ca

L. Etienne
Department of Planning and Environment, University of Tours, Tours, France

L. P. Hildebrand et al. (eds.), *Sustainable Shipping in a Changing Arctic*, WMU Studies in Maritime Affairs 7, https://doi.org/10.1007/978-3-319-78425-0_6

Keywords Sense-making · Ice data · Canadian Arctic · Operational limit · POLARIS · Ship classification · Statistical aggregation

1 Introduction

Maritime traffic in the Canadian Arctic is expected to increase in coming years as northern communities grow, tourism increases, and large resource development projects enter into operation (Higginbotham et al. 2012). Potentially accelerating this growth is evidence of a decline in ice coverage in the Canadian Arctic (Vihma 2014). As activity increases, the number of vessels exposed to the navigational risks in Canada's Arctic will rise. Additionally, it is a rather self-explanatory fact that the vulnerability of a vessel to these risks depends heavily on the type and class of the ship, the training and experience of the crew, and access to high quality information to support decision making, both during the planning and execution phase of an operation.

Passage planning for Polar Regions involves navigational practices typically accepted as standard, with additional considerations based on the expectation of the presence of ice (Snider 2012). The Canadian Ice Service (CIS), a division of the Meteorological Service of Canada, is the leading authority for information about ice in Canada's navigable waters. Currently, the CIS provides open-access to digital Arctic regional sea ice charts for marine navigation, climate research, and input to the Global Digital Sea Ice Data Bank (GDSIDB) (Canadian Ice Service 2009). CIS sea-ice charts provide information on the ice concentration, stage of development, and form of ice within the Canadian Arctic. Open-access weekly sea-ice charts covering the Canadian Arctic are available from the National Snow and Ice Data Centre (NSDIC). Weekly sea-ice charts are stored in the standard World Maritime Organization (WMO) ice chart archive vector format, Sea Ice Grid (SIGRID-3). Originally proposed in 1981 and adopted by the WMO, the SIGRID format was designed to meet larger scale climate requirements, providing a computer-compatible sea-ice data bank (National Snow and Ice Data Centre 2015a). The CIS SIGRID-3 vector format provides information about ice conditions in a specific geographic area. It can handle three different forms of ice (Fa, Fb, Fc), the stage of development (Sa, Sb, Sc), and the concentration (Ca, Cb, Cc) for each location (World Meteorological Organization 2004; Etienne and Pelot 2013).

This chapter will show how archived SIGRID-3 ice data can be used to evaluate ship operational limits in ice. Operational limitations were assessed using the Polar Operational Limitations Assessment Risk Indexing System (POLARIS), which is currently being considered for inclusion in the new International Code for Ships Operating in Polar Water (Polar Code) for use in both the Arctic and the Antarctic (Maritime Safety Committee 2014a, b). The study period covers Canadian Arctic ice conditions observed from 2007 to 2014. The Area of Interest (AOI) is largely driven by the Canadian Shipping Safety Control Zones delineated by the Shipping Safety Control Zones Order of the Arctic Waters Pollution Prevention Act, with the addition of a new Zone 0 to include the southernmost extent of Hudson Bay

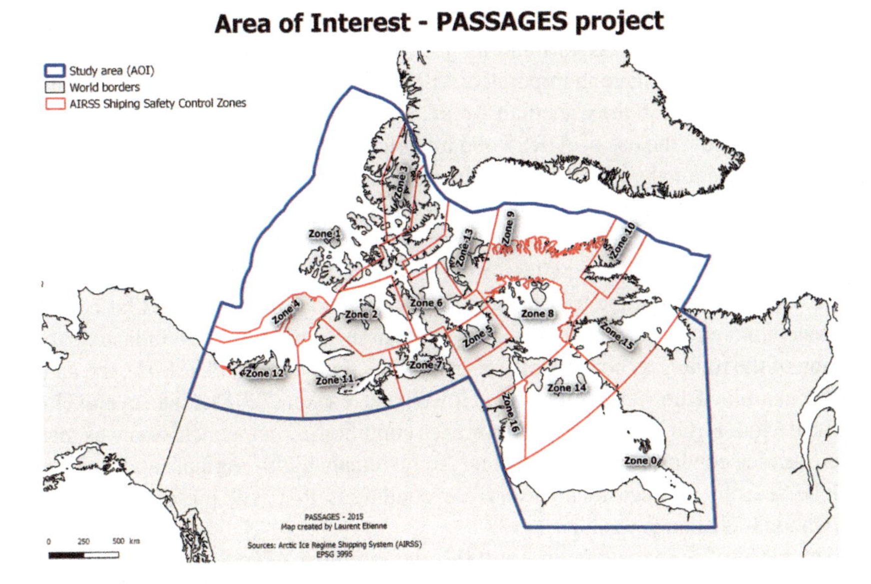

Fig. 1 Study AOI based on the Transport Canada Shipping Safety Control Zones with the addition of a Zone 0 to include Hudson Bay

(Transport Canada 1998, 2003). Figure 1 provides a geographical outline of the study AOI.

A total of 3744 POLARIS scenario risk maps were generated for the study AOI: for each of 12 ship types (PC1-7, IA, IA Super, IB, IC, no ice strengthening (NOT IS)), using 6 statistical aggregations (minimum, 1st quartile, median, average, 3rd quartile, maximum) of POLARIS results over our study period (2007–2014), and calculated for each of the 52 weeks of the year. Discussion of the results is aided by an example use case that clearly demonstrates how POLARIS scenario risk maps can be used to support the strategic appraisal of ice conditions and ice risk across a large AOI.

2 Assessing Ship Operational Limits in Ice

Transport Canada has provided mariners with a quantitative method to characterize the relative risk which various ice conditions pose to the structure of different ships, referred to as the Arctic Ice Regime Shipping System (AIRSS) (Transport Canada 1998). AIRSS is a widely used maritime framework to assess navigation safety in a given ice regime as a function of ice conditions (see Sect. 2.1) and the structural and engineering capabilities of a particular vessel class (Smith and Stephenson 2013). The system uses a basic algorithm to determine a result called the Ice Numeral (IN). A positive IN indicates that a vessel can proceed under the given operational conditions; a negative number indicates that a vessel cannot proceed under the given operational conditions (Snider 2012). Transport Canada (1998) provides a complete description of AIRSS and its use to assess ship operational limitations.

Many researchers have incorporated AIRSS into modeling and simulation studies to examine shipping in the Canadian Arctic. The efforts of Howell and Yackel (2004) demonstrated how the use of AIRSS and historical CIS digital ice charts could be used to assess navigational variability over three defined Western Arctic transit routes from 1969 to 2002. Transit routes were sampled at 5 km spacing and assigned an IN from a pre-computed IN raster grid. The results were used to visualize the spatial variability of the IN for a Canadian Arctic Class 3 (CAC3) ship throughout their study AOI and along each Western Arctic transit route. Each route was also examined based on the INs encountered during transit. A route containing a negative IN indicated that a section of the route was not suitable for operations. In Wilson et al. (2004), the authors further examined the AOI presented in Howell and Yackel (2004) using several Global Climate Models (GCMs) to predict future ice conditions. Their conclusion was that the future sea-ice conditions in the Canadian Arctic remain highly variable and there could still be seasons of occasional heavy ice conditions that will present a significant navigational challenge to ships.

The work of Somanathan et al. (2007) incorporated AIRSS into a simulation to compare the relative economics of shipping through the Northwest Passage and shipping through the Panama Canal. Using historical ice regime data for the Canadian Arctic they prepared probabilistic ice regimes that could be used in AIRSS. IN and ship type where then used to calculate transit speeds through the Northwest Passage in their simulation. The major conclusion of their study was that for the ice conditions in the Canadian Arctic for the period of 1999 through 2003 the use of the Northwest Passage is economically favored over the traditional route through the Panama Canal. The results were for a CAC3 ship where transit time was the main economic consideration. They qualify this result by saying that the difference is not very compelling, especially given the uncertainty and risk associated with transiting the Northwest Passage.

More recently, Etienne and Pelot (2013) presented a simulation tool that can be used to determine feasible paths throughout the Canadian Arctic based on historical sea ice conditions. The feasibility of a shipping path was determined using AIRSS, where historical ice data and ship classification are used to calculate the IN and resulting "Go" or "No Go" ship limit along a given route. In Smith and Stephenson (2013), the authors provide a similar analysis of shipping route feasibility using AIRSS. Instead of focusing on the use of historical sea ice data, the authors considered several leading Global Climate Models (GCMs) to construct future ice regimes for use in AIRSS. AIRSS was then used to determine the feasibility of trans-Arctic routes for Polar Class (PC) 6 and open-water (OW) vessels. A feasible navigational route consisted of a least-cost path (minimum total voyage time) from a start point to an end point, avoiding areas where the IN would obstruct a particular vessel class (IN < 0). A major conclusion of their paper was that by mid-century, September sea-ice conditions will have changed sufficiently in the Northwest Passage (NWP) such that trans-Arctic shipping to/from North America could commonly capitalize on the approximately 30% geographic distance savings that this route offers over the Northern Sea Route (NSR) which follows the Russian coastline.

2.1 Polar Operational Limitations Assessment Risk Indexing System (POLARIS)

POLARIS, a proposed risk assessment framework for determining ship operational limits in ice, was produced by the International Association of Classification Societies (IACS). Similar to the Transport Canada AIRSS system, POLARIS provides a risk assessment framework to assess navigation safety in a given ice regime, using observed or historical ice conditions and concentration and the vessel classification (Maritime Safety Committee 2014b). An ice regime is used to describe an area with a relatively consistent distribution of a number of ice types, including open water. The concentration of each ice type within an ice regime is reported in tenths. Also, for each ice type there is an associated ice type score defined for each ship ice classification.

In POLARIS, the ice type score is referred to as a Risk Value (RV), and a collection of RVs that correspond to a particular ice regime is referred to as a Risk Index Outcome (RIO). Using POLARIS, RIO is determined by summing the RVs for each ice type present in the ice regime encountered, multiplied by its respective concentration:

$$RIO = C_1RV_1 + C_2RV_2 + \ldots + C_nRV_n$$

where C_1, C_2, ..., C_n are the concentrations (in tenths) of ice types within the ice regime and RV_1, RV_2, ..., RV_n are the risk values corresponding to each ice type and for a given ship ice class classification. The resulting RIO value is then evaluated for either independent operations or icebreaker escorted operations to determine the appropriate operational limitation (see Table 1).

2.2 Ice Risk Visualization Using POLARIS Scenario Risk Maps

To facilitate the calculation and visualization of risk within the study AOI an Archimedean (uniform) tessellation was used to produce a rectangular mesh grid of the study AOI (Okabe et al. 1992). This tessellation was chosen for its ease of calculation and simplicity of the resulting data structure. The resulting quantized AOI contains approximately 16 million grid cells at a 1 km^2 resolution. The quantized AOI was further processed using a vector layer of the Canadian shoreline to delete grid cells that are outside of the AOI or inland. These spatial processing steps yielded an AOI containing four million grid cells. Next, CIS SIGRID-3 sea ice information was filtered to only include ice polygons within the study AOI and the resulting sea ice information was associated with the 1×1 km grid cells of our quantized AOI. For illustration purposes, Fig. 2 contains the Bellot Strait area of the study AOI with the associated gridded overlay at a 1 km^2 resolution.

Table 1 POLARIS evaluation criteria and associated Risk Index Outcome (RIO) condition, as defined in Tables 1.1 and 1.2 from the Maritime Safety Committee (2014b)

Evaluation criteria	Group A (PC1–PC5)	Group B (PC6–PC7)	Group C (IA Super—IA)	Group C (Below IA)
Operations not permitted	$RIO_{ship} < -10$ or $RIO_{escorted} + 10 < -10$	$RIO_{ship} < -10$ or $RIO_{escorted} + 10 < -10$	$RIO_{ship} < -10$ or $RIO_{escorted} + 10 < -10$	$RIO_{ship} < 0$ or $RIO_{escorted} + 10 < -10$
Escorted operations permitted—limited speed	$-10 \leq RIO_{escorted} + 10 < 0$	$-10 \leq RIO_{escorted} + 10 < 0$	$-10 \leq RIO_{escorted} + 10 < 0$	Not permitted
Escorted operations permitted	$RIO_{escorted} + 10 \geq 0$	$RIO_{escorted} + 10 \geq 0$	$RIO_{escorted} + 10> = 0$	$RIO_{escorted} + 10 \geq 0$
Limited speed operation permitted	$-10 \leq RIO_{ship} < 0$	$-10 \leq RIO_{ship} < 0$	Not permitted	Not permitted
Operations permitted	$RIO_{ship} > 0$	$RIO_{ship} > 0$	$RIO_{ship} > 0$	$RIO_{ship} > 0$

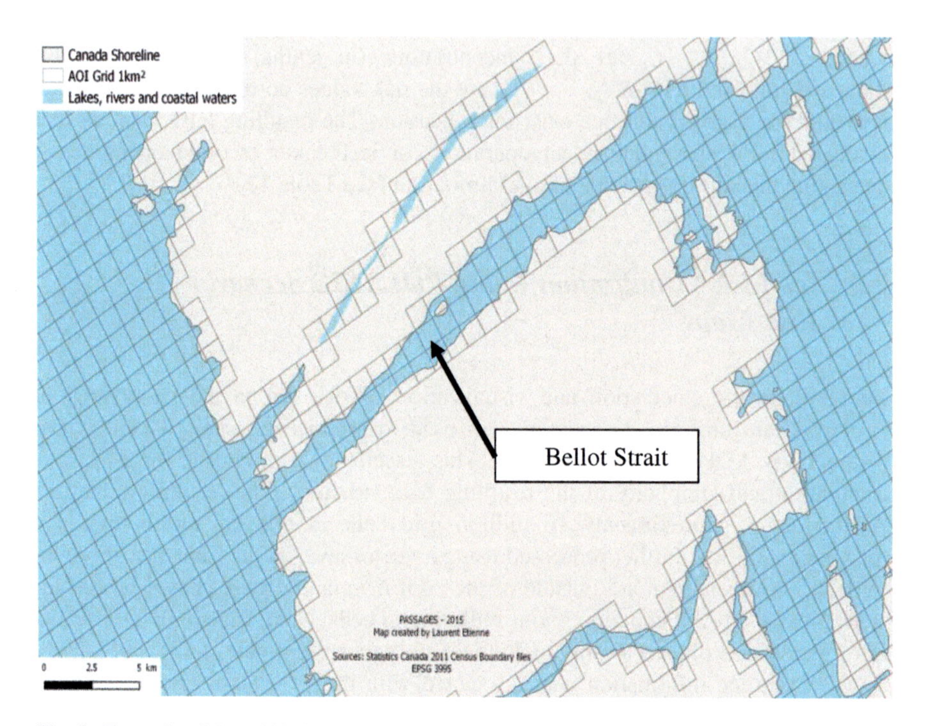

Fig. 2 Example of the gridded (tessellated) overlay produced for the study AOI. The area shown contains the Bellot Strait area with a rectangular grid resolution of 1 km^2

Table 2 POLARIS/SIGRID-3 ice type conversion table

Polaris ice type	SIGRID-3 codes	SIGRID-3 stage of development
Ice free	00, 80	Ice free
New ice	81, 82	New ice < 10 cm
Grey ice	84	Grey ice (10–15 cm)
Grey white ice	83, 85	Young ice (10–30 cm)
Thin first year ice, 1st stage	88	Thin first year ice (stage 1)
Thin first year ice, 2nd stage	86, 87, 89	Thin first year ice (>30 cm) or Brash ice
Medium first year ice 2nd stage	91	Medium first year ice (70–120 cm)
Thick first year ice	93	Thick first year ice (>120 cm)
Second year ice	96	Second year ice
Light multi year ice	95	Old ice
Heavy multi year ice	97, 98	Multi year ice

In order to use CIS SIGRID-3 ice regime information within POLARIS, a processing step was required to convert SIGRID-3 ice stage into POLARIS ice type. The Maritime Safety Committee (2014a) defines 12 ice types that are used by POLARIS. Table 2 provides the mapping of POLARIS ice type to the CIS SIGRID-3 ice types used in this study [N.B. *The mapping of POLARIS and CIS SIGRID-3 ice types provided in Table 2 is intended for academic purposes only*].

Using the POLARIS ice type and ship class information, a RIO for each grid cell in the study AOI can be calculated. The last step is to use the evaluation criteria for operational limitations (see Table 1) and apply an ordinal color scheme to convey the following evaluation results: (1) Operation Permitted (GREEN), (2) Limited Speed Operations Permitted (YELLOW), (3) Escorted Operations Permitted (ORANGE), (4) Escorted Operations Permitted—Low Speed (DARK ORANGE), and (5) Operations Not Permitted (RED).

Figure 3a summarizes the results of the POLARIS risk assessment for a IA class vessel as a thematic map, hereafter referred to as a POLARIS scenario risk map. For comparison, Fig. 3b provides an AIRSS scenario risk map for a Type B vessel to highlight the differences between POLARIS and AIRSS evaluations. Table 3 provides the vessel classification mapping that was used in this study to compare the results from AIRSS and POLARIS for Category C ships (below PC7). The Maritime Safety Committee (2014b) discusses modifications that could offer better alignment of the AIRSS system with the Polar Code and Polar Classes. These discussions, and potential impact on the mapping provided in Table 3, are outside of the scope of this study [N.B. *The vessel classification mapping provided in Table 3 is intended for academic purposes*].

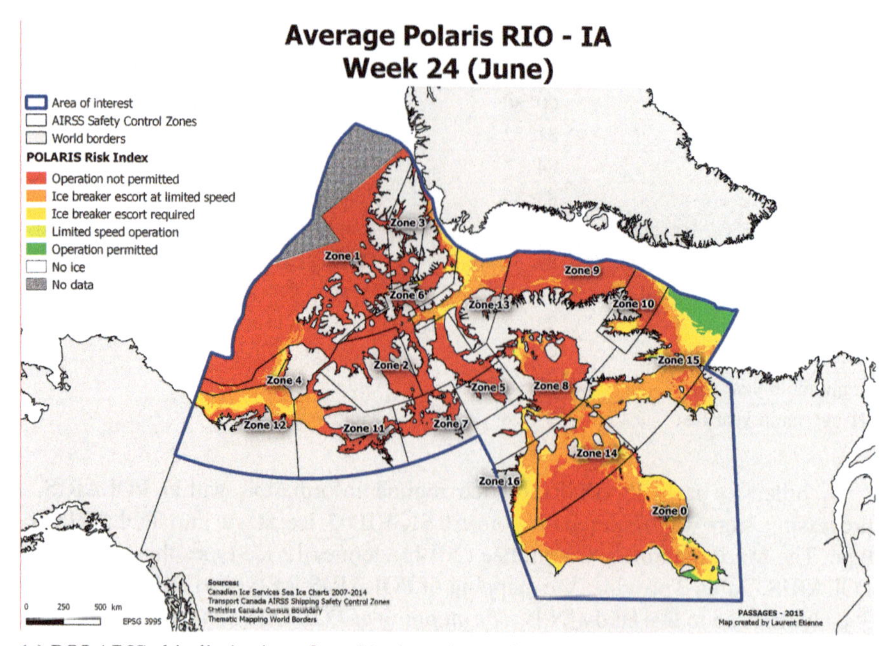

(a) POLARIS ship limitations for a IA classed vessel operating in the AOI during week 24

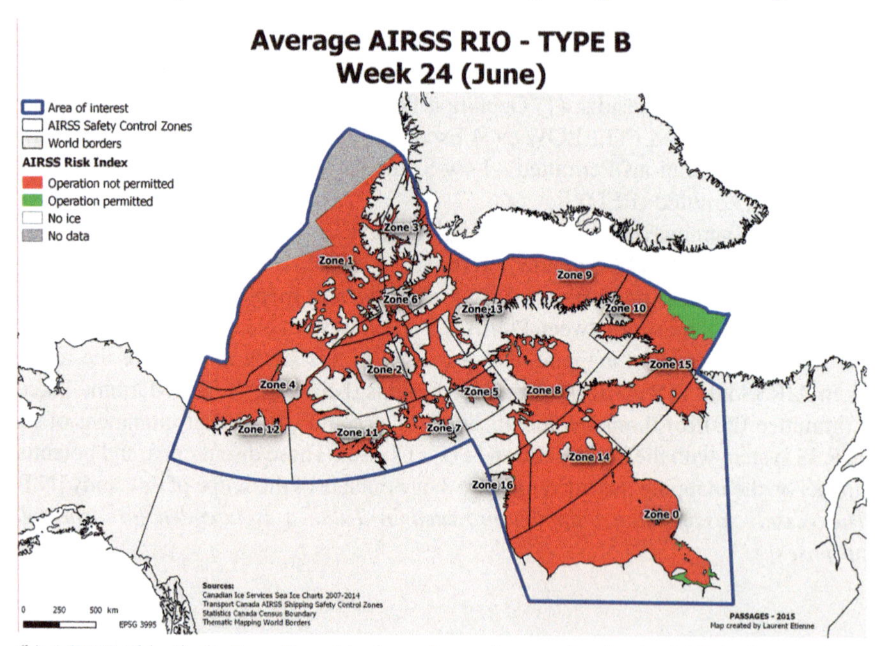

(b) AIRSS Ship limitations for a IA classed vessel operating in the AOI during week 24

Fig. 3 (**a**) Summary plot of the POLARIS ship limitations for a IA vessel operating in the study AOI during week 24 using the 2007–2014 average Risk Index Outcome (RIO). For comparison, (**b**) shows the AIRSS ship limitations for an equivalent classed vessel (Type B) during week 24 using the 2007–2014 average Ice Numeral (IN)

Table 3 Vessel classification mapping for Polar Code Category C (below PC7) classed vessels

Type of ship (Canada)	Lloyd's register of shipping	Finnish-Swedish ice class rules
Type A	100 A1 Ice Class 1 LMC or 100A1 Ice Class 1A Super LMC	IA Super
Type B	100 A1 Ice Class 1 LMC or 100A1 Ice Class 1A LMC	IA
Type C	100 A1 Ice Class 2 LMC or 100A1 Ice Class 1B LMC	IB
Type D	100 A1 Ice Class 3 LMC or 100A1 Ice Class 1D LMC	IC/II
Type E	100 A1 LMC	II

3 Example Use Case: Strategic Appraisal of Ice Conditions

The strategic appraisal of ice conditions involves the use of all information sources to give the most complete picture of the ice conditions possible (Canadian Coast Guard 2012). CIS SIGRID-3 ice data was discussed in Sect. 1 as a high quality source of historical ice conditions suitable for strategic appraisal of ice conditions and that can be easily accessed. While historical ice charts and knowledge of historical ice conditions can be combined towards the creation of a picture of the sea ice conditions possible in an area, the use of POLARIS improves sense-making for ship operators by visualizing the expected impact of ice on ship operations. In order to illustrate the use of the POLARIS scenario risk maps three potential uses for strategic appraisal are discussed, including the effect of: (1) varying polar ship classification when ice regime remains constant; (2) varying the transit/study period when polar ship classification and statistical aggregation of the RIO remains constant; and (3) varying the statistical aggregation of the RIO when ship classification and transit/study period remains constant.

3.1 Use 1: Varying Ship Classification

POLARIS scenario risk maps can be used to visualize the effect of varying polar ship classification in the AOI. The obvious result is that the more ice capable a ship is, the greater the freedom to maneuver it will have within the AOI. What remains of interest is the comparison of multiple ship classifications over the entire AOI. POLARIS RIO provides a consistent method to compare the capabilities of ships of varying polar ice classification within an AOI. The POLARIS scenario risk maps provide a visually appealing method to examine spatio-temporal phenomena within the AOI that vary in response to changing ship class. Figure 4 provides some insight to how the POLARIS risk index changes when the Polar Ship classification is varied within the study AOI.

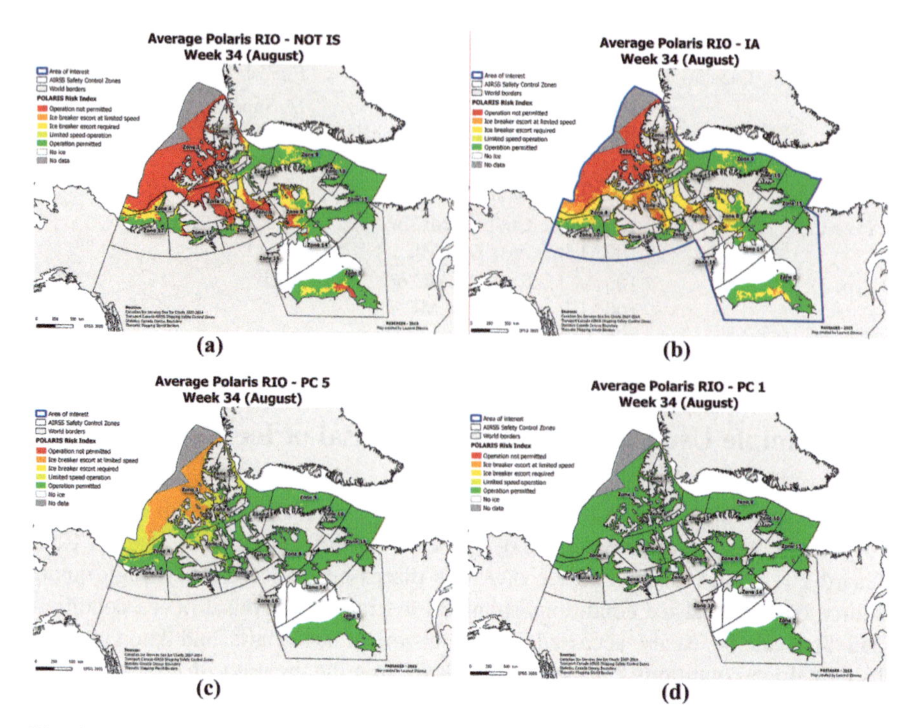

Fig. 4 Average POLARIS RIO results for four ship classifications operating in the study AOI during week 34, using the average RIO result from 2007 to 2014. (**a**) Average RIO for Not Ice Strengthened vessel in week 34; (**b**) Average RIO for IA vessel; (**c**) Average RIO for PC5 vessel; (**d**) Average RIO for PC1 vessel

3.2 Use 2: Temporal Variation of RIO

The second use examines the temporal variation in RIO for a given ship class and ice regime over a 4-week period in the AOI. Ice loss typically quickens in June with the largest loss rate occurring in July, the warmest month of the year (National Snow and Ice Data Center 2015b]. The computational results shown in Fig. 5 illustrate the temporal variation in average RIO results during the July time frame (week 27 to week 30) for a PC 7 vessel. By isolating Transport Canada Shipping Safety Control Zone 13, temporal variation in ship limitations, based on the percentage (%) of the Zone's total surface area corresponding to a particular ship type's operational limitation, can be examined. With the help of Fig. 6, it is evident that there is a steady increase in the extent of operations permitted starting around week 29 and gradually declining towards week 44 for an IA class vessel.

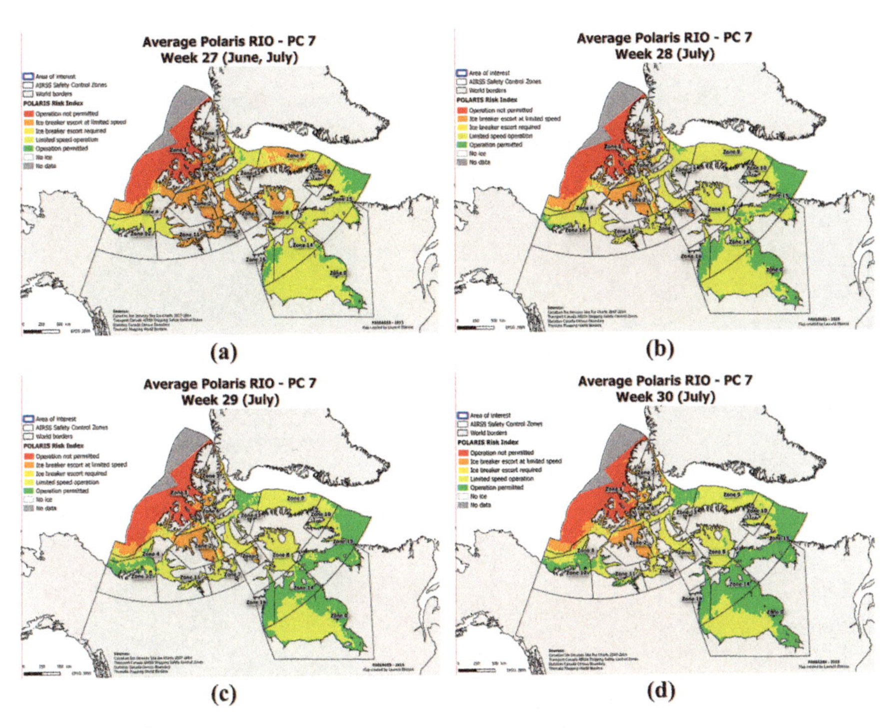

Fig. 5 Polaris risk index outcome (RIO) for a PC 7 class vessel operating in the defined Area of Interest (AOI) for week 27 to week 30 (month = June/July) using the 2007–2014 average RIO result. (**a**) Average RIO for a PC 7 vessel in week 27; (**b**) Average RIO for a PC 7 vessel in week 28; (**c**) Average RIO for a PC 7 vessel in week 29; (**d**) Average RIO for a PC 7 vessel in week 30

3.3 *Use 3: Impact of Statistical Aggregation on RIO*

The last use examines the impact of six different statistical aggregations of the RIO results from 2007 to 2014. Six different statistical aggregations are considered: (1) maximum RIO, (2) 1st quartile RIO, (3) average RIO, (4) median RIO, (5) 3rd quartile RIO, and (6) minimum RIO. The selection of the best aggregation is a question that cannot be easily answered; minimum and maximum statistical aggregations indicate the worst case and best case RIO results, while the average and percentiles are useful to understand the distribution of RIO scores over the study period (2007–2014). Further research and validation with expert Ice Navigators could shed light on the best statistical aggregation for route planning and evaluation.

The POLARIS scenario risk maps that correspond to each of the six chosen statistical aggregation methods are shown in Fig. 7. The maximum RIO score corresponds to the least severe ice conditions, while the minimum RIO score corresponds to the most severe ice conditions. The area of the study AOI most affected by the selection of statistical aggregation method is the Western entrance to the Northwest Passage. In the Maximum RIO score scenario, the entrance is open

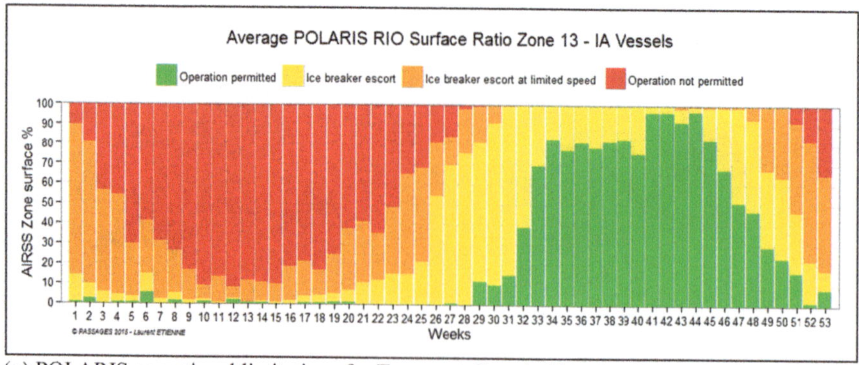

(**a**) POLARIS operational limitations for Transport Canada Shipping Safety Control Zone 13

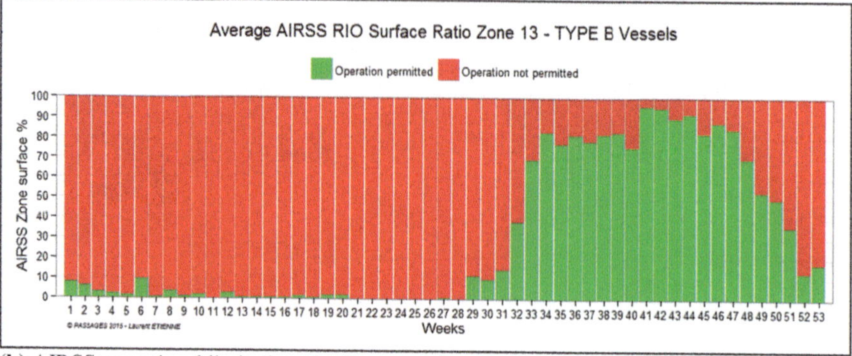

(**b**) AIRSS operational limitations for Transport Canada Shipping Safety Control Zone 13

Fig. 6 (**a**) Shows the temporal variation in ship operational limitations given by POLARIS for a IA vessel operating in Zone 13. Operational limitations are given as a percentage of the Zone 13 total surface area. (**b**) Shows the temporal variation in ship operational limitations given by AIRSS for an IA equivalent Type B vessel (see Table 3) operating in zone 13

and operations are permitted. For the Minimum RIO score scenario (Fig. 7f), the entrance is closed and operations are not permitted.

4 Discussion

The preceding section illustrated how POLARIS and CIS SIGRID-3 ice information can be used to support the strategic appraisal of ship operational limitations over a large AOI. Polar ship classification was identified as a variable that significantly influences ship operational limitations throughout the AOI, as expected. These results could be used by decision makers to examine the selection of a particular polar classification for a planned route, or the feasibility of operations in a particular region of the Canadian Arctic during a specific time of the year. Varying the transit period/study week also significantly impacts ship operational limitations. Figure 5 was used to visualize a four-week period within our study AOI for a PC7 vessel.

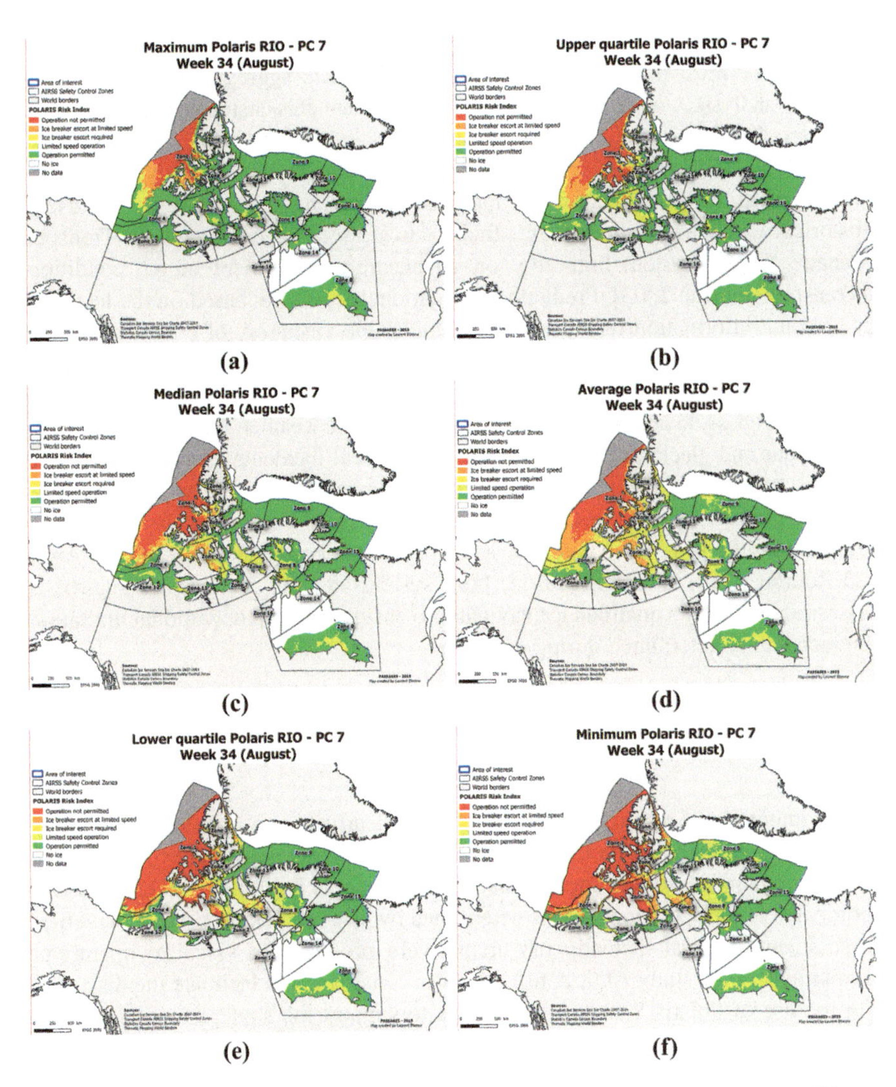

Fig. 7 RIO results for a PC7 vessel operating in the AOI during week 34 using six different statistical aggregation methods, including (**a**) Maximum RIO, (**b**) 75th Percentile RIO, (**c**) 50th Percentile/Median RIO, (**d**) Average RIO, (**e**) 25th Percentile RIO, and (**f**) Minimum RIO

Figure 6 extended this analysis and examined ship operational limitations over a 52-week period for a particular Transport Canada Shipping Safety Control Zone of interest (Zone 13), using both POLARIS and AIRSS. An example of this use is demonstrated in Stoddard et al. (2016), comparing an actual vessel trip route with the POLARIS rating along the route, and noting the implications of altering the trip timing to avoid anticipated navigation problems. Lastly, six different statistical aggregation methods of RIO results for the study AOI were compared. Figure 7 provided a visualization for each of the statistical aggregation methods. Based on a visual analysis of Fig. 7, the operational limitations in Zone 1 (Western Arctic) appear to be most influenced by the chosen statistical aggregation method. More detailed analysis would be required to substantiate this statement and could be considered in future studies.

The POLARIS scenario risk maps developed as part of this study served as a useful tool to support the strategic appraisal of ice conditions; however, the use of historical ice data inherently limits their tactical use. POLARIS and the Transport Canada AIRSS system both rely on an accurate assessment of ice conditions [Transport Canada 2003]. Predicting navigational feasibility based on the historical ice regime information remains difficult due to the vast array of other explanatory atmospheric and oceanic factors (Howell and Yackel 2004). Some other major factors that influence passage planning in the Canadian Arctic include operating area remoteness, locations of nearby support, extreme weather, daylight, presence of multi-year and glacial ice, and location of historical incidents (Snider 2012). With respect to sea ice, the conditions in the Canadian Arctic are characterized by substantial inter-annual variability (Lasserre 2011). The resulting unpredictability of ice conditions is often cited as a major hindrance for shipping in the Arctic (O'Rourke 2010). For this reason, both systems require that the ice regime be assessed in-situ by a qualified ice navigator to ensure that ship operational limitations are accurately determined during execution.

5 Conclusions

The computational results and analysis at hand clearly demonstrate how open-access SIGRID-3 ice data can be used to support the strategic assessment of ship operational limits in ice. Where appropriate, ship operational limits using POLARIS were compared to the operational limits determined by using the Transport Canada AIRSS assessment. Several thematic risk maps were produced to visualize operational limitations in the study AOI. A use case was constructed to facilitate the discussion on the use of Polaris Scenario Risk Maps to support the strategic appraisal of ice conditions and associated ship operational limitations. POLARIS has been clearly demonstrated as an excellent risk assessment framework to assess and visualize ship operational limits over large geographic areas. In addition to assessing ship operational limits, the POLARIS scenario risk maps and underlying data presented in this chapter could be used by decision makers to examine fleet mix, ship allocation, or ship scheduling and routing within a large AOI.

Acknowledgments Funding for this research was provided by the Natural Sciences and Engineering Research Council of Canada, exactEarth Ltd., the German Ministry of Economy and Technology, and Airbus Defence & Space.

References

Canadian Coast Guard. (2012). *Ice navigation in Canadian waters: Navigation in ice covered waters*. Ottawa: Government of Canada.

Canadian Ice Service. (2009). *Canadian Ice Service Arctic regional sea ice charts in SIGRID-3 format, (2007–2014)*. Boulder, CO: National Snow and Ice Data Center.

Etienne, L., & Pelot, R. (2013, June). Simulation of maritime paths taking into account ice conditions in the Arctic. In R. Devillers, C. Lee, R. Canessa, & A. Sherin (Eds.), *CoastGIS conference 2013: Monitoring and adapting to change on the coast. 11th international symposium for GIS and computer cartography for coastal zone management* (p. 116). Victoria, BC: University of Victoria.

Higginbotham, J., & Charron, A., & Manicom, J. (2012, November). *Canada – US Arctic Marine Corridors and Resource Development* (Policy Brief No 24). The Centre for International Governance Innovation. Retrieved from www.cigionline.org/publications

Howell, S. E. L., & Yackel, J. J. (2004). A vessel transit assessment of sea ice variability in the western Arctic, 1969–2002: Implications for ship navigation. *Canadian Journal of Remote Sensing, 30*, 205–215.

Lasserre, F. (2011). Arctic shipping routes: From the Panama myth to reality. *International Journal, 66*(4), 793–808.

Maritime Safety Committee. (2014a). *POLARIS – proposed system for determining operational limitations in ice*. International Association of Classification Societies (IACS), International Martime Organization.

Maritime Safety Committee. (2014b). *Technical background to POLARIS*. International Association of Classification Societies (IACS), International Maritime Organization.

National Snow and Ice Data Centre. (2015a). *Format for gridded sea ice information (SIGRID)*. Boulder, CO: National Snow and Ice Data Center.

National Snow and Ice Data Centre. (2015b). *Downwardly mobile. Arctic sea ice news and analysis*. Retrieved July 8, 2015.

O'Rourke R (2010) *Changes in the Arctic: Background and issues for congress* (7-5700, R41153, p. 13). Congressional Research Service, Retrieved July 8, 2015.

Okabe, A., Boots, B., & Sugihara, K. (1992). *Spatial tessellations: concepts and applications of Voronoi diagrams*. Chichester: John Wiley and Sons.

Smith, C. L., & Stephenson, S. R. (2013). New trans-Arctic shipping routes navigable by midcentury. In E. S. Mosley-Thompson (Ed.), *Proceedings of the National Academy of Sciences of the United States of America (PNAS)* (Vol. 110, p. E1191). Retrieved July 28, 2015, from www.pnas.org

Snider, D. (2012). *Polar ship operations: A practical guide*. London: The Nautical Institute.

Somanathan, S., Flynn, P. C., & Szymanski, J. K. (2007). Feasibility of a sea route through the Canadian Arctic. *Maritime Economics and Logistics, 9*, 324–334.

Stoddard, M. A., Etienne, L., Fournier, M., Pelot, R., & Beveridge, L. (2016). Making sense of Arctic maritime traffic using the polar operational limits assessment risk indexing system (POLARIS). In *IOP Conference series: Earth and environmental science* (Vol. 34, 012034, p 9).

Transport Canada. (1998). *Arctic Ice Regime Shipping System (AIRSS) Standards* (TP 12259E). Ottawa: Government of Canada.

Transport Canada. (2003). *Arctic Ice Regime Shipping System (AIRSS): A pictorial guide* (TP 14044E). Ottawa: Government of Canada.

Vihma, T. (2014). Effects of Arctic sea ice decline on weather and climate: A review. *Surveys in Geophysics, 35*, 1175. https://doi.org/10.1007/s10712-014-9284-0.

Wilson, K. J., Falkingham, J., Melling, H., & De Abreu, R. (2004). Shipping in the Canadian Arctic: Other possible climate change scenarios. In *Proceedings of IGARSS'04: Geoscience and Remote Sensing, Anchorage, Alaska*, September 2004 (Vol. 3, p. 1853). New York: IEEE International.

World Meteorological Organization. (2004). *SIGRID-3: A vector archive format for sea ice charts* (JCOMM Technical Report No. 23, WMO/TD-No. 1214). Retrieved October 10, 2016, from ftp://ftp.wmo.int/

Vessel Tracking Using Automatic Identification System Data in the Arctic

Torkild Eriksen and Øystein Olsen

Contents

Abstract Satellite AIS data collected with AISSat-1 and AISSat-2 represent more than 5 years of maritime traffic data from the Arctic (north of 67°N). The number of ships observed per month shows large seasonal variations, as well as an annual growth. In August 2014, 2272 ships were observed, in December the number was 1563 ships. The annual growth rate in number of ships per month varies with month; between 113 ships/year for November and 201 ships/year for July. Some of the increase in the number using Class A equipment can be explained by the Control

T. Eriksen (✉) · Ø. Olsen
Defence Systems Division, FFI, Kjeller, Norway
e-mail: Torkild.Eriksen@ffi.no; Oystein.Olsen@ffi.no

L. P. Hildebrand et al. (eds.), *Sustainable Shipping in a Changing Arctic*, WMU Studies in Maritime Affairs 7, https://doi.org/10.1007/978-3-319-78425-0_7

Regulation (EC 1224/2009) that has required AIS to be applied to fishing vessels above 15 m since 31 May 2014. To what extent the remaining increase in reporting vessels is due to higher activity or a higher number of ships using AIS is not studied. Considering geographic sectors of 45° longitude, the most trafficked sector is the 0–45°E sector where typically 75% of the ships in the Arctic are present. The 67.5°-longitude sector has the largest annual relative growth with of 46% or 695 ship-months in the same period. Looking at ship types, the peak numbers recorded in any month are 600 fishing vessels, 430 cargo ships, 120 tankers, 100 passenger ships, and 100 tugs. The seasonal variation most prominent is the fishing vessel peak in winter, while the remaining ship types peak in various summer months. Ship tracks for selected months illustrate the variation of the activity; August has a high number of tracks all around the Arctic as well as to the North Pole. As winter approaches the tracks fistly disappear in Alaska and Canada, then west of Greenland and lastly in the eastern part of the Northeast Passage in February. Activity is seen all year in the Norwegian Sea, the Barents Sea and the Kara Sea. In August 2014 the number of ships observed per day reached 1200 vessels. The number of position updates per ship per day was typically 13 and the largest daily time gap was typically less than 6.7 h. It should be noted that the variation over the Arctic region is large.

Keywords Satellite AIS · Ship tracks · Number of ships · Ship traffic trends · Ship observations · Ship updates · Arctic

1 Introduction

1.1 Norwegian SAT-AIS

The Norwegian satellite AIS (SAT-AIS) project, aimed at maritime surveillance in the High North, is a partnership between the Norwegian Coastal Administration, the Norwegian Space Centre, and the Norwegian Defence Research Establishment (FFI). SAT-AIS data collected with the AISSat-1 and AISSat-2 satellites launched 12 July 2010 and 8 July 2014, respectively, represent more than 5 years of maritime traffic data from the Arctic.

FFI started the development of SAT-AIS in 2004, at the same time as the network of coastal AIS stations was put up, and was the prime contractor and operator of AISSat-1 and AISSat-2 from the start. As a developer of technologies and functionality for the next generation maritime surveillance systems, FFI has been engaged in the further development of the AIS standard as well as critical technologies, and in the evaluation of SAT-AIS systems and the quality of service.

1.2 Purpose of This Chapter

The purpose of the analyses presented here is to look at geographic variations over time in number of ships in the SAT-AIS data, as well as to show some of the capabilities provided by SAT-AIS data. Even though the examples are relevant for operational maritime surveillance as well as planning, they only represent the contribution from SAT-AIS to the situational awareness, which also is based on several other data sources and systems. The examples are made for research and illustration purposes.

The Arctic is here considered as the region north of 67°N. For the trend and growth analysis, the count of "number of ships per month" is the count of unique MMSI[1] numbers per month. The study covers geographic variation in sectors of 45° longitude, and annual variation in categories according to ship type and flag state. Examples show ship tracks and densities, as well as the quality of the tracking service.

1.3 AIS Requirements, Equipment Classes and Message Types

The AIS requirements are outlined in Chapter V of the International Convention for the Safety of Life at Sea (SOLAS) of the International Maritime Organization (IMO 1974/2000). The carriage requirement for Class A shipborne mobile equipment include all passenger ships, tankers and other ships of 300 tons engaged in international voyages, as well as all ships of 500 tons or more in national voyages. In addition to the IMO requirements, there are regional requirements. For EU fishing vessels, article 10 of Council regulation (EC) 1224/2009 (the Control Regulation) has required AIS to be applied to vessels above 15 m since 31 May 2014 (European Commission 2009).

The AIS data used to analyse ship traffic using AIS Class A transponders are the position reports from message types 1, 2 and 3 defined by the International Telecommunication Union (2010). For the ship type, information from message type 5 is used.

Vessels that not have AIS Class A requirements, may also be equipped with such equipment, or be equipped with AIS Class B transponders that are intended for use on smaller vessels and pleasure craft. The Class B equipment has its own message types with longer reporting intervals and lower transmitter power than the Class A equipment and is therefore more difficult to detect from space. Ships using Class B are analysed using message type 18.

[1]MMSI—Maritime Mobile Service Identity—A unique identifier for ships (and base stations) broadcasting on the AIS system.

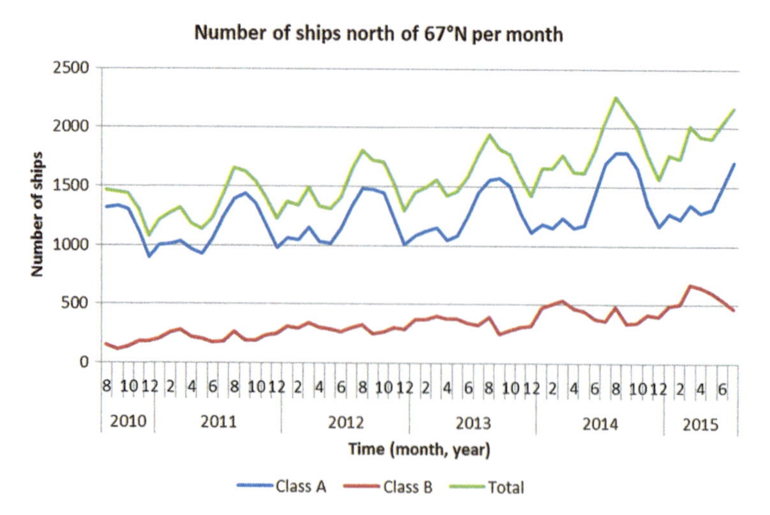

Fig. 1 Count of number of ships per month with AIS Class A, Class B, and in total, north of 67°N from August 2010 to July 2015

2 Trends in Number of Ships Observed in the Arctic

2.1 Number of Ships Per Month

Figure 1 shows the number of ships per month north of 67°N with Class A- and B equipment, as well at the total, for the period August 2010–July 2015. To illustrate typical monthly values data from 2014 are used: The curve for Class A shows seasonal variations from a minimum of 1148 ships (February) to a maximum of 1791 ships (August), Class B vary between 344 ships (September) and 532 ships (March), and the total vary between 1563 ships (December) and 2272 ships (August). It can be noted that whereas Class A has a minimum in winter, Class B has a maximum, giving the total local maximums in winter and annual peaks in summer.

2.2 Annual Growth Rates

The curves in Fig. 1 indicate a growing number of ships observed per month on an annual basis. The growth rate is calculated individually for each month applying trend lines to the data for the 4 years 2011–2014 month by month. The slopes of the 12 trend lines give the average annual growth by month over the 4 years. The results are shown in Fig. 2.

For Class A the median growth by month is 74 ships/year. The growth is below the median in January–April, and November–December, and above in the 6 months

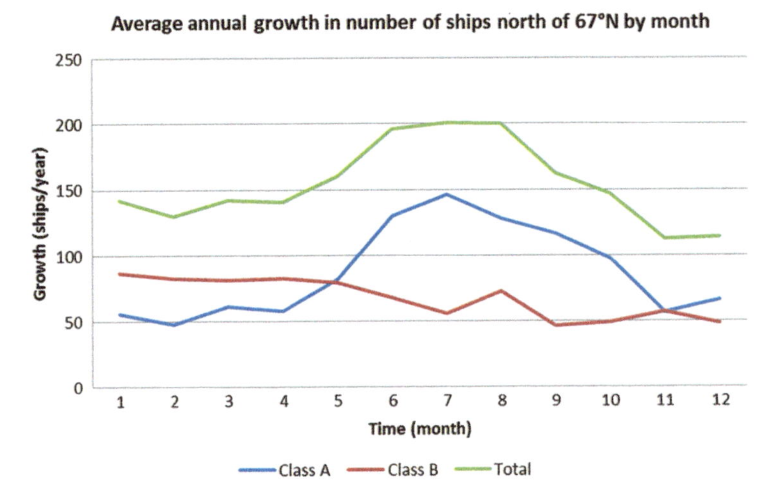

Fig. 2 Average annual growth in number of ships with AIS Class A, Class B, and in total from 2011 to 2014

from May–October. The maximum, 146 ships/year, is in July, and the minimum, 48 ships/year, is in February. While the peak of the counts of Class A is highest in August and September, see Fig. 1, the growth is largest in June and July, indicating a slight shift towards higher numbers earlier in summer.

For Class B the median growth by month is 70 ships/year. The growth is below the median in June–July and September–December, and above in January–May and August. The minimum is in September, 46 ships/year, and the maximum is in January, 87 ships/year. While the peak of the counts of Class B is highest in March, see Fig. 1, the growth is slightly larger in January; the winter months show the largest growth in numbers, with a weak indication of a shift towards larger values in January.

For the total number of ships the median growth by month is 144 ships/year. The growth is below the median in January–April, and November–December, and above in May–September. The growth is largest in the summer; as the Class A and Class B data are added, the growth becomes almost equal in June, July and August, with 200 ships/year for each month. The minimum is in November, 113 ships/year.

2.3 *Observations and Predictions for 2015*

The observed numbers of ships for the first 7 months of 2015 were not used to make the estimates of the growth. Figure 3 shows the plot of the observed number of ships per month from January–July as solid lines together with the 95% prediction interval for new values as dashed lines. Data from 2011 to 2014 are used to make predictions

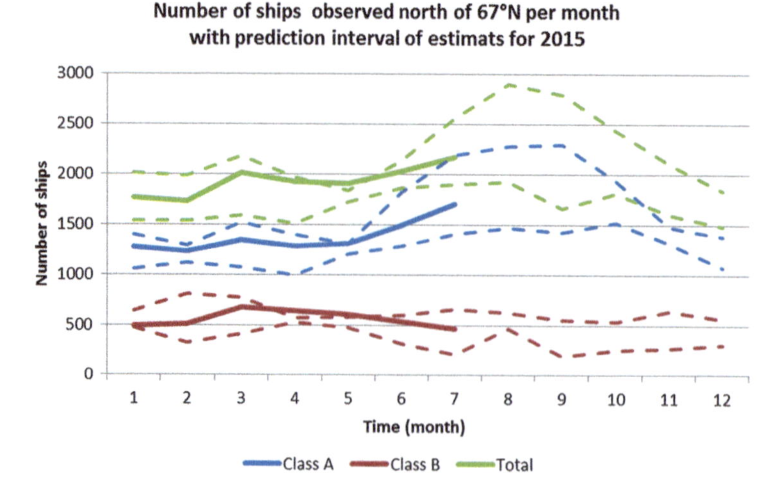

Fig. 3 Observed number of ships per month up till July 2015 (solid lines) and 95% prediction interval based on 2011 to 2014 data

for 2015. The prediction intervals are obtained using a least squares fit of a linear trend (Morrison 2014).

The Class A observations are between 27 and 87 ships above the predictions for the first 5 months, but 54–96 ships below for respectively June and July. The observed numbers are all within the 95% prediction interval, even for the month of May where the width of the prediction interval is at its lowest; ±50 ships. The maximum width is ±401ships for August.

The Class B observations show a different behaviour, being 63 and 56 ships below the prediction in January and February, but between 34 and 95 ships above the prediction for the following 5 months. The observations are mostly within the 95% prediction interval that has a width of between ±22 (April) and ±241 ships (February). The observed values are however higher than the maximum value for the prediction interval for April and May, where the observations are 645 and 605 ships and the prediction interval 551 ± 22 and 529 ± 53 ships, respectively. The values outside the prediction interval for these months means that the observed numbers are extremely high relative to the trend for the month, whereas the narrow prediction interval means that the observations of the previous years have followed the trend line very closely. Hence, the calculation shows an increase of the growth for the months relative to the steady growth of the previous years.

The observed increase of 165 Class B ships from February to March is the highest increase seen for the 5 years. The number of ships for March, 674 ships, is still within the 95% prediction interval. The ships are all in the 22.5°-longitude sector (0–45°E, 67–90°N), see Sect. 3.1.

The total curve shows the sum of the number of ships of Class A and B. The numbers are mostly within the 95% prediction interval, but slightly higher for May where the width of the prediction interval is only ±55 ships, the observed value is

1910 ship and the predicted value 1780 ships, giving a difference of 130 ships between predicted and observed value. The deviation occurs in a month in which both Class A and B, and hence the total, has earlier shown a very steady growth, and the deviation makes the annual variation of this month more similar to the variation of the other months.

The maximum width of the prediction interval for the total is for September having ±570 ships. The prediction gives a new peak number of 2417 ships in August 2015, 145 ships more than in August 2014.

Altogether 36 linear trends are investigated for the three data series'. For each estimate, the standard deviation is calculated as part of the estimate of the prediction interval. For a month with a narrow prediction interval, using May as example, the ratio of the standard deviation of the slope to the slope is typically 0.03. The coefficient of determination (R-squared) is 1.00, showing that the linear model is very well suited for the estimation. For a month with large uncertainty, September, the ratio for the slope is typically 0.25. The coefficient of determination is close to 0.89, indicating that the linear model is still well suited for the estimation.

2.4 Count of Ships by Ship Type

AIS message type 5 (Ship static and voyage related data) transmitted by Class A stations has the "Type of ship and cargo type" parameter that is used to group the ships by type (cargo, fishing, tanker, ..., other), but not the ship position (International Telecommunication Union 2010). To count the ships by type, the position reports of message type 1, 2 and 3 must be matched to message type 5 using the MMSI number. If no message type 5 is found, the ship is counted in the category "other". "Other" may also contain survey and research vessels, military vessels, and all ship types other than those named in the plot legend of Fig. 4. The plot shows the variation of the number of the different ship types from August 2010 to May 2015, which continues the time series presented earlier by Olsen et al. (2014) by 15 months. The plots are for Class A equipped ships in the entire Arctic area, but as the number of ships is significantly higher in the 22.5°-longitude sector than in any other sector, these ships dominate the result.

The three major ship types represented in descending order are fishing vessels, cargo ships, and other ships. The next three types are tankers, passenger ships and tugs. At the bottom is the pleasure craft that choose to use AIS Class A. Different seasonal variations are seen for the different ship types. Fishing vessels have peaks in winter as well as late summer, and a minimum in spring. All other types have their peaks in summer; the passenger traffic has its maximum earlier than cargo, tankers and tugs.

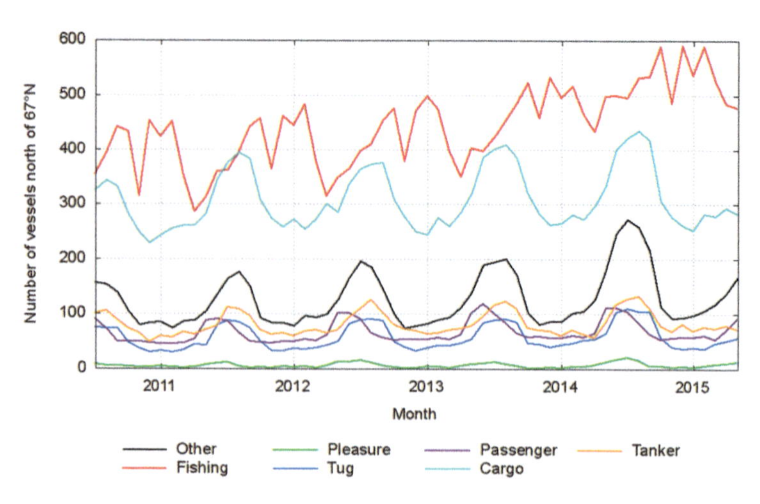

Fig. 4 Count of number of ships per ship type per month using AIS Class A equipment north of 67°N

2.5 *Discussion*

The SAT-AIS data show a clear annual growth in the number of ships. For the total, the growth rate over 4 years is 113 ships/year at minimum in November and 201 ships/year at maximum in July.

One can assume increased activity as a result of the reduction of the sea ice extent in parts of the Arctic waters, as a result of fish migration—and therefore the fisheries—to more northern latitudes, as well as national and commercial interest in claiming the right to natural resources as well as to the continental shelves and the North pole.

However, some or all of the increase in the number of detected fishing vessels is likely caused by regulatory changes rather than a real increase in the number of vessels. The Control Regulation (European Commission 2009) for fishing vessels apply to all activities covered by the common fisheries policy carried out on the territory of Member States or in Community waters or by Community fishing vessels or by nationals of Member States. The regulation states that the requirement for AIS equipment shall gradually apply to ships according to overall length:

- As from 31 May 2012: all vessels above 24 m
- As from 31 May 2013: all vessels above 18 m
- 31 May 2014: all vessels above 15 m.

These changes increase the total number of AIS equipped Norwegian fishing vessels in the Arctic with approximately 100 (numbers retrieved from the vessel statistics by groups of length; Norwegian Directorate of Fisheries 2014), and it explains most of the increase seen in Fig. 4. Assuming a similar effect for Icelandic and Greenlandic fishing vessels, the apparent trend in the number of fishing vessels may be explained by the Control Regulation.

3 Geographic Distribution

3.1 *Number of Ships in Sectors*

To analyse the geographic distribution of the activity, the total number of ships (Class A + B) per month north of 67°N is counted in eight sectors of 45° longitude, each being referred to by their longitudinal centre coordinate. Figure 5 shows the number of ships per month in the sectors. The 157.5°W sector is labelled −157.5 in the plot legend and referred to as the −157.5° sector in the text, and so on for the other sectors.

The 22.5° sector (0–45°E, 67–90°N) has by far the largest number of ships at any time of year. Here typically 75% of the ships in the Arctic are present; 1723 of 2272 ships in August 2014. The area encompasses the Norwegian Sea and the Barents Sea as well as the Svalbard archipelago. It is no surprise that the number of ships is highest in this sector, as part of it has a mild climate thanks to the Gulf Stream. Among the ships in this sector it is a significant number of vessels in coastal traffic in northern Norway as well as ships sailing to Murmansk and other ice-free ports in northern Russia. Seasonal variations are seen, with peaks in March and August (1710 and 1723, 2014 figures) and minimums in May and December (1513 and 1471, 2014 figures). The 165 ships that caused the rise of the Class B curve in Fig. 3 are found in this sector.

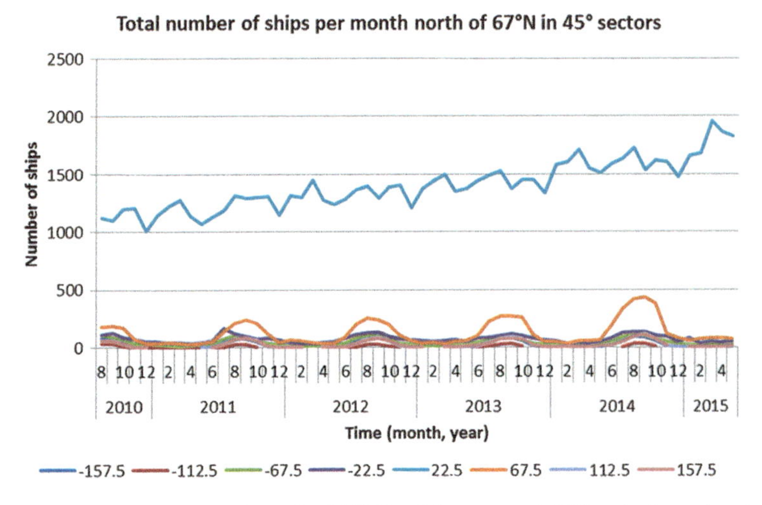

Fig. 5 Number of ships per month north of 67°N in eight sectors of 45° longitude, each referred to by their longitudinal centre coordinate

Table 1 Difference in number of ships per month in the sectors, as well as the total annual and sectorial figures, between 2013 and 2014

	−157	−112	−67	−22	22.5	67.5	112.5	157.5	Total
Jan	1	1	3	−2	204	9	0	1	217
Feb	0	1	6	−10	163	−10	1	2	153
Mar	−2	0	3	−15	218	29	−2	0	231
Apr	1	1	4	−25	204	7	3	3	198
May	1	−1	9	−7	141	9	8	−2	158
Jun	4	−2	10	1	142	85	8	3	251
Jul	−10	−5	23	36	144	108	9	12	317
Aug	2	5	18	29	195	149	14	27	439
Sep	4	2	15	11	166	157	23	27	405
Oct	−1	1	11	1	164	117	−6	1	288
Nov	−3	0	16	12	153	4	0	−1	181
Dec	−1	3	5	−12	138	31	0	−2	162
Total	−4	6	123	19	2032	695	58	71	3000

The area with the second highest numbers is the 67.5° sector (45–90°E, 67–90°N) that encompasses the Kara Sea and the western part of the Northeast Passage (NEP). Also in this sector it is activity all year, but strong seasonal variations are seen in the numbers. The peak is typically in September, not in August as for the 22.5° sector, and numbers range from 240 ships in 2012 to 429 ships in 2014. The minimum is typically in March, numbers range from 27 ships (2013 figure) to 56 ships (2014 figure).

The remaining sectors all show strong seasonal variations, with peak numbers in summer and low or even no activity in winter.

3.2 *Growth in the Sectors*

The growth in number of ships per month (and per year) is here studied as the difference from year to year, rather than as the long-term variation given by the trend line. Table 1 shows the difference between the number of ships in 2013 and 2014. Both in the monthly count (ships/month) and the sum of ships per month over the year (ship-months/year), growth is most commonly seen. Figure 6 shows the annual growth in the number of ship-months; the point at 2012 represent growth from 2011 to 2012, and correspondingly for the other years.

The 22.5° sector shows increase in every month, with a peak in March and a local maximum in August. The total annual increase was 2032 ship-months, which is a 12% growth relative to the 17,085 ship-months of 2013. The 67.5° sector, with growth in all months except February, has the strongest relative growth; 695 ship-months give a relative growth of 46%.

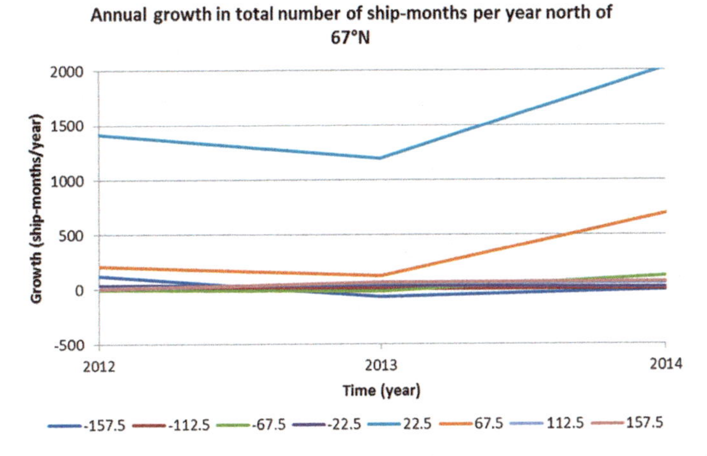

Fig. 6 Annual growth in the total number of ship-months for each sector from 2011 to 2014

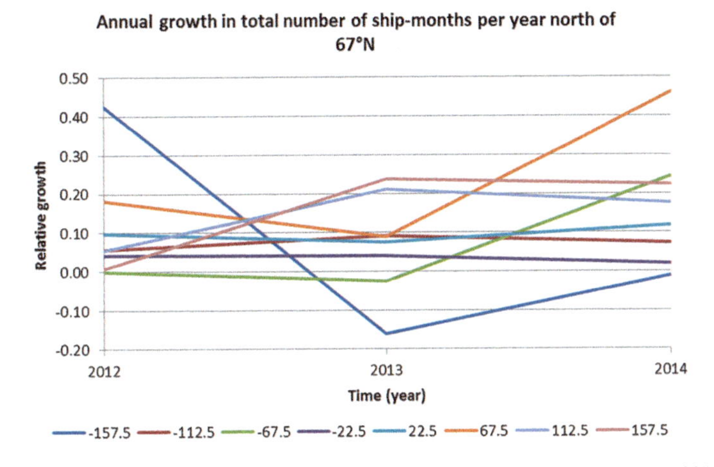

Fig. 7 Annual relative growth in the total ship-months for each sector from 2011 to 2014

The $-157.5°$ sector is the only sector that shows a decrease in the annual total; 4 ship-months reduction in the total, mainly due to a decrease of 10 ships in July. The $-22.5°$ sector shows a decrease of 71 ship-months in the wither months, but ends up at an annual growth of 19 ship-months thanks to the high growth in summer.

The annual growth in the total ship-months for a year relative to the sum of ship-months of the year before is shown in Fig. 7. The high growth in numbers of the 22.5° sector is seen as a relatively steady relative growth of 10%. The relative growth of the other sectors shows much larger fluctuations, partly due to low number of counts. The sectors that show the strongest relative growth are the eastern sectors of 67.5°, 112.5° and 157.5°. Among the sectors to the west, the $-22.5°$ and $-112.5°$ sectors show a growing trend, whereas the $-67.5°$ and $-157.5°$ sectors have both positive and negative figures.

3.2.1 Recent Growth of Class A and Class B

There was a significant increase in the number of Class A counts in the summer 2014. The peak in the 22.5° and 67.5° sectors were 1320 and 402 ships, respectively in August and September, representing a relative growth of 13% and 53% compared to 2013 values. The relative annual growth of the total ship-months in the two sectors was 9% and 42% from 2013 to 2014. The two other eastern sectors both had an annual growth of 17% in the same period. The sectors to the west show more up-and-down variation in the annual numbers. Only the −22.5° sector has had a steady year-by-year growth of 2%, 6%, and 3% from 2011 to 2014. From 2013 to 2014, the first 5 months showed a monthly reduction of up to 40% in that sector, but June and July had up to 46% growth, and in total the number of ship-months ended at a growth of 3%. The remaining three sectors have had an annual variation of up to ±10% the last 4 years, the −157.5° shows the largest variation from 94 ship-months increase in 2012 to 41 and 27 ship-months decrease in 2013 and 2014, respectively.

Class B equipped ships had a significant increase in March 2015 relative to 2014. While the January and February numbers were almost at the same level as in 2014, March showed an increase of 144 ships compared to March 2014 in the 22.5° sector. Ninety-nine percentage of the Class B counts are in the 22.5° sector in winter, but reduced to 80% in summer. The 67.5° sector had a peak in summer 2014; while the months from June to October never had more than 9 ships in earlier years, the number for September and October 2014 was 27. Also the 157.5° sector had a peak in summer 2014; while only one ship a month has been observed earlier, between 4 and 6 ships were observed from July to October 2014. The −22.5° sector has had a decrease of 20% and 6% for 2013 and 2014, respectively.

3.3 *Count of Ships by Flag State*

The Maritime Identification Digits (International Telecommunication Union 2015), abbreviated MID, are used to count the number of ships from various flag states. The MID is the first three digits of the MMSI number, which is included in every AIS message, and uniquely identify the flag state of the ship.

Figure 8 shows the distribution of total number of ships in the Arctic on flag states for August 2014. The data labels representing the flag state use the two-letter country codes defined in ISO 3166—the International Standard for country codes (International Organization for Standardization 2015), the numbers are the count of ships for the respective flags. The label "noFl" (no Flag) is used for the MMSI numbers with MIDs that do not represent flag states (such as AIS stations in aircraft and aids to navigation) or ships with invalid MMSI numbers.

Altogether 58 different flag states are identified, 39 of which have less than 11 counts and are gathered together in the label "Other" for readability of the annotation. The amount of Norwegian ships is 1002, or 44%, of the 2272 ships in

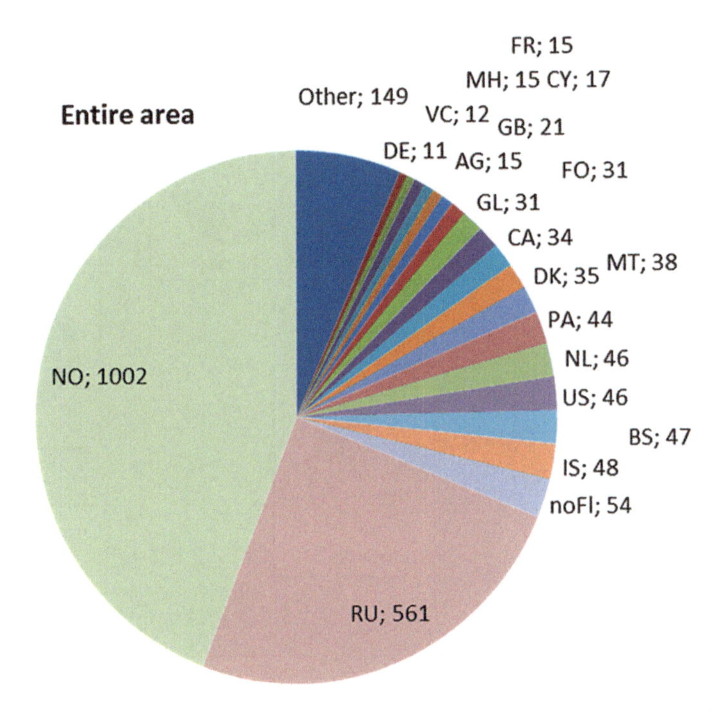

Fig. 8 Number of ships counted by flag state in August 2014 in the Arctic. The annotation shows two-letter country codes and the number of ships for each flag state. The total for all flags is 2272 ships, including 54 ships that did not have a valid MID number

total; the Russian constitutes 561 or 25%; the proportion with no flag constitutes 54 or 2%. Other countries that also have 2% are Iceland, the Bahamas, the United States of America, the Netherlands, Panama, Malta and Denmark. Further, 49 other flag states are present with less than 35 ships, equivalent to a ratio of less than 2%.

The count by flag state is also made for the eight sectors. The results for the sectors as well as for the entire Arctic in August 2014 are shown in Fig. 9. Note again that the total count for the entire area, being 2272 ships, is less than the sum of the sectors, being 2710 ship-months, which means that some ships are counted in more than one sector.

Like for the largest number of vessels, the largest number of flag states is also found in the 22.5° sector: there are 49 different flags among the 1723 ships. The sector has by far the largest part of the Norwegian ships, and also the largest part of the Russian and of several other flags. This Norwegian and Russian dominance is no surprise, as the sector comprises the Norwegian Sea and part of Norwegian territory, the Barents Sea and the port of Murmansk as well as the entrance of the White Sea. The large number of flags shows a large international activity in the area.

Russian dominance is found in the next three sectors with the Kara Sea, the Laptev Sea, and the East Siberian Sea, which are all part of the NEP. The 67.5° sector has 25 flags and 419 ships, the 112.5° sector has 6 flags and 98 ships, and the

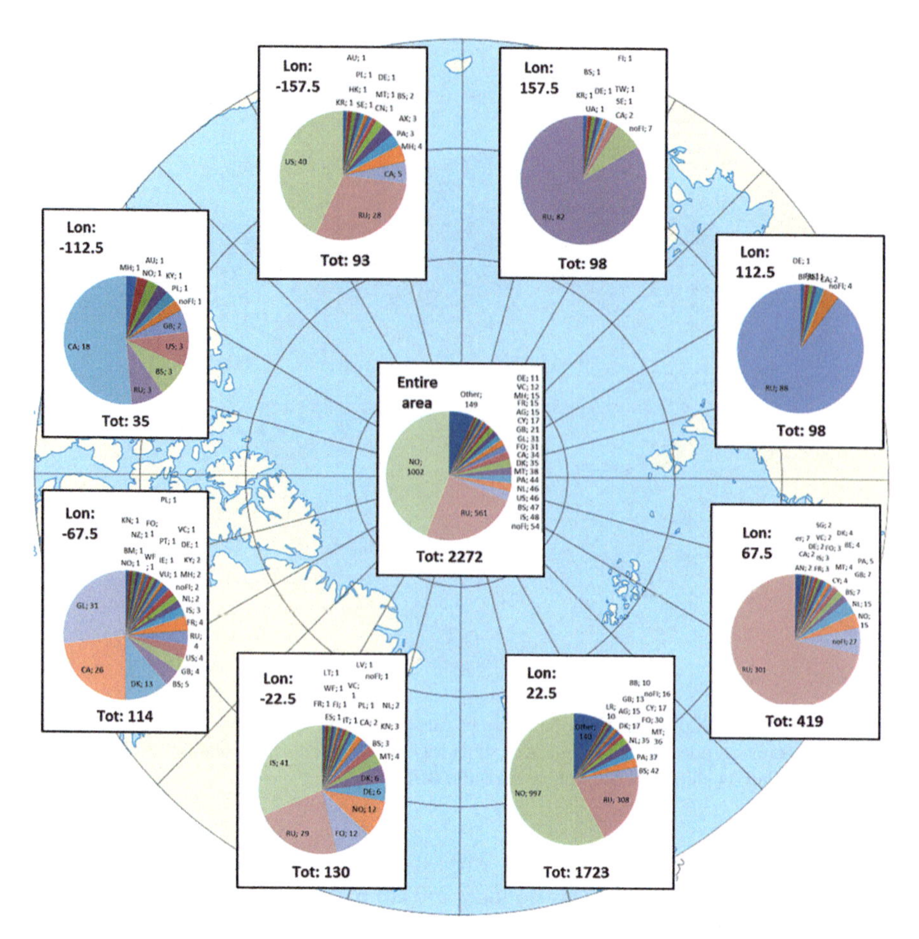

Fig. 9 Number of ships counted by flag state in August 2014 in the eight sectors as well as in the entire area

157.5° sector has 9 and 98 ships. It is in these sectors the ratio of MMSIs with no valid MID code is largest, hence it gives a number of ships labelled "noFl".

Crossing the dateline, there is US dominance in the −157.5° sector with the Chukchi Sea, the Beaufort Sea and the Bering Strait, but the Russian percentage is high also here. The number of flags present is 15 for 93 ships.

The −112.5° sector with Victoria Island and several of the Queen Elisabeth Islands has Canadian dominance. The number of flags is 10 and the number of ships only 35.

The −67.5° sector with Baffin Bay and the Davis Strait is dominated by Greenland and Canada, and also Denmark is significant. The number of flags is 24 and the number of ships 114.

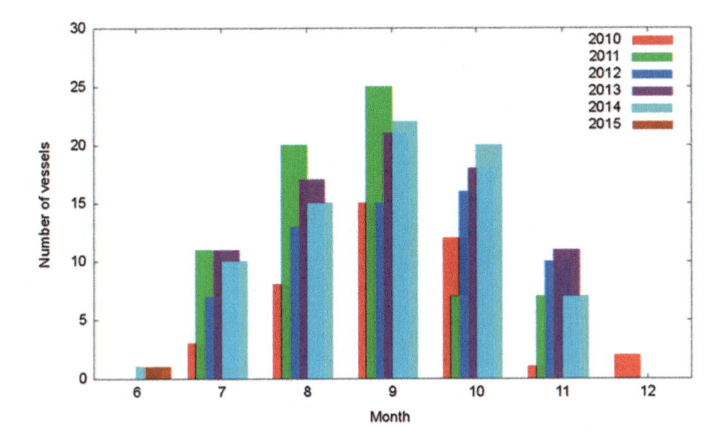

Fig. 10 Number of ships per month that has passed through the NEP

The −22.5° sector with the Greenland Sea is dominated by Iceland and Russia, and also the Faroe Islands and Norway is significant. The number of flags is 20 and the number of ships 130.

It should be noted that activity near the pole may easily be counted in several sectors as they all meet at 90°N.

3.4 Count of Ships in the Northeast Passage

The number of ships that has passed through the NEP has been counted for each month. A vessel is counted if it was seen in the Barents Sea either at least 10 days before it arrived at, or it left, the Bering Strait. The trip was not allowed to take more than 6 weeks. The trips were confirmed manually by generating tracks for each counted vessel.

The results for the period July 2010–June 2015 is shown in Fig. 10. It can be seen that the route has been used from June to December, but approximately 75% of the crossings is in the three middle months with a peak in September. No clear trend can be seen in the annual data.

Investigating the growth using monthly trend lines on the data from 2011 to 2014, values of between −1 (August) and 4 (October) ships/year are found, but no trend. Analysing the total annual ship-months from 2011 to 2014; being 70, 61, 78, and 75, and predicting the result for 2015 applying the same method as in Sect. 2.3, we find that the slope is 3 ship-months/year and a predicted value for 2015 of 78 ship-months, but with a prediction interval of 27–131 ship-months. The coefficient of determination is as low as 0.3, indicating that the linear model is not well suited to make predictions. Even though a small increase in annual numbers has been seen, the analysis shows that there is no clear trend.

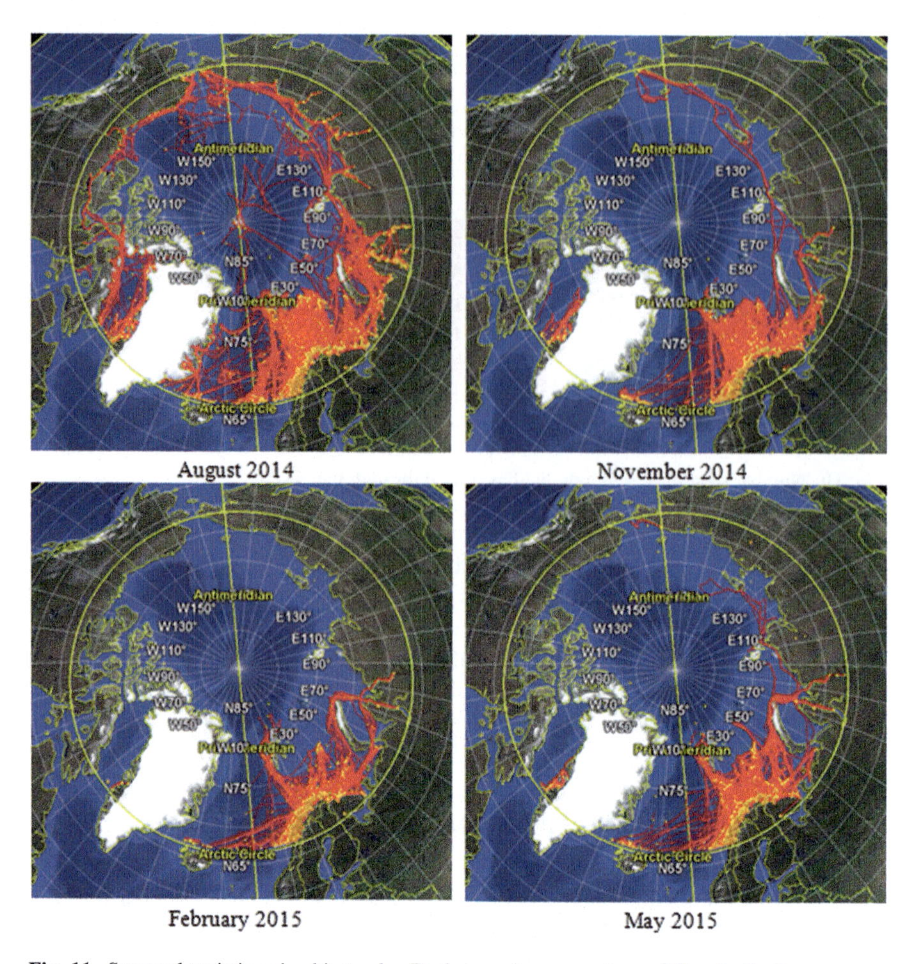

Fig. 11 Seasonal variations in ship tracks. Background map courtesy of Google Earth

4 Seasonal Variations in Ship Tracks

The activity in the Arctic is illustrated by the ship tracks from August and November 2014, and February and May 2015 as shown in Fig. 11. The yellow symbols represent the ship position at the end of the respective month; the red lines represent the track for the month, linear interpolation is used between the reported positions. The four selected months show the seasonal variations in the activity. Tracks are seen all around the Arctic in August. In November the activity has ended outside Canada and Alaska, and moved away from the coast in Russia. In February only the North Atlantic, the Barents Sea and the Kara Sea have tracks, as well as some limited activity near Greenland. In May the activity increases again along parts of the Russian coast and near Greenland.

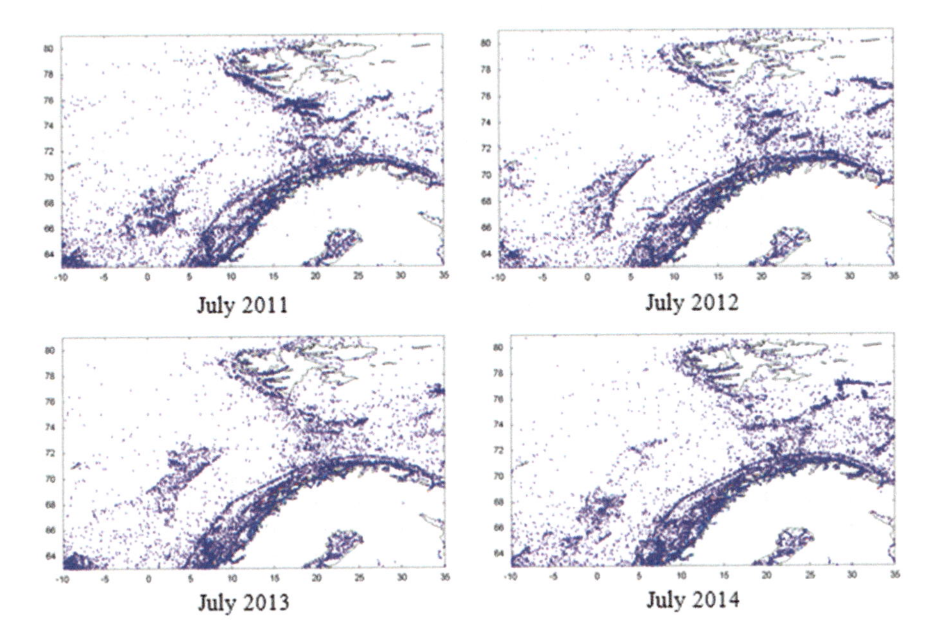

Fig. 12 Annual variations in ship densities in July 2011–2014

5 Annual Variations in Ship Density

The variation in the geographic distribution of ships in July over a period of 4 years is shown in Fig. 12, showing the expected mean ship density calculated applying one ship position per day on a grid of $1' \times 1'$ angular resolution, adding the daily results and dividing by the number of days. The resulting grid point values range from 1/31 to approximately 10. The plots are for July month in the years 2011–2014. Large variations are seen in the Banana Hole (Smutthavet) west of Norway, fewer vessels are seen at the end of the period west of Svalbard as well as north of Bear Island, whereas more vessels are seen in the eastern part of the Barents Sea.

6 Quality of Service

The contribution from SAT-AIS data to the maritime situational picture (MSP) is estimated as the number of detected ships, the number of observations per day, and the update interval. The analyses are based on Class A data received by the two satellites AISSat-1 and AISSat-2. The example shown in Fig. 13 shows the ship positions received coloured according to the number of observations for each ship on 14 August 2014. The date is chosen to illustrate ship detection and updates in more or less the entire Arctic region. An "observation" is defined as an essential update of the ship position, one per satellite pass in which the ship is detected.

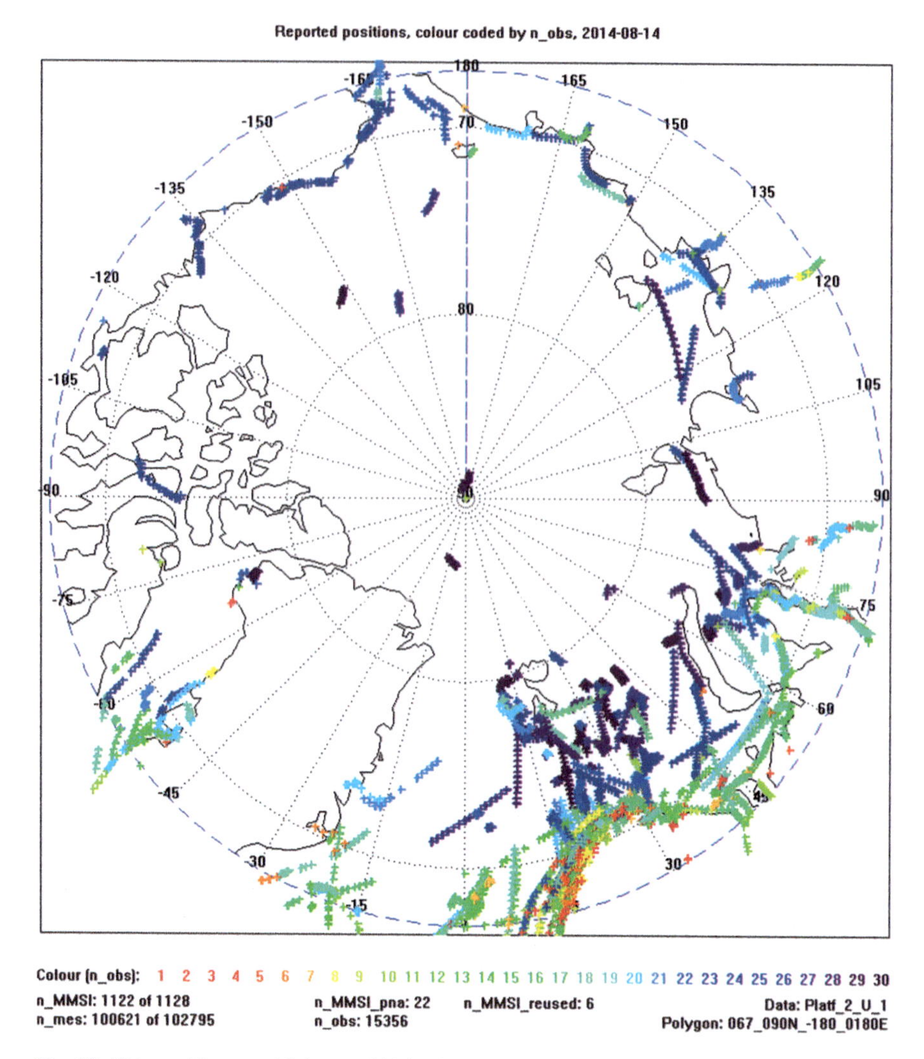

Fig. 13 Ship positions on 14 August 2014 coloured by the number of observations for each ship

The quality of service is studied on a daily basis in terms of number of received messages and observations as well as time between messages for each ship for Class A equipped vessels. The following performance figures are calculated for each ship, for each day:

- Number of messages per ship per day (n_mes): the count of position reports from SAT-AIS.
- Number of observations per ship per day (n_obs): an observation is attributed to the first message in each satellite pass.

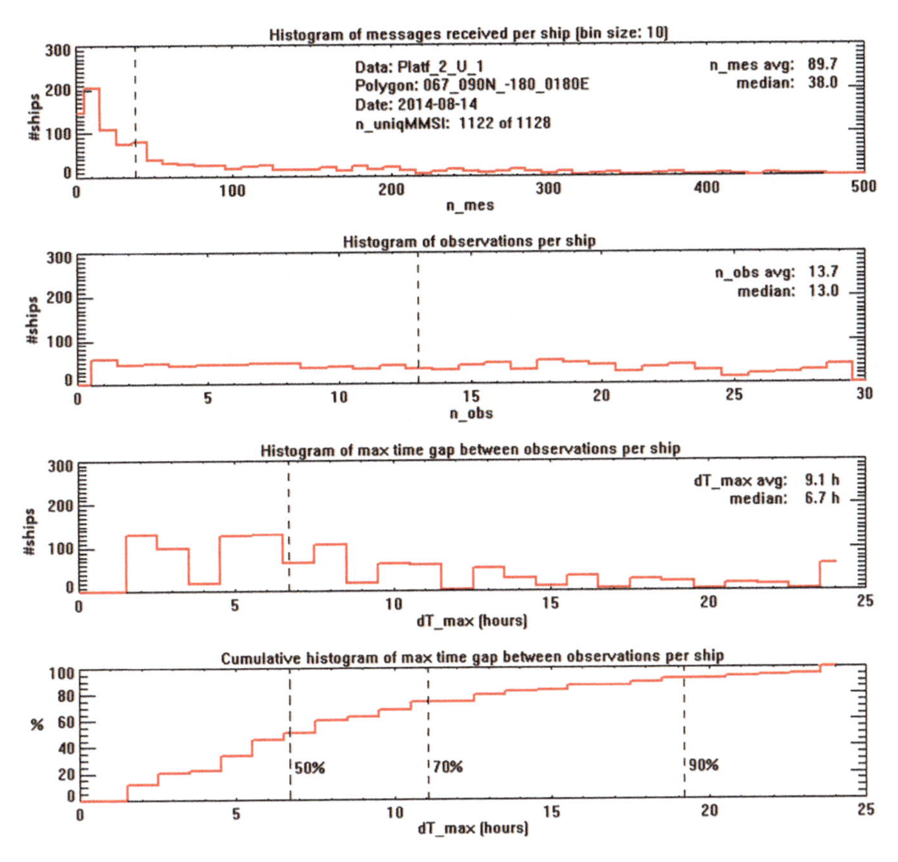

Fig. 14 Quality of service parameters showing the distribution for 1122 ships on 14 August 2014

- Maximum time gap (dT_max): the largest time gap between consecutive messages from any satellite for each ship.

The distributions of the parameter value for all ships are plotted in Fig. 14.

The two satellites are in polar low Earth orbits and pass over the areas north of 75°N every orbit. The periods of these orbits are approximately 97 min. Hence, an area north of 75°N is observed at most 30 times per day. Further south the maximum number of observations decreases due to the lower number of satellite passes. In some regions the number of observations is also reduced due to interference that make decoding of the AIS signals difficult. Further, ships in ports or at anchor reduce the reporting rate and hence become less visible. Also, some ships have poor AIS installations and transmit a weak signal, and finally some ships turn off the AIS.

The number of messages received per ship per day peaks at low values, with a median value of 38 as illustrated by the dashed line. The median number of observations is 13, but the distribution of the number of observations per ship per day is very wide; all numbers between the minimum being one and the maximum being 29 are found with more or less the same frequency. The maximum time gap between consecutive messages from any satellite is dominated by values in the lower end, with a median of 6.7 h. The cumulative distribution shows that 70% of the ships are update more often than every 11th hour, and 90% are update more often than every 19th hour.

The difference between two and one satellites is approximately a factor of two in number of messages and observations, median numbers for 14 August 2014 being 20 and 7 for AISSat-1 alone. The temporal performance for one satellite is a median of 9.7 h, 70% of the ships more often than every 13th hours, and 90% more often than every 22nd hour. The number of ships detected is 49 ships higher for two satellites than with one satellite, all of these are in coastal waters or ports. From the navigational status reported of the additional ships, it is found that 26 ships report that they are at anchor or moored, which implies a low reporting rate and hence lower probability of detection. For the density plots in Fig. 12, the availability of AISSat-2 data as well as AISSat-1 data for July 2014 increase the values along the coast of Norway a little, whereas on the high-seas no difference should occur in the one-position-per-day numbers.

7 Summary

Satellite AIS data collected with AISSat-1 and AISSat-2 (launched 12 July 2010 and 8 July 2014, respectively) represent more than 5 years of maritime traffic data from the Arctic. Whereas SAT-AIS gives a significant contribution to the maritime situational picture, especially in remote areas, it should be noted that the complete picture is made from data from other satellites as well as in situ observations.

Counting the AIS messages from Class A and Class B equipped vessels north of 67°N on a monthly basis, the total number of ships varies from a winter minimum of 1073 per month (December 2010) to a summer maximum of 2272 per month (August 2014). The annual growth rate in number of ships per month varies with month; from 113 ships/year at the minimum in November to 201 ships/year at the maximum in July.

Counting the number of ships in geographic sectors of 45° longitude, the significantly highest numbers are found in the 0–45°E sector. Here typically 75% of the ships are present; 1723 of 2272 ships in August 2014. The annual growth in the sector has been around 10% the last 4 years. The sectors with the highest relative growth rate are the following three eastern sectors, having an average growth between 15 and 24% per year.

The monthly ship-type counts give 600 fishing vessels, 430 cargo ships, 120 tankers, 100 passenger ships, 100 tugs and 280 other ships (ships that are not of the listed types or missing ship type information). The numbers represent 2014 peak-month values. Peaks occur in different months for the different types: Fishing vessels numbers peak in winter as well as late summer, the other types have their peaks in summer, passenger traffic earlier than cargo, tankers and tugs.

The most resent ship tracks for August, November, February and May illustrate the variation of the activity in Arctic through the seasons; August has a high number of tracks all around the Arctic as well as to the North Pole. Later on the tracks first disappear in Alaska and Canada (November), then west of Greenland and lastly in the eastern part of the Northeast Passage in February. Activity in the Norwegian Sea, the Barents Sea and the Kara Sea is seen all year.

Plots of ship densities for July month from 2011 to 2014 show that the activity west and south of Svalbard has decreased, while the activity east of Svalbard has increased. Also, the activity in the Banana Hole (Smutthavet) shows large variation.

In August 2014 the number of ships observed by SAT-AIS reached the highest value so far. 2272 ships were observed during 1 month. Up to 1200 vessels were observed per day, of which 1100 were using AIS Class A and 100 Class B. Using median values for 14 August 2014 as an example of the contribution from SAT-AIS data to the maritime situational picture, the number of position updates per ship per day from different satellites and passes were typically 13 and the largest daily time gaps typically smaller than 6.7 h. It should be noted that the variation over the Arctic region is large.

The analyses shown here are only based on the SAT-AIS data from AISSat-1 and AISSat-2, hence gives an insight into the trends and locations of ships using AIS. It is not studied in detail to what extent the observed variations over time is due to an actual increase of activity, or a more wide use of AIS onboard the ships. Further studies can both aim at discovering the source and type of activity, as well as the geographic location and regional differences of the activity.

References

European Commission Council regulation (EC) 1224/2009, the Control Regulation. (2009). Retrieved July 29, 2015, from http://eur-lex.europa.eu/legal-content/EN/TXT/?qid=1397742834654&uri=CELEX:32009R1224

International Maritime Organisation, International Convention for the Safety of Life at Sea "SOLAS". (1974). Chapter V "Safety of Navigation", Regulation 19. Although the original SOLAS Convention is from 1974, AIS was only introduced by amendment in December 2000.

International Organization for Standardization. (2015). Country Codes - ISO 3166. Retrieved July 29, 2015, from http://www.iso.org/iso/country_codes

International Telecommunication Union. (2010). *Technical Characteristics for a Universal Shipborne Automatic Identification System using Time Division Multiple Access in the VHF Maritime Mobile Bad*. ITU-R M.1371-4.

International Telecommunication Union. (2015). *Table of maritime identification digits*. Retrieved July 29, 2015, from http://www.itu.int/online/mms/glad/cga_mids.sh?lang=en. Updated 18 May 2015.

Morrison, F. A. (2014, September 25). *Obtaining uncertainty measures on slope and intercept of a least squares fit with Excel's LINEST*. Houghton, MI: Department of Chemical Engineering, Michigan Technological University. Retrieved August 6, 2015, from http://www.chem.mtu.edu/~fmorriso/cm3215/UncertaintySlopeInterceptOfLeastSquaresFit.pdf

Norwegian Directorate of Fisheries. (2014). Fartøy, totalt og på fylkesnivå t.o.m. 2014. Retrieved August 6, 2015, from http://www.fiskeridir.no/Yrkesfiske/Statistikk-yrkesfiske/Fiskere-fartoey-og-tillatelser/Fartoey-i-merkeregisteret. Updated 9 April 2015.

Olsen, Ø., Skauen, A., & Helleren, Ø. (2014, January 23). Predicting near future vessel traffic conditions in the Arctic using data from AISSat-1. *Arctic Frontiers*. Retrieved August 11, 2015, from http://www.arcticfrontiers.com/downloads/arctic-frontiers-2014/conference-presentations-3/thursday-23-january-2014/part-iv-arctic-search-and-rescue-sar-1/548-04-oystein-olsen/file

Detection and Tracking of Ships in the Canadian Arctic

Steven Horn

Contents

Abstract The Canadian Arctic is becoming increasingly important as climate change and economic pressures stimulate increasing activity in the region. The number of transits, cruise ships, and adventurer expeditions in this area is on the rise. Ensuring environmental, economic, archeological, defence, safety and security responsibilities in this challenging area has resulted in many recent investments including the Arctic Offshore Patrol Vessels and the RADARSAT Constellation Mission. This chapter will explore the challenges in detection and tracking of ships in the Arctic from perspectives including: ship-ice discrimination in remote sensing, sparse data tracking, effects of constrained navigation, and operational decision aids.

Keywords Arctic · Surveillance · Situational awareness · Detection · Sparse data

1 Introduction

The Canadian Arctic is a vast and remote area which is becoming increasingly accessible due to changing environmental conditions. There is currently a wave of investment in new immediate and future capabilities for the Canadian Arctic on land, in sea and in space (Canada's Northern Strategy 2013). New facilities are being

S. Horn (✉)
Defence Research and Development Canada, Centre for Operational Research and Analysis, Ottawa, ON, Canada
e-mail: Steven.Horn@forces.gc.ca

L. P. Hildebrand et al. (eds.), *Sustainable Shipping in a Changing Arctic*, WMU Studies in Maritime Affairs 7, https://doi.org/10.1007/978-3-319-78425-0_8

constructed, new ships are being built, and satellites are being launched. The responsibility for the North spans across many Canadian government departments. The Canadian Coast Guard provides a significant service to the Arctic through the provision of icebreakers, monitoring, regulation, and search and rescue. The Department of National Defence (DND) also plays an important role in the defence of the Arctic, and will be receiving a new fleet of Arctic Offshore Patrol Ships (AOPS) to patrol the north. Furthermore, remote sensing capabilities, like the privately-owned RADARSAT-2 satellite that is used to detect ships operationally by DND's Polar Epsilon project, and the future RADARSAT Constellation Mission (RCM), in addition to increasing commercial satellite sources, provides a potential means to monitor vessel activity, environmental impact, and ice within Canadian Arctic waters. Much of this northern development is being supported by new facilities such as the Nanisivik Naval Facility on Baffin Island to refuel ships, and the Canadian High Arctic Research Station (CHARS).

The Arctic is also an important economic resource for Canada. Industries in fishing, natural resources, and tourism in the Arctic are some examples of this economic value. All of this activity also carries a risk and responsibility, such as providing search and rescue. For example, Arctic adventurers navigate the Arctic waters in various pleasure craft which are at risk from the environmental conditions and can lead to a search and rescue event.

Monitoring the activity in the Canadian Arctic is an important maritime safety and security challenge. The increasing seasonal accessibility of the Arctic opens up this northern approach for potential criminal or adversarial exploitation. To address these threats, whether through prevention or deterrence, two integral components are situational awareness in the Arctic and subsequent response capability.

This chapter focuses discussion on the situational awareness challenges and capabilities vice response capabilities. Section 2 presents some of the Canadian surveillance capabilities in the Arctic, and Sect. 3 presents some examples of Arctic situational awareness achieved through these capabilities, as well as some future avenues of research for analysis and operational decision support.

2 Present and Future Capabilities

Sources of ship position information in the Arctic include the Northern Canada Vessel Traffic Services Zone Regulations (NORDREG), Space-based Automatic Identification System (S-AIS) and to a lesser extent a few Terrestrial AIS stations, Long Range Identification and Tracking (LRIT), Space-based Synthetic Aperture Radar (SAR), and open source reporting.

Managed by the Canadian Coast Guard, NORDREG is a regulation requiring vessels greater than 300 gross tonnes, vessels towing or pushing with a combined 500 gross tonnes or more, or vessels with pollutants or dangerous goods to periodically report their position and status (Canadian Coast Guard 2013). This regulatory reporting provides information which helps to ensure the safety and security of Arctic vessels, and also serves as a means to protect the Arctic environment.

Automatic Identification System (AIS) technology is an International Maritime Organization (IMO) mandated vessel safety system which is mandatory for vessels with passengers or greater than 300 gross tonnage, mandated by the International Convention for the Safety of Life at Sea (SOLAS), however, AIS can also be voluntarily used by others. AIS operates via Very High Frequency (VHF) radio transmissions at 161,975 and 162,025 MHz and there are also two classes of AIS transceivers: class A (with a minimum 12 Watts transmission power), and class B (with a maximum of 2 Watts transmission power). Class B transmitters are intentionally limited in their transmission power to prevent saturation of the available radio bandwidth, and are used by non-mandated vessels for the primary purposes of safety and navigation. Class A transmitters provide position updates every three minutes or up to every 2 s when maneuvering while class B transmitters typically transmit every 30 s. Notably, many Arctic adventurers carry either class B transponders and/or other satellite transponder systems despite not being required to do so.

In order to receive these AIS radio messages, only an antenna and decoder are required. While initially envisioned to be used for local area communication of ship positions, coastal AIS receiving antenna networks provide a means to monitor traffic within radio range of antennas effectively in real time. By placing receiving antennas on aircraft or satellites, the area of coverage for AIS monitoring is greatly increased. While satellite-based AIS receivers provide a wide area of coverage, two drawbacks are in the reduced persistence of sensing as the satellite orbits out of a monitoring area (this drawback is being addressed by increasing the number of satellites in orbit), and a drawback in the degradation of detection performance due to the nature of the AIS protocols (Cervera and Alberto 2008).

The nature of the detection performance degradation in S-AIS has been estimated as a geospatial function (Papa et al. 2012), and as a function of the number of ships in the satellite field of view (Tunaley 2011). However, even with the suboptimal detection capabilities, the availability and relatively low cost of an AIS transceiver means that many non-mandated vessels can also provide their positions via AIS. Specifically, the use of class B AIS means that small participating vessels can be tracked via S-AIS.

The LRIT system is an IMO global vessel monitoring system for SOLAS mandated vessels, which provides periodic updates of participating vessel positions and status when within 1000 nautical miles of a nation's coastline. The LRIT system is different from the AIS system in the sense that it uses real-time satellite communications to provide the positions of vessels. Every LRIT participating ship provides position updates every 6 h, but more frequent updates are possible by request of a nation. In the Polar Regions, many satellite communication systems are not as readily available since most communication satellites focus on serving regions at range from the equator, and so LRIT tracking is typically achieved via the Iridium constellation, which provides service in the Arctic region.

Active sensing is defined here as a capability which can detect non-cooperative or non-emitting vessels. Active sensing therefore provides a benefit over the aforementioned voluntary, regulatory, and passive capabilities. This enhances maritime security by being able to detect vessels which are either difficult to track passively, or may be attempting to evade detection for illicit purposes. Specifically, the RADARSAT-2 satellite, and the future RCM provide a means to achieve active sensing. These satellites have a sun-synchronous polar orbit, which means that they have frequent access to the Arctic region as access is constrained by the satellite duty cycle (Canada Space Agency 2015).

While SAR satellites provide a tantalizing opportunity to detect ships in the Arctic, there are also some significant challenges to overcome in order to exploit SAR ship detection in this environment. The primary challenge is the discrimination of ships from icebergs. Other related challenges, not discussed in this chapter, include ship detection performance (missed detections), and false detections. In order to achieve ship-ice discrimination, there are multiple techniques which can be used. The most basic of which is the association of SAR imagery with information from other systems such as AIS. Vachon et al. (2014) describe the SAR-AIS Association System (SAAS), developed by Defence Research and Development Canada (DRDC) which achieves this association. In this way, SAR detections which are truly ships can be readily identified. The RADARSAT-2 satellite does not have an on-board AIS receiver therefore SAAS requires alternate sources of AIS ship detections. The RCM, however, will include on-board AIS receivers. In the case that a ship is not providing its position via AIS or other reporting means, other image processing techniques for ship-ice discrimination in the SAR imagery must be used.

Ship-ice discrimination can be enhanced through increased imagery resolution, and the configuration of transmitted and received radar polarizations (Howell et al. 2004, 2008). Ice and ships have different sensitivities to the radar polarizations and result in different polarizations on the reflected radiation. Transmission and reception polarization can be varied to transmit in horizontal (H) and/or vertical (V) polarization, and receive in H and/or V polarization. In quad polarization modes, the radar transmits both H and V polarized radiation and receives both H and V polarized returns. The combinations of these signals can be used to implement detectors that can better discriminate between ice and ships. Howell et al. (2004, 2008), for example, report ship-iceberg discrimination performance accuracy using HV and HH polarizations on the order of 92–96% for large vessels in images with 30 m resolution and swath widths between 56 and 105 km.

Other enhancements to ship detection in SAR are also achieved via special maritime surveillance beam modes. Vachon et al. (2014) presents two RADARSAT-2 beam modes tuned for maritime surveillance, under the title Maritime Satellite Surveillance Radar (MSSR). The Detection of Vessels, Wide swath, Far incidence angle (DVWF) mode is specially designed for vessel detection, and the Ocean Surveillance, Very wide swath, Near incidence angle (OSVN) mode is tuned for general ocean surveillance, including ice detection and oil spill detection. The DVWF mode operates using a single polarization, and the OSVN using dual polarization.

To fully address the issue of maritime security, it is not sufficient to just develop and employ additional sensing capabilities. Future maritime Command and Control (C2) systems will have to support the processing and exploitation of greater quantities, varieties, and more complex information. For example, due to the non-persistent nature of the space-based SAR detections, and the vast area of surveillance with relatively few ships to detect in comparison to other parts of the world, any ship detections are spatially sparse in their nature. Generating effective situational awareness from this temporally and spatially sparse data presents an interesting research challenge. By fusing the available information, a clearer picture of maritime activities can be generated. Nonetheless, use of RADARSAT-2 in the Arctic for ship detection remains a practical challenge in terms of performance constraints due to the relatively low densities of traffic and high clutter due to land and ice.

The DRDC project for the next generation maritime C2 systems has as one component focusing on support for maritime and coastal Intelligence, Surveillance and Reconnaissance (ISR). To deliver a holistic Canadian solution for C2, one must consider the unique Canadian Arctic aspects in the ability to achieve maritime ISR in the Arctic which includes consideration of the types of information and data currently available in the existing C2 systems, and consideration for unique requirements to enhance and exploit Arctic situational awareness.

3 Situational Awareness

While navigable accessibility in the Arctic is increasing, in-situ sensing and communications remains a challenge due to the harsh environmental conditions. Satellite based transponder systems (e.g. LRIT and S-AIS) are the primary sources of information to track ships and active remote sensing capabilities (e.g. RADARSAT-2) are contributing to an increasing extent.

Figure 1 shows all of the open-source and unclassified ship position reports captured in the Royal Canadian Navy's (RCN) Global Position Warehouse (GPW) from May 1, 2011 to July 1, 2015. GPW is a database which archives ship position reports which were processed by the Navy command and control system. There is no guarantee of correctness in this dataset, but it does record reported (as received) position, time, and any available identifying information. Shown also is the ice extent for the week of September 17, 2014, which represents the minimum ice extent for 2014. This ice data was retrieved from the United States National Oceanic and Atmospheric Administration (NOAA) Optimum Interpolation (OI) Sea Surface Temperature (SST) V2 dataset (method of Reynolds et al. 2002) and was retrieved via the National Centers for Environmental Information.

It is clear from the spatial distribution of position contacts that there is a significant amount of activity in the Arctic, and one can begin to observe potential patterns and activities from just the basic positional observations shown in Fig. 1. Not shown in Fig. 1 is the number of transits and many of the position reports could be from a smaller subset of ships making repeated journeys. The density of these observations is presented in Fig. 2, which highlights the areas of higher density of position reports. In interpreting Fig. 2, one should be careful to consider the convolution of increased reporting frequency due to sensor locations, and increased

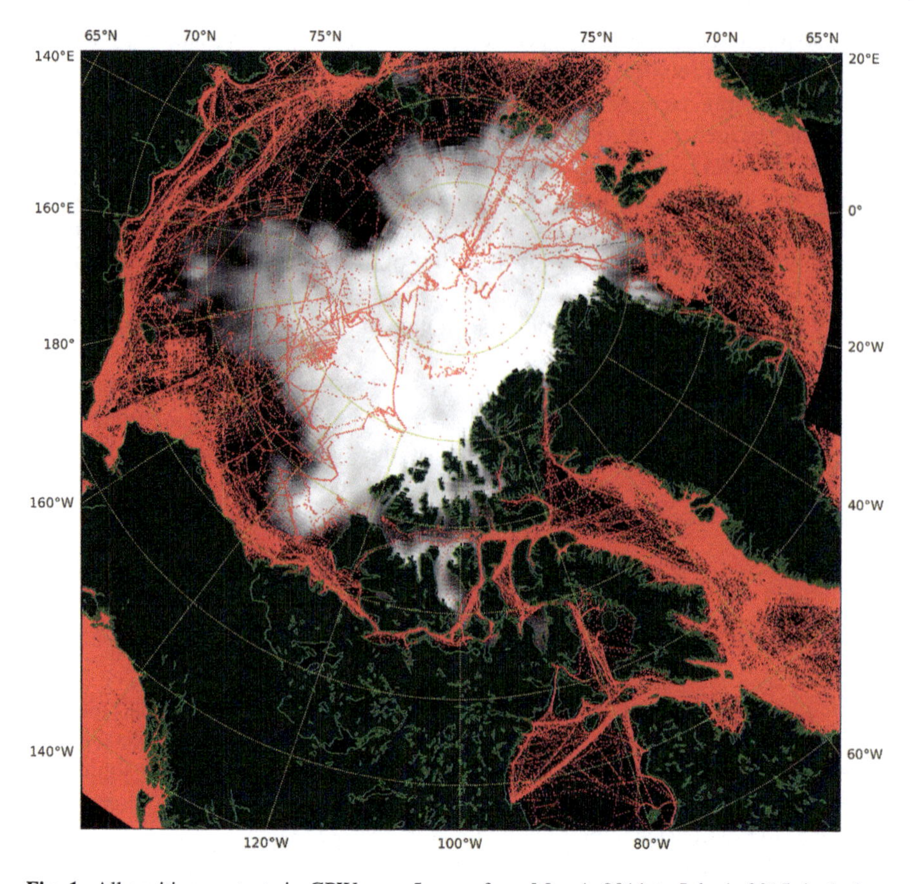

Fig. 1 All position contacts in GPW over 5 years from May 1, 2011 to July 1, 2015, inclusive, plotted as red dots. For spatial reference, the ice extent shown is from the week of September 17, 2014

reporting due to actual traffic density. This means that while the density in Fig. 2 is representative of the density of traffic, it should not be taken as an absolute value for traffic density (as it is conditioned also on sensor persistence and update rates).

In Fig. 2, three primary active maritime approaches to the Canadian Arctic are labelled: the western approach from the Beaufort Sea along the coast of Alaska into Canadian waters; the eastern approach south of Greenland through the Davis Strait and into Baffin Bay (and potentially the North West Passages); and the eastern approach through Hudson Strait into Hudson Bay or into the North West Passages via Fury and Hecla Strait The investigation of this dataset presented next will investigate the eastern Arctic approaches.

One of the labels captured by GPW is the category of ship in terms of merchant (commercial) vessels, government (i.e. Navy or Coast Guard) vessels, or fishing boats. Figure 3 presents colour coded detections in the Baffin Bay area to illustrate the visible vessel patterns in the Arctic.

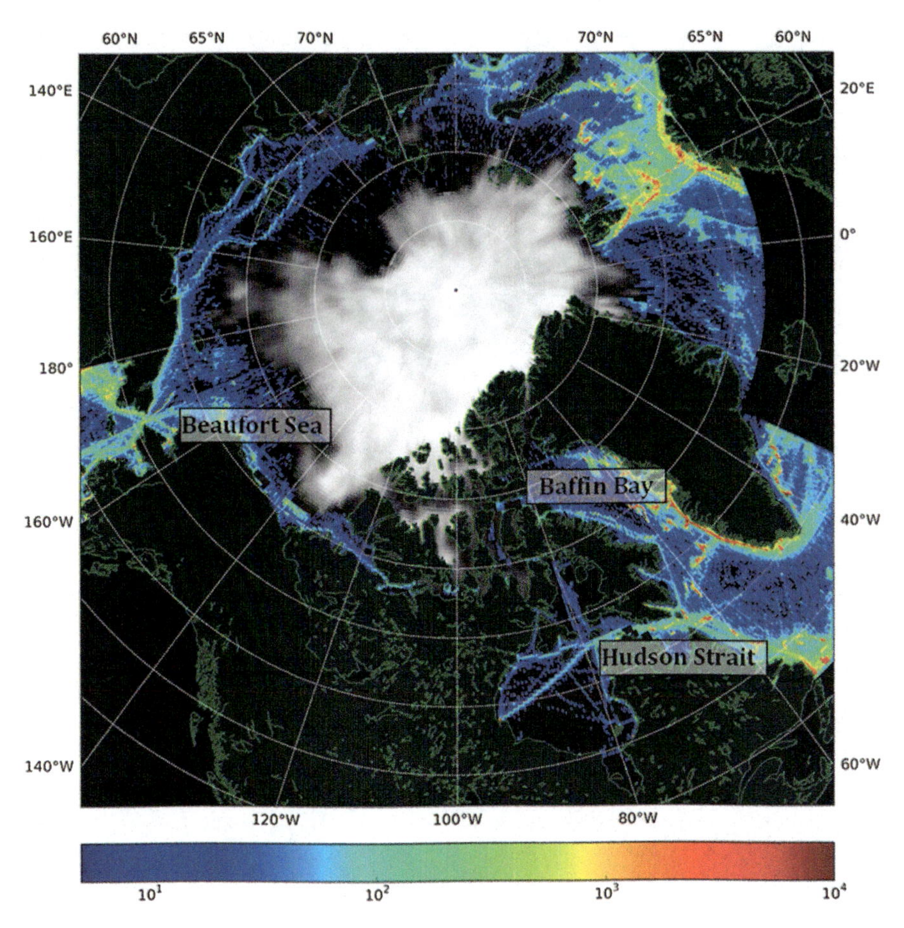

Fig. 2 Density map of position contacts with a grid size for the density layer of one degree latitude by one degree longitude, and logarithmic color scale. Each of the three major maritime approaches to the Canadian Arctic are labelled

The data in Fig. 3 for fishing boats is further analyzed to learn about their pattern of life. The pattern of life in the Arctic is a valuable context when evaluating sparse data. One of the more recent methods available for generation of pattern of life from large datasets is by using automated machine learning algorithms. For example, by using the DBSCAN algorithm (Ester et al. 1996) on the fishing data in Fig. 3, historical fishing zones can be extracted (among other items of interest such as locations of ports and harbours, stopping areas, and even transit corridors). Figure 4 shows the generation of clusters (reported speed between zero and one knots, minimum 10 observations, with a 20 km Euclidean neighbourhood threshold) for the fishing data using the DBSCAN algorithm. The clusters generated in Fig. 4 are also shown overlaid on bathymetry data obtained from the ETOPO1 dataset provided by NOAA (Amante and Eakins 2009). One can observe that the seemingly

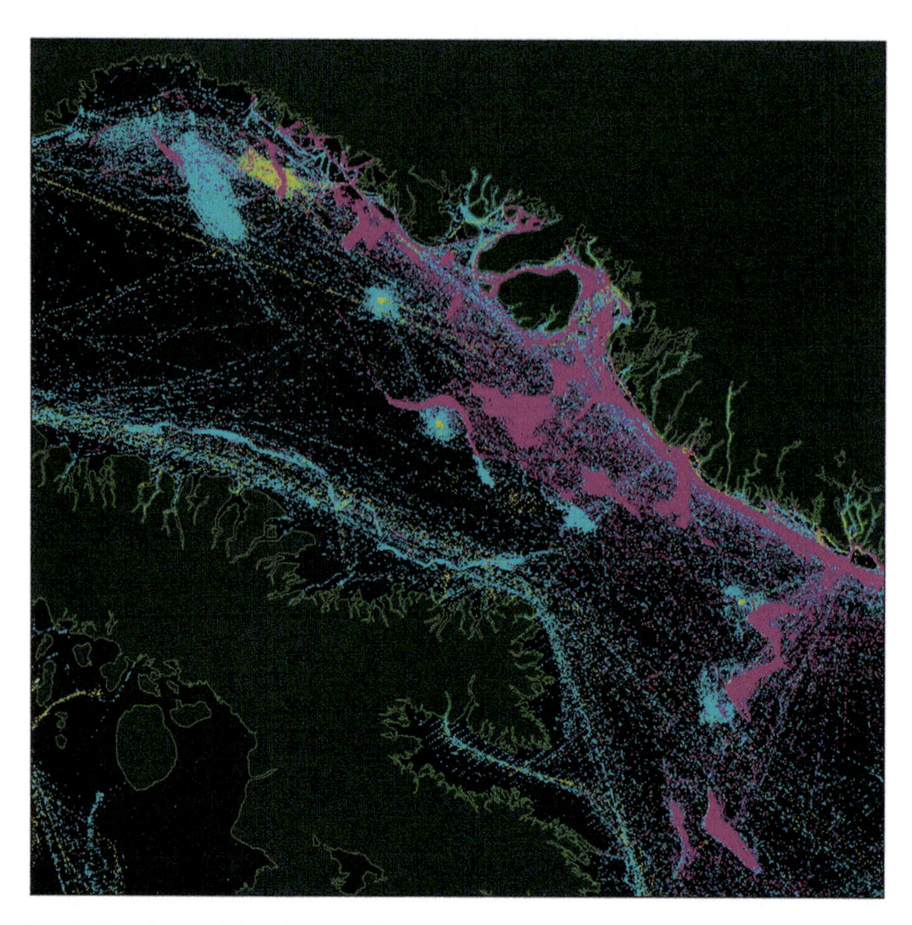

Fig. 3 Close-in plot of detections in Baffin Bay with reported fishing boats as magenta, commercial ships as cyan, and government ships as yellow

odd-shaped high-density regions of fishing activity are aligned to bathymetric features, which are no doubt linked to the occurrence of the resources being fished.

While the type of context generated by the analysis of this dataset is useful for general situational awareness, it is also a valuable piece for the enhancement to the detection and tracking of maritime threats or other vessels of interest. Pallotta et al. (2013) presents a powerful technique to automatically learn pattern of life activity from AIS observations using automated machine learning. Adapting their type of analysis to Arctic data would provide an atlas of "normal" pattern of life, which can then be used for threat analysis, anomaly detection, and decision making. Another application of pattern of life information has been shown by Mazzarella et al. (2015) where knowledge of normal shipping activities improves the ability to associate SAR detections with temporally asynchronous AIS detections. Improved association of SAR ship detections with pattern of life can help disambiguate the detection of

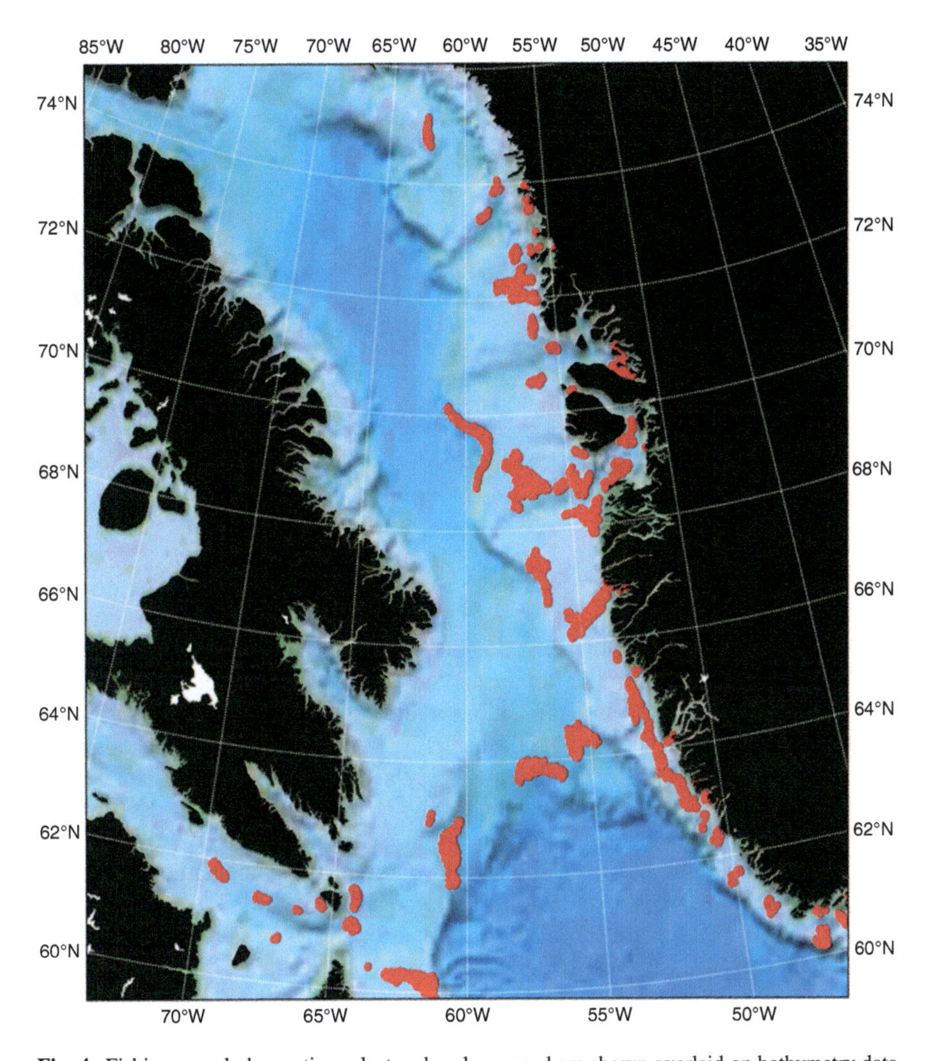

Fig. 4 Fishing vessel observations clustered on low speed are shown overlaid on bathymetry data

ships from icebergs and improve the tracking of ships which is currently highly dependent on AIS.

Another application being explored to support decision making is the enhanced prediction of vessels of interest by using known transit activities (Pallotta et al. 2014). Here, the authors found that an Integrated Ornstein-Uhlenbeck model describes the growth of uncertainty in a predicted position for generally open-water transits. The Ornstein-Uhlenbeck model is a stochastic model where the statistics of state changes over a time series are described using normal distributions for the range of possibilities combined with a mean-reverting tendency. One can think of this as an approximation of a driver aiming to steer straight, but the actual

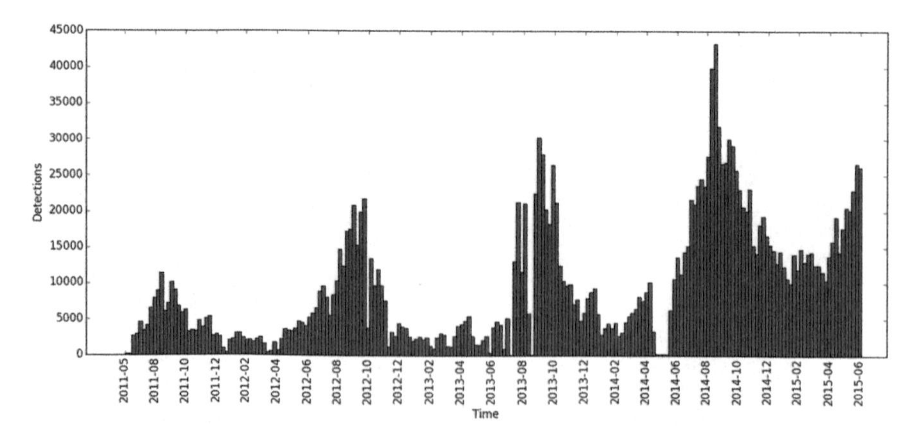

Fig. 5 Histogram of all ship detections in GPW north of 60 degrees latitude and between 170 West and 40 West degrees longitude with time in the x axis is indicated as year and month, and weekly histogram bins

path may have deviations to either side of the ideal line of transit. The driver is applying a mean reverting force.

While this Integrated Ornstein-Uhlenbeck model has been recently shown to work well for describing traffic in unrestricted open water, it is not certain whether this same model applies to ship predictions in the Arctic environment. It is arguably unlikely to be as effective in the Arctic due to the significant navigational constraints from land and ice. Therefore, for effective decision support in the Arctic, a prediction model for constrained or semi constrained navigation is required for vessel prediction in regions such as the Canadian Arctic. Hammond (2014) proposes one approach using graphs, however, additional work to reduce computational complexity, and validation against real data has yet to be achieved for this approach.

The foundations for a new paradigm of maritime ISR are being developed. Of relevant interest here is the use of large datasets to enhance the use of sparse or noisy sensor data. The Arctic trend in both the amount of traffic and information available in the Arctic is clear from the histogram of ship position reports over time, presented in Fig. 5. From May 2011 to July 2015, the seasonality of the traffic report quantities is evident in the periodic rise during the summer months and fall during the winter months. However, the trend to draw attention to in Fig. 5 is the ever-increasing quantity of reports over time. This increasing amount of data, in combination with automated machine learning algorithms and enhanced active remote sensing capabilities, enables new approaches to detect and track maritime threats in the Canadian Arctic.

4 Conclusions

The Government of Canada continues to invest in the Canadian Arctic, and the capability to detect and track vessels in the remote and challenging Canadian Arctic maritime environment is continuously increasing. The future RADARSAT Constellation Mission is one example capability which has the potential to enhance Arctic situational awareness and improve maritime security.

Future work by DRDC in developing the requirements for the next generation maritime C2 system will investigate the combination of unidentified ship detections (i.e. from RADARSAT-2 or RCM) against patterns of life in order to cross-cue surveillance capabilities. Furthermore, with the inclusion of the pattern of life data and navigational constraints, new and existing applications for operational decision support such as vessel of interest reconnaissance tools can be improved.

Acknowledgements Gratefully acknowledged is Paris W. Vachon of DRDC Ottawa Research Centre, who provided helpful comments, discussion, and reference material related to space based SAR and S-AIS. The author also acknowledges the Arctic dataset provided by the Royal Canadian Navy's Global Position Warehouse developed and maintained by the team of Scott Syms and Andrew DeBaie. The operational ship detection capabilities of RADARSAT-2 and the RADARSAT Constellation Mission are implemented via the Department of National Defence Polar Epsilon and Polar Epsilon 2 projects, MacDonald, Dettwiler and Associates, and DRDC Ottawa Research Centre.

References

Amante, C., & Eakins, B. W. (2009). *ETOPO1 1 Arc-Minute Global Relief Model: Procedures, Data Sources and Analysis. NOAA Technical Memorandum NESDIS NGDC-24*. National Geophysical Data Center. NOAA. https://doi.org/10.7289/V5C8276M

Canada Space Agency. (2015). *RADARSAT Constellation Mission description*, online. Retrieved July 21, 2015, from http://www.asc-csa.gc.ca/eng/satellites/radarsat/description.asp

Canada's Northern Strategy. (2013) *Our north, our heritage, our future*, online. Retrieved July 21, 2015, from http://www.northernstrategy.gc.ca

Canadian Coast Guard. (2013). *Vessel traffic reporting Arctic Canada traffic Zone (NORDREG)*, online. Retrieved July 21, 2015, from http://www.ccg-gcc.gc.ca/eng/MCTS/Vtr_Arctic_Canada

Cervera, M., & Alberto, G. (2008). On the performance analysis of a satellite-based AIS system. In *IEEE 10th International Workshop on Signal Processing for Space Communications*.

Ester, M., Kriegel, H. P., Sander, J., & Xu, X. (1996). A density-based algorithm for discovering clusters in large spatial databases with noise. In *Proceedings of the 2nd International Conference on Knowledge Discovery and Data Mining* (pp. 226–231). AAI Press.

Hammond, T (2014) Applications of probabilistic interpolation to ship tracking. In *Proceedings of Joint Statistical Meetings* (pp. 1–19).

Howell, C., Power, D., Lynch, M., Dodge, K., Bobby, P., Randell, C., et al. (2008). Dual polarization detection of ships and icebergs – recent results with ENVISAT ASAR and data simulations of RADARSAT-2. In *Proceedings of IEEE International Geoscience and Remote Sensing Symposium (IGARSS)* (pp. 206–209).

Howell, C., Youden, J., Kane, K., Power, D., Randell, C., & Flett, D. (2004). Iceberg and ship discrimination with ENVISAT multipolarization ASAR. In *Proceedings of IEEE International Geoscience and Remote Sensing Symposium (IGARSS)*, (pp. 113–116).

Mazzarella, F., Vespe, M., & Santamaria, C. (2015). SAR ship detection and self-reporting data fusion based on traffic knowledge. *IEEE Geoscience and Remote Sensing Letters, 12*(8), 1685–1689.

Pallotta, G., Horn, S., Braca, P., & Bryan, K. (2014). Context-enhanced vessel prediction based on Ornstein-Uhlenbeck processes using historical AIS traffic patterns: Real-world experimental results. In *Information Fusion (FUSION), 2014 17th International Conference on* (pp. 1–7).

Pallotta, G., Vespe, M., & Bryan, K. (2013). Vessel pattern knowledge discovery from AIS data: A framework for anomaly detection and route prediction. *Entropy, 15*(6), 2218–2245.

Papa, G., Horn, S., Braca, P., Bryan, K., & Romano, G. (2012) Estimating sensor performance and target population size with multiple sensors. In *Information Fusion (FUSION), 2012 15th International Conference on* (pp. 2102–2109).

Reynolds, R. W., Rayner, N. A., Smith, T. M., Stokes, D. C., & Wang, W. (2002). An improved in situ and satellite SST analysis for climate. *Journal of Climate, 15*, 1609–1625.

Tunaley, J. K. E. (2011). *Space-based AIS performance* (Technical Report 2011-05-23-001). London Research and Development Corporation.

Vachon, P. W., Kabatoff, C., & Quinn, R. (2014). Operational ship detection in Canada using RADARSAT. In *Proceedings of IEEE International Geoscience and Remote Sensing Symposium (IGARSS)* (pp. 998–1001).

Knowledge Discovery of Human Activities at Sea in the Arctic Using Remote Sensing and Vessel Tracking Systems

Michele Vespe, Harm Greidanus, Carlos Santamaria, and Thomas Barbas

Contents

Abstract Adequate knowledge of human activities in the Arctic is fundamental to support safe and secure maritime operations and sustainable development in the area. Such knowledge is often incomplete in terms of activities, geographic area and spatial resolution. For example, in the specific case of the transits over the Arctic shipping routes, such information can be accessed through domain expert knowledge, open source statistics or data from ship reporting systems. Offshore energy and exploration, fishing, and shipping activities can be monitored and/or mapped using surveillance tools such as satellite based remote sensing (e.g. Synthetic Aperture Radar—SAR) and vessel tracking systems (e.g. Automatic Identification Systems—AIS, and Long Range Identification and Tracking—LRIT), supplemented by knowledge discovery approaches. Such data-driven methodology, combined with meteorological and oceanographic information, enables a high level of situational awareness that is otherwise often difficult to access, hard to update or challenging to extract. In this chapter we analyse ways to understand and characterise activities

M. Vespe (✉) · H. Greidanus · C. Santamaria · T. Barbas
European Commission, Joint Research Centre (JRC), Ispra, Italy
e-mail: Michele.VESPE@ec.europa.eu

L. P. Hildebrand et al. (eds.), *Sustainable Shipping in a Changing Arctic*, WMU Studies in Maritime Affairs 7, https://doi.org/10.1007/978-3-319-78425-0_9

and discover their trends in the Arctic. This new information will assist policy makers and operational authorities when conducting Maritime Spatial Planning and the evaluation of new routing systems and impact assessments of Marine Protected Areas.

Keywords Knowledge discovery · Maritime situational awareness · Maritime surveillance · Maritime transport

1 Introduction

This chapter discusses ways to monitor human activities at sea in the Arctic. In particular, it endeavour to acquire knowledge of such activities by analysing and combining data sources on ship traffic and environment that are becoming available in the Arctic as an output of satellite surveillance.

Human activity in the Arctic seas is increasing, due to the profound reduction in the sea ice cover, advances in technologies to cope with the Arctic maritime environment, and continued acquisition of resources such as oil, gas, minerals and fish. To ensure safety, security and sustainability in these developments, authorities need to be well aware of the nature, locations and extent of the activities. A comprehensive marine shipping assessment was conducted by the Arctic Council and presented together with future scenarios in the AMSA report (2009). As a practice, shipping is used as a proxy for human activities as all human activities at sea are either ship or platform-based, and all platforms are accessed by ships. This chapter offers a data-driven methodology to integrate and enrich the available information.

This study therefore aims to use available data on ship traffic, in combination with available data on the geophysical environment and other auxiliary information, to produce knowledge on human activities that can be used by authorities in their assessments of the needs to regulate or intervene with the view to the mentioned criteria of safety, security and sustainability.

2 Data

We can usefully categorise the data types that are used in (1) ship data leading to information on individual movements and activities at sea; (2) geophysical data providing context information on oceanographic features and meteorological conditions; (3) auxiliary data providing additional informative layers over the area of interest.

2.1 Ship Data

Many ships nowadays carry automatic self-reporting systems, that report the ship's identity and position as derived from the on-board GNSS (Global Navigation Satellite System) position, plus possibly additional information such as speed, course, navigation status and destination. The main systems are AIS (Automatic Identification System) and LRIT (Long Range Identification and Tracking). Carriage of these two systems is globally mandated by the United Nations' International Maritime Organisation (IMO) for specific classes of ships, roughly ships of 300 GT and up. The VMS (Vessel Monitoring System) is another such system specifically required for fishing vessels and implemented by many States and regional programmes. While LRIT and VMS data are government-owned (by the Flag State) and not openly accessible, AIS data is more readily available, even with nearly global coverage as a result of the use of networked coastal AIS receivers and receivers deployed on satellites. Vessel Traffic Systems (VTS) along coastal states may send information about vessels which are not carrying AIS and which are tracked only by coastal radars, via the AIS to vessels equipped with AIS.

In addition, satellites that image the Earth (sea) surface can be used for ship detection to also find non-reporting ships—down to a certain size and over a limited area as determined by the satellite sensor's properties. Satellite imagers operate in optical and radar wavelengths, the latter being able to penetrate clouds. Satellite imaging capacity is however orders of magnitude too small to ensure continuous global tracking of the ship traffic, whereas AIS can already attain the necessary data for the reporting ships.

2.2 Geophysical Data

Geophysical data includes data that are more static such as coastlines and bathymetry, and more dynamic data such as winds, waves and ice coverage. The static data types are available from open source data sets, although in the Arctic sometimes with limited accuracy. The met/ocean and ice data are produced by models that are driven with observations, a large part coming from earth observation (remote sensing) satellites. Some of these satellites are the same as the ones that can be used for ship detection. For example, the Sentinel-1 imaging radar satellite of the EU's Copernicus program can be used for ship detection and sea ice mapping—albeit in each case with a limited accuracy only. It is worth noting that Arctic observation systems currently cover national or regional areas, although there are initiatives aiming at developing an integrated observation system (e.g. the H2020 INTAROS—Integrated Arctic Observing System project).

2.3 Auxiliary Data

Under this header can be lumped a diverse collection of data that can be used for better understanding of the maritime situation, such as maritime boundaries (EEZ, fisheries convention area limits, traffic separation schemes, etc.), positions of platforms, lighthouses, buoys, pipelines, etc., but also background information on ships such as ownership, infringements history, etc. In addition, government polar icebreakers can be used to track and relay pertinent auxiliary data along various routes through the arctic. More icebreakers have been commissioned and more are in the exploratory phase as well.

3 Methodology

Ship position reports from AIS and LRIT are collected within a certain geographic area (or even globally) and over a period of time.[1] Different AIS sources can be used in terms of networks and satellites (Høye et al. 2008; Eriksen and Olsen 2015), and likewise, LRIT and VMS can be sourced from its different owner governments. This is of significant importance because the government-owned data are not easily given access to due to legal and security restrictions; and the commercial data only come at a price. The position data of a certain ship can be collated from across the various sources, and strung together on a timeline, which produces the ship's track. As more data sources are accessible for use, the track will contain more points and fewer gaps. In this process, the ship's identity has to be unified across the various labels used (MMSI number for AIS, IMO number for LRIT). Also, outlier data points—that are regularly present in AIS data—must be recognised and removed. The resulting irregularly time-sampled track is then interpolated to a regular time grid, leaving open only gaps that are too long for reliable interpolation.

A thorough quantitative analysis of trends of number of ships in the Arctic and their geographic distribution over a period of 5 years can be found in (Eriksen and Olsen 2015). In this chapter, we aim at analysing ship behaviours in order to build the contextual awareness of activities at sea (Alessandrini et al. 2014). A certain ship will typically show a particular speed distribution. For instance, a cargo ship or tanker will spend most of its time at its cruising speed and the remainder manoeuvring at slow speed or at rest, while a fishing ship could be seen at some preferred slow speed when engaged in fishing but at another, higher speed when transiting. On the basis of the ship's type and speed, portions of its track can in this way be attributed to a specific activity.

For the present study, the geophysical data used was limited to ice coverage, with the aim to compare ship traffic with ice extent. Three types of ice cover data were used: monthly maps of the marginal ice zone; extent of 80% ice covered waters, both

[1] VMS data can also be included but were not used in this study.

Fig. 1 Sentinel-1 image from 19 Apr 2015, 01:28:13 UTC, EW mode, HH polarisation. Land is in the top right corner and in the bottom centre and left corner, the rest of the image is the ice-covered Yenisei Gulf. © *Copernicus 2014, 2015*

with Arctic-wide coverage from U.S. National Ice Center (NIC)[2]; and ice coverage as interpreted from Sentinel-1 satellite radar images in some particular locations in the Arctic.

The resulting ship tracks are then displayed on a GIS viewer, on a map background and with selectable layers that represent ice coverage. For this, the JRC's Blue Hub[3] viewer was used, presently implemented on GeoServer.

4 Results

4.1 Monitoring a Local Area

The Yenisei Gulf in Siberia (72.4N, 79.6E) was used to explore the correlation between AIS ship tracks and tracks visible in the ice in Sentinel-1 radar images. Figure 1 shows part of a Sentinel-1 image over the Yenisei Gulf taken on 19 Apr 2015; a prominent feature in the image is a bright line running diagonally across. Figure 2 maps the AIS positions over that area collected between 1 Jan and 22 Apr 2015 in which a total of 5088 messages from 12 unique vessels were recorded. Although the image is instantaneous and the AIS positions cover a 3.5 month period, it is clearly evident that the AIS positions coincide with the bright line in the Sentinel-1 image. This leads to a conclusion that the bright line in the radar image is a cut through the ice cover and that the ships keep using this opening in the ice, while the ice remains fixed in place during this months-long period. The few AIS

[2]http://www.natice.noaa.gov/.

[3]https://bluehub.jrc.ec.europa.eu.

Fig. 2 AIS positions collected between 1 Jan and 22 Apr 2015 over the same area as Fig. 1

positions outside the bright line are on the straight route into and out of the Gulf, which remained in use until 19 January 2015. After that day all the traffic moved to the slightly more winding route seen in Figs. 1 and 2.

Sentinel-1 operates following a pre-established observation plan and will routinely collect data in pre-defined modes and areas, the Arctic being one such area. Furthermore, the Copernicus program has adopted a free and open data policy, and Sentinel images are available in the Sentinels Scientific Data Hub.[4] Sentinel-1 acquired 67 images of the Yenisei Gulf area between 9 October 2014 and 20 April 2015, with an average revisit time of 2.9 days. Revisit time is defined here as the time difference between consecutive images. The revisit time will shorten as the sensor increases its imaging rate and as the second satellite in the Sentinel-1 constellation enters operation in 2016. The revisit time in the Arctic is expected to be around 1 day when the two-satellite constellation is in full operational mode (Sentinel-1 User Handbook 2013). This shorter revisit time will enable more frequent observations.

4.2 Analysis of an Individual Ship Track

An example of the speed distribution of a fishing ship and its relation with the ship's geographic location is shown in Fig. 3. The histogram shows the ship's speed distribution collected over 40 days of observation. The multi-modal speed distribution of fishing vessels has been recently analysed using AIS data by Mazzarella et al. (2014) and Natale et al. (2015). The speed histogram indeed shows three peaks: one around 0, interpreted as at rest; one around 2.5 knots, interpreted as engaged in fishing; and one around 11.5 knots, interpreted as transiting. When the ship's track is plotted on a map and colour-coded by these speed intervals, it is verified that the

[4]https://scihub.esa.int.

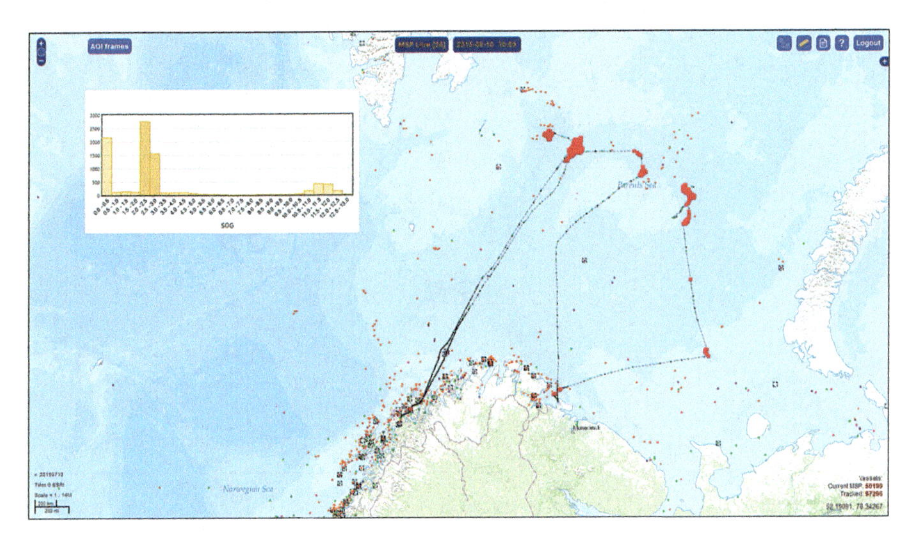

Fig. 3 Track of a fishing ship during 40 days plotted on a map, with positions coloured red when the ship's speed is near 2.5 knots. The scattered coloured dots are positions of other ships at a certain point in time. Inset: speed histogram of the ship during this period, with three peaks, at 0, 2.5 (highest peak) and 12.5 knots

transit legs contain the high speed points, whereas the low speed points around 2.5 knots cluster in the middle of the sea at the far ranges of the track (red), and are therefore likely the fishing grounds.

4.3 *Analysis of All Ship Tracks*

When, along these lines, the tracks of many ships are plotted together, the results look like Fig. 4, where the density of AIS messages (top) is broken down into fishing ships at slow speed in red, while non-fishing ships at slow speeds are coloured green, and the medium- and high-speed positions of all ships are in blue (bottom). This distribution, inasmuch as it is not close to the coast, is then interpreted as indicating fishing activity in red, exploration activity in green, and other ship presence including transport and transit in blue.

4.4 *Arctic-Wide Seasonal Changes*

It is possible to make a synoptic overview of the entire Arctic in this way, aggregating data in monthly periods, and comparing with the ice extent.

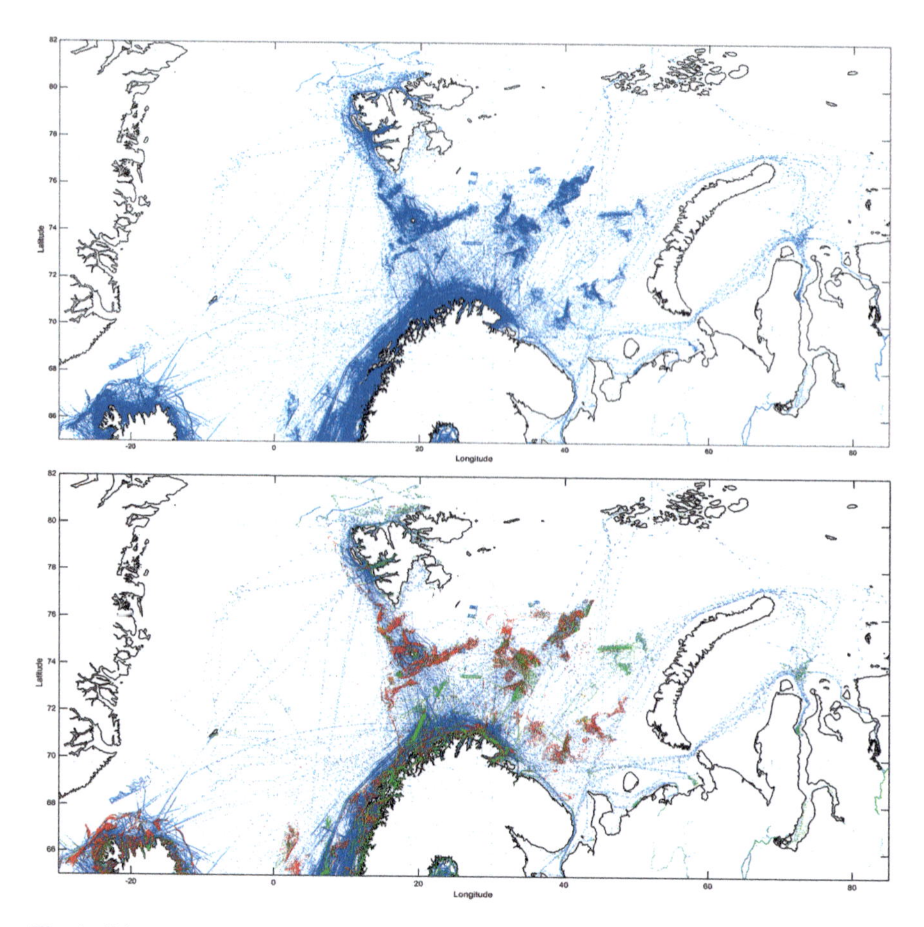

Fig. 4 Ship positions collected over a one month period plotted on a map around North Norway (top) and output of the behavioural analysis (bottom). Positions of fishing ships at slow speeds are coloured red; positions of non-fishing ships at slow speeds are coloured green; all other positions are coloured blue

This is shown in Figs. 5 and 6 for two sample periods. During the period from mid-August to mid-September 2014 (Fig. 5), the ice extent allowed shipping (blue), exploration (green), icebreaking (magenta) and fishing (red) activities in many areas of the Arctic. The classification is obtained from ship type information in the AIS messages. Conversely, in the period between mid-February and mid-March (Fig. 6) the ice extent is at maximum coverage and does not permit activities in the majority of the Arctic. It can clearly be seen that the seasonal retreat of the ice is mirrored in the expansion of human shipping activity.

The variability of fishing and other activities can clearly be observed in Fig. 7, where two seasons over two consecutive years can be inspected. In particular, given the bathymetry of the area between Iceland, Jan Mayen Island and Norway, the relevant fishing activities in July of both years (column of red dots above the

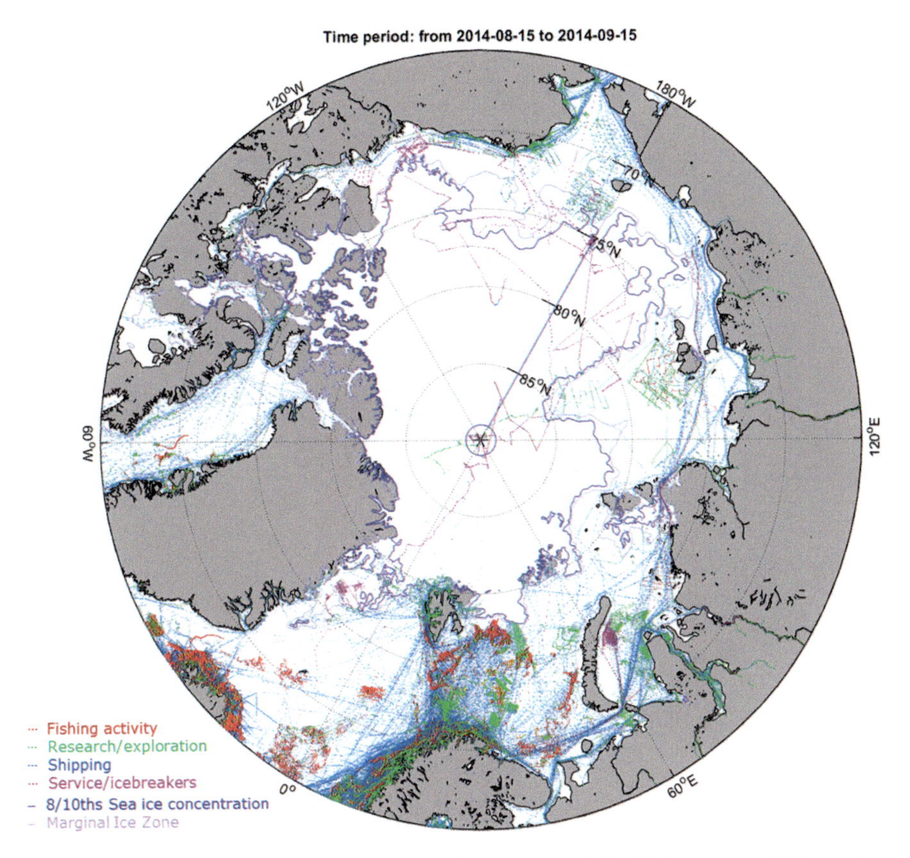

Fig. 5 Human activities in the Arctic and ice extent for the period mid-August to mid-September 2014, period of minimum extent of Arctic sea ice. Intense activities can be observed in many Arctic areas. The Northern Sea Route is *also clearly mapped, predominantly covered by re-supply vessels for the Arctic community*

0° longitude marker) are related to mid-water pelagic catches (Olsen et al. 2010). As expected, other activities such as research and exploration have a higher degree of variability.

5 Discussion

There is lack of available knowledge related to Arctic activities at sea, including fishing, shipping, exploration and tourism. However, this chapter has demonstrated the possibility to map such activities using vessel tracking data (AIS) and remote sensing data (satellite Synthetic Aperture Radar). Satellite AIS systems such as the ones operated by the Norwegian authorities are very powerful tools to monitor shipping in the Arctic.

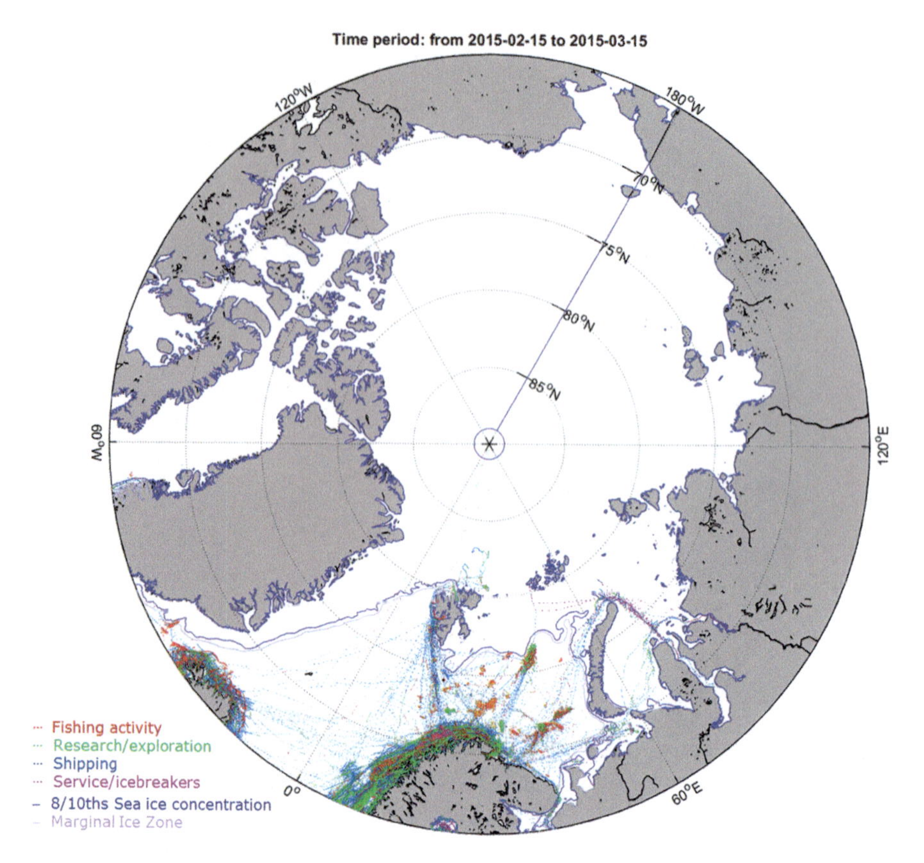

Fig. 6 Human activities in the Arctic and ice extent for the period mid-February to mid-March 2014. The activities at sea are confined to the Norwegian Sea and Barents Sea due to the seasonal ice extent. Icebreaking activities can also be observed to create maritime routes in the Kara Sea to the Yamal Peninsula and the port of Dudinka that services the industrial complex at Norilsk

The information obtained is essential for the better understanding of the Arctic human activity dynamics. Moreover, the knowledge of maritime activities is fundamental for a better planning of maritime uses and infrastructures, and is necessary for a safe and responsible development of the Arctic.

Still, there is a significant amount of Arctic information from earth observation, vessel tracking systems and other commercial services. Collaboration and information sharing are needed among the research community and operational authorities (e.g. regulatory compliance, law enforcement, Search and Rescue, emergency response) to bring together all available data, in many cases undisclosed or simply not used. This partnership would be highly beneficial to the same operational authorities as well as to policy makers.

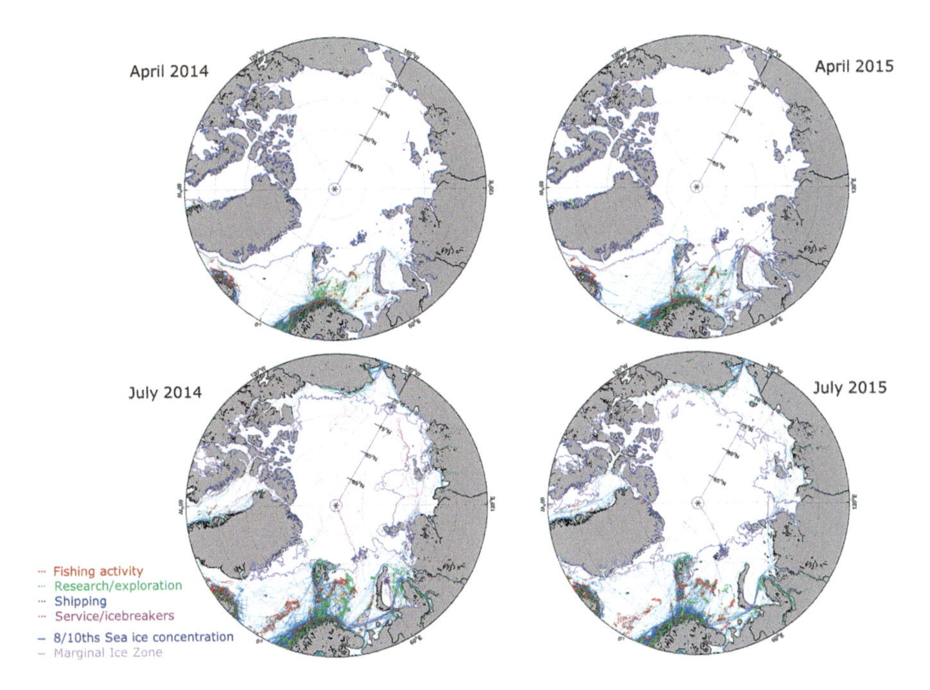

Fig. 7 Comparison of activities at sea in different months of two consecutive years, highlighting seasonal and yearly occurrences

Acknowledgments The Norwegian Defence Research Establishment (FFI) and the Norwegian Coastal Administration are thanked for providing access to AISSat-1 and AISSat-2 data that have proved very valuable for this study.

Further AIS data were obtained from MSSIS, courtesy of the Volpe Center of the U.S. Department of Transportation and the U.S. Navy.

Sentinel-1 data are © Copernicus 2014, 2015.

Ice maps were obtained from the U.S. National Ice Center.

The authors would like to thank the reviewers for their valuable contribution in improving this chapter.

References

Alessandrini, A., Argentieri, P., Alvarez, M. A., Barbas, T., Delaney, C., Arguedas, V. F., Gammieri, V., Greidanus, H., Mazzarella, F., Vespe, M. and Ziemba, L. (2014). Data driven contextual knowledge from and for maritime situational awareness. In *Context-Awareness in Geographic Information Services (CAGIS).*

AMSA Report. (2009). *Arctic Marine Shipping Assessment – Arctic Council.* Protection of the Arctic Marine Environment – PAME. Retrieved January 17, 2017, from http://www.pame.is/images/03_Projects/AMSA/AMSA_2009_report/AMSA_2009_Report_2nd_print.pdf

Eriksen, T., & Olsen, Ø. (2015). Vessel tracking using automatic identification system data in the Arctic. In *Proceedings of the ShipArc2015 conference "Safe and Sustainable Shipping in a Changing Arctic Environment".*

Høye, G. K., Eriksen, T., Meland, B. J., & Narheim, B. T. (2008). Space-based AIS for global maritime traffic monitoring. *Acta Astronaut, 62*, 240–245.

Mazzarella, F., Vespe, M., Damalas, D., Osio, G. (2014). Discovering vessel activities at sea using AIS data: Mapping of fishing footprints. In *17th International Conference on Information Fusion (FUSION).*

Natale, F., Gibin, M., Alessandrini, A., Vespe, M., & Paulrud, A. (2015). Mapping fishing effort through AIS data. *Plos One, 10*(6), e0130746.

Olsen, E., Aanes, S., Mehl, S., Holst, J. C., Aglen, A., & Gjøsæter, H. (2010). Cod, haddock, saithe, herring, and capelin in the Barents Sea and adjacent waters: A review of the biological value of the area. *ICES Journal of Marine Science, 67*(1), 87–101.

Sentinel-1 User Handbook. (2013). *GMES - S1OP – EOPG – TN – 13 – 0001, 66–67.* Retrieved January 17, 2017, from https://sentinel.esa.int/documents/247904/685163/Sentinel-1_User_Handbook

Part III
Arctic Governance

The Place of Joint Development in the Sustainable Governance of the Arctic

Buba Bojang

Contents

Abstract As the ice continues to melt away unabated, access to the areas of the Arctic, hitherto inaccessible, becomes real. The coastal States bordering the Sea have since laid claims to the continental shelf of what they believe is their legal entitlement, in order to exploit the resources of the seabed particularly oil and gas. Those who claim under the 1982 United Nations Convention on the Law of the Sea (UNCLOS), the relevant provisions thereof will be triggered, and for those outside UNCLOS, the Geneva Convention on the Continental Shelf and the rules of Customary International Law. Overlapping claim areas and the presence of oil and gas resources that transcends international boundaries are highly possible. For these reasons, the Arctic is referred to as another untamed place of the world, where the competition for resources in disputed areas or of a transboundary nature, without an established legal framework, could mar the geopolitical landscape and ultimately leading to confrontation. This may prove detrimental to the marine environment, shipping, and other peaceful uses of the sea.

The existing international legal regimes that regulate the activities in the Arctic, include, the Geneva Conventions of 1958; United Nations Convention on the Law of

B. Bojang (✉)
IMO International Maritime Law Institute, Msida, Malta
e-mail: r01bb13@abdn.ac.uk

L. P. Hildebrand et al. (eds.), *Sustainable Shipping in a Changing Arctic*, WMU Studies in Maritime Affairs 7, https://doi.org/10.1007/978-3-319-78425-0_10

the Sea of 1982, and the Polar Code, among others. However, none of these regimes is explicit on the rules for the exploitation of transboundary oil and gas resources or those found in overlapping claim areas/disputed areas. The delimitation of maritime boundary may not be effective, where states based their respective claims on different rules, or where oil and gas resources transcend international boundary or boundaries to an extend that same resources forms part of a single geologic unit and is exploitable from either side of the divide. The economic imperative that motivates states to venture into offshore oil and gas development hold same for the Arctic states too, especially with the findings of the US Geological Survey on the hydrocarbon potentials of the Arctic (USGS). This must be balanced with the social imperative of management.

Joint Development appears to be the alternative option for the Arctic States. Its role has expanded from the traditional development and apportionment of shared oil and gas resources to other aspects of ocean governance, including but not limited to the protection and preservation of the marine environment and the conservation of the living resources. However, its status (whether a provisional arrangement pending maritime boundary delimitation, or an alternative thereto) and the legal basis for states venturing into it, remains a discourse and sometimes elusive as an international rule of law.

The contribution of this chapter to the above-mentioned discourse is to examine whether joint development is in fact the best option for a truly Arctic governance and will seek to determine the legal basis for the Arctic states to enter into such an arrangement. It will also look at whether the Arctic Council could play a leading role in instituting joint development in the Arctic, through a multilateral treaty regime, rather than leaving it to the bilateral will of the states. Further, the chapter will critically analyse the Polar Code to determine whether it could secure a successful Arctic governance on its own. The chapter will then recommend, in addition to Joint Development, the development and adoption of 'the Arctic Natural Resources Development Code. The interaction of these arrangements will not fail to achieve the aspirations of the Arctic stakeholders. This chapter will conclude that a holistic ocean management, through joint development and the adoption of a natural resources development code will not fail to achieve a sustainable Arctic governance, including the protection and preservation of the marine environment. This will also institute a mechanism for the service of collective interest in the Arctic, through cooperation, rather than rivalry and confrontation.

Thus, a brief recount of the Arctic region and its special treatment under UNCLOS will be given. This will be followed by the analysis of the legal regimes governing the conduct of the coastal States in the Arctic and the limitation if any of such regimes. Identification will be made of the need for the development of appropriate mechanisms to fill in the gap.

Keywords UNCLOS · Joint development zone · Arctic legal regime · Sustainable governance · Arctic

1 Introduction

The North Pole, the Arctic Ocean and the area demarcated by the Arctic Circle are the constituents of the Arctic region. The Arctic Ocean, the main emphasis of this chapter, comprises of the following water bodies: The Barents; Kara; Laptev; East Siberia, and Chukchi Seas, among others. It has a continental shelf of about four and half million square kilometers, half of which is the shallow extensions of the land mass of the coastal States bordering the Sea, whilst the other half plunges into the deep ocean floor (Hober 2012). Ice has been the main characteristic, which differentiates it from other oceans or seas, the thickness of which ranges between 10 and 40 feet in layered format. It is an enclosed sea in that it is surrounded by eight States and connected to other seas through a sea lane (United Nations Convention on the Law of the Sea 1982). The Northwest Passage, which traverses the Arctic Ocean, along the northern coast of North America via waterways through the Canadian Arctic Archipelago and finally connecting the Atlantic and Pacific Oceans. There are five major coastal States bordering the Arctic: United States (US), through Alaska; Russia; Canada; Norway, and Denmark, through Greenland. Other States with territories in the Arctic but without a frongate in the arctic ocean are Sweden, Finland and Iceland.

With the seemingly unstoppable transformation of the Arctic, with the erosion of the ice cap, blamed on global warming, with its long extended continental shelf, signalling huge commercial potential, pertinent issues of concern will arise, not just for the Arctic coastal States, but for the international community as a whole (Lindsay 2012). Prominent among such issues is the determination of the maritime boundaries amongst the coastal States and the exploitation of the natural resources (living and non-living) in the Arctic, as overlapping claims or disputed areas abound. Other issues will include the protection and preservation of the marine environment and safety of international shipping as new routes will open up with the melting of the ice. The combination of these issues present emergent opportunities, but also a fertile ground for conflict in a similar manner as observed in the South China Sea. However, the concern of this chapter is more on how to effectively manage the exploitation of the resources, where maritime boundary delimitation may be unhelpful. With the unique character of the Arctic and the number of coastal States having legitimate interest in that enclosed area and recalling that it has been inaccessible for ages, one may inquire into the efficacy of the legal regimes that regulate access to the Arctic continental shelf and its natural resources.

2 The Arctic Legal Regime

The Applicability of particular legal norm (s) to a region may sometimes be dependent on the whether or not, the States of the region have subscribed to that norm (s). In the Arctic, all major stakeholders, but one, are parties to UNCLOS.

Some of them are still bound by both UNCLOS and the 1958 Geneva Convention on the Continental Shelf and Customary International Law.

2.1 The United Nations Convention on the Law of the Sea

The most widely accepted international legal regime that regulates maritime boundary disputes is UNCLOS. It establishes a legal framework for the peaceful resolution of disputes relating to its application to the world's oceans including the Arctic (Lindsay 2012). However, in cases of maritime boundary disputes, UNCLOS provides various options to parties entangled in such disputes. These options include bilateral agreement through negotiation (United Nations Convention on the Law of the Sea 1982). and where such an option fails, a dispute resolution framework under Part XV is resorted to, subject to the choice of parties expressed during signing, ratifying or acceding to UNCLOS and in doing so, may exercise their right to remove certain categories of disputes, such as those concerning maritime boundary delimitation, from the reach of such a framework (United Nations Convention on the Law of the Sea 1982). Further UNCLOS does not apply to States that are not parties thereto or sovereignty claims over disputed territories (Islands).

Firstly, it must be noted that US is not a State Party to UNCLOS. Therefore, the Geneva Convention on the Continental Shelf, 1958 and rules of Customary International Law applies, but only in relation to a State party under the same treaty, that is, Canada and Russia. For the other Arctic neighbours, the application of the provisions of UNCLOS relevant to the third party dispute settlement in relation to maritime boundary delimitation in the Arctic will be very difficult if not impossible, as the concern States have made differing declarations on how they want to resolve their respective maritime frontier disputes. For example, Canada made a declaration pursuant to article 298 (1) of UNCLOS that it does not accept any of the procedures provided for in Part XV, section 2, with respect to disputes concerning the interpretation or application of articles 15, 74 and 83 relating to sea boundary delimitations (Division for Ocean Affairs and Law of the Sea 2004). Russia, too, made similar declaration to the same effect. The respective declarations of Denmark and Norway appear to further complicate the issue. They do not accept an arbitral tribunal constituted in accordance with Annex VII as a third party dispute resolution mechanism in respect of maritime boundary dispute. The declarations of the other Arctic neighbours are equally worth mentioning. Both Finland and Sweden did not make a declaration under article 298, they however, choose ICJ and ITLOS, and ICJ respectively, as their forum for the settlement of disputes concerning the interpretation and application of UNCLOS and Part XI. Iceland on the other hand, declares that it will submit any dispute concerning the continental shelf to Conciliation under Annex V, section 2. This already looks like a treaty jigsaw in the interaction of these States.

The Arctic has seven international boundaries that merit delimitation, four of which already have delimitation agreements and two of the remaining un-delimited boundaries are governed by the UNCLOS regime whilst the remainder is to be governed by the Geneva regime on the Continental Shelf and Customary International Law (Hawker et al. 2012). Furthermore, the US Geological Survey reveal the existence of a significant deposit of oil and gas in about twenty-five provinces in the Arctic, seven of which are believed to be located in disputed areas or straddle across already delimited boundaries (United States Geological Survey 2008). Applying the above treaty relations to the current situation in the Arctic will only revealed the limitations of the existing legal regimes in resolving the highly potential maritime boundary disputes. Resultantly, it appears from the above declarations and statements that none of the third party adjudicatory bodies listed in UNCLOS may have the opportunity to exercise any form of jurisdiction over Arctic maritime frontier disputes. There are areas where delimitation remains a thorny issue, either in whole or in part, whilst in other areas the same has been achieved with ease, through bilateral treaties. This will present more challenges due to the presence or suspected presence of hydrocarbon deposits. In this regard, it must be noted that other than jurisdictional clarity and apportionment of maritime space, delimitation of maritime boundary alone does not in any way resolve the issue of a straddling resource. In essence, the Arctic is faced with two challenges, maritime boundary delimitation, in some instance compounded by territorial claims and trans-boundary oil and gas resources, a non-respecter of international boundaries.

The application of UNCLOS and the Geneva Convention on the Continental Shelf, among other international legal regimes to the Arctic Ocean may not deliver it from the scourge of human activities incidental to the exploitation of the natural resources. Delimiting the respective maritime frontiers of these Arctic neighbours will no doubt be dramatic on paper and nearly impossible in fact other than through cooperation, as the difference is not just about the mechanism but also the forum and even where such an exercise (maritime boundary delimitation) were to be successful, the intricacies of transboundary resources will continue to rear its ugly head. The 2008 Conference of the coastal States of the Arctic Ocean (Canada, Denmark, Norway, Russia and the United States of America) adopted the Ilulissat Declaration, acknowledging first that '(t)he Arctic Ocean stands at the threshold of significant changes. Climate change and the melting of ice have a potential impact on vulnerable ecosystems, the livelihoods of local inhabitants and indigenous communities, and the potential exploitation of natural resources.' (The Ilulissat Declaration 2008). The Declaration also recalled on the applicability of what it regards as an "extensive international legal framework" (UNCLOS) to the Arctic Ocean, in terms of the rights and obligation of State parties concerning delimitation and delineation, the protection and preservation of the marine environment and other peaceful uses of the sea. It concluded that there is 'no need to develop a new comprehensive international legal regime to govern the Arctic Ocean.' However, the Conference did undertake to keep abreast with the developments in the Arctic to adopt appropriate rules when the need arises.

With the current flux of activities in the Arctic, it appears that the above declaration, particularly its conclusion needs to be revisited. To the contrary, there is the need for Arctic States to develop a comprehensive Arctic management framework in order to cover the existing limitations and gaps in the current legal regimes. This position is confirmed by the following germane observation:

> The Arctic region is not currently governed by any comprehensive multilateral norms and regulations because it was never expected to become a navigable waterway or a site for large-scale commercial development. Decisions made by Arctic powers in the coming years will therefore profoundly shape the future of the region for decades (Borgerson 2008).

The question to ask would therefore be, what then is required in the Arctic to ensure its sustainable governance and to deliver it from the impending threats to the harmonious exploitation of hydrocarbon resources and its marine environment? Suggestions have been proffered, that the Arctic Ocean needs a wide-ranging treaty regime similar to the 1961 Antarctica treaty in order to resolve the conflicting claims by instituting a framework for joint governance (Watson 2008–2009). Others are of the view that the creation of an international treaty that would allow the establishment of an international sector or park, along the lines of the Limpopo Trans-Frontier Park in Southern Africa and parallel to the Antarctica arrangement, or the creation of an authority to place a moratorium on natural resource exploitation and development in the Arctic will not fail to provide the panacea (Dubner 2005). The Above suggestions, while being applauded for their wisdom, must be treated with caution if they are suggestive of a regime that would place an embargo or a moratorium on the right of the Arctic States to undertake economic activities in their respective or claimed continental shelf. Further, whilst the Antarctica and the Limpopo Park are terrestrial, the Arctic is a sea, governed by UNCLOS, albeit with gaps, which has given coastal States the sovereign right to explore and exploit the natural resources of the continental shelf appertaining to them (United Nations Convention on the Law of the Sea 1982).

2.2 *Joint Development*

This chapter argues that the Arctic needs an international regime that will strike a balance between the economic imperative of natural resource exploitation and the social obligation to manage the environment that harbours such resources. It thus proposes that since this balancing act has been the hallmark of Joint Development as it expands its role, from mere resource exploitation and the apportionment of the proceeds therefrom, to a more sophisticated mechanism for zonal management, it appears to be the most appropriate under the circumstances. It is not a provisional arrangement pending maritime boundary delimitation, as argued, but rather an alternative thereto.

Joint Development is a cooperative mechanism used by States for the harmonious exploitation and apportionment of natural resources that exist in more than one jurisdiction. Such natural resources include, international rivers, transboundary fisheries and transboundary oil and gas. The concept could be traced to the joint utilisation of international rivers and the conservation and utilisation of transboundary fish stocks. Therefore, the subjects of JD are known as shared natural resources and it existed as a mechanism prior to UNCLOS. Its application to transboundary oil and gas resources, could be traced to the early practice of States bordering the North Sea, as instructively referred to by the ICJ in the North Sea Continental Shelf Cases. The Court described the North Sea and why joint development was the most appropriate option in the following:

In a sea with the particular configuration of the North Sea, and in view of the particular geographical situation of the Parties' coastlines upon that sea, the methods chosen by them for the purpose of fixing the delimitation of their respective areas may happen in certain localities to lead to an overlapping of the areas appertaining to them. The Court considers that such a situation must be accepted as a given fact and resolved either by an agreed, or failing that by an equal division of the overlapping areas, or by agreements for joint exploitation, the latter solution appearing particularly appropriate when it is a question of preserving the unity of a deposit.

The Court's recommendation for the institution of joint development in the North Sea was due largely to the semi-enclosed nature of the area. This description could also be a true representation of the geographical fact of the Arctic. From that singular cooperative role of natural resource management, to a more expanded one. In recent cases, the cooperative arrangement establishes a zone of cooperation out of the disputed area, either wholly or in part, where the concerned States would have rights and responsibilities in and towards that zone, as agreed in the arrangement (Mensah 2006). Further, in contrast with the early joint development arrangements, recent joint development treaties have witnessed the inclusion of marine environmental protection provisions within it and placed such responsibility on the body responsible for the management of the designated zone. For example, article 23 of the UK/Norway Frigg Field Agreement requires the parties to ensure, either jointly or severally, that the exploitation of oil and gas and other incidental operations, and other peaceful uses of the sea (shipping and fishing) shall not cause pollution of the marine environment. Further, the Treaty between Nigeria and Sao Tome and Principe on joint development of their common offshore resources included a provision on the prevention of pollution and protection of the marine environment. Further, the Protocol to the Management and Cooperation Agreement between the Government of the Republic of Senegal and the Government of the Republic of Guinea-Bissau, for the joint exploitation of a designated zone of their continental shelf and exclusive economic zone, created an Agency saddled with the responsibility of managing the agreed zone. In the sphere of the protection of the marine environment from the activities in the zone, Article 23 (2) states that:

> In accordance with article 11, subparagraphs (i), (k), (*l*), (m) and (n), of this Protocol, the Agency shall lay down regulations to protect the marine environment in the Area. It shall establish an emergency plan or management plan to combat pollution and any degradation arising from resource prospecting, exploration and exploitation activities in the Area

Unlike the above treaty regimes, the 2010 Barents Sea Treaty between Norway and Russia did not create a designated zone and thus no established body for the administration of a designated zone. However, in the matter for the protection of the marine environment from the exploitation of transboundary hydrocarbon resources, parties are required, under the treaty, to consult each other in respect of environmental measures to be adopted as required by their national legislations.

3 Proposed Regime Contents

What is required in the Arctic is the concerted and uniform approach of the States, through the Arctic Council, towards the sustainable management of the Ocean, by regulating access to shared resources as well as the exploitation of such resources, especially to protect and preserve the marine environment from pollution resulting from hydrocarbon exploration and exploitation. This action should not be left at the convenience of or for the national laws of respective States, as seen in the Norway/Russia Treaty. A more elaborate regime (international joint development arrangement), is thus required at the level of the Arctic Council in a form of draft articles or a treaty (Crawford 2014). This will not fail to ensure the cooperative utilisation of hydrocarbon resources in the contested waters of the Arctic or even those of a transboundary nature. This cooperation will instil in the parties the duty to adhere to the environmental protection and preservation standards instituted by such a regime. The Arctic States are encouraged to consider the following principles for inclusion into the contents of such an international cooperative mechanism.

3.1 Joint Development Zone (Designated Zone)

The creation of a zone for the exercise of joint authority by countries entangled in maritime boundary delimitation in cases of overlapping claims or face with the existence of trans-boundary hydrocarbon resource is key to a successful implementation of a joint development mechanism. This is because such a zone, otherwise a zone of the contest, will now become the basis for cooperation. The respective claimants will exercise their rights and obligations in that zone, in unison. This is the case with Senegal/Guinea Bissau, and Nigeria/Sao Tome Agreements, mentioned above.

3.2 Body Corporate/Institution (Agency/Authority/ Commission)

The creation of a body corporate, responsible for the management of the zone is another important facet in a successful implementation of a joint development arrangement. Such a body, whatever called, should be composed of officials from concern States and be given an independent juridical personality in such States. The relevance of this body is that, since the respective States are unable to act severally, they could do so jointly, through this body.

3.3 The Constitution of the Body Corporate/Institution

Joint development arrangement must also establish a working tool for the body responsible for the administration of the designated area. It is a very important document that guides the Institution in its functions of regulating both exploitation of the resources, taking measures to prevent pollution of the marine environment and other matters incidental to hydrocarbon extraction. When it comes to pollution from oil extraction activities, such measure should cover the entire life cycle of oil production (exploration, exploitation and decommissioning).

3.4 Arctic Natural Resources Development Code

In addition to joint development arrangement, a code should be developed to regulate the exploitation of hydrocarbon resources in areas not covered by joint development. This code may embody rules, regulations and procedures for the exploration and exploitation of oil and gas resources in the Arctic, undertaken by respective States.

4 Conclusion

From the expanding role of joint development, it should be noted that such a role does not only ensure the peaceful utilization of hydrocarbon resources in overlapping or contested claims area, but also ensure the protection and preservation of the marine environment from such resource extraction activities. It is a functional institutional framework, establishing rules of engagement that offers opportunity for States to overcome intractable challenges of maritime boundary delimitation.

The interaction between the Code and the Joint Development Arrangement will not fail to contribute to the sustainable development of the Arctic, particularly with regard to oil and gas activities. Achieving this will be not difficult for the Arctic States, as a number of them already have relevant precedents (States practice) on joint development of common or transboundary hydrocarbon resources.

References

Borgerson, S. G. (2008). *Arctic meltdown: The economic and security implications of global warming*. Retrieved August 2, 2015, from https://www.foreignaffairs.com/articles/arctic-antarctic/2008-03-02/arctic-meltdown

Crawford, J. (2014). *State responsibility: The general part* (p. 43). Cambridge: Cambridge University Press.

Dubner, B. H. (2005). On the basis for creation of a new method of defining international jurisdiction in the Arctic Ocean. *Missouri Environmental Law and Policy Review, 13*(1), 11.

Hawker, E. E., Loftis, J. L., & Tyler, T. J. (2012). Gaps in the Ice: Maritime boundaries and hydrocarbon field development in the Arctic. *OGEL, 2*. www.ogel.org

Hober, K. (2012). Territorial disputes and natural resources: The melting of the ice and Arctic dispute. *Oil, Gas and Energy Law Intelligent, 10*(2).

Ilulissat Declaration. (2008). *Arctic Ocean Conference*. Retrieved August 2, 2015, from http://www.oceanlaw.org/downloads/arctic/Ilulissat_Declaration.pdf

Lindsay, T. J. (2012). (Un) Frozen frontiers: A multilateral dispute settlement treaty for resolving boundary disputes in the Arctic. *Oil, Gas and Energy Law Intelligent, 10*(2).

Mensah, T. A. (2006). Joint development zones as an alternative dispute settlement approach in maritime boundary delimitation. In R. Lagoni & D. Vignes (Eds.), *Maritime delimitation* (p. 147). Leiden: Martinus Nijhoff Publishers.

North Sea Continental Shelf Cases (Federal Republic of Germany v. Denmark; Federal Republic of Germany v. Netherlands), I.C.J. Reports 1969 3.

Protocol to the Agreement Between the Republic of Guinea-Bissau and the Republic of Senegal Concerning the Organization and Operation of the Management and Cooperation Agency Established by the Agreement of 14 October 1993 (adopted and entered into force in 1995) 1903 UNTS 1145.

Treaty between the Kingdom of Norway and the Russian Federation concerning Maritime Delimitation and Cooperation in the Barents Sea and the Arctic Ocean (adopted in 2010 and entered into force 2011) Annex II arts 10 and 11. Retrieved September 8, 2015, from https://www.regjeringen.no/globalassets/upload/ud/vedlegg/folkerett/avtale_engelsk.pdf

Treaty between the Kingdom of Norway and the Russian Federation concerning Maritime Delimitation and Cooperation in the Barents Sea and the Arctic Ocean. (2010). Retrieved August 2, 2015, from https://www.regjeringen.no/globalassets/upload/SMK/Vedlegg/2010/avtale_engelsk.pdf

United Nations Convention on the Law of the Sea. (1982). 1833 U.N.T.S. 397.

United Nations, Division for Ocean Affairs and Law of the Sea. (2004). Bulletin No.53. 15. Retrieved July 30, 2015, from http://www.un.org/depts/los/doalos_publications/LOSBulletins/bulletinpdf/bulletin53e.pdf

United Nations, Division for Ocean Affairs and Law of the Sea, Declarations and Statements. Retrieved July 30, 2015, from http://www.un.org/depts/los/convention_agreements/convention_declarations.htm

United States Geological Survey. (2008). *Circum-Arctic appraisal: Estimates of undiscovered oil and gas north of the Arctic circle*. Retrieved July 24, 2015, from http://pubs.usgs.gov/fs/2008/3049/

Watson, M. (2008–2009). An Arctic treaty: A solution to the international dispute over the polar region. *Ocean and Coastal Law Journal, 14*, 307, 226.

Arctic Strategies of the EU and Non-Arctic States: Identifying Some Common Elements

Henning Jessen

Contents

Abstract The national Arctic strategies of the eight Member States of the Arctic Council serve as important domestic policy guidelines to pursue long-term national objectives in Arctic matters. As part of an evolving process, several non-Arctic States have developed such policy guidelines as well. Most of these nations are recurring observers to the Arctic Council (i.e. on a "non-ad hoc" basis). Their national Arctic strategies outline the driving factors for active research engagement and other objectives in the region. Moreover, the European Union (EU) is in process of defining its major policy objectives in the Arctic as well. The EU's goals are evidenced by a series of publications from different EU institutions, developing further an "EU Integrated Arctic Policy". This chapter first provides a summary and reference guide on the EU's general policy objectives in international (marine) environmental law and ocean governance, including statements on the evolution of an Integrated EU Arctic policy since 2008. It is supplemented by some references on the German national Arctic strategy (first published in 2013) which represents an exemplary policy document of a non-Arctic State with a comprehensive interest in Arctic matters. The chapter also identifies some further common elements of national Arctic strategies of other non-Arctic States.

H. Jessen (✉)
World Maritime University (WMU), Malmö, Sweden
e-mail: hj@wmu.se

L. P. Hildebrand et al. (eds.), *Sustainable Shipping in a Changing Arctic*, WMU Studies in Maritime Affairs 7, https://doi.org/10.1007/978-3-319-78425-0_11

Keywords Arctic Council · Arctic governance · Non-Arctic States · Observers to the Arctic Council · Integrated EU Arctic Policy

1 Introduction: Non-Arctic States "Queuing Up" at the Arctic Council

The Arctic Council has emerged to be the most important political coordination forum for Arctic governance (Baker 2013, pp. 275–279; Weidemann 2014, p. 49; Vigeland Rottem 2015, pp. 50–59; Schram Stokke 2007, pp. 164–184). The majority of discussions on the Arctic Council are traditionally centred on the vital role of the five littoral "*Arctic inner circle States*" (Offerdal 2011, p. 862). This is especially true for debates relating to the opening of Arctic shipping routes and to the possible use and extraction of resources within the Exclusive Economic Zone (EEZ) and on the (outer) Continental Shelf (CS/OCS). Generally, any matters relating to the EEZ, the CS and the High Seas are legally governed by UNCLOS, i.e., the United Nations Convention on the Law of the Sea (done at Montego Bay, 10 December 1982, U.N. T.S., vol. 1833, 3, entry into force on 16 November 1994), which shall, however, not be the centre of the discussion of this chapter. Nevertheless, in 2008, the "Arctic five" have confirmed the overall importance of the law of the sea, in their "Ilulissat Declaration" which—*inter alia*—states that UNCLOS provides a "solid foundation for responsible management by the five coastal States *and other users* of this ocean" (Dodds 2015, p. 48).

The recent years have evidenced a continuous enlargement of the Arctic Council forum—in relation to its observers. Observers to the Arctic Council can be governmental, intergovernmental and non-governmental "non-Arctic" entities. As of 2016, there were twelve non-Arctic States, nine Intergovernmental and Inter-Parliamentary Organizations and eleven Non-governmental organizations (NGOs) which had previously been accepted as recurring observers to the Arctic Council. Their recurring appearance at Arctic Council meetings is usually based on a unanimous approval of the eight Arctic Council Members. When addressing State observers to the Arctic Council, the term "recurring observers" (in the sense of "*non ad hoc*" observers) should be preferred. In any case, it would be imprecise to use the term "permanent observers" instead. There are no legal grounds for a *permanent* State observer status in Arctic Council matters (Knecht 2015: "*There is no Permanency in Observer Status.*"). In fact, the legal status of a recurring State observer could be revoked at any time on the request of only one Member of the Arctic Council (Knecht 2015: "*There Should be no Permanency in Observer Status.*").

At their 2013 Kiruna Ministerial Meeting, the eight Arctic Council Members adopted an "*Arctic Council Observer Manual*" for the first time. The manual (available online, Arctic Council website, Observers 2015) formalized the requirements and procedures for granting recurring observer status to applicant entities. It thus summarizes the criteria for admittance as recurring observers to the Arctic

Council (i.e. on a "non ad hoc basis") and the corresponding rights and obligations of those observers. However, from the perspective of "traditional" Arctic stakeholders, an "interest" in the Arctic does not equate with a right to make decisions as to how the Arctic is governed (The Gordon Foundation 2011). Consequently, the Arctic Council grants no active voting rights or any other comparable instruments or powers to its recurring observers. Effectively, their procedural rights at Arctic Council high-level meetings are reduced to "sitting in the back" and to listening to the statements of the Arctic Council members. Nevertheless, five Asian countries have been granted the highly-desired recurring observer status in 2013, i.e., China, Japan, South Korea, Singapore and India (Nong Hong and Dey Nuttall 2014).

This political development could demonstrate the political desire of various non-Arctic stakeholders to participate more actively in emerging Arctic governance. Examples and commercial realities of long-standing business ties and further expansion opportunities between different Asian and Russian companies in the area of Arctic offshore oil and gas exploration and production do exist. For example, the Russian company Gazprom-Neft actively cooperates with the State-owned company PetroVietnam in the Dolginskoye field in the Russian Pechora Sea.

Although the acceptance as a recurring observer to the Arctic Council is purely an act of political symbolism, it can be expected that the long observer list is even going to be extended in the future. Applicants for new observers will definitely "queue up" as evidenced, e.g., by the pending applications of Turkey, Greece, Switzerland and Mongolia to be accepted as a future "non ad hoc" observer to the Arctic Council (Knecht 2015). To give another example, admittedly referring to a far less institutionalized forum, a high level Arctic *Circle* Conference (held in 2015 in Iceland) listed several "country sessions" of non-Arctic States (e.g. of Brazil, China or Germany) during which these nations explained their Arctic (and/or Polar) policy interests to a wider international public. China, in particular, has published a "first edition" of its National Arctic Policy in 2018. This strategy document also serves to explain and justify China's future Arctic interests.[1]

In fact, as evidenced by the multilateral Antarctic Treaty System (ATS) for decades, geographical proximity to the Polar regions of the world is not a prerequisite for States, international organizations and other entities to express a strategic interest in those areas, either directly or indirectly (on the ATS: Elferink et al. 2013, pp. 12 et seq. and 390 et seq.; Koivurova 2013, pp. 443 et seq.). Nevertheless, the question remains whether Arctic strategies of non-Arctic States are predominantly an inward-oriented tool, i.e. helping to prevent non-Arctic states from missing an "Arctic connection" for the benefit of national economic stakeholders, or whether they really represent wider foreign policy objectives which also contribute to more "institutionalized" global Arctic governance.

[1]China's Arctic Policy is available online: http://english.gov.cn/archive/white_paper/2018/01/26/content_281476026660336.htm (last visited: 1 April 2018).

2 Foundations on *International Environmental Law,* the EU and the Arctic

As a first practical example for the efforts of a non-Arctic stakeholder to contribute to Arctic governance, the most recent developments at the European Union (EU) level shall be discussed. Originally, the EU's interests in developing its own Arctic Policy were primarily motivated by geopolitical considerations. In particular, Russia planting its flag on the sea bottom beneath the North Pole on 1 August 2007 had raised some political concerns in Brussels (Offerdal 2011, p. 863). It is a fact that this unilateral symbolic act (also largely geared towards the global media and allegedly sponsored privately) coincides with the EU starting to evaluate its own role in the Arctic more systematically.

Nevertheless, the reasons for the EU Commission to deploy more manpower on Arctic issues are multi-layered and are composed of a number of different motivations. Norway first motivated the EU in a proactive way (i.e. to discover the Arctic at all as a dormant policy area) while later having to slow down the European Commission and taking a more defensive approach (Offerdal 2011, pp. 861–877). At the forefront of the original EU motives are issues commonly associated with the buzzword "*sustainable development*", in particular relating to global climate change but also other environmental concerns (Hossain and Koivurova 2012).

On an ad hoc basis, the EU has already been an observer to various Arctic Council meetings before. However, in contrast to 32 other different stakeholders, it is still not officially among the list of its *recurring* observers. The Arctic Council had already received the application of the EU for observer status affirmatively. However, the Members have deferred the final decision on its implementation until the Arctic Council Ministers can agree on this request by consensus. The underlying understanding is that the EU may observe Arctic Council proceedings until such time as the Arctic Council feels ready to take a final decision on the EU's application (European Commission 2014, p. 15). As a result, the ultimate decision on the EU's request to be officially "upgraded" to a recurring Arctic Council observer is still adjourned. Quite paradoxically, three EU Members (Denmark, Finland, and Sweden) are Arctic Council Members and a number of other EU Members (France, Germany, Italy, the Netherlands, Poland, Spain and, at the time of this writing still, the United Kingdom) are officially among the group of its recurring observers.

The peculiar situation for the EU has several political reasons. One of those reasons has been commonly attributed to the EU's treatment of Canadian seal products and the related disputes at the World Trade Organization (WTO, European Communities—Certain Measures Prohibiting the Importation and Marketing of Seal Products) prompting Canada to temporarily frustrate the EU's political desire to institutionalize its observer relation to the Arctic Council (Wegge 2013, pp. 255–273). In particular, after the successful conclusion of the Comprehensive Economic and Trade Agreement ("CETA") between the EU and Canada this will most probably be remembered as a diplomatic "interlude". The most difficult diplomatic challenge for the EU will neither be Canada nor Denmark but, rather, the EU-Russian relations (Offerdal 2011, pp. 870 and 877). Obviously, the technical argument that the EU has legal personality itself and that it is also a party to UNCLOS will not suffice. Nevertheless, even in times of deteriorating diplomatic relations and economic sanctions applied between the EU and Russia, the question is not "if" but rather "when" the step of granting recurring observer status to the EU in the Arctic Council will be politically acceptable to all of its Members.

In order to focus on the possible "Arctic relevance" of key EU legal acts this chapter includes only a compressed reference guide on internationally accepted legal principles of (marine) environmental law and the applicable EU law. Key principles of international environmental law have been gradually integrated into the European Treaties, above all, into the Treaty on the Functioning of the European Union (TFEU) itself. Some internationally accepted principles of environmental law have now been "upgraded" to become primary sources of EU law. Article 191 TFEU sets out that the EU's policy on the environment shall contribute to the pursuit of a number of objectives, stating in the first sentence of the provisions' second paragraph: *"Union policy on the environment shall aim at a high level of protection taking into account the diversity of situations in the various regions of the Union. It shall be based on the precautionary principle and on the principles that preventive action should be taken, that environmental damage should as a priority be rectified at source and that the polluter should pay."*

This provision refers explicitly only to "*the regions of the Union*" and the first sentence of Article 191(4) TFEU also concedes that "*within their respective spheres of competence, the Union and the Member States shall cooperate with third countries and with the competent international organisations*". Thus, the direct effect of EU legal measures is, of course, generally confined to the EU itself. The EU is cautious not to give an outside impression of being a self-appointed global environmental regulator. Nevertheless, an explicit intra-EU endorsement of international environmental law principles—which has also been confirmed on various occasions by the Court of Justice of the EU (CJEU) (Case 240/83 Procureur de la Republique v ADBHU (1985) ECR 531 and Case C-379/92 Re Peralta [1994] ECR I-3453; Case T-13/99 Pfizer v European Commission [2002] ECR II-3305)—is of fundamental importance when it comes to EU actions and policies in a setting which potentially transcends the EU borders.

Thus, the additional reference to international environmental law principles can be helpful for the intra-EU persuasiveness of initial legislative drafts of the European Commission, for example, in case some EU Members do not see an immediate necessity for legislative action. A good practical example is Directive 2008/56/EC, better known as the "*Marine Strategy Framework Directive*", which had been agreed late in 2007 and which was formally adopted in 2008 (Markus et al. 2011, pp. 59–90; Long 2011, pp. 1–44). Effectively, the Directive serves to implement the precautionary principle within the EU in a comprehensive setting of marine governance: As part of the EU's overall "*Integrated Maritime Policy*" the Commission had proposed the adoption of the Directive already in 2005 to implement a broad thematic strategy and to be able to address marine pollution through a long term programme of diagnosis and action carried out by competent authorities in the Member States and under the European regional seas conventions.

Thus, while "*Marine Strategy Framework Directive*" did not address the environmental impacts of maritime transport or other uses of the sea specifically, it has triggered governance mechanisms which over time generate new EU actions having direct implications for any marine-related sector. As a result, the "*Marine Strategy Framework Directive*" has now emerged to be the environmental pillar of the EU's "*Integrated Maritime Policy*". It promotes and applies several internationally-accepted environmental principles, such as

- the principle of sustainable development,
- the principle of environmental integration,
- the precautionary principle, and
- the ecosystem approach (Long 2014, pp. 699–726).

To give another practical example, one of the EU's recent legal activities in the area of the safety of offshore oil and gas operations is explicitly based on Article 191 TFEU as well: Recital (1) of Directive 2013/30/EU of the European Parliament and of the Council of 12 June 2013 on safety of offshore oil and gas operations and amending Directive 2004/35/EC (OJ L 178/66 of 28 June 2013) reads: "*Article 191 of the Treaty on the Functioning of the European Union establishes the objectives of preserving, protecting and improving the quality of the environment and the prudent and rational utilisation of natural resources. It creates an obligation for all Union action to be supported by a high level of protection based on the precautionary principle, and on the principles that preventive action needs to be taken, that environmental damage needs as a matter of priority to be rectified at source and that the polluter must pay.*"

Moreover, Article 11 TFEU broadly states that "*[. . .] environmental protection requirements must be integrated into the definition and implementation of the Union's policies and activities, in particular with a view to promoting sustainable development.*" Quite obviously, the term "*sustainable development*" can be charged with an endless variety of political objectives. However, it is also a legal principle at the forefront of the environmental policies of the EU and of many nations (Long 2014, p. 716). Global cooperation in all environmental matters is—indisputably—an essential policy objective of the EU, as evidenced also by Article 3(3) and (5) of the Treaty on the European Union (TEU).

In particular, Article 3(5) TEU states that: "*In its relations with the wider world, the Union shall uphold and promote its values and interests and contribute to the protection of its citizens. It shall contribute to peace, security, the sustainable development of the Earth, solidarity and mutual respect among peoples, free and fair trade, eradication of poverty and the protection of human rights, in particular the rights of the child, as well as to the strict observance and the development of international law, including respect for the principles of the United Nations Charter.*"

The EU's evolving Integrated Arctic policy is a good practical example for the inclusion of a global dimension in EU instruments (in this case relating to the

"*Northern Dimension*"). The EU has, e.g., evaluated its own Arctic carbon footprint in order to gain a more precise view on its own environmental impact in the Arctic (Ecologic Institute 2010; Offerdal 2011, p. 872). From a more "globalized" perspective, it also becomes more why the EU seeks to finally have more "institutionalized" observer position within the Arctic Council. From the EU perspective itself, this position will be a supplement to the already existing EU participation in the Barents Euro-Arctic Council and in other regional fora for coordinating Arctic matters. In this context, the "*Opinion of the European Economic and Social Committee on 'EU Arctic Policy to address globally emerging interests in the region — A view of civil society*" (European Economic and Social Committee 2013, para. 2.4) stated that: "*[...] The EU should have a stronger position in the [Arctic] Council, because this would allow it to better contribute to the Council's work and to boost the Council's influence through its participation. The EU has a lot to contribute to cooperation. One possible way to strengthen EU's position is to become an observer entity and the Arctic EU member States should take into account also EU views in the Council. The EU should also endeavour to strengthen cooperation in the Barents Euro-Arctic Council (and Barents Regional Council), because they play a key role in cross-border interaction amongst the 13 member regions (in Norway, Sweden, Finland and Russia) of the resource-rich Barents region [...]*".

The EU's role in promoting sustainable "*Ocean Governance*"—primarily concentrated on international shipping and also the politically difficult area of fisheries activities but (inter alia) also including offshore extractive industries—has already been analysed elsewhere (Long 2014, pp. 699–726; on the broad topic of "*Ocean Governance*" itself: Freestone 2014, pp. 729–752; Freestone 2009, pp. 44–49; Freestone 2008, pp. 385–391). Generally, just like in any other regulatory field, the EU applies its own unique legal instruments, i.e., secondary legislation imposed on its Members in accordance with Article 288 TFEU. This legislation takes the form of legally binding Regulations and (more flexible) Directives, to further the EU's primary policy objectives. In sum, since 2005, the EU is following a long-term, principle-fuelled and goal-based (marine) environmental policy. In achieving "good environmental status" of marine waters by 2020, the EU is fully aware of the fact that it cannot create legal obligations for third (i.e. non-EU) parties. However, both the sustainability approach and the cooperative elements of the EU's (marine) environmental policy also have a global dimension. These elements extend to areas beyond national jurisdiction and thus far beyond the formal territorial and aquitorial boundaries of the EU itself, including the Arctic, where the EU has no geographic ties anymore: Although associated with the EU Member State Denmark, since 1 February 1985, Greenland is not part of the EU (then EEC) territory anymore, following the results of a referendum of 1982. Rather, Greenland is included in the list of overseas countries and territories set out in Annex II to the TFEU. In accordance with Article 198 TFEU, the purpose of the association of the overseas countries and territories with the EU is "*to promote the economic and social development of the overseas countries and territories and to establish close economic relations between them and the EU as a whole*".

Since 2007/2008, the emerging EU Arctic policy is made up of a continuously evolving network of "soft law" instruments. These years also mark the time in which the EU started to highlight its own global responsibilities in environmental matters. This has been done, for example, by specifying the EU's "*Integrated Maritime Policy*" in more detail, e.g., by passing the "*Marine Strategy Framework Directive*". About ten years later, it seems quite logical that the EU has begun to develop and to substantiate its own "*Integrated Arctic Policy*". All EU instruments discussed below are available online at the internet presence of the European Commission (Sea Basin Strategy: Arctic Ocean 2016).

3 EU Council Conclusions on the Arctic

By the end 2016, the EU had passed three Council Conclusions (by the Foreign Affairs Council) in 2009, 2014 and 2016 on Arctic matters. Though legally non-binding, multiple official Council Conclusions potentially have a preparatory function and character. They represent a coordinated position of the Commission and the EU Member States, setting out the EU's policy objectives for the future. From a policy perspective, Council Conclusions have a potential of facilitating the track to adopt future legally-binding intra-EU instruments. It simply adds to the political persuasiveness when the Commission can refer to a series of Council Conclusions. In turn, it can get harder for reluctant EU Member States to obstruct entering into a new "hardened" phase of EU policy integration.

Between 2009 and 2014, the title of the Arctic-related Council Conclusions has transformed from merely addressing an incoherent variety of "*Arctic Issues*" to a more institutionalized "*Developing a EU Policy towards the Arctic Region*" in 2014. However, in 2016, the coordinated position of the Commission and the EU Member States was reduced in ambition again by setting out the EU's policy objectives simply as "*Council Conclusions on the Arctic*". Nevertheless, there is a visible gradual evolution towards a more coordinated (i.e. "integrated") policy on Arctic issues and to address the related EU interests and responsibilities.

4 Joint Communications on the Arctic

The year 2012 will once be remembered as the turning point of EU Arctic Policy. First in 2012 and in 2016, the European Commission and the High Representative of the EU for Foreign Affairs and Security published ambitious "*Joint Communication*" on the EU and the Arctic. In contrast to EU Council conclusions, a Joint Communication (and also a Communication issued solely by the Commission) does not reflect a coordinated approach between the EU Commission and the EU Member States. Rather, a Commission communication represents a vision of its originator (s) setting out the details for the most important cornerstones of a certain EU policy area. It has been rightly pointed out that sometimes such documents are mistaken for representing official positions of the EU as a whole (Offerdal 2011, p. 862). However, this is a wrong assumption and it is also a reason why Commission Communications can be far more extensive and detailed as compared to Council Conclusions.

The 2012 Arctic Joint Communication was built on a broader 2008 joint chapter (of the same originators) on climate change and international security and on an earlier 2008 Commission Communication on "*the EU and the Arctic Region*". Four years later, the offspring of those two rather general documents of 2008, included a total of 28 action points adopting the strapline of "*knowledge, responsibility, engagement*" (Chuah 2012, pp. 251–252). The 2012 Arctic joint communication stressed the considerable financial engagement and contribution of the EU to Arctic research and the EU's support for a sustainable use and management of Arctic resources. The document avoided addressing politically contentious issues (like the earlier idea of creating a completely new "Arctic Treaty System", broadly based on the legal role model of the Antarctic Treaty System). The 2016 Joint Communication clearly builds on its 2012 predecessor, stressing now more precisely that the EU's primary objective is an "*integrated*" policy for the Arctic. The explicit reference to the term "*integration*" emphasizes that it is a now a policy goal of the EU to ensure more effective synergies between the various EU funding instruments in the Arctic region. That is also why the slogan "*knowledge, responsibility, engagement*" will continue to serve as the three cornerstones of an "Integrated EU Arctic Policy". In particular, the EU will continue to highlight three key policy objectives to:

- protect and preserve the Arctic environment in cooperation with the people who live there, and in particular relating to climate change,
- promote sustainable use of resources in and around the Arctic, and
- to foster international cooperation on Arctic issues, emphasizing enhanced scientific cooperation.

Specifically on the third policy objective, the EU has devoted financial resources to create and develop Arctic observatory networks, and to facilitate access to research facilities in the Arctic to scientists from Europe and beyond. This is done by funding projects such as INTERACT, a multi-disciplinary network of 58 land-based Arctic and northern research stations, building capacity throughout the Arctic for environmental monitoring, research, education and outreach (Interact 2016). The EU will also initiate a new five-year project (2016–2021) coordinated by Norway to develop an Integrated Arctic Observing System (INTAROS). This project will involve scientists in 14 European countries as well as in a number of countries elsewhere in the world and has a €15.5 million budget. Furthermore, the EU will initiate two new projects to understand the impact of the changing Arctic on the weather and climate of the Northern Hemisphere. The projects APPLICATE (standing for: "Advanced Prediction in Polar regions and beyond: modelling, observing system design and LInkages associated with a Changing Arctic climate", 2016–2020, €8 million budget) and Blue-Action (2016–2021, €7.5 million budget) will involve scientists of 13 European countries as well as in a number of countries elsewhere in the world (The White House, Fact Sheet 2016).

5 Resolutions of the European Parliament on the Arctic

Finally, the European Parliament had expressed original interest in Arctic issues and in shaping the interrelationship with EU policies (Offerdal 2011, p. 873). The group has pushed forward the European Parliament to pass (again: non-binding) Resolutions, in particular, an Arctic-specific Resolution of 2014 (European Parliament Resolution 2014) which is based on an earlier and broader Resolution of 2011 on a sustainable EU policy for the High North (European Parliament 2011).

However, though generally more progressive, the European Parliament as a whole (i.e. unlike individual parliamentarians) has also eschewed to address contentious political issues (e.g., the question of contested Fisheries Zones and the "*Svalbard issue*" Churchill and Ulfstein 2005, p. 20). Consequently, the latest Resolution of the European Parliament had also been criticized as serving an alibi-function by simply reiterating the Parliament's own consultative importance in the EU-Arctic policy-making process (Raspotnik and Østhagen 2014).

6 The German Example of an Arctic Strategy by a Non-Arctic State

Germany acknowledges that the Arctic Council is the most promising forum to tackle the overarching policy challenges in the Arctic region. Moreover, like many other non-Arctic national States, Germany has repeatedly expressed its vital interest in efficient, environmentally-friendly maritime traffic in the Arctic as a region in massive transition. This includes a German (political) contribution to the work of the Arctic Council as one of its recurring observers (i.e. on a "non ad hoc" basis).

The German political agenda emphasizes the sustainable use of Arctic marine resources by actively promoting the enforcement of the highest environmental standards, including the establishment of protected areas to maintain Arctic biodiversity and the application of the "polluter pays" principle. Thus, the German Arctic strategy—entitled "*Assuming Responsibility, Seizing Opportunities*" stresses, above all, the protection of the Arctic environment (German Arctic Strategy 2013). Comparable to the EU policy chapters, the German Arctic strategy promotes a comprehensive active application of the precautionary principle both by Arctic States and by non-Arctic States. To explain the use of the term "*comprehensive*", the German Arctic strategy addresses political, legal and economic topics via

- promoting environmental standards and principles, (German Arctic Strategy 2013, p. 7)

- accepting economic opportunities for the global community (German Arctic Strategy 2013, p. 6)
- guaranteeing the freedom of navigation and the freedom of scientific research (German Arctic Strategy 2013, p. 8)

Notably, and emphasizing its domestic interests in international shipping, Germany is actively "*campaigning*" for freedom of navigation in the Arctic Ocean in accordance with high safety and environmental standards. It also seeks to promote security and stability in the Arctic by achieving "*horizontal coherence*" (German Arctic Strategy 2013, pp. 9–10). Apart from supporting the successful negotiation of the IMO's Polar Code, Germany also supports the IMO in furthering the highest standards of maritime surveillance, infrastructure expansion, and Arctic search and rescue (SAR) capabilities.

In particular, Germany repeatedly refers to its sophisticated and traditional national profile in polar research (Steinicke 2014, p. 120). For example, the "*Alfred Wegener Institute*" and the "*Helmholtz Centre for Polar and Marine Research*" are leading international research institutes based in Germany. Their research focus relates to issues of climate change, changes in sea ice and biological diversity as well as to oceanographic, biological and geological changes in the Polar Regions. Moreover, together with France, Germany operates its own Arctic research base in the Norwegian Arctic (on the island of Spitsbergen). Finally, Germany fully supports the emerging EU Integrated Strategy for the Arctic which would also include a "promotion" of the EU to ultimately and officially become another recurring observer to the Arctic Council.

As a consequence and in sum, Germany places itself as being an indispensable partner to the littoral Arctic nations. It stresses the existence of its vast expert knowledge in the areas of research, technology and adherence to environmental standards and as a provider of specialized technology and know-how. This potentially generates new business opportunities for the German industry. In this context, Germany seeks to find a good balance between environmental responsibility and geo-economic opportunities while it also recalls that coastal States in general are legally obliged to foster international cooperation in accordance with Articles 242 et seq. UNCLOS.

7 Common Features of Arctic Strategies of Other Non-Arctic States

There is already vast (and constantly growing) academic literature available on the Arctic policy interests of several non-Arctic States, e.g., (with further references and without any claim for completeness) on:

- Asian countries in general (Schram Stokke 2014, p. 770; Nong Hong and Dey Nuttall 2014; Solli et al. 2013, p. 253);

- China and Japan (Tonami 2014, p. 105);
- China (Nong Hong 2014, p. 271; Peng and Wegge 2014, p. 287; Beck 2014, p. 306; Chao 2013, p. 467);
- Japan (Tonami and Watters 2012, p. 93);
- South Korea (Bennett 2014, p. 886; Kim 2014, p. 917);
- India (Lackenbauer 2013, p. 1; Chaturvedi 2014);
- Singapore (Watters and Tonami 2012, p. 104);
- Poland (Graczyk 2012, p. 139);
- the United Kingdom (Depledge 2012, p. 130) and
- Scotland (Johnstone 2012, p. 114).

Generally, a basic literature review confirms that non-Arctic States' interests are predominantly sectoral and/or resource-based. At the outset, national Arctic strategies or political statements of both Arctic States and non-Arctic States are often centred on the management of global challenges, in particular relating to climate change but also to Arctic shipping (The Gordon Foundation 2011). All non-Arctic States mentioned above agree—for obvious reasons—that the freedom of navigation in Arctic waters must be maintained without any undue restrictions by coastal States. In this context, several non-Arctic States explicitly endorse the entry into force and implementation of the IMO's Polar Code as an instrument to be adhered to while exercising the right to freedom of navigation. Consequently, there are common policy interests of non-Arctic States, relating above all to freedom of navigation as well as the necessity of Polar research, but also to peace and security in the region and the fight against global warming.

What is more striking, however, is the fact that there seems to be no visual policy coherence and political coordination *among* non-Arctic States. There might be "backroom talks" and informal consultations. However, non-Arctic States did never join diplomatic forces so far. For example, they could have proposed an "*Advisory Body to the Arctic Council*". The advantage of such an instrument would be to serve as a possible tool to coordinate policy positions of non-Arctic States prior to Arctic Council meetings. As such, an "*Advisory Body to the Arctic Council*" could possibly also have a right to express a common position at High-Level Arctic Council meetings (The Gordon Foundation 2011). Thus, at the current state, a "*Non-Arctic States Advisory Body to the Arctic Council*" seems to be an illusionary policy vision. This is true even though it could probably channel the shared political interests of non-Arctic States more efficiently and it could probably even address the challenge of institutional overlaps and multi-layered governance.

However, at this stage, non-Arctic States have not even asked for enhanced participation rights at the Arctic Council. And there is no real case for enhanced individual rights of non-Arctic States at the Arctic Council as long as there is also only ad hoc, sporadic appearance of representatives of non-Arctic States in various specialized technical working groups which operate under the institutional realm of the Arctic Council. However, apart from overarching comprehensive issues like maintaining freedom of navigation and fighting global climate change, non-Arctic States seem to be very often only interested in some sectoral issues of specific Arctic challenges. Their actual presence and possible active contributions seems to be often conditioned on a specific national competence, resulting in a national interest in a particular topic. For example, with 16 out of 17 attendances, the Netherlands had a visual concentration of its past activities in the Arctic Council working group of the "Arctic Monitoring and Assessment Programme" (AMAP) while other representatives of non-Arctic States had almost no or even zero attendances in those meetings (Knecht 2015).

To give another example, for obvious reasons South Korea is highly interested in the ship-building requirements for ice-breakers or ice-capable freight vessels, even including oil and LNG tankers. As a result, South Korea has concentrated its strategic activities particularly in this area without, of course, giving the impression that other issues would be neglected. But doubtlessly, ship-building will remain to be the key commercial driver for South Korea's emerging Arctic interests (Bennett 2014, p. 886; Kim 2014, p. 917).

As a result and quite paradoxically, non-Arctic State recurring observers to the Arctic Council all seem to attach utmost political importance to attending the highest level meetings of the Arctic Council. For them, this is part of a political "great power competition" and they seek to prevent "missing the connection". As stated above, their procedural rights at the high-level meetings are reduced to "sitting in the back of the room" (if they are allowed to attend the meeting at all). In contrast to that, the technical Working Group level would offer far better opportunities to influence any future decision-making processes of the Arctic Council. However, this level seems to be significantly underutilized. If at all, recurring non-Arctic State observers seem to concentrate their expertise and resources in the work of one or two Arctic Council Working Groups (Knecht 2015).

8 Conclusions and Outlook

The continuously intensified efforts of the EU to establish a full-fledged Arctic strategy did not generate any EU hard law yet (in the form of Directives or even directly binding Regulations). Rather, it is still an evolving EU policy which highlights, more and more, the necessity of integrating different policy aspects in a more coherent way, such as scientific research, climate mitigation and adaptation strategies or sustainable innovation and investment. As a result, this political process is characterized by a visible shift from an initial EU approach which was more focussed on geopolitics to a more innovation-centred and research-related attitude.

As it is often the case, bringing the EU to the diplomatic table could bring the risk of creating an even more complex picture in an area which is already characterized by institutional overlaps and multi-layered governance. However, a recent answer given by the High Representative of the EU for Foreign Affairs and Security on behalf of the Commission to a question of a Member of the European Parliament can serve as "diplomatic showcase" for the current inoffensive state of EU affairs in Arctic matters. The parliamentarian simply asked (inter alia): "*Does the EU have specific interests in the Arctic region?*" (European Parliament, Question for written answer E-002847/14 to the Commission). The written reply of the High Representative of 24 June 2014 first referred to the 2012 Arctic joint communication as the key guiding document and to the other existing EU soft law instruments, as discussed above. Under the joint chapter's notion of EU "*knowledge, responsibility, engagement*", the High Representative continued to stress that: "*The EU is stepping up its engagement with its partners to jointly meet the challenge of safeguarding the environment while ensuring the sustainable and peaceful development of the Arctic region through investment in knowledge, promoting responsible approach to arising commercial opportunities and constructive engagement with Arctic partners. [...] The majority of today's known resources are within the boundaries of the 200-mile zones and/or continental shelves of the Arctic coastal states and are uncontested.*"

More generally, all recurring non-Arctic observers to the Arctic Council will definitely have an important role to play in the future. Despite all problems of logistics caused by the rising number of participants, allowing non-Arctic States to participate opens far more possibilities, for example:

- to propose and partly fund new projects,
- to disseminate information and contribute relevant expertise and input to subsidiary body meetings,
- to contribute to the production of public goods (like knowledge about the state and development of the Arctic environment),
- to let them serve as "global" multipliers for Arctic Council initiatives, and
- to distribute results of environmental assessments and policy recommendations into the public sphere and other national and international venues with political authority (Knecht 2015).

References

Books and Book Chapters

Chao, J. K. T. (2013). China's emerging role in the Arctic. In H. N. Schreiber & J.-H. Paik (Eds.), *Regions, institutions, and the law of the sea* (pp. 467–491). Boston/Leiden: Brill Academic Publishers.

Churchill, R., & Ulfstein, G. (2005). *Marine management in disputed areas – The case of the Barents Sea*. Routledge.

Ecologic Institute. (2010). *EU Arctic footprint and policy assessment: Final report*. Berlin.

Elferink, A. G. O., Molenaar, E. J., & Rothwell, D. R. (2013). The regional implementation of the law and the sea and the Polar regions. In E. J. Molenaar, A. G. O. Elferink, & D. R. Rothwell (Eds.), *The law of the sea and the Polar regions* (pp. 1–16). Boston/Leiden: Brill Academic Publishers.

Freestone, D. (2008). Principles applicable to modern ocean governance. *International Journal of Coastal and Marine Law, 23*(3), 385–391.

Freestone, D. (2014). Governing the Blue: Governance of areas beyond national jurisdiction in the twenty-first century. In C. Schofield, S. Lee, & M.-S. Kwon (Eds.), *The limits of maritime jurisdiction* (pp. 729–752). Boston/Leiden: Brill Academic Publishers.
Long, R. (2014). Principles and normative trends in EU ocean governance. In C. Schofield, S. Lee, & M.-S. Kwon (Eds.), *The limits of maritime jurisdiction* (pp. 699–726). Boston/Leiden: Brill Academic Publishers.
Nong Hong, & Dey Nuttall, A. (2014). Emerging interests of non-Arctic countries in the Arctic. In R. W. Murray & A. Dey Nuttall (Eds.), *International relations and the Arctic: Understanding policy and governance* (Chapter 19). Cambria Press.
Weidemann, L. (2014). *International governance of the Arctic marine environment.* Heidelberg/New York et al.: Springer International Publishing.

EU Documents

European Commission and the High Representative of the EU for Foreign Affairs and Security, Climate change and international security, Chapter from the High Representative and the European Commission to the European Council, S113/08 of 14 March 2008.
European Commission, Communication from the Commission to the European Parliament and the Council: The European Union and the Arctic Region, Brussels, 20 November 20, COM(2008) 763 final.
European Commission and the High Representative of the EU for Foreign Affairs and Security, Joint Communication to the European Parliament and the Council: Developing a European Union Policy towards the Arctic Region: progress since 2008 and next steps, Brussels, 26 June 2012, JOIN(2012) 19 final.
European Commission and the High Representative of the EU for Foreign Affairs and Security, Joint Communication to the European Parliament and the Council: An integrated European Union policy for the Arctic, Brussels [2016], JOIN(2016) 21 final.
European Commission, Roadmaps for international cooperation, Commission Staff Working Document, Brussels, 11 September 2014, SWD(2014) 276 final.
European Commission, Sea basin strategy: Arctic Ocean: http://ec.europa.eu/maritimeaffairs/policy/sea_basins/arctic_ocean_en (last visited: 1 November 2016).
European Council, Council Decision No. 136/2014/EU of 20 February 2014 laying down rules and procedures to enable the participation of Greenland in the Kimberley Process certification scheme, OJ L84/99 of 20 March 2014.
European Economic and Social Committee, Opinion of the on 'EU Arctic Policy to address globally emerging interests in the region — A view of civil society, OJ C 198/26 of 10 July 2013.
European Parliament, Resolution of 20 January 2011 on a sustainable EU policy for the High North (2009/2214(INI)).
European Parliament, Question for written answer E-002847/14 to the Commission, Sergio Paolo Francesco Silvestris (PPE), The race for the Arctic: European prospects (11 March 2014), OJ 2014/C 326/01.
European Parliament, Resolution of 12 March 2014 on the EU strategy for the Arctic (2013/2595 (RSP)).
Interact 2016, available for download at http://www.eu-interact.org/ (last visited: 1 November 2016).

Journal Articles (Including Online Articles and Digital Object Identifiers)

Baker, B. (2013). Offshore oil and gas development in the Arctic: What the Arctic Council and International Law can – and cannot – do. *ASIL Proceedings*, 275–279.
Beck, A. (2014). China's strategy in the Arctic: A case of lawfare? *The Polar Journal, 4*(2), 306–318.
Bennett, M. M. (2014). The Maritime Tiger: Exploring South Korea's interests and role in the Arctic. *Strategic Analysis, 38*(6), 886–903. https://doi.org/10.1080/09700161.2014.952935.
Chaturvedi, S. (2014, July). India's Arctic engagement: Challenges and opportunities. *Asia Policy, 18*, 73–79.
Chuah, J. (2012). The development of an EU Arctic Policy? Perhaps Not.... *Journal of International Maritime Law, 18*, 251–252.
Depledge, D. (2012). The United Kingdom and the Arctic in the 21st century. *Arctic Yearbook*, 130–138. http://www.arcticyearbook.com/images/Articles_2012/Depledge.pdf (last visited: 1 November 2016).
Dodds, K. (2015). The Ilulissat Declaration (2008): The Arctic States, "Law of the Sea," and Arctic Ocean. *SAIS Review of International Affairs, 33*, 45–55. https://doi.org/10.1353/sais.2013.0018.
Freestone, D. (2009). The modern principles of high seas governance. *The Legal Underpinnings, International Environmental Policy and Law, 39*(1), 44–49.
Graczyk, P. (2012). Poland and the Arctic: Between science and diplomacy. *Arctic Yearbook*, 139–155. http://www.arcticyearbook.com/images/Articles_2012/Graczyk.pdf (last visited: 1 November 2016).
Hossain, K., & Koivurova, T. (2012). Hydrocarbon development in the offshore Arctic: Can it be done sustainably? *Oil Gas and Energy Law (OGEL)*, 2. www.ogel.org/article.asp?key=3258 (last visited: 1 November 2016).
Johnstone, R. L. (2012). An Arctic strategy for Scotland. *Arctic Yearbook*, 114–129. http://www.arcticyearbook.com/images/Articles_2012/Johnstone.pdf (last visited: 1 November 2016).
Kim, J. D. (2014). Overview of Korea's Arctic policy development. *Strategic Analysis, 38*(6), 917–923. https://doi.org/10.1080/09700161.2014.952939.
Knecht, S. (2015). *New observers queuing up: Why the Arctic Council should expand - and expel.* The Arctic Institute. http://www.thearcticinstitute.org/2015/04/042015-New-Observers-Queuing-up.html (last visited: 1 November 2016).
Koivurova, T. (2013). Multipolar and multilevel governance in the Arctic and the Antarctic. *Proceedings of the Annual Meeting (American Society of International Law), 107*, 443–446.
Lackenbauer, P. (2013). Whitney, India's Arctic engagement: Emerging perspectives. *Arctic Yearbook*, 1–19. http://www.arcticyearbook.com/images/Articles_2013/LACKENBAUER_AY13_FINAL.pdf (last visited: 1 November 2016).
Long, R. (2011). The Marine Strategy Framework Directive: A new European approach to the regulation of the marine environment, marine natural resources and marine ecological services. *Journal of Energy and Natural Resources Law, 29*, 1–44.
Markus, T., Schlacke, S., & Maier, N. (2011). Legal implementation of integrated ocean policies: The EU's Marine Strategy Framework Directive. *The International Journal of Marine and Coastal Law, 26*, 59–90.
Nong Hong. (2014). Emerging interests of non-Arctic countries in the Arctic: A Chinese perspective. *The Polar Journal, 4*(2), 271–286. https://doi.org/10.1080/2154896X.2014.954888.
Offerdal, K. (2011). The EU in the Arctic. *International Journal* (Autumn), 861–877.
Peng, J., & Wegge, N. (2014). China and the law of the sea: Implications for Arctic governance. *The Polar Journal, 4*(2), 287–305.
Raspotnik, A., & Østhagen, A. (2014). To Svalbard and beyond – The European Parliament is back on its Arctic Track, March 17, 2014. http://www.thearcticinstitute.org/2014/03/to-svalbard-and-beyond-european.html (last visited:1 November 2016).

Schram Stokke, O. (2007). International institutions and Arctic governance. In O. Schram Stokke & G. Hønneland (Eds.), *International cooperation and Arctic governance* (pp. 164–184). London/New York: Routledge.
Schram Stokke, O. (2014). Asian stakes and Arctic governance. *Strategic Analysis, 38*(6), 770–783. https://doi.org/10.1080/09700161.2014.952946.
Solli, P. E., Wilson Rowe, E., & Lindgren, W. Y. (2013). Coming into the cold: Asia's Arctic interests. *Polar Geography, 36*(4), 253–270. https://doi.org/10.1080/1088937X.2013.825345.
Steinicke, S. (2014). *A slow train coming: Germany's emerging Arctic Policy*. http://www.kas.de/upload/Publikationen/2014/Perceptions_and_Strategies_of_Arcticness_in_Sub-Arctic_Europe/Perceptions_and_Strategies_of_Arcticness_in_Sub-Arctic_Europe_steinicke.pdf (last visited: 1 November 2016).
The Gordon Foundation. (2011). *Interests and roles of non-Arctic states in the Arctic*. http://gordonfoundation.ca/publication/418 (last visited: 1 November 2016).
Tonami, A. (2014). The Arctic policy of China and Japan: Multi-layered economic and strategic motivations. *The Polar Journal, 4*(1), 105–126. https://doi.org/10.1080/2154896X.2014.913931.
Tonami, A., & Watters, S. (2012). Japan's Arctic Policy: The sum of many parts. *Arctic Yearbook*, 93–103. http://www.arcticyearbook.com/images/Articles_2012/Tonami_and_Watters.pdf (last visited: 1 November 2016).
Vigeland Rottem, S. (2015). A note on the Arctic Council agreements. *Ocean Development & International Law, 46*, 50–59.
Watters, S., & Tonami, A. (2012). Singapore: An emerging Arctic actor. *Arctic Yearbook*, 104–115. http://www.arcticyearbook.com/images/Articles_2012/Tonami_and_Watters_Singapore.pdf (last visited: 1 November 2016).
Wegge, N. (2013). Politics between science, law and sentiments. Explaining the European Union's ban on trade with seal products. *Environmental Politics, 22*, 255–273.

Other Online Documents

Arctic Council, Observers, http://www.arctic-council.org/index.php/en/about-us/arctic-council/observers (last visited: 1 November 2016).
Arctic Council, PAME (2009 and 2014). http://www.pame.is/index.php/projects/offshore-oil-and-gas and http://www.pame.is/index.php/projects/offshore-oil-and-gas/systems-safety-management-and-safety-culture (last visited: 1 November 2016).
German Arctic Strategy 2013, available for download at: http://www.auswaertiges-amt.de/cae/servlet/contentblob/658714/publication File/185871/Arktisleitlinien.pdf (last visited: 1 November 2016).
The White House, Fact Sheet 2016, https://www.whitehouse.gov/the-press-office/2016/09/28/fact-sheet-united-states-hosts-first-ever-arctic-science-ministerial (last visited: 1 November 2016).

Legal Regime of Marine Insurance in Arctic Shipping: Safety and Environmental Implications

Proshanto K. Mukherjee and Huiru Liu

Contents

P. K. Mukherjee
World Maritime University, Malmö, Sweden

Juris Dr. Honoris Causa, Lund University, Lund, Sweden

Dalian Maritime University, Dalian, People's Republic of China
e-mail: pkm@wmu.se

H. Liu (✉)
World Maritime University, Malmö, Sweden
e-mail: P1507@wmu.se

L. P. Hildebrand et al. (eds.), *Sustainable Shipping in a Changing Arctic*, WMU Studies in Maritime Affairs 7, https://doi.org/10.1007/978-3-319-78425-0_12

Abstract The primary driver of Arctic shipping and maritime operations is the development of natural resources in the Arctic. The melting of Arctic sea ice as a result of global warming and climate change is providing greater maritime access and potentially longer navigation seasons. This combined with advancements in ice-breaking technology benefitting commercial carriers are translating into shorter sea routes and reduced shipping costs. This chapter is concerned with safety and environmental implications of insurability and indemnifiability of enhanced Arctic shipping risks attributable to the presence of ice. The specific focus is on examination of relevant provisions in marine insurance contracts subject to English marine insurance law and the Nordic Marine Insurance Plan (NMIP). The discussion begins with a synoptic overview of the fundamental principles of marine insurance and then moves on to the regulatory dimension of Arctic shipping as manifested in the IMO Polar Code. Whether a violation of the Code leads to a potential breach of the implied warranty of seaworthiness entrenched in the UK Marine Insurance Act 1906 (MIA) operating through the Institute warranties is examined together with the corresponding elements in the NMIP regime. A comparative analysis is carried out of the MIA and Institute Clauses and Warranties on the one hand, and the NMIP legal framework on the other, which includes premium considerations, and ice class and notification requirements. The environmental implications for Arctic shipping are examined through the concept of protection and indemnity (P&I) insurance focusing on compulsory cover for third party liability required under relevant ship-source pollution conventions. Finally, the emerging concept of environmental salvage particularly in relation to Arctic waters is discussed peripherally.

Keywords Arctic shipping · Marine insurance · Polar code · P&I clubs · STCW convention

1 Introduction

1.1 Background and Preliminary Observations

Shipping in the Arctic is a contemporary issue of increasing significance in the current maritime milieu. The primary driver of Arctic shipping and maritime operations is the development of natural resources in the Arctic. Global warming and consequential climate change is reportedly the cause of Arctic ice melting rapidly to the delight of some and chagrin of others. The warming of Arctic waters and consequential retreat of sea ice is opening up new waterways hitherto ice-bound and un-navigable providing greater maritime access and potentially longer navigation seasons. This, combined with growing ice-breaking navigational technology particularly of benefit to commercial carriers, is translating into shorter sea routes and dramatic decreases in shipping costs. All these point to positive prospects for the shipping industry. As well, increased fishing opportunities in northern waters are a welcome development for communities where fisheries resources are vital for their economies and the fishing industry in the Arctic region.

On the other side of the spectrum, however, there are the downsides; the safety of ships and the detriment suffered by the pristine Arctic environment. The shift in the fragile ecological balance due to the changing configuration of ice infestations compounded by the increasing intrusion of vessels can result in irreparable environmental harm. The international maritime community would wish a reconciliation of the two extremes of relative benefit and detriment and an optimum balance being reached by the utilization of scientific and technological tools at its disposal. To that end, Arctic shipping must involve multi-disciplinary action including law and economics in addition to science and technology.

At the centre of all the above observations and concerns lies the element of risks for shipowners operating in the Arctic. While it is conceded that shipping as a maritime adventure is inherently a risky business, the risks associated with the Arctic are more because of the presence of ice, remoteness, increased navigational difficulties in the high North, and lack of salvage equipment. Maritime risks are protected through marine insurance which consists of three basic components, namely, hull and machinery (H&M), cargo and third party liability (P&I). There are the safety-associated risks pertaining to the shipowner, that is, damage suffered by the hull and machinery of the ship and the cargo carried on board covered by H&M and cargo insurance. There is also the risk of damage caused to the Arctic marine environment which comprises liability of the shipowner as the polluter to third party claimants who are those who have legal rights in respect of the polluted environment. These risks are covered by P&I insurance.

1.2 Purpose

The principal purpose of this chapter is to examine the marine risks attributable to increased shipping in the Arctic and discuss the legal implications pertaining to their insurability and indemnifiability through the application of norms and practices of marine insurance. In specific terms, the chapter focuses on examining relevant provisions in marine insurance contracts pertaining to risks caused by Arctic ice through a comparative analysis of the so-called Institute Clauses under English marine insurance law and the corresponding provisions of what is currently known as the Nordic Marine Insurance Plan (NMIP) which is the law which prevails in Norway. Clauses developed by the Institute of London Underwriters are used widely throughout the maritime world. The UK legislation governing marine insurance is the Marine Insurance Act 1906 (MIA). Recently the UK Insurance Act, 2015 has been enacted which in effect has resulted in some changes to the MIA 1906. Even so, the 1906 Act is still the model legislation for numerous common law jurisdictions worldwide. The NMIP is the Norwegian version of a regime common to all the Nordic countries. From the time of its establishment in 1996 it was known as the Norwegian Marine Insurance Plan. In 2013 the name was changed to Nordic Marine Insurance Plan. The "Plan" has the status of legislation corresponding to the MIA.

These two jurisdictions have been selected for discussion because they represent two major systems of law and practice in the field of marine insurance. The English law in substance has considerable international application through the national legislation of virtually all common law jurisdictions while the Norwegian law represents the common platform used in all the Nordic states. The aspect of cargo insurance in Arctic marine transportation will not be addressed in this discussion mainly because it involves a substantial amount of law which falls much beyond the intended scope of this work.

It is hoped that this chapter will stimulate continuing discussion in the international maritime arena and instigate further consciousness of a highly dynamic and contemporary subject. Needless to say, this effort is aimed at fostering a regime of marine insurance for the Arctic consistent with safe and sustainable shipping.

1.3 Structure

Following this brief introduction, a synoptic review of the historical evolution of marine insurance and its fundamental precepts is presented. It is deemed necessary to provide at least a sketchy backdrop to the detailed discussion to follow for the benefit of a multi-disciplinary readership and audience; to afford an adequate appreciation of the peculiarities of marine insurance in general and issues pertaining to Arctic shipping. The discussion then moves on to introducing the concept of risks, their insurability and indemnifiability and the basic principles and doctrines of marine insurance law. It is pointed out that marine insurance has important commercial implications. Next, the central theme of the chapter is brought into focus. The insurance implications for Arctic shipping are discussed in light of enhanced risks in terms of safety and environmental protection issues. The specific clauses in the UK marine insurance regime mainly governed by Institute Clauses and the relevant corresponding provisions in the NMIP are examined through a focused comparative analysis. The very topical issue of environmental salvage is then addressed synoptically given its relevance to the environmental dimension of indemnifiability for pollution damage caused in ice-infested Arctic waters. A summary is presented in conclusion together with comments touching on the need for consistency and uniformity in the marine insurance regime in consideration of the changing parameters relating to shipping safety and the need for sustainability of the Arctic environment.

2 Marine Insurance and Regulation of Arctic Shipping

2.1 *Evolution and Precepts of Marine Insurance*

The origins of marine insurance are said to be "veiled in antiquity and lost in obscurity" (Gold et al. 2003, p. 297). The progenitor of the maritime mortgage and marine insurance is the notion of bottomry and its cargo counterpart known as respondentia pursuant to which a shipowner in ancient times pledged the security of his ship or cargo in exchange for a loan for the purpose of prosecuting the ship's voyage. If the ship failed to return from its voyage the loan did not have to be repaid; in effect therefore, the lender became the insurer of the pledged ship or cargo. The practice of bottomry prevailed in ancient Babylon as recorded in the Code of Hammurabi dating back to the period 2000-1600 BC and in Hindu India around 900 BC as recorded in the Manu Samhita (Mukherjee 2002, pp. 11–12; Reddie 1841, pp. 482 and 493). The Greeks also practiced bottomry and respondentia during the turn of the first millennium and contracts of foenus nauticum or usury, considered specifically as precursors of the marine insurance contract of subsequent times, were in vogue which were enforced by the Athenian courts (Mukherjee 2002, pp. 11–12; Reddie 1841, pp. 482 & 493). Modern marine insurance practice started with the Jewish merchants of Lombardy and Florence in Italy moving to London and setting up business in the shipping district of the city near the Tower of London. In the late seventeenth century, Edward Lloyd's coffee house in Tower Street evolved as a venue from where marine insurance business began to be transacted. Even though it has long since disappeared physically, it is famous as the original forerunner of the present day Lloyd's marine insurance market (Gold et al. 2003, pp. 298–300). Incidentally, there is at present a coffee house at the bottom of the Lloyd's Building at 1 Lime Street in London where insurers, brokers and others still meet to transact business.

The conceptual essence of marine insurance which came before any other kind of insurance was ever perceived, and which has existed since time immemorial, is the notion of "spreading of risk" in a common maritime adventure so that upon the collection of a relatively small premium from multiple shipowners, each known as an assured, a singular entity known as the insurer contracts to indemnify each assured in the event of a loss caused by an insured peril to the extent to which the subject matter (ship, cargo or freight) is insured (Gold et al. 2003, p. 308). Today, however, in many instances there is one ship owner (the assured) with one premium but multiple insurers sharing the risk by providing cover for the shipowner's whole fleet. This is the norm for hull and machinery, and cargo insurance. Notably, the concept of small premium paid by multiple ship owners mainly prevails in the field of protection and indemnity (P&I) insurance discussed under Sect. 3.4 of this chapter. In this context, risk is indubitably associated with safety. In contrast, third party liability insurance provided by the P&I regime, in large part relates to the risk of environmental damage caused by ship-source pollution, injury to crew, passengers and third parties, and wreck removal. As may be gleaned from the discussion so far, in legal terms marine insurance involves a contractual relationship between insurer and assured manifested through an instrument known as the "policy". Marine insurance is an integral feature of commercial shipping and aside from the contractual element is also governed by legislation. There is no international convention and it is unlikely that it will ever be necessary to have one. The significance of marine insurance in relation to Arctic shipping is the presence of ice as a navigational and environmental risk which does not exist in other commercially navigable waters although it is recognized that there are ice-infested areas in waters south of the Arctic region where ships regularly operate but the Arctic is unique in several respects. Other risks include remoteness, increased navigational difficulties in the high northern latitudes and lack of salvage equipment.

2.2 *Regulation of Arctic Shipping*

The Arctic region is shared by several countries[1] which all have interests at stake in terms of safety and pollution concerns instigated by enhanced navigational risks in the marine environment which is at once potentially hostile and fragile, resulting from the melting of ice and consequential increase in maritime traffic. Of the eight states which are members of the Arctic Council five are Arctic Coastal States, namely, the US, Canada, Russia, Norway, and Denmark (Greenland). It is therefore expected and inevitable that Arctic states will impose regulatory controls over ships traversing through national waters overlying the purely private contractual legal regime of marine insurance which, as mentioned above, is not governed by any international convention instrument, and it is recognised that the commercial viability of marine insurance in the Arctic region is heavily dependent on adequate and stringent regulatory standards properly enforced by the Arctic states (Johannsdottir and Cook 2014, p. 3; Sarrabezoles et al. 2014, p. 2). National regulatory regimes naturally impact on indemnifiability of losses arising from insured risks and perils under private law contractual relationships between the assured and insurer. Any tension between national regulatory regimes and the private law legal framework of marine insurance is potentially detrimental to the smoothness and functional efficiency of commercial shipping operations in the Arctic. In view of this verity, harmonization of national safety standards is invariably of utmost importance and

[1]Canada, Denmark, Finland, Iceland, Norway, Russia, Sweden and the United States are the eight members of the Arctic Council. See "Member States, Arctic Council", available at http://www.arctic-council.org/index.php/en/about-us/member-states, accessed on 22 September 2016 and "Arctic Council", available at http://en.wikipedia.org/wiki/Arctic_Council, accessed on 22 September 2016. Notably, 12 non-Arctic states have permanent observer status in the Council. They are China, France, Germany, India, Italy, Japan, South Korea, Netherlands, Poland, Singapore, Spain and United Kingdom are observer states in the Council. Turkey and the European have ad-hoc observer status. See "Observers, Arctic Council", available at http://www.arctic-council.org/index.php/en/about-us/arctic-council/observers, accessed on 22 September 2016.

virtually indispensable for the maintenance of sustainability. It seems the quest for harmonization has finally been achieved through the emergence and adoption of the International Code for Ships Operating in Polar Waters (Polar Code) under the auspices of the International Maritime Organization (IMO) following lengthy and protracted negotiations.

2.3 *Salient Features of the IMO Polar Code*

The principal object of the Polar Code, is to standardize the technical requirements for ships operating in waters that fall within the definition of Polar Waters, as well as implementing environmental requirements, crew training requirements, and search and rescue provisions The Code is being implemented by way of amendments to the IMO's International Convention for Safety of life at Sea (SOLAS) in respect of the safety requirements in the Polar Regions, the International Convention for the Prevention of Pollution from Ships (MARPOL) in respect of the environmental provisions, and the International Convention for Standards of Training, Certification and Watch-Keeping (STCW) Convention in respect of the crew training provisions. The safety aspects of the Polar Code were adopted at the 94th session of the Maritime Safety Committee (MSC) in November 2014 through amendments to SOLAS (Resolution MSC.385 (94)). The amendments entered into force on 1 January 2017 through the tacit acceptance procedure of the convention. The amendments to MARPOL related to the Code were adopted by the Marine Environment Protection Committee (MEPC) in May 2015 and the date of their entry into force is aligned with that of SOLAS, that is, 1 January 2017, again through the tacit acceptance procedure of MARPOL (Resolution MEPC.264 (68)). The Code is a mandatory instrument for state parties to SOLAS and MARPOL in respect of commercial ships operating in polar waters. The marine insurance industry together with the classification societies are the entities responsible for evaluation of risks for ships operating in polar waters for whom the Code ostensibly provides uniform standards through a regulatory framework for enhanced maritime safety and environmental protection (Brigham 2013).

The first point of observation is that the application of the Code is not confined to the Arctic but rather is applicable to "Polar Waters", a term that is not specifically defined in the Code but its import may be gleaned from the words in preambular clause 6 which states, inter alia, that "the Code is intended to apply as a whole to both Arctic and Antarctic...". Notably, the Code in its "Introduction" following paragraph 4—Structure, contains figures illustrating the applications of maximum extents of the Antarctic area and Arctic waters as those terms are defined in various provisions in the relevant Annexes of SOLAS and MARPOL. The goal of the Code as expressed in paragraph 1 of the Introduction is "to provide for safe ship operation and the protection of the polar environment". The manner in which this goal is to be achieved is reflected in the words "by addressing risks present in polar waters and not adequately mitigated by other instruments...". Ship categories A, B and C are defined in paragraph 2 of the Definitions by reference to design for operation in young ice, first year ice, medium first year ice and old ice defined in temporal terms and according to thickness. The Code refers to hazards in terms of "elevated levels of risk", "increased probability of occurrence" of those risks and "more severe consequences" which recognizes the likelihood of enhanced risks in the Arctic. This is provided in paragraph 3.1 of the Introduction to the Polar Code. It is also recognized that risk levels may vary according to geographical location, amount of daylight according to the time of year and ice-coverage. As well, measures for mitigating specific hazards can be different.

The concept of "Polar Class" (PC) is introduced which is basically the ice class assigned to a ship by its flag state Administration or the ship's classification society recognized by the Administration based upon IACS Unified Requirement as provided in paragraph 1.2.10 of Part I-A bearing the title "Safety Measures".

The International Association of Classification Societies (IACS) has developed and adopted a set of unified requirements for polar class ships designated as classes ranging from PC 1 to PC 7 based on the times of operation of a ship in the run of the year in ice-infested waters (Brigham 2013, pp. 3–4). It is notable in this context that the table of IACS polar class ship categories contains ice descriptions taken from the sea-ice nomenclature of the World Meteorological Organization.

In the usual manner of IMO instruments, there are requirements for surveys and certification. Each ship to which the Code applies must have on board a Polar Ship Certificate issued by the flag state Administration or an organization recognized by it which in most cases is a classification society after an initial or renewal survey has been carried out in accordance with Regulation 1/1 of Chapter XI of SOLAS, and must be in the prescribed form as provided in Chapter 1 paragraph 1.3 of the Polar Code. A Code-compliant ship must also carry on board a Polar Water Operational Manual (PWOM) as per Chapter 2 of the Polar Code.

In Chapter 1 Part 1–A of the Code, it appears that there is no specific provision relating to what ships or ship types are subject to the Code. However, the Code being a part of SOLAS, it would apply to ships engaged on international voyages as provided in Regulation 1 of Chapter 1 Part A, including cargo ships of 500 tons or greater and passenger ships as defined in SOLAS, i.e., ships carrying more than 12 passengers as provided in Regulation 2 of that Chapter. SOLAS does not apply to inter alia, fishing vessels and pleasure yachts not engaged in trade as per Regulation 3 of that chapter. The Code is also associated with MARPOL which applies to "ships" which in turn carries a broader definition in Article 2 of the Convention. Thus, in the Code itself, there is no clear application clause in respect of ships. There is a reference to operational limitations being established taking account of hazards referred to in paragraph 3 of the Introduction and additional hazards if they are identified. This is provided in Chapter 1 paragraph 1.5 of the Polar Code. It is anticipated that such hazards will affect the insurability of ships operating in polar waters. Chapter 2 provides for an operational manual; namely, the PWOM referred to above, which sets out functional requirements in relation to regulations in paragraph 2.3 of this Chapter. They essentially provide that specified risk-based procedures must be included in the manual to facilitate compliance with the functional requirements relating to specific procedures to be followed for normal operations of ships. The procedures include measures to be taken in the event the ship encounters conditions exceeding the ship's specific capabilities and procedures or operational limitations in respect of which assessments have been made, and procedures to be followed when ice breaker assistance is used. Chapter 2 paragraphs 2.2 and 2.3 of the Polar Code provides for this. Chapters 3 to 6 deal respectively with ship structure, subdivision and stability, watertight and weathertight integrity and machinery installations. The contents of these Chapters provide for structural and safety matters. In terms of Chapter 3, additional guidance provided in Part I-B for determination of equivalents of ice class for categories A and B, new and existing ships, which involves comparisons of other ice classes with IACS polar class. Notably, for category C ice classes, additional information regarding comparisons of strengthening levels are available in the annex to HELCOM Recommendation. 25/7. It appears that some classification societies have developed user-friendly tools for determining compliance with IACS polar class structural requirements. Some Administrations and other third parties have done the same. Paragraph 4 "Additional Guidance to Chapter 3" of Part I-B bearing the title "Additional Guidance regarding the Provisions of the Introduction and Part I-A" of the Polar Code contains the details to be followed.

Chapters 7 to 11 deal with fire safety protection, life-saving appliances and arrangements, safety of navigation, communication and voyage planning. Among these, safety of navigation is of crucial importance in the context of this work in view of the fact that the prescriptions for equipment may potentially affect the seaworthiness requirements of the ship under a marine insurance contract. Chapter 12 deals with manning which also has an impact on seaworthiness as an extension of the issue of safety. Masters and navigation officers must be suitably qualified in accordance with the STCW Convention and STCW Code appropriate for ship operations in polar waters. Part II–A of the Code addresses pollution prevention measures; it consists of Chapters 1 to 5 which correspond to the five Annexes of MARPOL, namely, prevention of pollution relating to the five types of ship-source pollutants. They are oil, noxious liquid substances, packaged harmful substances, sewage and garbage. Air pollution dealt with in Annex 6 of MARPOL is not addressed in the Polar Code. These provisions are obviously preventive measures against pollution damage that may be suffered by polar waters falling within the jurisdictions of polar states. In the context of this chapter, these are the Arctic coastal states that are members of the Arctic Council.

The Polar Code is indeed a laudable effort on the part of the IMO and its member states to supplement existing IMO Conventions, SOLAS, STCW and MARPOL relating to safety and environmental protection. In that sense, the Code strengthens and supplements the relevant international conventions in terms of their application to polar waters including the Arctic which is the subject of this chapter. It is assumed that through the instrumentality of this Code, Arctic states will give effect to these specialized technical and regulatory requirements in their national legal regimes. Although no such objective is expressly stated in the Code itself, the fact that the Code is now in force is ample evidence that it has gained universal acceptance. No doubt it serves to create harmonization in this emerging and contemporary issue in shipping despite some perceived lack of clarity. Hopefully the instrument will be revisited so that it can get as close to perfection as possible and its acceptability reaffirmed as a uniform and non-discriminatory set of international standards providing a level playing field for all marine and maritime activities in polar waters.

2.4 Principles of Marine Insurance

Aside from the basic precepts of marine insurance referred to above, there are a number of legal principles usually referred to as the doctrines of marine insurance. These are set out below in introductory synoptic form for the benefit of those who are uninitiated in this highly specialized field of commercial maritime law; for a fuller understanding and clear perception of the discussion on specific provisions pertaining to cover for enhanced risks in Arctic shipping addressed in this chapter.

The starting point of the discussion must be the notions of insurability and indemnifiability of risks and losses. Insurability refers to the condition whereby the subject matter to be insured such as a ship, and the risks against which the insurance is to be effected meet the basic requirements of the law and are insurable. Insurability occurs before the fact; it must be determined before insurance is placed, in other words, before the parties enter into a contractual relationship. By contrast, indemnifiability signifies the condition under which following a loss or damage suffered by the subject matter such as the ship, the insurer becomes liable to indemnify the assured. Thus, the concept of indemnifiability operates ex post facto or after the fact. In light of these fundamental propositions, the basic principles or doctrines of marine insurance must be viewed and appreciated.

The first principle is the notion of indemnity. A marine insurance contract is a contract of indemnity which means that the insurer is liable to pay the assured only to the extent of the value of his pecuniary loss resulting from an insured peril up to the level of indemnity which may not always be the full value of the property. This is relatively simple in the case of a valued policy where the parties agree on the value of the subject of insurance at the time they enter into the insurance contract. But where the policy is unvalued, the loss must be evaluated by a professional valuer or loss adjuster after the occurrence of the loss or damage and the insurer pays accordingly. The essence of the doctrine of indemnity is that the assured cannot make a profit from the loss he suffered and the insurer is not obliged to pay beyond the amount equivalent to that loss.

The second principle is that the assured must have an insurable interest in the subject to be insured recognized at law. It is related to insurability of the subject of insurance mentioned above. Where a person stands in an equitable position in relation to a maritime adventure or to any insurable property at risk such as a ship so that he stands to benefit from its safe arrival or is prejudiced by its loss, damage or detention, he is said to have an insurable interest in the subject of the insurance. (See Canadian Marine Insurance Act, 1993, s. 8, S.C. 1993, c.22.) Aside from the shipowner, entities such as the cargo owner, charterer, master and crew and the mortgagee have insurable interests in a ship. In the absence of an insurable interest the transaction amounts to a wager or gamble which is prima facie unlawful. (See Marine Insurance Act, s. 4.)

Mutual disclosure of material facts between insurer and assured otherwise referred to as the doctrine of utmost good faith or uberrimae fidei, is a remarkable feature of marine insurance law. The rationale for this requirement applicable to both the insurer as well as the assured is that the information to be disclosed is solely in the possession of one or the other party to the marine insurance contract, which is why "utmost" good faith and not merely good faith, as required in contracts generally, has been deemed to be necessary in the legal regime of marine insurance. Disclosure of material facts and representations allows the insurer to decide whether to accept the assured for providing insurance, and if so, to determine what premium to charge. The assured for his part needs disclosure from the insurer to decide whether he should contract with the particular insurer or choose some other on the basis of the insurer's reputation and track record. While it is not necessary to disclose non-material facts, materiality must be determined by objective criteria. The traditional position under English law has been that any fact deemed to be material that the underwriter would have taken into account, even if he would not have changed this decision to insure that risk, that becomes subsequently known to the insurers gives him the right to avoid the insurance policy as if it never existed. Information that is readily available in the public domain or a fact that is a commonly known verity is not necessarily subject to disclosure even if it is material to the transaction. In recent times, further to papers and reports published by the English and Scottish Law Commissions, there has been considerable discussion regarding whether the doctrine of uberrimae fidei should be retained in the current milieu of marine insurance. The upshot is that by virtue of section 14 of the new U.K. Insurance Act 2015, non-observance by one party of the requirement of utmost good faith is no longer a ground for avoidance of the contract by the other although a marine insurance contract remains based on that doctrine under section 17 of the MIA. Incidentally, uberrimae fidei as depicted in the original MIA continues to operate in its totality in other jurisdictions.

Subrogation is a concept peculiar to insurance law in general and features significantly in marine insurance as a fundamental doctrine. It is best described by the expression "stepping into the shoes" of the assured by the insurer to pursue the perpetrator of the wrong that caused the loss or harm suffered by the assured after the assured is indemnified. (See Castellain v. Preston (1883), 7 A.C. 333 (Q.B.)) Subrogation gives to the insurer only a right of action in the name of the assured and does not vest in him any proprietary right or interest (Gold et al. 2003, p. 327). Closely related to subrogation, sometimes featuring as an outfall thereof is the concept of windfall. This occurs when for reasons such as, for example, a drastic alteration of a currency exchange rate the proceeds of insurance turn out to be greater than the contracted amount. The question then arises as to who benefits from the windfall, insurer or assured. (See Yorkshire Insurance Co. v. Nisbet Shipping Co., [1961] 2 All E.R. 487.)

Sometimes confused with subrogation is the doctrine of abandonment which stems from the nature of a loss. In marine insurance law, losses are of the partial or total variety. Partial losses are subdivided into particular average losses arising from fortuitous incidents at sea and general average losses which are basically voluntary acts such as sacrifices made in the interest of saving the whole maritime adventure. Total losses are also divided into two groups, namely, actual total loss (ATL) and constructive total loss (CTL). An ATL takes place when the assured is irretrievably deprived of the subject of insurance or it undergoes a change in specie so that it is no longer a thing of the kind insured. A CTL happens when recoverability of the property is unlikely or the cost of retrieval is grossly prohibitive being far in excess of its market value. In such situation, the doctrine of abandonment becomes applicable under which the assured abandons the physical property and its associated legal rights in favour of the insurer after the latter has paid for a total loss. To effectuate this, the assured must issue a notice of abandonment, acknowledgement or acceptance of which the insurer may decline, to avoid inheriting liabilities associated with the property such as wreck removal obligations. The distinction between abandonment and subrogation is that abandonment results in the insurer acquiring a proprietary interest in the property but in subrogation he only acquires a right of legal action (Khurram 1994, p. 95).

3 Arctic Shipping, Enhanced Risks and Marine Insurance Implications

3.1 Marine Insurance in Arctic Shipping: General Issues and Considerations

The reduction of sea ice cover in the Arctic is invariably a great incentive for increased economic and shipping activity there. Although the volume of traffic in the Northern Sea Route has decreased notably between 2014 and 2016, it is relatively minor in global terms. Although shipowners are free to assume the risk of trading on Arctic routes subject to obtaining suitable marine insurance (Østreng et al. 2013, p. 331), it is the norm for marine insurance contracts to exclude or limit the coverage of Arctic marine perils through imposing navigating limits on the ship because of the extraordinarily high risks, manifested by the Institute Warranties 1976 (See Appendix 1) and the International Navigating Limits (INL) in the International Hull Clauses (IHC) 2003 (Bennett 2006, p. 553). The INL are an alternative to the Institute Warranties 1976 made by the Institute of Chartered Underwriters. They became effective as Clause 32 of IHC 2003 (See Appendix 2). Needless to say, there are limitations on navigation outside the UK regime as evident from Clause 3-15 and the Appendix to the 2013 NMIP (See Appendix 3) (Liu 2016, pp. 83–88).

A pertinent observation is that marine insurance with respect to the Arctic is disadvantaged by limited knowledge and information exemplified by a lack of adequate empirical data which makes it rather difficult for insurers to determine appropriate assessment parameters for the risks likely to be encountered in the various shipping routes. General risks in shipping are known to insurers but with respect to those associated with Arctic transportation there remains a dearth of understanding, particularly regarding their identification (Østreng et al. 2013, p. 332; Tamvakis et al. 1999, pp. 221–280).

Underwriters usually establish premiums based on accident statistics Thus they are prompted to carry out a strict scrutiny of a ship's track record in relation to safety, and expect to see evidence that it is capable of sound performance in Arctic waters (Sarrabezoles et al. 2014, p. 4; Østreng et al. 2013, p. 332). As a corollary to the concerns of insurers, shipowners expect reliable and consistent rules, standards and provision of services from their insurance providers (Chircop 2009, pp. 355–380). They face rigid conditions imposed by insurers such as increased deductibles for ice-related damage which can be quite onerous. But insurers consider this to be quite justified given that the Arctic is a high risk market to cover in the face of uncertainty and lack of reliable data (Johannsdottir and Cook 2014, p. 21). Because of the lack of information at present, save in respect of very experienced operators, insurers currently consider Arctic risks on a case by case basis.

The insurance industry perceives a whole range of multifarious risk factors; the principal one being the presence of ice whether it is concealed or visible. Perils of the sea in Arctic waters in a legal context are addressed in academic literature. (See Liu 2016, pp. 88–92) To deal with this, ice-breaker assistance is viewed as a solution recognizing the fact that most ships presently operating in the Arctic and those under construction are designed to operate independently of icebreaker support. An exception is the Russian mode of transportation where ice-breaker escort and movement in convoys remains prevalent. Other risks include inadequacy of search and rescue (SAR) and salvage facilities, hydrographic parameters including bathymetric and seabed mapping information, unreliable communication networks at the higher latitudes, crew competence, including Arctic knowledge and experience, and language deficiencies. Ice class requirements are a must and the industry is not prepared to support any dilution of requirements by national authorities in the Arctic region on the basis of season and amount of ice. However, certain cruise ships and cargo ships (under Category C of the Polar Code) are capable of operating in ice-infested waters in the same manner as in clear waters without having to comply with any additional Polar Code requirements, if following an assessment no additional requirements are deemed to be required. Additionally, any reduction of safety-related regulatory measures by such authorities is viewed from the perspective of insurers as hugely increasing the risk factor (Normann and Mikkelsen 2014; Sarrabezoles et al. 2014, p. 2).

It is evident that the international marine insurance market is prepared to underwrite shipping risks in the three Arctic corridors, namely, the Northeast Passage (NEP), Northwest Passage (NWP) and Transpolar Passage (TPP). However quite understandably, the premiums charged by them would be different. Also, it appears that the premiums for the NWP and TPP would be higher than those for the NEP based on the variously differentiated risk factors involved. Premiums are difficult to calculate due to complexities relating to, inter alia, the vagaries of the Arctic environment. Every risk is different and must be determined on a case-by-case basis depending on such factors as the nature of the operation, time of the year, etc. In addition, a vessel venturing into the high Arctic may need to obtain a separate policy for each separate passage (Østreng et al. 2013, p. 331; Johannsdottir and Cook 2014, pp. 23–24; Lajeunesse 2012, pp. 521–537). In so far as the NSR is concerned it seems premiums have been more or less settled. Even so, given that the essential factors governing the determination, premiums are subject to international as well as national law and policy. The associated uncertainties including the bearing of costs of necessary infrastructure and the investments relating to those costs are a cause of anxiety for the insurance industry (Normann and Mikkelsen 2014).

3.2 English Marine Insurance H&M Regime

3.2.1 Institute Clauses and International Hull Clauses in H&M Policies

Marine insurance policies contain standard form clauses covering varieties of risks. As such, this discussion must of necessity begin with an explanation of the terms "Institute Clauses" and "Institute Warranties" for the benefit of those who are unfamiliar with the typical jargon of English marine insurance law and practice. These terms refer to clauses and warranties articulated by what was originally known as the Institute of London Underwriters (ILU), designed to be used in conjunction with the old Lloyd's Ship and Goods (SG) policy and subsequently with the Lloyd's Marine policy (MAR 91). They appeared primarily as cargo, voyage and time clauses in respect of hulls and freight; and war and strikes clauses associated with the foregoing. The ILU subsequently evolved into the International Underwriting Association (IUA) and the International Hull Clauses were formulated initially in 2002 and amended in 2003 as a modernized version of the Institute Hulls Time Clauses under its auspices. In 2009, the Institute Cargo Clauses and War Clauses were revised. Also a new IUA Marine Policy was formulated for accommodating the IHC 2003 and the other modernized Institute Clauses (Rose 2012, pp. 3–4 and 695–840).

H&M policies are subject to limitations on navigation so that cover is denied in the event that the insured vessel enters certain specified waters. Such limitations have traditionally been expressed as promissory warranties (Bennett 2006, p. 553). By way of explanation it is useful to note that in the MIA, a warranty is referred to as a "promissory warranty" and is "a warranty by which the assured undertakes that some particular thing shall or shall not be done, or that some condition shall be fulfilled, or whereby he affirms or negatives the existence of a particular state of facts" (Rose 2012, p. 188).

While there are no navigating limits in the Institute Time Clauses-Hulls (ITC-H) 1983 and 1995, both sets of clauses refer to the Institute Warranties 1976 which exclude perils of the seas in certain areas and contain a list of warranties relating to geographical limits of navigation. These limits cover most of the Arctic waters.[2] Under these Institute warranties, the assured warrants (promises) that his vessel will not navigate in areas prohibited under the contract. These prohibitions are stipulated by the insurer and apply to certain months of the year within certain areas the perimeters of which are defined by geographical coordinates. According to section 33 of the MIA, the insurer is automatically discharged from liability under the contract from the time of a breach of such warranty. In contrast, section 10(2) of the 2015 Insurance Act stipulates that a breach of such warranty only suspends rather than discharges the insurer's liability from the time of the breach, until the breach is remedied; that is, until the vessel ceases to navigate in the prohibited area. In such case, the insurer is not liable for anything which occurs, or which is attributable to something occurring, during the period of suspension.[3] However, pursuant to sections 16 and 17 of the 2015 Insurance Act the insurer can contract out of such consequences of the breach of warranty; in other words, if he contracts out, he may be discharged from liability.

From the shipowner's perspective, to cover Arctic marine risks as the assured, he must have in his favour "held covered" provisions in the contract to avoid consequences of breaching the warranty of navigation limits (Liu 2016, p. 94). The inclusion of such provisions requires special notification to and arrangements with the insurer. Thus, a "held covered" clause is almost always used in conjunction with a trading limits warranty. Such a clause gives to the assured the right to retain cover when operating within the prohibited area, subject to prompt notification to the insurer and payment of additional premium (Hodges 1996, p. 120). The notion is illustrated by Clause 3 of the Institute Time Clauses-Hulls 1983 and 1995 which is worded—"[H]eld covered in case of any breach of warranty as to cargo, trade, locality, towage, salvage services or date of sailing, provided notice be given to the Underwriters immediately after receipt of advices and any amended terms of cover and any additional premium required by them be agreed". The "held covered" clause

[2]The Polar Code does not contain a definition of Arctic Waters although in the original draft the following definition was available. "Arctic waters" means those waters which are located north of a line from latitude 58°00′0 N and longitude 042°00′0 W to latitude 64°37′0 N, longitude 035°27′0 W and thence by a rhumb line to latitude 67°03′9 N, longitude 026°33′4 W and thence by a rhumb line to Sørkapp, Jan Mayen and by the southern shore of Jan Mayen to the Island of Bjørnøya, and thence by a great circle line from the Island of Bjørnøya to Cap Kanin Nos and hence by the northern shore of the Asian Continent eastward to the Bering Strait and thence from the Bering Strait westward to latitude 60°N as far as Il'pyrskiy and following the 60th North parallel eastward as far as and including Etolin Strait and thence by the northern shore of the North American continent as far south as latitude 60°N and thence eastward along parallel of latitude 60°N, to longitude 56°37′1 W and thence to the latitude 58°00′0 N, longitude 042°00′0 W". See Regulation 1—Definitions in "Draft International Code for Ships Operating in Polar Waters", Annex 2 in Guidance on Arctic Navigation in the Northeast Route published by the Maritime Safety Administration of China.

[3]UK Parliament. (2015). Explanatory Notes, Insurance Bill [HL], para.88, available at http://www.publications.parliament.uk/pa/bills/lbill/2014-2015/0039/en/14039en.htm, accessed on 22 September 2016.

leads to compulsion on the part of the assured to notify and negotiate with the insurer if the ship is going to enter into Arctic waters.

Notably, "held covered" clauses fall under the concept of post-contract duty of good faith. In effect, a "held covered" clause in an existing policy represents an extension or variation of the contract in respect of which additional premium may be payable by the assured where such an extension or variation is found to be well-established. There is a duty of disclosure applicable to the assured which is effectuated through his duty to provide prior notification to the insurer as mentioned above. However, a failure of this duty does not vitiate the whole contract (Rose 2012, pp. 108–109).

In so far as the IHC 2003 is concerned, its Clause 10 provides that the ship must comply with any provisions as to locality mentioned in Clause 32 (an additional clause in Part 2 of the IHC), which prohibits entry into inter alia, the Arctic (depicted as north of 70 degrees north latitude) unless agreed by the underwriters. Clause 11 releases underwriters from liability in the event of non-compliance with Clause 10 but they would be liable if they were given notice under Clause 11. Clause 32 in turn cross-refers to Clause 33 which provides that the ship may breach Clause 32, in which case Clause 11 is not applicable so long as prior permission of the underwriters is obtained and the requisite additional premium is agreed. The net effect of the foregoing maze of provisions is that it is incumbent on the assured to obtain permission from the underwriters before engaging in shipping in the Arctic. Incidentally, unlike the warranty in Institute Time Clauses, Clause 11 read in conjunction with Clause 32 in the IHC 2003 indicates that the navigating limits including Arctic waters are no longer warranties but rather are "suspensive conditions" (Bennett 2006, p. 553).

3.2.2 Seaworthiness

Seaworthiness is a concept peculiar to maritime law and consists of a public law component, usually referred to as statutory seaworthiness, and a private law component within the sphere of commercial maritime law, namely, carriage of goods under bills of lading and charterparties, and marine insurance (Schroeder et al. 2006, pp. 21–28). Violation of statutory seaworthiness leads to sanctions as illustrated in sections 457 and 463 of Merchant Shipping Act 1894 and in section 457 under which sending an unseaworthy ship to sea is a misdemeanor, whereas a breach of seaworthiness in commercial maritime law including marine insurance leads to available remedies in the relevant law. While safety and seaworthiness are correlated, the relationship in legal terms is somewhat tenuous. The term "unsafe" as distinguished from "unseaworthy" was used in section 459 of the Merchant Shipping Act 1894 although the virtual meaning of unsafe is almost the same as that of unseaworthy, namely, "... defective condition of her hull, equipments, or machinery, or by reasons of her under manning, or by reason of overloading or improper loading, unfit to proceed to sea without serious danger to human life, having regard to the nature of the service for which she is intended, ...". It is further notable that an unsafe ship as described above is subject only to detention. The reason for the change of terminology is unclear since the definitions for both terms are the same. The editors of Temperley's Merchant Shipping Acts note that the degree of danger associated with sending a ship to sea in an unseaworthy state is lower than the standard required to make a vessel an unsafe ship. Its only practical implication is that in the case of the former the burden of proof is at a lower threshold (Thomas and Steel 1976, p. 130).

Furthermore, maritime safety is unquestionably regulatory in scope whereas seaworthiness primarily has private law connotations, the definitions of which can be gleaned from case law jurisprudence reflected in seminal texts on commercial maritime law. In UK legislation, that is, sections 94 and 95 of the Merchant Shipping Act 1995, the term "unsafe ship" has been changed to "dangerously unsafe ship" following the recommendations made in the Donaldson Report relating to the official investigation of the sinking of the Herald of Free Enterprise. It is noteworthy that whereas seaworthiness is mentioned in Article 94 of UNCLOS, there is no such reference in SOLAS which exclusively deals with safety. The reason for pointing this out is that the foregoing discussion is relevant to any conclusion regarding the status of the Polar Code, or at least the provisions relating to safety, manning and environmental protection. The object here is to discuss whether non-compliance by a ship operating in the Arctic is likely to render a ship unseaworthy for the purposes of the implied warranty in a marine insurance contract.

In this context it would be instructive to look at the legal positions of the ISM and ISPS Codes in terms of seaworthiness by analogy. In respect of the ISM Code notably IHC 2003 clause 13 requires ships to be classed and shipowners to comply with Chapter IX of SOLAS. Non-compliance with this provision automatically terminates the insurance under clause 13.2. With regard to the ISPS Code it is instructive to look at relevant academic literature (Andrewatha and Stone 2004, p. 370) where the authors refer to deficiency in ISPS compliance, in particular, lack of crew security training, deficient ISPS documentation and master or crew negligence as possibly constituting unseaworthiness.

In the context of the present discussion, it is incumbent on us to first look carefully at the issue of the implied warranty in marine insurance contracts tempered by relevant statutory provisions (Schroeder et al. 2006, pp. 21–28). In this vein, it can be stated that seaworthiness itself is a cornerstone element in a marine insurance contract in that it is an implied warranty which is explained in detail in section 39 of the MIA as follows:

(1) In a voyage policy there is an implied warranty that at the commencement of the voyage the ship shall be seaworthy for the purpose of the particular adventure insured.
(2) Where the policy attaches while the ship is in port, there is also an implied warranty that she shall, at the commencement of the risk, be reasonably fit to encounter the ordinary perils of the port.
(3) Where the policy relates to a voyage which is performed in different stages, during which the ship requires different kinds of or further preparation or

equipment, there is an implied warranty that at the commencement of each stage the ship is seaworthy in respect of such preparation or equipment for the purposes of that stage.

(4) A ship is deemed to be seaworthy when she is reasonably fit in all respects to encounter the ordinary perils of the seas of the adventure insured.

(5) In a time policy there is no implied warranty that the ship shall be seaworthy at any stage of the adventure, but where, with the privity of the assured, the ship is sent to sea in an unseaworthy state, the insurer is not liable for any loss attributable to unseaworthiness.

Sub-sections (1), (2) and (3) are statements of law pertaining essentially to a voyage policy and include the notion of the so-called "stages" doctrine in relation to seaworthiness (Soyer 2001, pp. 92–95). Sub-section (5) is also a statement of law in relation to a time policy to state that there is no implied warranty but the ship must be seaworthy at any stage of the maritime adventure; and the insurer is not liable if the assured sends the ship to sea in an unseaworthy state. Perhaps most importantly, sub-section (4) provides a statutory definition of seaworthiness.

Judicially, a seaworthy ship has been described as one that is "...in a fit state as to repairs, equipment, crew and in all other respects, to encounter the ordinary perils of the sea of the voyage". (See Dixon v. Sadler (1839), 5 M & W 414.)

Another authoritative definition of "seaworthiness" is found in the case of F.C. Bradley & Sons Ltd. v. Federal Steam Navigation Co. ((1926), 24 Ll.L.R. 446 at p. 454) where approving a statement in the classic text Carver on Carriage by Sea, the court held that "[T]he ship must have that degree of fitness which an ordinary careful owner would require his vessel to have at the commencement of her voyage having regard to all probable circumstances of it." It has also been held that "[S]eaworthiness is not an absolute concept; it is relative to the nature of the ship, to the particular voyage or even to the particular stage of the voyage on which the ship is engaged." (The Fjord Wind [1999] 1 Lloyd's Rep. 307 at p. 315 per Moor-Bick J. approved by Clark J. in the Court of Appeal in [2000] 2 Lloyd's Rep. 191 at p. 197.) The standard for the implied warranty of seaworthiness is based on whether or not factually a ship is reasonably fit in all respects to encounter the ordinary perils of the sea, which is a condition which must be determined on a case-by-case basis and is not something that can be considered as a constant parameter. In this context, the question arises as to whether or not the presence of ice in Arctic waters is an "ordinary peril of the sea". Arguably, sea ice may be considered an ordinary peril if it exists in the normal course of a voyage in the Arctic. But it may not be considered an ordinary peril if it exists only in certain "ice-infested" waters posing extraordinary unpredictable navigational hazards. In Popham and Willet v. St. Petersburg Insurance Co., ((1904), 10 Com Cas 31 at p 34) it was held by Walton J. that "the obstruction by ice was accidental and unexpected" causing "extraordinary difficulty and danger". He reached the conclusion that "the obstruction and danger and difficulty from the ice ... was a peril of the sea ... covered by the policies". Notably, an alternative opinion is expressed by one author (Torrens 1994, pp. 91–92). But it can be equally argued that if a ship is equipped for ice navigation in accordance with the Polar Code, then for its purposes, the presence of ice could be an ordinary peril in that context.

Flowing from the above discourse, for a vessel entering Arctic waters, can it be concluded that non-compliance with the Polar Code constitutes unseaworthiness? The answer should be in the affirmative if non-compliance with the Code leads to the vessel not being fit in all respects for its intended purpose, namely, traversing the waters of the Arctic. As mentioned earlier, several crucial elements of the Polar Code reflect requirements of SOLAS and MARPOL in respect of structural integrity and equipment of a ship as well as STCW in respect of safe manning and seafarer qualifications. Inasmuch as the ISM Code is an extension of Chapter IX of SOLAS, if non-compliance with SOLAS requirements leads to unseaworthiness, then non-compliance with the ISM Code would also constitute the same. While the ISM Code does not set the standards for seaworthiness; it does for safety management. The Code was not designed to address unseaworthiness arising from fortuitous circumstances. But if there is proper maintenance and management in accordance with the requirements of SOLAS and the ISM Code, undoubtedly a vessel would be better equipped to maintain its seaworthy condition and counter the negative effects of any fortuitous circumstances (Honka 2004, p. 2). The standards for seaworthiness must be derived from case law and by reference to safety-related instruments such as SOLAS. Unfortunately, there is no clear or definitive judicial opinion in this matter. The ISM Code features only peripherally in The Patraikos 2. ([2002] 4 SLR 232.) In The Eurasian Dream ([2002] 1 Lloyd's Rep. 719) liability was imposed on owners for failure of "due diligence" under the Hague-Visby Rules not on non-compliance with the ISM Code. One commentator has opined that failure to comply with the ISM Code is by no means conclusive evidence of a breach of the seaworthiness requirement (Soyer 2001, p. 122).

Be that as it may, the ISM Code is highly relevant as an evidentiary tool and is of probative value for a claimant seeking to prove breach of seaworthiness by the shipowner. In light of the above observations, it cannot be stated with any degree of certainty whether non-compliance with the Polar Code may amount to a breach of the implied warranty of seaworthiness for a vessel purporting to operate in the Arctic. If there is such a breach, the insurer will be relieved of his liability to indemnify the assured shipowner. It has been stated that ISM Code compliance has the effect of making ships safer as a consequence of which it is likely that there will be a reduction in allegations of unseaworthiness (Soyer 2001, p. 131).

In the view of the present authors, in the first instance, the same can be said about the Polar Code given that its object is to ensure, inter alia, ship safety in Polar waters including the Arctic. Having said that, the Polar Code provides for the issuing by the flag state Administration or a recognized organization (RO) of a Polar Ship Certificate after a an initial or renewal survey of a ship purporting to operate in polar waters meets the relevant requirements of the Code (paragraph 1.3 of Chapter 1—General in PART 1-A Safety Measures). A ship not in possession of a valid Polar Ship Certificate may prima facie be deemed to be unseaworthy for purposes of obtaining marine insurance cover, but whether or not a ship was seaworthy for purposes of indemnifiability by the insurer will remain subject to the various factors discussed above and the applicable marine insurance regime.

3.3 Nordic Marine Insurance Regime: NMIP Clauses

Unlike the Institute Clauses, the NMIP 2013 constitutes quite a different marine insurance regime without warranties and held-covered clauses but bears similar stipulations regarding the coverage of Arctic marine risks. Under Clause 3-15 of NMIP 2013 (See Appendix 3), the trading areas are divided into into three types, namely, the ordinary, conditional and excluded trading areas. Sub-paragraph 1 of Clause 3-15 provides that the assured shall notify the insurer before the ship proceeds beyond the ordinary trading limit. According to sub-paragraph 2, the ship is allowed to continue to sail in the conditional trading areas but the insurer may require an additional premium and also may stipulate other conditions. Based on sub-paragraph 3, if the ship proceeds into an excluded trading area, in principle, the insurance ceases to have effect, unless the insurer has given permission in advance. The Appendix to Clause 3-15 sets out the conditional and excluded trading areas geographically associated with relevant charts and maps, in which Arctic waters are included. Therefore, if the assured wishes to embark on Arctic shipping under NMIP 2013, he has to notify the insurer and enter into a special agreement with him, as in the case of Institute Time Clauses or the IHC.

Notably, sub-clause 1 of Clause 3-22 (See Appendix 4) provides by way of definition that "[A] safety regulation is a rule concerning measures for the prevention of loss, issued by public authorities, stipulated in the insurance contract, prescribed by the insurer pursuant to the insurance contract, or issued by the classification society". Sub-clause 3 (See Appendix 4) in turn provides that "[T]he rules prescribed by the classification society regarding ice class constitute a safety regulation under sub-clause 1." The breach of such safety regulations, as pointed out in Clause 3-25, can produce the result that the insurer is only "...liable to the extent that the loss is not a consequence of the breach, or that the assured has not breached the safety regulation through negligence." Needless to say, only rules issued by a public authority or classification society that are binding on the assured will qualify as safety regulations under Clause 3-22. Public authorities may adopt rules, recommendations or guidelines similar to classification society ice class rules for application in ice-infested waters within their jurisdiction. Whether they are binding on the assured in terms of the insurer's obligation to indemnify for losses depends on whether they are made mandatory. It appears that Finnish and Swedish ice-surveillance authorities issue recommendations but they are not mandatory. Ships are free to not comply with them. Thus, they would not be binding and not be considered as safety regulations for insurance purposes. But if a ship does not comply with the recommendations, it will not be entitled to assistance from a state-owned icebreaker if stuck in ice (Cefor 2013).

In the above context, the official commentary of The Nordic Association of Marine Insurers (Cefor) is pertinent. It states—Sub-clause 3 is new in the 2007 version. The classification society's ice class is a voluntary supplementary classification. Consequently, it is doubtful whether these rules qualify as safety regulations under sub-clause 1. Sub-clause 3 therefore expands the definition of a safety regulation in sub-clause 1 to include ice classes (Cefor Archive 2013).

In light of the above comment, it is the view of the present authors that while ice class notation is voluntary, it becomes a safety regulation once a ship obtains it from the classification society and the ice class rules then become mandatory. Thus, the entry of a vessel into ice-infested waters may lead to a safety regulation contravention if the ship in question has ice class and is not in compliance with its ice class requirements under Clause 3-15 in NMIP 2013. (See Appendix 3).

Given that ice class is acquired voluntarily, class designations do not include ice class notation. Such notation is additional; it simply identifies the vessel and documents the fact that it is designed to operate in specifically described conditions of ice. The ice class is dictated by the thickness of the ice which essentially means that a higher ice class vessel is designed to operate in ice of commensurate thickness. How a vessel is to be operated in ice-infested waters is not regulated as such by the rules of the classification society (Cefor 2013).

In view of the doubt expressed in the official commentary cited above, changes have been effectuated in the 2016 version of NMIP 2013. Cefor has introduced a new clause as an amendment to NMIP clause 3-15 together with removal of sub-clause 3 of clause 3-22 as part of the 2016 revision of the Plan (See Appendix 5). Incidentally, the new wording has been in place since 26 November 2013. The removal of sub-clause 3 of clause 3-22 means that whether the ice class prescribed by the classification society constitutes a safety regulation requiring compliance by the assured depends on how "safety regulation" is defined. Whether the public authorities require the ship to have a certain ice class to allow it to pass through their jurisdiction or the insurer has consented to the ship trading in a conditional trading area subject to a certain ice class, the requirement of ice class will constitute a safety regulation that will apply in addition to any safety regulation that might apply by virtue of Clause 3-22, sub-clause 1 of the Plan.

Following the entry into force of the Polar Code on 1 January 2017, a ship will require the appropriate ice class to enable it to enter into Arctic waters. In such case the ice class requirement will surely fall within the scope of "safety regulation" under the NMIP. In this regard it must be noted that in the Polar Code guidance is provided to facilitate the determination of national equivalency standards using a simplified approach by reference to the newly developed IACS Polar Classes as provided in Paragraph 4.2 of Part I-B of the Polar Code.

Also, the insured must refer to an Ice Regime System on their Polar Ship Certificate explaining how the system or methodology has been applied in their Polar Waters Operation Manual. It must be a system such as the Canadian AIRS System, the Russian Ice Passport System or the newly developed IMO Polar Operational Limitation Assessment Risk Indexing System (POLARIS), recognized by the relevant flag state authority. There is no doubt that the CEFOR guidelines will be amended to reflect this, as will the Lloyd's Market Association Guidelines for Insurers.

Incidentally, in the NMIP, until 2007, there were specific provisions in § 45 relating to unseaworthiness but there was no definition of that term. However, a definition in the Seaworthiness Act 1903 of Norway is stated in negative terms relating to conditions that are virtually the same as those found in English law (Torrens 1994, p. 92). It is important to note that the NMIP no longer has any provisions relating to unseaworthiness because the term is no longer used in the new Norwegian Ship Safety and Security Act of 2007. Amendments to that effect were made to the Plan in 2007.[4]

As discussed above, a violation of the Polar Code under the present legal regime will certainly be considered a violation of a safety regulation in NMIP 2013 which could lead to both sanctions in public law as well as a breach of obligations in the context of marine insurance. The clarity of the position regarding provisions of the Polar Code relating to prevention of loss ostensibly characterized as safety regulations under the NMIP regime is in contrast to the uncertain position of the Code in terms of seaworthiness under English law which, in the view of the present authors, is unfortunate.

3.4 *Third Party Liability in Respect of Environmental Risks and P&I Cover*

In view of the fragility of the Arctic environment, P&I insurance is of particular significance in relation to various risks relating to non-H&M cover, not the least of which is third party liability of the shipowner for pollution damage causing harm to the marine environment (Østreng et al. 2013, p. 331). In the cited publication other risks covered by P&I insurance are mentioned as well. The principal claimants are state entities whose claims emanate from their rights and interests founded in the international law of the sea. This is the essence of environmental risks in the Arctic that the shipowner has to bear if he chooses to venture into Arctic waters. In that vein, he also has to consider the issue of navigating limits from an insurance viewpoint.

Similar to H&M insurance, there are also implications involving navigating limits in respect of P&I cover. Any restrictions imposed on the type of trade or geographical limits of an entered ship are generally endorsed on the member's Certificate of Entry rather than in the rules of the P&I Club but the endorsements must be construed in light of the Club rules. Again, as in the case of H&M insurance, navigation limits are usually regarded as promissory warranties, a breach of which

[4]Preface to the Norwegian Marine Insurance Plan of 1996, Version 2007, item (10), available at http://www.nordicplan.org/Documents/Archive/Plan-2007/Norwegian%20Plan%20of%201996,%20Version%202007%20-%20English.pdf, accessed on 11 October 2016.

relieves the Club from all liability based on the relevant provisions of the MIA (Hazelwood and Semark 2010, p. 202). As mentioned earlier, the Insurance Act 2015 which entered into force in August 2016 has altered this position. Therefore, if the shipowner plans to take his ship into Arctic waters, he has to notify not only his H&M insurer but also his P&I Club.

3.4.1 P&I Clubs

The P&I Clubs are at the centre of third party liability insurance and the UK legislation, the MIA 1906 conspicuously governs P&I insurance as much as it governs H&M insurance. By contrast, it is notable that the Norwegian Clubs are less affected by the NMIP compared with the H&M insurance. It is particularly notable that the Gard Club has certain club rules relating to cover and limits of exposure that are fairly autonomous. The English Clubs are greatly tempered by the MIA because of the Act's legal status but they are free to articulate their own rules provided they are not contrary to the MIA as the governing legislation. As a result, whether the clubs are of English or Norwegian vintage, there is reasonable opportunity for them to create "custom built" cover for shipping in the Arctic sea routes to meet the needs of the shipping industry (Torrens 1994, pp. 129–130).

A fundamental characteristic of P&I clubs is that they operate on a non-profit basis as distinguished from H&M insurance providers who operate in the insurance market. Mutuality is the cornerstone of the P&I system simply because the shipowners as assureds are themselves the members of P&I clubs. The so-called "omnibus rule" is an important feature of P&I insurance which allows a claim to be covered even if it does not fit squarely into the list of specific risks which would allow a member to invoke a P&I claim even if he encountered a situation in an Arctic sea route which was unanticipated. In effect P&I insurance is not bound by any rules of precedent as may be the case with H&M insurance. If there is a dispute between a member and the club in which a certain ship owned by him is entered, dispute resolution would in most cases be carried out through arbitration which would often be in accordance with the so-called "Scott v. Avery rule" (1856), 5 H.L. Cas. 811; 10 E.R. 1121 (H.L.) pursuant to which arbitration would be a condition precedent to court proceedings taken by a party to the agreement. In Norwegian clubs such as Skuld or Gard the rules would in all likelihood stipulate that arbitration would be subject to Norwegian law. Even though there is a sizeable amount of case law in the field of arbitration, cases involving Arctic sea routes would be dealt with through the application of analogous situations (Østreng et al. 2013, p. 331). Notably, in the field of P&I insurance there are no differences in principle between the treatment of claims in Arctic waters and non-Arctic waters. The distinctions would lie in the manner in which liability would arise and the quantum of indemnification involved in the claims. Indeed, it is at present uncertain whether P&I calls in respect of Arctic shipping would be higher than their counterparts in relatively southern waters (Østreng et al. 2013, pp. 131–132). At any rate, there is no evidence that potential liability exposure would be significantly higher in respect of sea routes in the Arctic as compared with other types of cover. Thus, insurers at present view third party liability cover for shipping in the Arctic routes as more of an academic issue although with increases in traffic, this position may be subjected to change; the crucial question being whether the liability risks involved will be more conspicuously enhanced. While it is a matter of speculation, the risks will likely remain static because Arctic traffic will only increase with improved maritime infrastructure which at present is grossly deficient. One characteristic of mutuality in this regard is that the assessment of risk in Arctic sea routes would have to be assessed by all P&I clubs under the auspices of the International Group of P&I Clubs and also the measure of the risk. In other words, the group will assume a risk taken on by an individual member of the group. Finally, it is important to note that with regard to shipping in Arctic sea routes a condition precedent for P&I insurance cover is that in the first place there must be full value H&M cover with respect to the subject ship (Østreng et al. 2013, pp. 132–133).

3.4.2 Indemnification of Pollution Liability

While this work is not concerned with the role of P&I insurance with regard to varieties of non-H&M risks including collision liability, it is pertinent to observe that pollution liability often arises out of a collision, and if it happens in the Arctic, it will concern the present subject matter. It is well known that P&I insurance at least within the English marine insurance regime, has traditionally covered only 25% as the residual portion of collision liability, the lion's share of 75% being covered by H&M insurance under the so-called "running down clause (RDC)". It is evident, however that P&I Clubs are increasingly bearing all 100% of collision liability as a matter of choice as liability in respect of matters falling outside the scope of the RDC such as the costs of salvage of wrecks and wreck removal, and most importantly in the context of this chapter, pollution damage.[5] It is evident that among others, the English and Norwegian Clubs cover oil pollution damage done to other vessels and their cargo as well as other property on board such vessels. This, for example, is provided in Gard Rule 38 (See Gold 2002, p. 455).

Unlike H&M insurance involving the first party proprietary interests of the shipowner, it is unlikely that P&I insurance with its indemnification mechanism for environmental pollution will rely on voluntary action of shipowners. Thus, mandatory insurance, usually in the form of P&I cover, is required in both international and the domestic laws of numerous jurisdictions in respect of potential liability of the shipowner for ship-source pollution damage in respect of Arctic shipping. International conventions concerning such pollution liability apply in Arctic waters

---

[5]Since the 1989 grounding of the Exxon Valdez in the sub-Arctic waters of Alaska, there have been several serious shipping casualties in the Arctic region involving sinkings and groundings and collisions of the Maxim Gorky (1989) near Svalbard the Clipper Adventurer in the NWP and the Nanny in the Canadian Arctic (both in 2010), and the Nordvik (2013), respectively.

falling within the geographical scope of application of these conventions that require ships to carry compulsory third-party liability insurance. The conventions in question are the CLC 1992 and the FUND 1992, Bunkers and HNS Conventions. At the domestic law level, Russia and Canada have enacted relevant laws giving effect to Article 234 of the United Nations Convention on the Law of the Sea, 1982 (UNCLOS) to control pollution in Arctic waters. Through Article 3 of the Federal Law of Shipping on the Water Area of the Northern Sea Route, 2012, the Merchant Shipping Code, 1999 of Russia was amended by adding Article 5.1.4 to the Code requiring vessels navigating in the area of the Northern Sea Route to have in their possession insurance or other financial security relating to civil liability for pollution damage caused by the vessel. The Arctic Waters Pollution Prevention Act, 1970 (AWPPA) of Canada in section 8(1)(d) requires the same before ships are permitted to enter Arctic waters within the Canadian jurisdiction.

The international regime of liability for oil pollution damage had its genesis in the adoption and eventual entry into force of the International Convention on Civil Liability for Oil Pollution Damage, 1969 revised in 1992 (CLC) and subsequently, the International Convention on the establishment of an International Fund for Compensation for Oil Pollution Damage, 1971 revised in 1992 (FUND Convention) as a companion instrument. As indicated, both conventions initially adopted in 1969 and 1971, respectively, were revised and upgraded in 1992 which is the year by which they are identified at present. The conventions were articulated in response to the infamous Torrey Canyon incident which occurred in 1967 when the Liberian tanker by that name ran aground on Seven Stones Reef off the west coast of England.

The CLC is designed to provide for liability and compensation for oil pollution damage on the part of the registered shipowner of the polluting vessel. By contrast, the Fund Convention has been adopted to provide for additional compensation beyond the limit of liability of the shipowner through the establishment of the International Oil Pollution Compensation (IOPC) Fund financed through the imposition of levies on importers of oil, essentially representing the oil industry. The CLC comprises a strict liability regime which means that the claimant as a pollution victim does not have to prove fault on the part of the polluting ship for its owner to be liable under the convention. This chapter is concerned exclusively with the compensation payable by the liable shipowner through a system of compulsory insurance (Hazelwood and Semark 2010, p. 393). As mentioned above, the state whose environment has suffered pollution damage and the cost associated with its mitigation is a major seeker of remedy together with various public and private entities. The convention requirement for compulsory insurance is to ensure that adequate compensation is made directly available to victims of pollution damage up to the shipowner's limit of liability. Viewed together, the object and purpose of the two conventions is to establish a uniform and comprehensive international liability and compensation regime for ship-source oil pollution victims (Xu 2013, p. 107). Notably, there are two other pollution liability conventions, namely, the International Convention on Liability and Compensation for Damage in connection with the Carriage of Hazardous and Noxious Substances by Sea, 1996 (HNS Convention, 1996) governing liability and compensation in relation to pollution damage from hazardous and noxious substances as defined by the convention and the International Convention on Civil Liability for Bunker Oil Pollution Damage, 2001 (Bunkers Convention, 2001) which provides for liability and compensation in respect of oil pollution damage from the bunkers of non-tankers.

Against the above brief background, it is of prime significance in the context of this section of this chapter that the compulsory insurance required by the CLC is provided by the polluting ship's P&I Club. Inevitably, in Arctic waters, the P&I clubs face payment of indemnification associated with enhanced risks due to the presence of ice which has the potential for engendering casualties including collisions and groundings and consequential pollution in an environment that is at once ecologically fragile and where the pollutant does not readily dissipate.

Most notably, the convention regimes mentioned above provide for two significant features. One relates to the requirement of certificates evidencing financial responsibility undertaken by the registered shipowner. In other words, in respect of each ship of a shipowner to which a particular convention such as the CLC applies, must be in possession of a certificate known as a certificate of insurance or security without which the ship may be prevented from entering a port or offshore installation of a convention state. In practical terms with respect to the CLC, the owner of a tanker will be issued a "Blue Card" upon application to the ship's P&I club which is evidence to a state party to the CLC that the ship is covered for pollution liability by insurance as required by the convention. This enables the flag state of the ship to issue the CLC certificate which is required by the convention. Usually a CLC state party will not issue a CLC certificate without a "Blue Card" from the P&I club (Hazelwood and Semark 2010, p. 398). Indeed, quite apart from the convention regimes, such certificates are required by virtue of the national legislation of a state. In the Arctic region, the states in question undoubtedly require such certificates regardless of whether they are parties to pollution conventions. As mentioned earlier, P&I clubs have traditionally covered their members for pollution caused by vessels, and therefore they would be the providers of such convention certificates or certificates required under national legislation (Hazelwood and Semark 2010, p. 398).

The other essential feature is the right of direct action against the insurer who provides the certificate of insurance or financial security required by the relevant convention. This means that a claimant who is the victim of pollution damage, be it cargo or bunker oil or HNS, can proceed directly against the insurer, that is the P&I Club, for payment of compensation. These two important elements apply to ships traversing the waters of the Arctic in the same manner as they would apply to ships in other waters. The combined effect of the provision of compulsory insurance together with issuing of a convention certificate and the direct action against the P&I Club is expressed in the following words:

> ... to the extent P&I clubs agree to provide such certificates, they relinquish their right to rely on the "pay-to-be-paid" rule under the first party indemnity principle and other defenses available to them under their Rules, or which may otherwise have been available in the event the member concerned had breached a condition or warranty (Hazelwood and Semark 2010, p. 398).

As in any other waters, if there is an oil spill in Arctic waters, as a practical matter, the P&I Club of the ship acting on behalf of the shipowner will, as soon as possible meet with the representative of the IOPC Fund to sort out the details regarding making compensation available to victims with immediate effect based on their respective liabilities under the conventions. Speed of functionality in respect of both entities would be crucial in the event of a catastrophe in the Arctic. It is therefore imperative that the responsible parties act in due haste in consonance with the operational teams which will be on site to mitigate the pollution damage and prevent it from spreading.

4 Environmental Salvage in the Arctic and Its Marine Insurance Implications

It is no intention of this chapter to delve deep into the legal regime of marine salvage but it is recognized that it is a crucial element associated with casualties. As such, salvage has an impact on the safety and environmental implications of shipping in the Arctic in a more pronounced way than in non ice-infested waters because salvage operations in the Arctic environment are more difficult and costly to conduct. As discussed above, violations of the Polar Code can lead to casualties which in turn would concern indemnifiability under marine insurance law. In this respect, the role of salvage is indispensable from the viewpoint of marine insurance especially in terms of the measure of indemnity. Thus, in the inter-link between the Polar Code and the relevant marine insurance contract, salvage is a significant element which cannot be overlooked.

In this chapter, in the context of salvage, the focus is on protection of the marine environment which involves the applications of relevant provisions relating to the environment, particularly Article 14 of the International Salvage Convention which deals with the phenomenon of special compensation payable to salvors to recover their costs in cases where they mitigate pollution damage but are unable to collect a salvage reward. While it is beyond the scope of this work to enter into a critical analysis of Article 14 and the associated case law, their importance must be noted in view of the emerging concept of environmental salvage. A good elaboration of the concept of environment salvage can be found in academic literature on the subject. (See for example, Bishop 2012a, b, pp. 65–105) Whether or not a new and separate regime of environmental salvage should be introduced is a matter of contemporary interest and is understandably conducive to debate and discussion. Whatever may be the final outcome, there will undoubtedly be implications for the indemnifiability of salvage charges in respect of Arctic waters given its particularly sensitive marine environment and the need to prevent and mitigate potential pollution damage from shipping accidents.

5 Summary and Conclusion

In this chapter the authors have attempted to provide an overview of marine insurance implications for shipping in the Arctic in relation to safety and environmental concerns. Given the expansive nature of the law and practice of marine insurance not all areas of marine insurance law, even if they may be relevant to Arctic shipping, have been within the contemplation of the authors. The scope of discussion has been limited to selected areas of focus to allow for as much narrow analytical treatment as possible. Thus, the law and practice regarding cargo insurance in Arctic transportation has been left unaddressed. In terms of a central theme, it has been considered expedient to examine from a comparative law perspective, how provisions pertaining to risks in the Arctic are dealt with under English and Norwegian law through the respective instrumentalities of the English Institute Clauses and Warranties in conjunction with the MIA and the corresponding Norwegian counterpart, the NMIP.

To begin with, in recognition of the antiquity of marine insurance, the fundamental age-old principles and legal doctrines governing this discipline have been introduced in synoptic form to provide a backdrop for the ensuing discussion. Next the emergence and recent adoption of the Polar Code under IMO auspices has been addressed with regard to its status as international regulatory law and its impact on indemnifiability of risks faced by ships navigating in Arctic waters. In particular, the thorny issue of seaworthiness is examined with a view to determining by analogy with the ISM Code whether non-compliance with the Polar Code constitutes unseaworthiness for the purposes of the implied warranty found in marine insurance contracts pursuant to the MIA in the English legal system and whether it amounts to a violation of a safety regulation under the Norwegian NMIP regime. No definitive conclusion can be drawn in this regard in terms of the English law in the absence or insufficiency of case law given the fact that neither the Codes nor their parent conventions actually set standards for seaworthiness.

The seaworthiness issue is discussed in the context of marine insurance implications for enhanced risks in Arctic shipping. The operations of the relevant Institute Clauses and Warranties and the IHC, and their counterpart provisions in the NMIP regime are examined in contextual detail and a comparative analysis of the two regimes is presented as the central focus of the chapter. The examination of the English marine insurance regime delves into the legal implications of "held-covered" clauses in view of limitations and prohibitions appearing in H&M policies in respect of entry of ships into Arctic waters. The NMIP regime does not have warranties or "held-covered" clauses as in the English regime but consists of numerous provisions pertaining to navigating or trading limits including the Arctic and safety regulations for prevention of loss. Sub-clause 3 therefore expands the definition of a safety regulation in sub-clause 1 to include ice classes. Doubt has been expressed in the Cefor Commentary regarding whether ice class rules qualify as safety regulations under Clause 3-22 of NMIP 2013. This has been dealt with by deleting the ice class requirement in the new Cefor Trading Areas Clause and NMIP

2016. The provisions are germane to vessels purporting to operate in Arctic waters and seeking marine insurance coverage for enhanced risks.

Following this detailed discussion, the chapter addresses the role of P&I Clubs and third party liability in respect of pollution damage centering on Arctic environmental peculiarities characterized by the presence of ice. An introduction to P&I insurance is presented followed by the requirement for compulsory insurance or evidence of financial security by relevant international ship-source pollution conventions as well as national jurisdictions exemplified by Canadian and Russian legislation. It is recognized that the role of the P&I Club is crucial to indemnification of pollution damage which is the second prong of the overall legal regime of marine insurance. The implications of salvage law in relation to marine insurance are briefly presented emphasizing the environmental dimension of salvage and mention is made of the emerging concept of environmental salvage which connects to the central focus of the chapter. It is envisaged that the efforts of the authors in providing an exposé of Arctic shipping that is at once contemporary and topical infused with the traditional subject matter of marine insurance, albeit one that is assuming new orientations, will evoke and instill interest in the minds of all involved in safety and sustainability in Arctic shipping.

Appendix 1: Institute Warranties 1976

1. Warranted no:-
   - (a) Atlantic Coast of North America, its rivers or adjacent islands,
     - (i) north of 52° 10′ N. Lat. and west of 50° W. Long.;
     - (ii) south of 52° 10′ N. Lat. in the area bounded by lines drawn between Battle Harbour/Pistolet Bay; Cape Ray/Cape North; Port Hawkesbury/Port Mulgrave and Baie Comeau/Matane, between 21st December and 30th April both days inclusive.
     - (iii) west of Baie Comeau/Matane (but not west of Montreal) between 1st December and 30th April both days inclusive.
   - (b) Great Lakes or St. Lawrence Seaway west of Montreal.
   - (c) Greenland Waters.
   - (d) Pacific Coast of North America its rivers or adjacent islands north of 54° 30′ N. Lat., or west of 130° 50′ W. Long.
2. Warranted no Baltic Sea or adjacent waters east of 15° E. Long.
   - (a) North of a line between Mo (63° 24′ N. Lat.) and Vasa (63° 06′ N. Lat.) between 10th December and 25th May b.d.i.
   - (b) East of a line between Viipuri (Vyborg) (28° 47′ E. Long.) and Narva (28° 12′ E. Long.) between 15th December and 15th May b.d.i.

(c) North of a line between Stockholm (59° 20′N. Lat.) and Tallinn (59° 24′N. Lat.) between 8th January and 5th May b.d.i.
(d) East of 22° E. Long, and south of 59° N. Lat. between 28th December and 5th May b.d.i.

3. Warranted not North of 70° N. Lat. other than voyages direct to or from any port or place in Norway or Kola Bay.
4. Warranted no Behring Sea, no East Asian waters north of 46° N. Lat. and not to enter or sail from any port or place in Siberia except Nakhodka and/or Vladivostock.
5. Warranted not to proceed to Kerguelen and/or Croset Islands or south of 50° S. Lat., except to ports and/or places in Patagonia and/or Chile and/or Falkland Islands, but liberty is given to enter waters south of 50° S. Lat., if en route to or from ports and/or places not excluded by this warranty.
6. Warranted not to sail with Indian Coal as cargo:-

(a) between 1st March and 30th June, b.d.i.
(b) between 1st July and 30th September, b.d.i., except to ports in Asia, not West of Aden or East of or beyond Singapore.

Appendix 2: Clause 32 Navigating Limits of International Hull Clauses 2003

Unless and to the extent otherwise agreed by the Underwriters in accordance with Clause 33 below, the vessel shall not enter, navigate or remain in the areas specified below at any time or, where applicable, between the dates specified below (both days inclusive):

Area 1—Arctic

(a) North of 70°N. Lat.
(b) Barents Sea

except for calls at Kola Bay, Murmansk or any port or place in Norway, provided that the vessel does not enter, navigate or remain north of 72°30′ N. Lat. or east of 35° E. Long.

Area 2—Northern Seas

(a) White Sea.
(b) Chukchi Sea.

Area 3—Baltic

(a) Gulf of Bothnia north of a line between Umea (63° 50′ N. Lat.) and Vasa (63° 06′ N. Lat.) between 10th December and 25th May.
(b) Where the vessel is equal to or less than 90,000 DWT, Gulf of Finland east of 28° 45′ E. Long. between 15th December and 15th May.

(c) Vessels greater than 90,000 DWT may not enter, navigate or remain in the Gulf of Finland east of 28° 45′ E. Long. at any time.
(d) Gulf of Bothnia, Gulf of Finland and adjacent waters north of 59° 24′ N. Lat. between 8th January and 5th May, except for calls at Stockholm, Tallinn or Helsinki.
(e) Gulf of Riga and adjacent waters east of 22° E. Long. and south of 59° N. Lat. between 28th December and 5th May.

Area 4—Greenland
Greenland territorial waters.
Area 5—North America (east)

(a) North of 52° 10′ N. Lat. and between 50° W. Long. and 100° W. Long.
(b) Gulf of St. Lawrence, St. Lawrence River and its tributaries (east of Les Escoumins), Strait of Belle Isle (west of Belle Isle), Cabot Strait (west of a line between Cape Ray and Cape North) and Strait of Canso (north of the Canso Causeway), between 21st December and 30th April.
(c) St. Lawrence River and its tributaries (west of Les Escoumins) between 1st December and 30th April.
(d) St. Lawrence Seaway.
(e) Great Lakes.

Area 6—North America (west)

(a) North of 54° 30′ N. Lat. and between 100° W. Long. and 170° W. Long.
(b) Any port or place in the Queen Charlotte Islands or the Aleutian Islands.

Area 7—Southern Ocean
South of 50°S. Lat. except within the triangular area formed by rhumb lines drawn between the following points

(a) 50° S. Lat.; 50° W. Long.
(b) 57° S. Lat.; 67° 30′ W. Long.
(c) 50° S Lat.; 160° W. Long.

Area 8—Kerguelen/Crozet
Territorial waters of Kerguelen Islands and Crozet Islands.
Area 9—East Asia

(a) Sea of Okhotsk north of 55° N. Lat. and east of 140° E. Long. between 1st November and 1st June.
(b) Sea of Okhotsk north of 53° N. Lat. and west of 140° E. Long. between 1st November and 1st June.
(c) East Asian waters north of 46° N. Lat. and west of the Kurile Islands and west of the Kamchatka Peninsula between 1st December and 1st May.

Area 10—Bering Sea
Bering Sea except on through voyages and provided that

(a) the vessel does not enter, navigate or remain north of 54° 30′ N. Lat.; and

(b) the vessel enters and exits west of Buldir Island or through the Amchitka, Amukta or Unimak Passes; and
(c) the vessel is equipped and properly fitted with two independent marine radar sets, a global positioning system receiver (or Loran-C radio positioning receiver), a radio transceiver and GMDSS, a weather facsimile recorder (or alternative equipment for the receipt of weather and routeing information) and a gyrocompass, in each case to be fully operational and manned by qualified personnel; and
(d) the vessel is in possession of appropriate navigational charts corrected up to date, sailing directions and pilot books.

Appendix 3: Clause 3-15 Trading Areas of Nordic Marine Insurance Plan 2013

The ordinary trading area under the insurance comprises all waters, subject to the limitations laid down in the Appendix to the Plan as regards conditional and excluded areas. The person effecting the insurance shall notify the insurer before the ship proceeds beyond the ordinary trading limit.

The ship may continue to sail in the conditional trading areas, the insurer may require an additional premium and may also stipulate other conditions. If damage occurs while the ship is in a conditional area with the consent of the assured and without notice having been given, the claim shall be settled subject to a deduction of one fourth, maximum USD 200,000. The provision in Cl. 12-19 shall apply correspondingly.

If the ship proceeds into an excluded trading area, the insurance ceases to be in effect, unless the insurer has given permission in advance, or the infringement was not the result of an intentional act by the master of the ship. If the ship, prior to expiry of the insurance period, leaves the excluded area, the insurance shall again come into effect. The provision in Cl. 3-12, sub-clause 2, shall apply correspondingly.

Appendix 4: Clause 3-22 Safety Regulations of Nordic Marine Insurance Plan 2013

A safety regulation is a rule concerning measures for the prevention of loss, issued by public authorities, stipulated in the insurance contract, prescribed by the insurer pursuant to the insurance contract, or issued by the classification society.

Periodic surveys required by public authorities or the classification society constitute a safety regulation under sub-clause 1. Such surveys shall be carried out before expiry of the prescribed time-limit.

The rules prescribed by the classification society regarding ice class constitute a safety regulation under sub-clause 1.

Appendix 5: Cefor Trading Areas Clause: Clause to Replace Clause 3-15 and Clause 3-22, Sub-Clause 3 of the Nordic Marine Insurance Plan of 2013

Instead of Clause 3-15—Trading areas the following clause shall apply:

The ordinary trading area under the insurance comprises all waters, subject to the limitations laid down in the Appendix to the Plan as regards conditional and excluded areas. The person effecting the insurance shall notify the insurer before the ship proceeds beyond the ordinary trading area.

The insurer may consent to trade outside the ordinary trading area and may require an additional premium. The insurer may also stipulate other conditions which shall constitute safety regulations cf. Cl. 3-22 and Cl. 3-25, sub-clause 1 of the Plan.

The vessel is held covered for trade in the conditional trading areas, but if damage occurs while the ship is in a conditional area with the consent of the assured and without notice having been given, the claim shall be settled subject to a deduction of one fourth, maximum USD 200,000. The provision in Cl. 12-19 of the Plan shall apply correspondingly. If claims arising out of ice damage are a result of the assured's failure to exercise due care and diligence, further reduction of the claim may be made based on the degree of the assured's fault and the circumstances generally.

If the insurer has been duly notified in accordance with sub-clause 1 of trade within the conditional trading areas, the insurance remains in full force and effect, subject to compliance with conditions, if any, stipulated by the insurer.

If the ship proceeds into an excluded trading area, the insurance ceases to be in effect unless the insurer has given his consent in advance, or the infringement was not the result of an intentional act by the master of the ship. If the ship, prior to expiry of the insurance period, leaves the excluded area, the insurance shall again come into effect. The provision in Cl. 3-12, sub-clause 2 of the Plan, shall apply correspondingly.

Cl. 3-22, sub-clause 3 of the Plan shall not apply.

References

Andrewatha, J., & Stone, Z. (2004). English maritime law update. *Journal of Maritime Law and Commerce, 35*, 369–404.

Bennett, H. (2006). *The law of marine insurance* (2nd ed.). Oxford: Oxford University Press.

Bishop, A. (2012a). Environmental salvage: Time for a change. In B. Soyer & A. Tettenborn (Eds.), *Pollution at sea; Law and liability*. London: Informa.

Bishop, A. (2012b). The development of environmental salvage and review of the London Salvage Convention 1989. *Tulane Maritime Law Journal, 37*(1), 65–105.

Brigham, L. (2013). *IMO Polar Code for ships operating in polar waters.* Produced by Arctic Climate Change Economy and Society (ACCESS), Policy Brief No.4. Retrieved October

11, 2016, from http://www.aeco.no/wp-content/uploads/2013/06/2015-04-ACCESS-PolicyBrief-Polar-Code.pdf

Cefor. (2013). *Commentary on Cefor Trading Areas Clause: Clause to replace Clause 3-15 and Clause 3-22, sub-clause 3 of the Nordic Marine Insurance Plan of 2013*. Retrieved October 11, 2016, from http://www.cefor.no/Documents/Clauses/Hull/CeforClause-TradinAreas-Commentary13-11-26.pdf

Cefor Archive. (2013). *Commentary to sub-clause 3 of Clause 3-22 in the Nordic Marine Insurance Plan of 2013*. Retrieved October 11, 2016, from http://archive.nordicplan.org/Commentary/Part-One/Chapter-3/Section-3/#-3-22

Chircop, A. (2009). The growth of international shipping in the Arctic: Is a regulatory review timely? *The International Journal of Marine and Coastal Law, 24*(2), 355–380.

Gold, E. (2002). *Gard handbook on P&I insurance* (5th ed.). Arendal: Assuranceforeningen Gard.

Gold, E., Kindred, H., & Chircop, A. (2003). *Canadian maritime law*. Toronto: Irwin Law.

Hazelwood, S. J., & Semark, D. (2010). *P&I clubs law and practice* (4th ed.). London: Lloyd's List.

Hodges, S. (1996). *The law of marine insurance*. London: Cavendish Publishing Limited.

Honka, H. (2004). Main obligations and liabilities of the carrier. *Transport Review, 27*, 278–283.

IMO Media Center. *"Shipping in Polar Waters"*. Retrieved October 11, 2016, from http://www.imo.org/MediaCentre/HotTopics/polar/Pages/default.aspx

Johannsdottir, L., & Cook, D. (2014). *An insurance perspective on Arctic opportunities and risks: Hydrocarbon exploration and shipping*. Working Paper for Institute for International Affairs of University of Iceland. Retrieved October 11, 2016, from http://ams.hi.is/wp-content/uploads/2015/04/An_Insurance_Perspective_PDF.pdf

Khurram, R. (1994). Total loss and abandonment in the law of marine insurance. *Journal of Maritime Law & Commerce, 25*, 95–118.

Lajeunesse, A. (2012). A new mediterranean? Arctic shipping prospects for the 21st century. *Journal of Maritime Law and Commerce, 43*(4), 521–537.

Liu, H. (2016). Arctic marine insurance: Towards a new risk coverage regime. *Journal of Maritime Law and Commerce, 47*(1), 77–100.

Mukherjee, P. K. (2002). *Maritime legislation*. Malmo: WMU Publications.

Normann, A. K., & Mikkelsen, E. (2014). *The role of insurance for use of the Northern Sea Route*. Northern Research Institute (Norut) Publication. Retrieved September 10, 2016, from http://www.arcticfrontiers.com/downloads/arctic-frontiers-2014/poster-presentations-3/part-iii-shipping-a-offshore-in-the-arctic-2/610-anne-normann-and-eirik-mikkelsen/file

Østreng, W., Eger, K. M., Fløistad, B., Jørgensen-Dahl, A., Lothe, L., Mejlænder-Larsen, M., et al. (2013). *Shipping in Arctic waters, a comparison of the Northeast, Northwest and trans polar passages*. Heidelberg: Springer-Praxis.

Reddie, J. (1841). *An historical view of the law of maritime commerce*. London: William Blackwood & Sons.

Rose, F. (2012). *Marine insurance law and practice* (2nd ed.). London: Informa.

Sarrabezoles, A., Lasserre, F., & Hagougn'rin, Z. (2014). Arctic shipping insurance: Towards a harmonisation of practices and costs? *Polar Record, 52*(4), 1–6.

Schroeder, J. U., Mejia, M. Q., Jr., Mukherjee, P. K., Manolis, F. M., & Dreessen, S. (2006). Potential consequences of imprecise security assessments. *IAMU Journal, 4*(2), 31–38.

Soyer, B. (2001). *Warranties in marine insurance*. London: Cavendish Publishing Ltd.

Tamvakis, M., Granberg, A., & Gold, E. (1999). Economy and commercial viability of the Northern Sea Route. In W. Østreng (Ed.), *The natural and societal challenges of the Northern Sea Route: A reference work*. Dordrecht: Kluwer Academic Publishers.

The Nordic Association of Marine Insurers (Cefor). (2007). *Norwegian Marine Insurance Plan of 1996, Version 2007, item (10)*. Retrieved October 11, 2016, from http://www.nordicplan.org/Documents/Archive/Plan-2007/Norwegian%20Plan%20of%201996,%20Version%202007%20-%20English.pdf

Thomas, M., & Steel, D. (1976). *Temperley's merchant shipping acts* (7th ed.). London: Stevens & Sons.

Torrens, D. L. (1994). *Marine insurance for the Northern Sea Routes pilot study*. INSROP Working Paper IV 3.3, NO.1-1994.

UK Parliament. (2015). *Explanatory Notes*. Insurance Bill [HL]. Retrieved October 11, 2016, from http://www.publications.parliament.uk/pa/bills/lbill/2014-2015/0039/en/14039en.htm

Xu, J. (2013). The international legal framework governing liability and compensation for ship-source oil pollution damage. In M.Q. Mejia, Jr. (Ed.), *Selected issues in maritime law and policy: Liber Amicorum Proshanto K. Mukherjee*. Hauppauge, NY: Nova Science Publishers.

Suggested Reading

Arctic Council. (2009). *Arctic marine shipping assessment 2009 report, second printing*. Retrieved October 11, 2016, from http://www.pmel.noaa.gov/arctic-zone/detect/documents/AMSA_2009_Report_2nd_print.pdf

European Commission Maritime Affairs and Fisheries. (2010). *Legal aspects of Arctic shipping: Summary report*. Luxembourg: Publications Office of the European Union.

Gold, E. (1989). Marine salvage: Towards a new regime. *Journal of Maritime Law and Commerce, 20*(4), 487–503.

Marsh Report. (2014). *Arctic shipping: Navigating the risks and opportunities*.

Mukherjee, P. K. (2006). Refuge and salvage. In A. Chircop & O. Linden (Eds.), *Places of refuge for ships*. Leiden: Martinus Nijhoff.

The Nordic Association of Marine Insurers (Cefor). (2013a). *Commentary on Cefor Trading Areas Clause: Clause to replace Clause 3–15 and Clause 3–22, sub-clause 3 of the Nordic Marine Insurance Plan of 2013*. Retrieved October 11, 2016, from http://www.cefor.no/Documents/Clauses/Hull/CeforClause-TradinAreas-Commentary13-11-26.pdf

The Nordic Association of Marine Insurers (Cefor). (2013b). *Commentary on Clause 3–22 of the Nordic Marine Insurance Plan 2013*. Retrieved October 11, 2016, from http://www.nordicplan.org/Commentary/

Arctic High Seas Governance of Biodiversity

Torsten Thiele

Contents

Abstract On June 19, 2015, following a long period of preparation, the UN General Assembly adopted Resolution A/69/L.65: 65 "Development of an international legally-binding instrument under the United Nations Convention on the Law of the Sea on the conservation and sustainable use of marine biological diversity in areas beyond national jurisdiction". A preparatory committee will develop draft recommendations in 2016 and 2017. The proposed new instrument will have important implications for the areas beyond national jurisdiction, including the Central Arctic Ocean and therefore for the Arctic governance regime overall. Key components of the "package" of measures discussed during the sessions of the Working Group were area-based management tools, including MPAs; marine genetic resources, including questions related to the sharing of benefits; environmental impact assessments and capacity-building and technology transfer. The potential implication of such a new legal instrument on areas beyond national

T. Thiele (✉)
Institute for Advanced Sustainability Studies e.V., Potsdam, Germany
e-mail: torsten.thiele@iass-potsdam.de; torsten@globaloceantrust.com

L. P. Hildebrand et al. (eds.), *Sustainable Shipping in a Changing Arctic*, WMU Studies in Maritime Affairs 7, https://doi.org/10.1007/978-3-319-78425-0_13

jurisdiction in the Arctic will be manifold. They will affect shipping and other marine operations. Arctic nations have expressed initial views on the proposed measures but it will in the end be a decision of the international community as a whole to decide on the details of the new Implementing Agreement which will then provide a binding regime for all High Seas areas, including the Central Arctic Ocean.

Keywords Arctic governance · High seas biodiversity agreement · United Nations general assembly resolution · Areas beyond national jurisdiction · Law of the sea · Central Arctic Ocean

Abbreviation

UNGA United Nations General Assembly

1 Introduction to the Governance of the Central Arctic Ocean Under the Law of the Sea

Global ocean governance is based on United Nations Convention on the Law of the Sea, 1982 (hereinafter referred to as UNCLOS), often described as the constitution for the ocean (Koh 2013). UNCLOS thus provides the fundamental governance regime for the Arctic Ocean (Young 2016). UNCLOS is evolving to address major new challenges, including those to marine ecosystems and their biological diversity resulting from human pressures (Druel and Gjerde 2014). The Central Arctic Ocean beyond 200 nautical miles lies outside national jurisdictions and falls therefore under the High Seas, regulated in Part VII. of UNCLOS.

Marine areas beyond national jurisdiction encompass almost half of the surface of the earth, their biodiversity is largely unprotected yet under threat (Global Ocean Commission 2014). The 7th Conference of Parties (hereinafter referred to as COP) of the Convention on Biological Diversity (hereinafter referred to as CBD) in 2010 agreed in Aichi Target 11 to conserve at least 10% of coastal and marine areas, this goal has not yet been reached and there is a need for larger ocean areas under protection. A very small percentage of High Seas areas is afforded protection at present (IASS 2013).

The CBD has specifically highlighted the multi-year ice and associated marine habitats of the central Arctic Ocean as Ecologically or Biologically Significant Areas (hereinafter referred to as EBSAs) and the Marginal Ice Zone and the Seasonal Ice-Cover Over the Deep Arctic Ocean as unique habitats (CBD 2016). Marine scientists working through the CBD EBSA-process have identified at least 30% of the global ocean as in need of special protection, taking into account connectivity and representativeness (Abdullah et al. 2014). This position was confirmed in the

"Promise of Sydney" declaration approved at the IUCN World Parks Congress in 2014 (IUCN 2014b) and again in motion 53 of the IUCN World Conservation Congress in 2016. Thus the Arctic marine environment is in particular need of good governance (Weidemann 2014).

Furthermore, as anthropogenic climate change effects the Arctic disproportionately, global civil society has similarly raised concerns for the fragile Arctic biodiversity, raising concerns about a potential race for Arctic resources and calling for a moratorium on activities in the central Arctic (http://www.arcticdeclaration.org).

Arctic nations through the Arctic Council, an intergovernmental forum, and its working group on the Protection of the Arctic Marine Environment (PAME) are undertaking a number of initiatives to understand the ecosystems of the Arctic better and to identify the challenges ahead (PAME 2016). PAME's mandate is to address policy and non-emergency pollution prevention and control measures related to the protection of the Arctic marine environment from both land and sea-based activities (PAME 2015).

Beyond the efforts of Arctic nations themselves to address these issues, the developing global environmental and legal governance regime plays a distinctive, relevant and important role in guiding the governance of the central Arctic Ocean (Gjerde et al. 2016). This is particularly crucial at a time where the link between ocean and climate change is increasingly clear (Poertner et al. 2014) and evidence of its consequences in terms of ocean warming and acidification are detected in the Arctic Ocean.

The Central Arctic Ocean (CAO), international waters more than 200 nautical miles from any coast (Pan and Huntington 2016).

2 UNGA Resolution A/69/L.95

Adoption of Resolution A/69/L.65: "Development of an international legally-binding instrument under the United Nations Convention on the Law of the Sea on the conservation and sustainable use of marine biological diversity in areas beyond national jurisdiction" concludes a 9-year process undertaken by the United Nations (hereinafter referred to as UN) General Assembly Ad Hoc Open-ended Informal Working Group (UN 2015) to identify the needs and requirements to close the gaps in existing global high-seas biodiversity governance (hereinafter referred to as HSA) (HSA 2014). Whilst it will take several more years to create the proposed legally-binding instrument now is the time to consider the potential impacts of such a new agreement on the Arctic and its large High Seas area.

Article 1 of the Resolution outlines the process to reach a future agreement: A preparatory committee working under a consensus approach and meeting in 2016 and 2017, which will be open to all States Members of the United Nations, members of the specialized agencies and parties to the Convention, with others invited as observers in accordance with past practice, will be working to make substantive recommendations to the General Assembly on the elements of a draft text of an international legally-binding instrument. This approach will allow interested states as well as relevant agencies such as the International Maritime Organisation and observers such as IUCN and others to participate, adding to the transparency and legitimacy of the process (Hubert 2015). The final text is due to be submitted to an international conference for negotiations as described in the letter from the Co-chairs (Kohona and Lijnzaad 2015).

Article 2 refers to the "package" agreed in 2011 in the Working Group, namely the conservation and sustainable use of marine biological diversity of areas beyond national jurisdiction, in particular, together and as a whole, marine genetic resources, including questions on the sharing of benefits, measures such as area-based management tools, including Marine Protected Areas (hereinafter referred to as MPA), environmental impact assessments and capacity-building and the transfer of marine technology. All of these aspects will need to be reflected in the final document.

Area-based management measures will require interaction of such new designations with existing arrangements, be they for fisheries through regional fisheries management organisations, or for shipping under designations undertaken by the International Maritime Organization (hereinafter referred to as IMO). The new instrument may clarify and elaborate the duty to cooperate enshrined in UNCLOS, for example through a duty to support conservation measures and governance principles agreed to under the new instrument, accompanied by reporting requirements (UNGA 2015).

Article 3 recognizes that the process should not undermine existing relevant legal instruments and frameworks and relevant global, regional and sectoral bodies. The IMO as the competent international body for the regulation of international shipping will continue to be the primary regulator of shipping activities and in May of 2015 adopted the environmental part of the Polar Code and associated International Convention for the Prevention of Pollution from Ships, 1973 (hereinafter referred to as MARPOL) amendments to make the Code mandatory. The IMO Polar Code will come into force on 1 January 2017 and the Central Arctic Ocean will be covered entirely under the Code's rules and regulations.

The new implementing instrument will be based on a consistent application of the precautionary principle. Thus the designation of marine protection will need to reflect ecosystem-diversity of all species, genes and habitats and connectivity, taking into account all stages of marine life, from larval through juvenile to adult (Gjerde et al. 2016).

3 Implications of the Proposed New Regime for Shipping and Marine Operations in the Central Arctic Ocean

Measures agreed under the new instrument, including area-based measures, could potentially affect the Central Arctic High Seas. The marine protected area approach could for instance be applied to the Central Arctic ocean (Delfour-Samama 2014).

Protecting an entire High Seas area of particular biodiversity value has also been proposed and can have many benefits (White and Costello 2014).

Similarly the discussion of environmental impact assessments, including strategic and cumulative assessments, will be of relevance for Arctic activities (IUCN 2014a). In addition to other impacts from commercial activities ocean noise is increasingly described as pollution, with serious impacts for instance on marine mammals (Zitterbart et al. 2013). Addressing potential cumulative environmental impacts in a comprehensive, transparent assessment, taking into account the views of all stakeholders, will be necessary to fully reflect the precautionary principle at the core the proposed agreement.

An emerging chapter of the negotiation package covers technology transfer, with a number of approaches under discussion (Thiele and Harden-Davies 2016). Training at regional and local levels in the Arctic will be needed to enable capacity building and technology transfer to strengthen the abilities of regional organizations (Rochette et al. 2015). Thus this aspect may be of particular benefit to indigenous communities and other remote settlements in the Arctic.

Marine genetic resources and the benefit sharing of information gathered through bio-prospecting is another complex issue, reflecting increased awareness of the potentially significant medical breakthroughs that could result from genetic analysis of extremophiles. Given the unique conditions in the Polar marine regions these areas could be of particular interest for this research.

The proposed new institutional framework may include a conference of parties and subsidiary bodies such as a scientific, environmental and technical advisory committee to address concerns arising due to the differences in existing governance regimes such as those for the maritime shipping. Several other agreements and institutions supplement the UNCLOS framework and have direct application to the Arctic (Becker 2010).

Thus the impact of the potential new governance regime on the High Seas of the Arctic is likely to be manifold and needs to reflect both regional and global concerns (Klein et al. 2015). In the following, a number of examples of such effects are discussed in more detail.

3.1 Oslo-Paris Convention (OSPAR) Example

A practical example of how shipping regulations may under the new instrument interact with areas covered under specific protection is provided by the interplay we see today between the IMO code and the OSPAR Convention, 1992 (Oslo-Paris Convention) (hereinafter referred to as OSPAR) (OSPAR 1992). Where the OSPAR Commission considers that an action for the protection and conservation of the North-East Atlantic is necessary in relation to questions concerning maritime transport, it draws those questions to the attention of the IMO (Johnson 2013). This approached is supported through an Agreement of co-operation between OSPAR and IMO. OSPAR Contracting Parties also cooperate among each other on such issues within the IMO (OSPAR 2010).

Area-based measures may then additionally be applied to protect identified regions (Molenaar and Oude Elferink 2013). The OSPAR High Seas MPAs offer interesting comparative cases. This reflects the complex legal and political characteristics governing the sites and might inform initiatives in other marine regions or at the global level (Rochette et al. 2014).

3.2 IMO Example

The shipping industry is of major and increasing relevance in the Arctic Ocean overall and is globally governed by the IMO (Johansson 2015; AMSA 2015a). One particular mechanism that the IMO already has in place to address marine environmental protection is the process to identify Particularly Sensitive Sea Areas (hereinafter referred to as PSSAs) (IMO 2006). Guidelines on designating a PSSAs are contained in IMO resolution A.982(24) Revised guidelines for the identification and designation of Particularly Sensitive Sea Areas (PSSAs). Whereas the establishment of MPAs under OSPAR has generally focused on the identification of potential damages first, the IMO's PSSA guidelines are designed to provide specific anti-pollution measures. A PSSA is an area that needs special protection through action by IMO because of its significance for recognized ecological or socio-economic or scientific reasons and which may be vulnerable to damage by international maritime activities (IMO 2006). When an area is approved as a PSSA, specific measures can be used to control the maritime activities in that area, such as routing measures, strict application of MARPOL discharge and equipment requirements for ships, such as oil tankers; and installation of Vessel Traffic Services (VTS) (IMO 2017).

An interesting example of the potential linkages between MPAs and the PSSA approaches is provided in the Mediterranean Pelagos MPA (Mayol et al. 2013). The Pelagos MPA aims to protect marine mammals.

The specific progress made by the IMO in terms of Arctic pollution control through the adoption of the Polar Code is covered elsewhere in this book.

3.3 An Example of the Regional Approach

A regional approach to addressing environmental challenges in the global ocean has also been advocated as an effective means of implementation (Toepfer et al. 2014). Regional initiatives can also be seen to act to support globally coordinated solutions (Visbeck et al. 2014).

An example for a regional initiative is provided in the Arctic by the Declaration Concerning the Prevention of Unregulated High Seas Fishing in the Central Arctic

Ocean (2015). The five Arctic Ocean coastal states as signatories "recognize the crucial role of healthy marine ecosystems" and "agree to promote international compliance" by "coordinating our monitoring, control and surveillance activities". The Declaration contains a number of non-legally binding commitments prior to the ultimately binding legal regime only UNCLOS can provide (Ryder 2015).

The uniform Arctic safety and environmental protection regulatory regime proposed by the Arctic Marine Shipping Assessment Recommendations (AMAP 2015a) and the binding obligations entered into under the 2011 Search and Rescue Agreement and the 2013 Agreement on Cooperation on Marine Oil Pollutions Preparedness and Response are further examples of regional initiatives. They show a commitment by the Arctic nations to address some these important environmental challenges (Berkman and Young 2009). However all these efforts require adequate infrastructure and financial support to be fully effective.

4 Implementing an Eco-System Governance Approach for the Central Arctic Ocean

The proposed area-based measures under the UN resolution aim to protect marine biodiversity on an ecosystem basis. They may therefore cover wider areas such as the Central Arctic, drawing on the Ecologically or Biologically Sea Areas criteria developed under the Convention for Biological Diversity (Dunn et al. 2014). Specifically, the new agreement will need to take ecosystem connectivity into account, as well as the different live stages of marine organisms. A lot of the science in this field is still emerging, in particular around species response to climate change. Other human stressors that have already been identified even in remote areas such as the Arctic include plastics and persistent organic pollutants (AMAP 2015b). A precautionary approach will need to take into account how the resilience of Arctic marine species can be protected in the light of these diverse pressures. The Arctic Council has consistently worked to apply an ecosystem approach to Arctic marine ecosystems and the Council's comprehensive study released in May 2013 at the Kiruna Ministerial Meeting entitled the Arctic Biodiversity Assessment states in Chapter 7: Ecosystem-based Management in the Arctic:

> (20) Arctic states should recognize, in accordance with the recommendations from the Arctic Council EBM Expert Group and the PAME lead Ecosystem Approach expert group, the importance of the following elements when implementing marine Ecosystem- based Management in the Arctic Council Working Groups: identification of the ecosystem, description of the ecosystem, setting ecological objectives, assessing the ecosystem, valuing the ecosystem and managing human activities.
>
> (21) The Arctic Council should promote common understanding and the mutual exchange of lessons learned by periodically convening Arctic Council- wide meetings on EBM to share knowledge and experiences with respect to management and science across Large Marine Ecosystems; and review information on integrated assessments. (PAME 2013).

4.1 Participation and Transparency

Some Arctic nations have expressed reservations as to the need for a new UNCLOS instrument (Haftendorn 2013). One way to address these concerns is for these nations to fully integrate their Arctic constituency into deliberations that are now taking place at the United Nations. Sufficient participation and widespread acceptance will be key to achieve a comprehensive global regime (Hubert 2015). The Aarhus Convention provides a relevant standard on how to address participation and transparency in the implementation of the proposed new agreement (Aarhus Convention 1998). A productive and ongoing discussion for Arctic governance solutions (Young 2010) at a regional level could provide an important complementary effort, providing local and regional stakeholders an opportunity to provide a crucial contribution to address Arctic challenges.

4.2 Monitoring and Enforcement

Monitoring and enforcement of measures over large areas will require modern technology, using satellites covering the region, sonar and radar, and sensors both in the water and on vessels (Kachelriess et al. 2014). These technical solutions are of particular relevance in areas that face tough climate conditions. Automatic Information Systems (AIS) technology already exists for the monitoring & surveillance of large ships and has been a requirement under the IMO Safety on Sea (Solas) rules since 1974. The IMO regulations have recently been tightened through revised guidelines (IMO 2015).

Progress is being made in monitoring vessel movements independently via satellites and there are a number of efforts for improved marine sensor technologies (Secades et al. 2014), including in Arctic waters. Developing an integrated Arctic ocean data management system would assist ecosystem-based measures and could impact the safety and reliability of Arctic shipping, delivering better route planning, weather information and other benefits (Schofield et al. 2013). Such a network should be fully integrated into the global system (McCurdy 2014).

An engagement of the shipping industry in the design and development of ocean solutions is desirable, drawing on existing expertise. This would allow the design a broader Arctic data infrastructure, reducing risks to the environment while providing operational data to interested parties.

A particular challenge is provided by black carbon, which mainly results of incomplete combustion of fossil fuels. Specific constraints on the type of fuel used in Arctic shipping have been proposed as a countermeasure. However shipping regulation on its own will be insufficient to control Arctic black carbon as the largest ground source for this black carbon is Arctic land based mining and air pollution from mid-latitude Asia (Allan et al. 2015). The shipping component in the Arctic is quite small and can be expected to remain so.

Funding for monitoring and enforcement measures could come from payments for the ecosystem services, including taxes, fines and fees from shipping, extractive industries and energy producers (Rogers et al. 2014). Fees are presently charged in Arctic waters generally relate to the delivery of specific services such as piloting and ice-breaking and this approach could similarly be applied in the Central Arctic Ocean.

4.3 Public-Private Partnerships

Public-private partnerships to develop ocean infrastructure such as remote sensing platforms could bring additional funders and reduce overall cost. Such an approach may also help to design partnerships that aim for multiple benefits for ocean and human health (Muller-Karger 2013). Including biodiversity monitoring in this structure would help Arctic science and the study of Arctic ocean resilience and allow for innovative financing (OECD 2013). Establishing a dedicated ocean finance institution to provide loans, guarantees, equity and debt instruments and to structure transactions and partner with new investors would be a way forward (Thiele 2015).

4.4 Interactions with Sectoral Approaches

The new legal instrument will address biodiversity and be complementary to efforts required by several sectors, including shipping, tourism and extractive industries. Issues relevant to Arctic shipping that may be addressed as a consequence of the new instrument could include limitations on the use and carriage of heavy fuel oil and on noise pollution. Specific measures may be required in relation to the prevention of the introduction of non-native species through hull fouling and ballast water in future Arctic MPAs such as the Central Arctic High Seas. Another prominent issue is the impact of shortlived climate pollutants on Arctic warming and biodiversity (AMAP 2015a).

Environmental impact assessments prior to any activities will be conducted based on processes defined by the proposed agreement. Until solid environmental baselines and robust processes are in place based on good science and best practices the new agreement will aim to help protect the Central Arctic as a place for research and exploration.

5 Conclusions

Whilst it remains to be seen when and in what form the new implementing instrument will emerge the UNGA resolution points the way forward to a new regime for large-scale marine ecosystems in High Seas areas such as the Central Arctic. It aims to deliver comprehensive, legally enforceable global governance regime, taking a holistic approach that takes into account cumulative impacts and recent science and fully integrates with existing legal sectoral regimes such as provided for shipping by the IMO. It is thus a crucial part of the global environmental governance regime (Young 2011). At the same time the new approach relies on regional ocean governance approaches to deliver effective marine protection. The Arctic Council has already developed a significant set of relevant approaches and principles in this regard. Arctic actors are therefore encouraged to actively engage with this important, ongoing UN process to make sure that the ultimate outcome delivers the most appropriate management regime for biodiversity in the Central Arctic Ocean.

Acknowledgement The author participated as an International Union for Conservation of Nature (hereinafter referred to as IUCN) observer in the final three sessions of the United Nations General Assembly (hereinafter referred to as UNGA) Ad hoc Informal Working Group on Biodiversity in Areas Beyond National Jurisdiction and at the first Preparatory Committee session for the proposed new agreement.

References

Abdullah, A., Obura, D., Bertzky, B., & Shi, Y. (2014). Marine World Heritage: Creating a globally more balanced and representative list. *Aquatic Conservation: Marine and Freshwater Ecosystems, 24*(2), 59–74.

Allan, J. D., Williams, P. I., Najera, J., Whitehead, J. D., Flynn, M. J., Taylor, J. W., et al. (2015). Iodine observed in new particle formation events in the Arctic atmosphere during ACCACIA. *Atmospheric Chemistry and Physics, 15*, 5599–5609.

Arctic Monitoring and Assessment Programme (AMAP). (2015a). *Summary for policy-makers: Arctic climate issues: Short-lived climate pollutants*. Arctic Council.

Arctic Monitoring and Assessment Programme (AMAP). (2015b). *Summary for policy-makers: Arctic pollution issues: Persistent organic pollutants*. Arctic Council.

Becker, M. A. (2010). Russia and the Arctic: Opportunities for engagement within the existing legal framework. *American University International Law Review, 25*(2), 225–250.

Berkman, P. A., & Young, O. R. (2009). Governance and environmental change in the Arctic Ocean. *Science, 324*(5925), 339–340. https://doi.org/10.1126/science.1173200.

Convention on Access to Information, Public Participation in Decision-Making and Access to Justice in Environmental Matters [Aarhus Convention]. (1998). 2161 UNTS 447, 38 ILM 517.

Convention on Biological Diversity [CBD]. (2016). https://chm.cbd.int/database/record?documentID=204088

Declaration Concerning the Prevention of Unregulated High Seas Fishing in the Central Arctic Ocean. (2015). Oslo.

Delfour-Samama, O. (2014). Review of potential legal frameworks for effective implementation and enforcement of MPAs in the high seas. *ICES Journal of Marine Science*, https://doi.org/10.1093/icesjms/fsu024

Druel, E., & Gjerde, K. M. (2014). Sustaining marine life beyond boundaries: Options for an implementing agreement for marine biodiversity beyond national jurisdiction under the United Nations convention on the law of the sea. *Marine Policy*. https://doi.org/10.1016/j.marpol.2013.11.023.
Dunn, D. C., Ardron, J., Bax, N., Bernale, P., Cleary, J., Cresswell, I., et al. (2014). The convention on biological diversity's ecologically or biologically significant areas: Origins, development, and current status. *Marine Policy, 49*, 137–145.
Gjerde, K., Reeve, L., Harden-Davies, H., Ardron, J., Dolan, R., Durussel, C., et al. (2016). Protecting Earth's last conservation frontier: Scientific, management and legal priorities for MPAs beyond national boundaries. *Aquatic Conservation: Marine and Freshwater Ecosystem, 26*(2), 45–60.
Global Ocean Commission. (2014). *From decline to recovery: A rescue package for the global ocean*. Oxford: Global Ocean Commission.
Haftendorn, H. (2013). *The case for Arctic governance: the arctic puzzle*. Institute of International Affairs/The Centre for Arctic Policy Studies.
High Seas Alliance. (2014). *The need for a new implementing agreement under UNCLOS on Marine biodiversity of the high seas*. HSA Briefing.
Hubert, A.-M. (2015). UN General Assembly Resolution to develop a new legally binding instrument on the conservation and sustainable use of marine biological diversity of areas beyond national jurisdiction. ABLAwg.ca.
Institute for Advanced Sustainability Studies (IASS)/Institute for Sustainable Development and International Relations (IDDRI). (2013). *Joint policy brief: Advancing governance of the High Seas*. www.iass-potsdam.de and www.iddri.org
International Maritime Organization (IMO). (2006). Revised Guidelines for the Identification and Designation of Particularly Sensitive Sea Areas. Res A.982(24), A/24/Res.982.
International Maritime Organization (IMO). (2015). Resolution A.1106(29) Adopted on 2 December 2015 (Agenda item 10) Revised guidelines for the onboard operational use of shipborne automatic identification systems (AIS).
International Maritime Organization (IMO). (2017). Retrieved January 6, 2017, from http://www.imo.org/en/OurWork/Environment/PSSAs/Pages/Default.aspx
IUCN. (2014a). *A strategy of innovative approaches and recommendations to enhance implementation of marine conservation in the next decade*. Gland, Switzerland.
IUCN. (2014b). The promise of Sydney: Innovative approaches for change. In *IUCN World Parks Congress*, Sydney, Gland, Switzerland. http://worldparkscongress.org/downloads/approaches/ThemeM.pdf
Johansson, T. (2015). *The shipping industry, ocean governance and environmental law in the paradigm shift*. Springer.
Johnson, D. (2013). Can competent authorities cooperate for the common good: Towards a collective arrangement in the North-East Atlantic. In P. A. Berkman & A. N. Vylegzhanin (Eds.), *Environmental security in the Arctic Ocean, NATO science for peace and security series C: Environmental security*. Netherlands: Springer.
Kachelriess, D., Wegmann, M., Gollock, M., & Petorelli, N. (2014, January). The application of remote sensing for marine protected area management. *Ecological Indicators, 36*, 169–177.
Klein, C. J., Brown, C. J., Halpern, B. S., Segan, D. B., McGowan, J., Beger, M., & Watson, J. E. M. (2015, December 3). Shortfalls in the global protected area network at representing marine biodiversity. *Scientific Reports, 5*(17539), 1–7.
Koh, T. T. B. (2013). *The Tommy Koh Reader: Favourite Essays and Lectures*. Singapore: World Scientific Publishing.
Kohona, P. T. B., & Lijnzaad, L. (2015). Letter dated 13 February 2015 from the Co-Chairs of the Ad Hoc Open-ended Informal Working Group to the President of the General Assembly, A/69/780, UNGA 69th sess, Item 74(a) (13 Feb).
Mayol, P., Labach, H., Couvat, J., Ody, D., & Robert, P. (2013). Particularly sensitive sea area (PSSA): An IMO status as an efficient management tool of Pelagos. In: *IMPAC 3*. Marseille.

McCurdy, A. (2014, March 20). *Deep ocean observing strategy - A global ocean observing system project report*. Consultative Draft, V2-1.

Molenaar, E. J., & Oude Elferink, A. G. (2013). Marine protected areas in areas beyond national jurisdiction: The pioneering efforts under the OSPAR convention. *Utrecht Law Review, 5*(1), 5–20.

Muller-Karger, F. E. (2013). Remote sensing applications to ocean and human health. System Monitoring. *Encyclopedia of Sustainability Science and Technology*. https://doi.org/10.1007/978-1-4614-5684-1_16

OECD. (2013). *Scaling up finance mechanisms for biodiversity*. Paris: Organisation for Economic Co-operation and Development.

OSPAR. (1992). Convention for the Protection of the Marine Environment of the North-East Atlantic [OSPAR Convention], 2354 UNTS 67, 32 ILM 1069.

OSPAR. (2010). Strategy of the OSPAR Commission for the Protection of the Marine Environment of the North-East Atlantic 2010–2020 (OSPAR Agreement 2010-3).

PAME. (2013). The Arctic Ocean Review Project, Final Report, (Phase II 2011–2013), Kiruna May 2013. Protection of the Arctic Marine Environment (PAME) Secretariat, Akureyri (2013).

PAME. (2015). Arctic Council website. Retrieved September 14, 2016, from http://www.arctic-council.org/index.php/en/about-us/working-groups/pame

PAME. (2016). DRAFT 10 Sept 2016 Briefing on EAIC 2016 for PAME II. https://oaarchive.arctic-council.org/bitstream/handle/11374/1825/EDOCS-3858-v1A-ACSAOUS203_Portland_2016_2-2_Ecosystem_Approach_Implementation_Draft-Report.PDF?sequence=1&isAllowed=y

Pan, M., & Huntington, H. P. (2016, January). A precautionary approach to fisheries in the Central Arctic Ocean: Policy, science, and China. *Marine Policy, 63*, 153–157. https://doi.org/10.1016/j.marpol.2015.10.015

Pörtner, H.-O., Boyd, P. W., Cheung, W. W. L., Lluch-Cota, S. E., Nojiri, Y., Schmidt, D. N., Zavialov, P. O. et al. (2014), Ocean systems. In *Climate Change 2014: Impacts, Adaptation, and Vulnerability. Part A: Global and Sectoral Aspects. Contribution of Working Group II to the Fifth Assessment Report of the Intergovernmental Panel on Climate Change.*

Rochette, J., Wright, G., Gjerde, K. M., Grieber, T., Unger, S., & Spadone, A. (2015). *A new chapter for the high seas?* Paris: IDDRI.

Rochette, J., Unger, S., Herr, D., Johnson, D., Nakamura, T., Packeiser, T. et al. (2014). The regional approach to the conservation and sustainable use of marine biodiversity in areas beyond national jurisdiction. *Marine Policy*. https://doi.org/10.1016/j.marpol.2014.02.005

Rogers, A. D., Sumaila, U. R., Hussain, S. S., & Baulcomb, C. (2014). *The High Seas and us: Understanding the value of High-Seas ecosystems*. Oxford: Global Ocean Commission.

Ryder, S. (2015). *The declaration concerning the prevention of unregulated High Seas fishing in the Central Arctic Ocean*. Posted 11/08/2015 on JCLOS blog.

Schofield, O., Glenn, S. M., Moline, M., Irwin, A., Chao, Y., & Arrott, M. (2013). Ocean observatories and information. In J. Orcutt (Ed.), *Earth system monitoring*. New York: Springer. https://doi.org/10.1007/978-1-4614-5684-1_1.

Secades, C., O'Connor, B., Brown, C., & Walpole, M. (2014). *Earth observation for biodiversity monitoring: A review of current approaches and future opportunities for tracking progress towards the Aichi biodiversity targets* (Technical Series No. 72, 183 p). Montréal, Canada: Secretariat of the Convention on Biological Diversity.

Thiele, T. (2015, November). Accelerating impact, the promise of blue finance. *Cornerstone Journal of Sustainable Finance & Banking*, 21.

Thiele, T., & Harden-Davies, H. (2016). Technology transfer: Policy brief. *Nereus Policy Brief.* http://www.nereusprogram.org/policy-brief-bbnj-technology-transfer/

Toepfer, K., Tubiana, L., Unger, S., & Rochette, J. (2014). Charting pragmatic courses for global ocean governance. *Marine Policy*. https://doi.org/10.1016/j.marpol.2013.12.004. United Nations Convention on the Law of the Sea [UNCLOS], preamble, 10 Dec. (1982), 1833 UNTS 397, 21 ILM 126.

UN document A/69/780 Annex 2/9 15-01992. (2015) Outcome of the Ad Hoc Open-ended Informal Working Group to study issues relating to the conservation and sustainable use of marine biological diversity beyond areas of national jurisdiction and Co-Chairs' summary of discussions.

United Nations General Assembly (UNGA). (2015). Development of an international legally-binding instrument under the United Nations Convention on the Law of the Sea on the conservation and sustainable use of marine biological diversity of areas beyond national jurisdiction. GA Res 69/922, 69th sess, A/RES/69/922 (2 June).

Visbeck, M., Kronfeld-Goharani, U., Neumann, B., Rickels, W., Schmidt, J., & van Doorn, E. (2014). A sustainable development goal for the ocean and coasts: Global ocean challenges benefit from regional initiatives supporting globally coordinated solutions. *Marine Policy*. https://doi.org/10.1016/j.marpol.2014.02.010

Weidemann, L. (2014). *International governance of the Arctic Marine environment*. Springer.

White, C., & Costello, C. (2014). Close the High Seas to fishing? *PLoS Biology, 12*(3), e1001826. https://doi.org/10.1371/journal.pbio.100182.

Young, O. (2010). Arctic governance - Pathways to the future. *Arctic Review on Law and Politics, 1* (2), 164–185. ISSN 1891-6252 A.

Young, O. (2011, December 13). Effectiveness of international environmental regimes: Existing knowledge, cutting-edge themes, and research strategies. *PNAS, 108*(50), 19853–19860.

Young, O. (2016, October). Governing the Arctic Ocean. *Marine Policy, 72*, 271–277. https://doi.org/10.1016/j.marpol.2016.04.038

Zitterbart, D. P., Kindermann, L., Burkhardt, E., & Bebel, O. (2013). Automatic round-the-clock detection of whales for mitigation from underwater noise impacts. *PLoS One*. https://doi.org/10.1371/journal.pone.0071217

The Legal Status of the Northwest Passage: Canada's Jurisdiction or International Law in Light of Recent Developments in Arctic Shipping Regulation?

Saied Satei

Contents

Abstract The recent adoption of the Polar Code relates to the Northwest Passage (the Passage) that connects the Atlantic and Pacific Oceans through the Canadian Arctic Archipelago. The Passage has not, however, been completely navigable due to the existence of Arctic sea ice. Arctic waters are however, increasingly becoming more accessible since sea ice, largely due to the effect of climate change, is thawing. This holds the potential of greater maritime activities in the Arctic waters including the Passage. It is consequently essential to ensure maritime safety and environmental protection. The question is, who has jurisdictional authority to govern such activities within the Passage? Canada claims that it is part of its historic internal waters and therefore, Canadian legislation is applicable. It also dismisses the notion that it is an international strait and/or may be used for innocent passage. There are two criteria for the qualification of a strait as international: Geographical situation connecting two parts of the high seas; and it is used for the purposes of international navigation. Moreover, littoral states do not have a right to prohibit innocent passage in time of

S. Satei (✉)
Robert Gordon University, Aberdeen, UK
e-mail: s.satei@rgu.ac.uk

L. P. Hildebrand et al. (eds.), *Sustainable Shipping in a Changing Arctic*, WMU Studies in Maritime Affairs 7, https://doi.org/10.1007/978-3-319-78425-0_14

peace. This is in conjunction with the 1982 United Nations Convention on the Law of the Sea and customary international law. Canada has the right to exercise jurisdiction over issues relating to marine pollution in the Passage waters. It simultaneously has the obligation to apply international rules such as the Polar Code.

Keywords The Northwest Passage governance · The Polar Code · UNCLOS · International navigation · Maritime safety · Environmental protection

1 Introduction

The International Maritime Organization (IMO) has recently adopted the Polar Code aiming to address growing naval operations in polar waters and to ensure safety of shipping and prevent maritime pollution. Increasing maritime activities in the Arctic, including the Passage, which is the main focus of this chapter, has resulted from climate change and receding of sea ice. The Passage, located within the Canadian territory connects the Atlantic and Pacific Oceans through the Canadian Arctic Archipelago and is currently one of the most viable waterways to the Arctic Ocean alongside the Northern Sea Route in Russia (Young et al. 2009). The legal issues relating to the Passage is discussed in Sect. 3. The extreme and rapidly changing Arctic environmental conditions have limited navigation through the Passage. The receding of sea ice is however, transforming shipping in the region and the Passage may be physically able to allow vessels to cross during the summer (Sakhuja 2014).

Since the majority of legal international trade is carried out by sea, international straits are significant as shortcuts for the shipping industry. Goods need to reach ports in a minimal time frame and with minimal costs. International straits have historically played key roles in global economy, transportation and trade relations between nations and are used as alternative routes to connect ports and harbours. They are therefore, significant as seaways for all nations. Among the best definitions of the concept of 'international strait' is that of Martin (2010). She defines it as "a natural maritime passage which entails a contraction of the waters no greater than double the width of the territorial sea of the respective coastal states, which separates two land masses, and communicates a high seas or EEZ area with another high seas or EEZ area, or a high seas or EEZ area with the territorial sea of another state or, possibly, with its interior waters or its archipelago waters, and is used for international navigation".

The Arctic is rich in living and non-living resources. Canada and the United States have been exploiting hydrocarbons from the Mackenzie Delta and Prudhoe Bay (Yenikeyeff and Krysiek 2007). States and companies have begun to invest and exploit other Arctic resources. They include Arctic and non-Arctic states (such as China, Japan and South Korea). The seabed of the Arctic Ocean has been projected to contain substantial amount of hydrocarbon deposits (Clark 2007). The Arctic waters encompassing the Passage waters may be used as an intercontinental sea route to transport such materials to the market. In November 2012, a large gas tanker, the Ob River, made the first winter journey from Norway to Japan (McGrath 2012).

This voyage is an example of trans-polar shipping through the Northern Sea Route noting that its economics are much different of the Passage. Although such developments provide opportunities there are challenges. Increased shipping activity may have an impact on the fragility of the Arctic environment. Such challenges have been the prime focus for the IMO, leading to the adoption of the Polar Code to address such issues and establish a mandatory international Code of safety for ships operating in the Polar waters.

As regards the legal status of the Passage, Canada strongly asserts that it is part of its historic internal waters and, therefore, Canadian law is applicable. Conversely, the United States (and some other states) argues that it is an international waterway and is subject to international law (Brubaker 2005; Lindsay 2012). Such disagreement between allies over the use of the Passage will continue to cause discomfort until this legal dispute is settled peacefully. The aim of this chapter is to tackle this dispute and answer the question: What law is suitable to provide jurisdictional authority to govern the growing maritime activities within the Passage? Such law will set rights and obligations as regards safety of shipping and marine environmental protection.

First, it provides an overview of the Passage and its significance. Second, it reviews the legal status of the Arctic from Canada's perspective. It also examines the existing criteria for a strait to qualify as international waterway and whether the Passage meets such criteria. The final section pulls together the findings of the preceding sections and offers some overarching conclusions.

The method used to achieve the objective is the comparative analysis of laws and regulations relating to the Passage and examine the legal status of that (micro legal analysis). It will also employ the IRAC mode as a generally accepted approach of performing legal study.

2 Brief Overview of the Northwest Passage

The Northwest Passage, as a shipping route, is a series of linked straits passages (Rothwell 1996). It connects the North Atlantic Ocean (via the Davis Strait) from the east through the Canadian Arctic Archipelago to the Beaufort and Chuckchi Seas and on to the North Pacific Ocean-the Bering Strait- (Rothwell 1993). Due to the presence of thick ice and shallowness in some areas, these potential shipping routes are not navigable for complete traverse without assistance by ice-breakers (Rothwell 1993). It is predicted, however, that as ice melts away this chokepoint will be open for navigation by all types of vessels.

The legal status of the Passage has been controversial for more than a century. In 1907 the Canadian Senate advised the Government to declare its possession of all of its Arctic territory including the lands and islands (Rothwell 1993). This controversy reached its peak when, in 1969, an American oil Tanker, the SS Manhattan, crossed through the Passage despite Canada's claim that the waters of the route were its internal waters.

Despite the US assertion that the traverse was innocent and that it was not blocked by Canadian authorities, it nonetheless, caused political and public rows inside Canada and demands that the Government claim sovereignty over the Passage (Dosman 1975). In 1970 Canada, in response, adopted three measures:

- The 1970 Arctic Waters Pollution Prevention Act (AWPPA) enactment: By passing this law, Canada extended its maritime boundary and regulated all navigation within 100 nautical miles from the low water line (Rostan 2009).
- Modification of Canadian relationship with the International Court of Justice (the ICJ): Canada amended its acceptance of compulsory jurisdiction of the ICJ declared in 1929. It aimed to avoid any international claims being made against Canada as regards themes relevant to the Arctic (Reid 1974).
- The extension of Canadian maritime boundary: Canada extended its territorial sea from three to twelve nautical miles spanning waters around the islands of the Arctic Archipelago. This resulted in foreign vessels navigating through the Passage become further the subject of Canadian jurisdiction under this Act (Rothwell 1996).

Dispute over the legal status of the Northwest Passage re-emerged in 1985 when the U.S. informed Canada of its plan to sail its icebreaker, the Polar Sea, through the Passage without seeking permission from Canadian officials. The Polar Sea completed its crossing from east to west with two Canadian Coast Guard captains on board as 'invited observers' (Byers and Lalonde 2009). The Canadian public once again protested against such action leading to the establishment of straight baselines around Canada's Arctic Archipelago whereby it enclosed the Arctic straits and to the public claim of full jurisdiction over the Passage based on claims of historic internal waters (Lolande 2004; Roach and Smith 1994).

In 1988, Canada and the U.S. agreed to cooperate on navigation and resource development in the Arctic. So far, such cooperation has not been fully successful and the issue still remains unsettled. One of the reasons is that the Agreement primarily deals with icebreakers navigation, not other vessels (Article 3). Moreover, the U.S. does not recognize the Passage waters as internal claimed by Canada although it pledges to undertake all navigation with the consent of Canada (Article 3).

2.1 *Strategic/Political*

The Northwest Passage is situated in the Arctic region which is a high geopolitical and distinctive area. Russia, Norway, Denmark (via Greenland), Canada and the U.S. (via Alaska) possess sovereignty in the Arctic Ocean. The physical geography of these countries encircling the Arctic Ocean has made the region important. The unique geographical formation of the area affects the maritime claims of the Arctic littoral states and leads to a number of maritime territorial disputes with a limited range of solutions. Sovereignty over the Northwest Passage as a viable Arctic seaway has also been disputed between Canada and the U.S. (and some of the international community). Free navigation for all vessels without restrictions could potentially pose risks to the interests of Canada in the Arctic. Terrorist threats are another issue that may affect the safety of shipping through the Passage waters.

2.2 Economy

The Northwest Passage is about 4000 miles shorter than routes via the Suez or Panama Canals connecting the Far East markets to the Northwest America, Canada and Northern Europe (Paulson 2009). If ice continues to recede, the Passage will be open for international commercial navigation (subject of course to resolving sovereignty disputes over the waterway) at least in the summer. Such a new shipping route may facilitate the transportation of living and non-living resources, particularly petroleum, with high practicality and lower costs. Canada currently navigates through Arctic waters including the Passage all year round. Fednav, as the only merchant shipping company operating in the Arctic Shipping Pollution Prevention Control Zone, is involved in a majority of mining projects transporting materials between the country's ports (Pelletier and Lasserre 2012).

2.3 Environment

Climate change has radically transformed the Arctic's environmental prospect. Arctic ice is receding at an alarming pace, bringing increased access to the region, particularly by ship. Conversely, it is important to protect the Arctic marine environment. The Arctic environment is fragile with unique flora and fauna which makes it susceptible to growing stress such as increased resource extraction, industrial development and pollution from outside sources (Nowlan 2001). Article 192 of UNCLOS states that "Sates have the obligation to protect and preserve the marine environment". Article 194 elaborates this further that states require taking all necessary measures to prevent, reduce and control marine pollution using their best practical means and capabilities. Article 197 also sets guidance on how to implement the environmental protection through cooperation on a global/regional basis in formulating international rules. This is important for Arctic environmental issues such as climate change and persistent organic pollutants which derive from sources beyond the Arctic.

The increased shipping traffic poses environmental (operational and search and rescue) concerns in the region. The consequences of any safety or pollution accident in the Passage waters, especially from oil tankers, will likely cause adverse harm to the ecosystem in the area and beyond. The United Nations Convention on the Law of the Sea (UNCLOS) lays out the rights and obligations of state parties in the Arctic Ocean. Part XII of UNCLOS specifically addresses issues concerning protection and preservation of the marine environment. Article 234 also confers on Arctic coastal states the right to extend their regulatory authority on environmental matters to ice covered areas of the Exclusive Economic Zone (EEZ). This Article allows Canada to extend its environmental laws and regulations mentioned in Article 192 to its EEZ with due regard to navigation. It is however, contrary to the claim made by Canada that Article 234 gives full jurisdiction over the Passage. It is noteworthy that this Article is the only provision in UNCLOS which addresses ice covered areas and applies to the EEZ, not to territorial or internal waters.

IMO has also developed the mandatory Polar Code following the non-binding 2002 and 2009 Guidelines for ships operating in polar waters. It aims to "cover the full range of shipping-related matters relevant to navigation in waters surrounding the two poles–ship design, construction and equipment; operational and training concerns; search and rescue; and, equally important, the protection of the unique environment and ecosystems of the polar regions". It will be applied to the Passage too if it is used for international navigation. The Polar Code is expected to enter into force on 1 January 2017.

Other treaties relevant to the Arctic passages include the 1972 London Convention on the Prevention of Marine Pollution by Dumping of Wastes and Other Matter (London Convention), the 1973 International Convention for the Prevention of Pollution from Ships (and its 1978 Protocol) (MARPOL 73/78), the 1989 Basel Convention on the Control of Transboundary Movements of Hazardous Wastes and their disposal (Basel Convention) and the 1992 Convention on Biological Diversity.

3 The Legal Status of the Northwest Passage: A Dormant Issue or a Resurfaced Dispute?

Canada claims full sovereignty over the Northwest Passage that is located in its maritime territory. Although the legal status of the Passage has been in dispute for almost half a century, it has remained a dormant legal issue for part of that time. Lack of international regime to govern the sea ice areas, the Cold War and harsh weather could be named as the reasons. Other problems reducing transport through the Passage include the high cost of insurance, the cost of designing new technologies for ships to strengthen them against ice, and operational and political risks. Thawing Arctic sea ice has, however, resurfaced such conflict.

3.1 Internal and Historic Waters

Canada's assertion over the Passage is primarily based on two legal concepts: internal waters and historic waters. The former is established in accordance with the normal baseline (the low-water line-Article 5 of UNCLOS) or the straight line (a line of the shortest distance between two points-Articles 7, 10 and 76(7) of

UNCLOS. Under UNCLOS, the provisions related to the archipelagic waters do not apply to Canada because it does not wholly consist of archipelagic islands. The naming of the Canadian islands as the Arctic Archipelago is a political definition, not a legal one. It would be however, correct to refer to the 'waters around the islands as part of Canada's internal waters'. Coastal state jurisdiction extends seaward: for the territorial sea (Article 2) and the EEZ (Article 44) are measured from coastal baselines and the continental shelf (Article 76) extends beyond the EEZ according to a geographic and geologic formula, but these areas do not constitute internal waters.

The latter is not defined by UNCLOS or the 1958 Convention on the Territorial Sea and the Contiguous Zone. The ICJ however, provides an appropriate meaning as: "By 'historic waters' are usually meant waters which are treated as internal waters but which would not have that character were it not for the existence of an historic title". To fill this deficit, the formation of any other historic title to territory may be applied by analogy. In the Fisheries case Norway asserted that the historic title could apply to any waters including straits, archipelago and bays. Thus, the state claiming historic waters must prove a well established intention to assert sovereignty over the strait (or archipelago or bay); and peaceful and unchallenged exercise of authority over the waters of the strait (or archipelago or bay) (in the form of effectiveness, continuity and notoriety) (Walker 2012). Such requirements have not yet been achieved as regards the Northwest Passage since the U.S. and international community opposed Canadian assertion when Canada established the baselines around the perimeter of its Arctic Archipelago.

In 1985, Canada drew straight baselines around its Arctic Archipelago which enclosed the Arctic straits and claimed full sovereignty based on historic internal waters. So far so good. What about Canada's sovereignty prior to 1985? Article 8 (2) states: "Where the establishment of a straight baseline in accordance with the method set forth in Article 7 has the effect of enclosing as internal waters areas which had not previously been considered as such, a right of innocent passage as provided in this Convention shall exist in those waters". Canada however, failed to meet the criteria for establishment of a title to such areas since the United States (and rest of the world) did not recognize Canada's assertion and therefore, cast doubts that the Arctic waters were previously considered as internal waters (see below). In other words, the Northwest Passage, from the United States perspective, is the subject of the transit passage and the principle of the freedom of navigation and is thus, regulated by international law (including the Polar Code). Furthermore, lack of attention or effort by the Canadian Government to pursue its full sovereignty over the Arctic waters by persistent actions or announcements since 1950s have adversely contributed to its failure (Lolande 2004).

Some authors suggest that enclosing straits by the claimant state as its historic internal waters will not jeopardize or will have only minor effects on the rights of other states for navigation (Symmons 2008). This may be done via a special arrangement (similar to the one signed in 1988 between Canada and the U.S.) between Canada and the international community.

3.2 *International Law and Qualification for International Straits*

In the 1949 Corfu Channel case the ICJ set out two criteria for the qualification of a strait as international: the geographical situation connecting two parts of the high seas; and the strait's use for the purposes of international navigation. It further concluded that littoral states do not have a right to prohibit innocent passage of an international strait in time of peace. This is in conjunction with Part III of UNCLOS, Section III of the 1958 Convention on the Territorial Sea and the Contiguous Zone and customary international law. The following will apply these tests to determine whether the Passage is considered as Canadian internal waters or it is an international strait as claimed by the U.S.

3.2.1 Geographical Location of the Northwest Passage

The Passage, as set forth above, is a series of routes through the Canadian Archipelago. Canada may argue (although it is not the strongest argument) that the Northwest Passage is a very long 'route/s' (about 900 miles) connecting one part of the high seas to another. By claiming that, Canada would be able to disqualify the Passage as a strait and may possibly enable it to pursue its assertion based on historic internal waters. In the Fisheries case Norway brought almost the same argument forward although it admitted that the Indreleia (the waters followed by the navigational route) was utilized, to a certain extent, for international navigation (Symmons 2008).

Freedom of the high seas is a generally accepted principle of international law. Rights in an international strait are however, far closer to those of innocent passage in the territorial sea than the high seas. Furthermore, non-suspendible innocent passage for all nations through a strait which joins two parts of the high seas (or even high seas to a territorial sea) is inherently a customary international law. Article 16(4) of the 1958 Convention on the Territorial Sea and the Contiguous Zone and UNCLOS (Art. 45) confirms that too. Even though the U.S. has yet to ratify UNCLOS, the principle of freedom of navigation and the concept of non-suspendible innocent passage are essential parts of customary international law (as well as treaty law and legal practice) and will thus, be hard for Canada to dismiss. Conversely, free navigation through the Passage will help worldwide economic growth and development but it is notable that the benefit to Canada would be no greater than would transit under Canadian legal jurisdiction.

It is worth noting that an international strait must be a natural waterway as opposed to artificial one (such as the Suez Canal). Given that the Northwest Passage is currently covered by ice this could affect the possible creation of artificial passage by breaking the ice during the voyage. This may in turn disqualify it as a natural waterway although the legal status of sea ice is unclear in international law (Molde 1982). This argument will however, become baseless while Arctic sea ice is melting away.

3.2.2 The Actual Use of the Northwest Passage for International Navigation

The practical application of the internal waters test is to determine an effective control of foreign shipping, that is, the degree and the nature of navigation (Symmons 2008). Pharand (2007) questions whether navigations of foreign ships (the total of 69 voyages between 1903 and 2005), particularly merchant ones, have constituted a sufficient historical precedent to make the Passage as international strait. This may be answered from two perspectives:

- Geographical Perimeter: Given the fact that the Northwest Passage is covered by sea ice for most of the year, it is predicted that the degree and type of navigational use of its waters will not be similar to other international straits. So the answer to the above question would be no. Conversely, according to international law a strait which connects two parts of the high seas and is used for international shipping may not be claimed as internal waters (Gross 1966). The Canadian argument is that the Passage is not used in international navigation (without Canada's approval) sufficiently to demonstrate that it is used in international navigation. However, the fact that it is used with the Canadian agreement at all appears to highlight the debate that it 'is' actually used for international navigation. With advancing technology and developed icebreakers they will be able to ply in and out of the Passage if Canada does not interfere and permit normal crossing. The answer thus, would be yes.
- The future of Sea Ice: It is predicted by scientists that the Arctic icecap covering the Northwest Passage will disappear as a result of climate change, at least in the summer, and therefore, increases shipping traffic through the Passage waters. In light of the current presence of sea ice, the answer to the above question may be "no", at least for now. This will, however, change in a few years' time and could fluctuate drastically depending upon severity of seasons.

4 Conclusion

This chapter reviewed the legal status of the Northwest Passage in Canadian territory and Canada's claims of sovereignty over it. The United States and other Governments including the member states of the European Union have objected to this claim. Their objection is in line with international laws that the right of innocent passage may not be hampered in time of peace let alone that the Passage is qualified for transit passage too with some reservations. Some countries don't object because they have no interest at stake, and some others such as Russia may have some shared interests with Canada. The Passage may not be completely navigable at present to meet one of the criteria in international law, but it will be in the near future when ice melts away, at least in the summer.

Conversely, UNCLOS as the main international legal instrument governing the Arctic Ocean confers Canada the right to exercise its jurisdiction for issues that relate to marine pollution and protection of the marine environment. This is manifested in Article 234 in which the coastal state may extend its environmental jurisdiction (as opposed to full sovereignty) over the EEZ with due regard to international navigation. Furthermore, Canada has to take into consideration its obligation to enact and/or enforce international rules and standards. That includes the Polar Code which has been developed and adopted by the IMO and will be able to adequately provide mandatory provisions for shipping in the Passage waters. It also appropriately addresses shipping safety and environmental protection. It therefore meets Canada's credible concerns on environmental issues which cause serious ecological challenges to the area and beyond when navigation increases. Furthermore, the IMO has a wide range of suitable legal instruments with high positive records of environmental protection for shipping operations and may provide sustainable navigation through the Northwest Passage. Internalisation or militarisation of the Passage is not an appropriate and viable solution.

Devising a special legal regime based on an agreement between Canada and the U.S. (and international community) similar to the Turkish Straits (the 1936 Montreaux Convention) may also provide an appropriate settlement for such a long standing territorial dispute. Such an agreement would need to be addressed through the IMO as the Competent International Organization if it is to be binding on all maritime parties and remain consistent with UNCLOS. Such arrangement could potentially end the current stalemate and adequately satisfy both sides by incorporating terms into that in order to achieve a tangible success. All they need to do is to cooperate and compromise although given the difficulty states, particularly democratic states, face in making compromises of claims to sovereignty over territory.

References

Journal Article

Byers, M., & Lalonde, S. (2009). Who controls the Northwest Passage. *Vanderbilt Journal of Transnational Law, 42*, 1156–1159.

Clark, M. (2007). Arctic: A tough nut to crack. *Petroleum Economist, 74*, 32.

Gross, A. (1966). Maritime boundaries of the states. *Michigan Law Review, 64*, 650.

Lindsay, T. (2012). (Un) Frozen frontiers: A multilateral dispute settlement treaty for resolving boundary disputes in the Arctic. *Oil, Gas & Energy Law Intelligence, 10*, 11.

Lolande, S. (2004). Increased traffic through Canadian waters. *Revue Ju-ridiue Themis, 38*, 74.

Molde, J. (1982). The status of ice in international law. *Nordisk Tidsskrift for International Ret, 51*, 165.

Paulson, J. (2009). Melting ice causing the Arctic to boil over: An analysis of possible solutions to a heated problem. *Indiana International & Comparative Law Review, 19*, 353.

Pelletier, S., & Lasserre, F. (2012). Arctic shipping: Future polar express sea-ways: Shipowners' opinion. *Journal of Maritime Law and Commerce, 43*, 558.

Pharand, D. (2007). The Arctic waters and the Northwest Passage: A final revisit. *Ocean Development & International Law, 38*, 5 and 29–30.

Reid, R. (1974). Canadian claim to sovereignty over the waters of the Arctic. *Canadian Yearbook of International Law, 12*, 124.
Rostan, M. (2009). The Northwest Passage's emergence as an international highways. *Southern Journal of International Law, 15*, 452–454.
Rothwell, D. (1993). The Canadian-U.S. Northwest Passage dispute: A reassessment. *Cornell International Law Journal, 26*, 352.
Sakhuja, V. (2014). The Polar Code and Arctic navigation. *Strategic Analysis, 38*, 803.

Book

Brubaker, R. (2005). *The Russian Arctic straits*. Leiden: Martinus Nijhoff.
Dosman, E. (1975). *The national interest: The politics of northern development, 1968–75*. Toronto: McClelland and Stewart.
Martin, A. (2010). *International straits: Concept, classification and rules of passage*. Heidelberg: Springer.
Nowlan, L. (2001). *Arctic legal regime for environmental protection*. Siegburg: IUCN.
Roach, A., & Smith, R. (1994). *Excessive maritime claims*. Newport: Naval War College.
Rothwell, D. (1996). *The polar regions and the development of international law*. Cambridge: Cambridge University Press.
Symmons, C. (2008). *Historic waters in the law of the sea: A modern re-appraisal*. Leiden: Martinus Nijhoff.
Walker, G. (2012). *Definitions for the law of the sea: Terms not defined by the 1982 Convention*. Leiden: Martinus Nijhoff.

Online Document

Arctic Council. (2009). *Arctic marine shipping assessment*. Arctic Council. Retrieved July 20, 2015, from http://www.arctic.noaa.gov/detect/documents/AMSA_2009_Report_2nd_print.pdf
IMO. (2010). *Guidelines for ships operating in polar waters*. IMO. Retrieved July 25, 2015, from http://www.imo.org/en/Publications/Documents/Attachments/Pages%20from%20E190E.pdf
IMO. (2015). Shipping in polar waters: Development of an international code of safety for ships operating in polar waters (the Polar Code). Retrieved July 21, 2015, from http://www.imo.org/en/MediaCentre/HotTopics/polar/Pages/default.aspx
McGrath, M. (2012). *Gas tanker Ob River attempts first winter Arctic crossing*. BBC News. Retrieved July 20, 2015, from http://www.bbc.co.uk/news/science-environment-20454757
Yenikeyeff, S., & Krysiek, T. (2007). *The battle for the next energy frontier: The Russian polar expedition and the future of Arctic hydrocarbons*. Oxford Institute for Energy Studies. Retrieved July 18, 2015, from http://www.oxfordenergy.org/wpcms/wp-content/uploads/2011/01/Aug2007-TheBattleforthenextenergyfrontier-ShamilYenikeyeff-andTimothyFentonKrysiek.pdf
Young, S., Minteer, K., Long, J., Hubach, C., & Carlton, J. (2009). The scramble for the Arctic: The United Nations convention on the law of the sea (UNCLOS) and extending national seabed claims. *Bepress Journal*. http://works.bepress.com/jondcarlson/28/

Case Law

UK v. Albania (the Corfu Channel case). International Court of Justice. 1949 Apr 09.
UK v. Norway (the Fisheries case). International Court of Justice. 1951 Dec 18.

Part IV
Protection and Response in the Arctic Marine Environment

Arctic Oil Spill Intervention: In Search of an Integrated Approach for the High Seas

Neil Bellefontaine and Tafsir M. Johansson

Contents

Abstract Article 86 of the United Nations Convention on the Law of the Sea (UNCLOS) defines the high seas as all parts of the sea excluding internal waters, territorial sea, exclusive economic zone and archipelagic waters belonging to an archipelagic state. The high seas, as such, are considered to be *res communis*, and can be enjoyed by any state (Through the freedoms to fish, navigate, lay submarine cables, research etc.). The notion of *res communis* has preceded today's concept of public domain and provides a sense of undisturbed entitlement to the shipping industry, which in recent years has translated into a dramatic increase in navigation and trans-Arctic shipping.

For the Arctic, shorter sea-routes and trans-Arctic shipping across the high seas of the Arctic raises significant governance issues. One such issue relates to oil spills and oil spill preparedness and response for the Arctic. Following the Torrey Canyon disaster in 1967, the shipping industry has witnessed a significant number of oil spills and severe damage to the marine environment. Owing to the fact that a coastal state's authority to regulate foreign shipping does not extend to the high seas, transiting ships would only be subject to international shipping safety and the environmental rules and standards (UNCLOS; Art. 211 (1)). For the high seas there exists a corpus of international law, i.e. the International Convention Relating

N. Bellefontaine · T. M. Johansson (✉)
World Maritime University, Malmö, Sweden
e-mail: nab@wmu.se; tm@wmu.se

L. P. Hildebrand et al. (eds.), *Sustainable Shipping in a Changing Arctic*, WMU Studies in Maritime Affairs 7, https://doi.org/10.1007/978-3-319-78425-0_15

to Intervention on the High Seas in Cases of Oil Pollution Casualties 1969 (Intervention Convention). But the inevitable question is to what extent can the Intervention Convention provide an effective framework to deal with oil pollution response in the Arctic high seas? Or do the Arctic high seas require an integrated approach, which can link together differing agendas and mandates of the Arctic States in trying to deal with the impacts of an oil spill disaster? This approach is analogous to the European Union initiatives reflected in various "macro-regional strategies", and would be similar to the North American (US-Can) joint preparedness agreements for oil spills response.

The operative word in the proposed paper is "intervention", which can be contrasted with "prevention" that usually runs to the conclusion of remediation efforts after an oil spill. Although related to response, intervention would occur the very moment a national authority is advised of an incident in progress that has the potential for a spill (e.g. a vessel in distress with a developing leak). This definition is guided by the fact that the Arctic is a pristine area, and for pristine areas there ought to be advanced intervention rules to stop all types of vessel-source oil pollution at the source. This goes beyond the given international oil spill prevention and response regime. The effort is to realize whether an integrated intervention plan for the Arctic high seas can bring the stakeholders together and form an alliance to save the pristine high sea areas from oil spill disasters in areas beyond national jurisdiction.

Keywords Arctic · High seas · Oil spill · Intervention · Integrated approach · Marine protection

1 Introduction

The Arctic from a global standpoint has been defined as the areas lying North of the Arctic Circle at 66°33′ North latitude. For the Arctic states situated in different geographical positions, this definition varies from one state to another. Article 234 of UNCLOS lays down a definition of "ice-covered areas" and the interpretation of the article relates to *lex Specialis* as it seeks to confine the coastal states jurisdiction in furnishing preventive and enforcement regulations;

> [c]oastal States have the right to adopt and enforce non-discriminatory laws and regulations for the prevention, reduction and control of marine pollution from vessels in ice-covered areas within the limits of the exclusive economic zone, where particularly severe climatic conditions and the presence of ice covering such areas for most of the year create obstructions or exceptional hazards to navigation, and pollution of the marine environment could cause major harm to or irreversible disturbance of the ecological balance... (UNCLOS, Article 234)

The areas within which the coastal states are permitted to adopt necessary measures do not extend to areas beyond national jurisdiction, commonly known as the high seas. The high seas are considered as a common heritage of mankind and are reserved for peaceful purposes and listed in the form of several rights preceded by the word "freedom" as incorporated Article 87 of UNCLOS (UNCLOS, Article 87).

An effort is also made by IMO, a specialized agency of the United Nations, to provide a definition of 'Arctic Waters' in provision G-3.3 and G-3.5 of the 'Guidelines for Ships Operating in the Polar Waters' (Guidelines for Ships Operating in Polar Waters 2009). G-3.3 is pertinent in the context of navigation and G-3.5 establishes that "Ice-covered waters means polar waters where local ice conditions present a structural risk to a ship". However, these ice-covered areas coincide with zones of essential importance for intra- and trans-Arctic navigation in the Northwest Passage and Northern Sea Route. Moreover, trans-Arctic shipping across the high seas of the Arctic raises significant governance issues due to the fact that a coastal state's authority to enforce regulations on foreign vessels does not extend to the high seas. The vessels engaged in trans-Arctic shipping are to be governed by and subject to public international law, which is an endeavor to establish cohesive environmental rules and standards adopted through IMO and to be strictly followed by the flag states involved in trans-Arctic shipping.

However, prior to dealing with the international law aspect in relation to the high seas of the Arctic, it is important to commence with a brief study on the international rules governing operational discharge in different areas within the maritime jurisdiction under relevant international instruments. The objective of the brief study is to understand the status quo of international regulations in different maritime zones within national jurisdiction and how they apply to ice-covered areas. The regulatory comparative analysis between "areas within national jurisdiction" and "areas beyond national jurisdiction" will help extract the varying binding-pattern of international law in the high seas and more specifically, the high seas of the Arctic and the much-needed measures to deal with the environmental vulnerability from vessel source pollution, both operational and accidental.

2 Arctic Maritime Zones and Operational Discharge Under UNCLOS and MARPOL 73/78

UNCLOS is viewed as a comprehensive legal instrument governing vessel source pollution and is aptly referred to as the "umbrella convention" as it is designed with the intention to serve as a unifying framework for a growing number of international agreements that address one or more particular ocean use (Kimball 2005). Part XII of UNCLOS underlines the general jurisdictional provisions for the regulation of operational discharge and narrows down the coastal states' prescriptive and enforcement jurisdictions (UNCLOS, Part XII). Article 194 (3) (b) provides for adopting measures against pollution from vessels by intentional or unintentional discharges and Article 194(5) gives special reference to assume measures to protect and preserve "rare or fragile ecosystems", a term which is sporadically used to describe the Arctic due to its undefined characteristics.

Article 211(6)(c) empowers coastal states to adopt additional laws in respect of vessel discharges for a common area and communicate it to the competent international organization. The enforcement aspect of those laws and regulations has been laid down in Article 217(1) through which flag states are under an obligation to ensure compliance with those applicable norms of international rules. Under the coastal state jurisdiction, as stipulated in UNCLOS, a state enjoys sovereignty in internal waters (UNCLOS, Article 2) subject to rendering due publicity to requirements for the prevention, reduction and control of marine pollution and communicating such requirements to the competent international organization (Article 211 (3)). Under Article 21(1)(f) and 211(4), the coastal states' prescriptive jurisdiction in the territorial sea has been confined to the obligation of providing innocent passage and to adopt laws and regulations in conformity with rules of international law for the prevention, reduction and control of marine pollution to the extent that it will not hamper such innocent passage. The enforcement jurisdiction of a coastal state in the territorial sea is limited only to circumstances where there is a clear case to believe that a vessel during innocent passage has violated international rules and standards (UNCLOS, Article 220(2)). As regards to prescriptive jurisdiction in the Exclusive Economic Zone (EEZ), Article 211(5) enunciates that the coastal state may adopt regulations in accordance with international rules and standards established by the competent international organization. Then again, Article 234 governs the coastal states' right to adopt and enforce laws and regulations for the prevention, reduction and control of marine pollution from vessels in ice-covered areas within the EEZ. Article 211(6) further establishes that in cases where international rules and standards are insufficient to meet special circumstances, a coastal state may in terms of the EEZ proceed to adopt mandatory measures subject to consultation with the competent international organization. Enforcement of regulations in the EEZ has been laid down in Article 220(3) and 220(5) in situations where the vessel in question is under an obligation to produce relevant documents and if there is a clear ground for suspecting a violation, the coastal state may conduct inspection of such vessel. Article 220(6) provides for more stringent measures, i.e. institute proceedings including arrest of vessel in cases where there is "clear objective evidence" as regards to the violation.

On the other hand, Article 211(2) of UNCLOS determines the prescriptive jurisdiction of the flag State to adopt laws and regulations for the prevention, reduction and control of vessel source pollution whereby, those "laws and regulations shall at least have the same effect" as generally accepted rules and standards established via competent international organizations. The competent international organization, in this regard, commonly refers to IMO, which plays a significant role in the steering of UNCLOS provisions and has attempted to harmonize requirements as regards to operational discharge standards. The requirements of operational discharge and pertinent standards and rules are clearly embedded in International Convention for the Prevention of Pollution from Ships, 1973 as modified by the Protocol of 1978 (MARPOL 73/78), which contains in its Annexes restrictions as regards to voluntary discharge from vessels. "Discharge" according to MARPOL 73/78 covers all voluntary releases with respect to disposal, spilling, leaking, pumping, emitting and emptying (MARPOL 73/78, Article 2). Compliance with MARPOL 73/78 can be traced back to the Arctic Environmental Protection Strategy

(AEPS[1]) in which reference is given to MARPOL 73/78, the objective of which was to ensure that states maintain the given standards while operating in Polar Waters.

Annexes I, II and V of MARPOL 73/78 pave the way for the possibility to establish areas designated as "Special Areas" and "SO_x emission control areas" where the particular sensitivity of Arctic waters justifies the application of more stringent discharge and emission standards. The discharge of oil from machinery spaces of all ships is regulated by regulation 15 (Annex I) and restricts the discharge of any amount in the Antarctic areas from ships less than 400 Gross Tonnage (GT) unless it complies with Regulation 15-C. However, the Arctic has been overlooked in terms of discharging restrictions despite the occasional implications of similarity in features between the two (Koivurova 2010). Moreover, Annex II (Discharge of Noxious Liquid Substance) and Annex V (Disposal of Garbage) of MARPOL 73/78 do not designate the Arctic as a "Special Area" and only provide restrictions in the Antarctic areas where operational discharges are unauthorized. Among the eight Arctic states (Canada, Denmark, Finland, Iceland, Norway, Sweden, the Russian Federation and the United States of America (US)), only Canada and the Russian Federation have adopted stringent regulations in compliance with Article 234 of UNCLOS in the Arctic North of 60° N latitude. Moreover, Canada has deliberately precluded MARPOL 73/78 for those areas. In addition to the provisions and standards of MARPOL 73/78, which are applicable to ice-covered areas, the Arctic states require compliance with stricter standards. Similar to Antarctica, there is a clear need to designate areas within national jurisdiction as "Special Areas" or "SO_x emission control areas" in order to obtain special protection under MARPOL 73/78. A designation as such would certainly add weight to the protection of the marine environment in Arctic ice-covered areas and could even be extended to "areas beyond national jurisdiction" where increased trans-Arctic shipping could increase the possibility of harmful discharge from commercial vessels.

When it comes to vessel operational discharge, it seems that there exists a range of interrelated and coinciding treaties, which are active at various levels. However, underneath all the layers of international regulations and regional co-operation lies the purely national layer of enactments, which gives effect to the former. The US has adopted the Act to Prevent Pollution from Ships (33 United States Codes §§ 1901–1903) (APPS), which is an enactment of MARPOL 73/78. Then again, the operation and response regime in the Arctic State of Alaska is governed by the Alaska Oil and Hazardous Substances Pollution control Act (AOHSPCA) and the Alaska Environmental Conservation Act (AECA) (Gold 2006). For Norway, the Seaworthiness Act of 1903 (which applies to Norwegian ships) containing pertinent regulations, has substantially incorporated MARPOL 73/78 and later been replaced by the Ship Safety and Security Act of 2007 (SSSA). Sweden as a part of the Arctic Council has executed legislation to give effect to its obligations pursuant to MARPOL 73/78. The Act Relative to Measures Against Pollution Caused by

[1]AEPS is the predecessor of the Arctic Council, established in Rovaniemi Finland in 1991. The Alta Declaration is the last AEPS declaration before the formal establishment of the Arctic Council.

Ships of 1980 (ARMAPCS) (Lag (1980:424) om åtgärder mot förorening från fartyg) embodies restrictions in the context of oil discharge in Sweden. The Arctic Waters Pollution Prevention Act of 1985 (AWPPA), subsequent regulations (Regulations include, Arctic Shipping Pollution Prevention Regulations (C.R.C., c. 353), Arctic Waters Experimental Pollution Regulations, 1978 (SOR/78-417), Arctic Waters Experimental Pollution Regulations, 1979 (SOR/80-9), Arctic Waters Experimental Pollution Regulations, 1982 (SOR/82-276), Arctic Waters Experimental Pollution Regulations, 1982 (Dome Petroleum) (SOR/82-832), Arctic Waters Pollution Prevention Regulations (C.R.C., c. 354) and policies enacted by Canada remains to date the most notable example of national enactments and can be considered as a functional approach via unilateral action since it promotes a "zero discharge" policy and stipulates that "no person shall deposit or permit the deposit of waste of any type in the Arctic waters" (Rothwell and Joyner 2000). The Danish regime of operational discharge relates to the statutory framework of Marine Environmental Protection Act of 1993) (MEPA). This act provides a blueprint for incorporating general and regional treaty instruments into Danish law and chapter 2 of this act implements the rules of MARPOL 73/78 concerning operational oil discharge.

3 Operational Discharge v. Accidental Oil Pollution

In retrospect, navigation in the Arctic waters was confined to supplying local communities in the summer season (Jensen 2007). Arctic shipping has been classified into many categories i.e. commercial vessels including tankers and fishing vessels, vessels for recreation and tourism, scientific research vessels, ice-breakers for re-supply and vessels engaged in offshore exploitation (Jensen 2007). Researchers and scholars have over the years laid down strong predictions on the traffic density in specific areas of the Arctic. The first of these areas include the Northwest Passage, which is the sea route connecting the Atlantic and the Pacific oceans through the archipelago of Canada where the islands of the archipelago are separated from each other and the continental mainland by a series of waterways (Jensen 2007). As to the Russian Arctic, the Northern Sea Route has become the focus of shipping as it extends approximately 2800 km along the Russian Arctic Coast from Novaya Zemlya to the Bering Strait (Jensen 2007). Finally, the Northeastern Passage connects the Atlantic with the Pacific Ocean, traversing the eastern part of the Arctic Ocean following Russia and Norway's coasts. The Northeastern Passage is considered as an alternative to the traditional route from Asia towards Europe through the Suez Canal and is actually 40% shorter compared to the one crossing the Indian Ocean (Marchenko 2014).

Under the international legal regime, international waters including the North Pole and the region of the Arctic Ocean surrounding it do not belong to any of the Arctic States. As such, the Arctic high seas are identified by a large central area surrounding the geographical North Pole, namely the polar cap (Cinelli 2011). Those areas of the high seas are commonly known as *res communis omnium*, a principle derived from the Roman private law upholding the notion of "the common heritage of mankind" and something that is not subject to the appropriation by any sovereign body. However, because of geographical complexities, there has been no determination to date as to whether the North Pole legally belongs to any one of the Arctic coastal states—or whether by delimiting the international seabed area around the North Pole it could be legally termed as *res communis omnium* (Cinelli 2011). The common understanding is that the areas beyond national jurisdiction have not received adequate attention and the international community that supports the "freedom of navigation", only seeks commercial advantages of a shorter sea route (Johansson and Donner 2014). The increase of both intra- and trans-Arctic shipping poses great pressures and risks in terms of impacts on the pristine Arctic marine environment, its living resources and its biodiversity, leaving the North Pole and the region of the Arctic Ocean in a vulnerable position (Johansson and Donner 2014).

When it comes to explaining vessel-source pollution and the vulnerable state of the Arctic Ocean, the two areas that surface with reference to environmental issues are mainly operational discharge and accidental pollution. Discharge of oil from shipping is mainly the result of deliberate operational discharges, which MARPOL 73/78 has explicitly covered. Moreover, the Convention on the Prevention of Marine Pollution by Dumping of Wastes and other Matter of 1972 (LC72) (also known as the London Dumping Convention, Annex I: The "Black List" prohibits the dumping of crude oil) has been developed and implemented to give effect to the provisions of The United Nations' Convention on the Law of the Sea of 1982 (UNCLOS[2]) whereby Article 210 addresses ocean dumping as an integral part of operational and voluntary vessel-source pollution. The five states bordering the Arctic (Canada, Norway, Russia, Denmark and the US) are parties to the LC72 and have implemented it domestically, but the Russian Federation and the United States of America are not parties to the protocol of 1996. Canada fulfills its international obligations, in part, through Part 7, Division 3 (Disposal at Sea) of the Canadian Environmental Protection Act of 1999 (CEPA). The US has adopted an Ocean Dumping Act, codified as titles I and II of the marine Protection, Research and Sanctuaries Act of 1972 (MPRSA), 33 U.S.C. (paragraph 1401 et seq.). Then again, Sweden's Law 1971:1154 on the Prohibition on Dumping of Wastes at Sea of 1971 (PDWA) (pp. 1–3) is a conforming national law in this regard. LC72 is applicable to all marine waters outside internal waters and sets a minimum standard for all States on the basis of categories of pollutants and a system of permits for those substances permissible for dumping. So, whether it is operational or deliberate dumping, discharge of oil and generic substance is governed by international conventions and a follow-up of national legislation in the form of "hard law".

In trying to understand whether the detrimental effects of operational discharge outweigh the effects of accidental oil pollution, whether the former precedes the other, or vice versa, has not been fully substantiated by any relevant literature. The

[2]Ocean dumping Articles refer to Articles 1(1) (5), 210 and 216.

degree of detriment varies from one situation to another. But from a general understanding, accidental discharges are said to occur when two seagoing vessels collide with each other or come in distress at sea, or where there is a blowout of an offshore oil well. Maritime scholars and researchers have left no stone unturned in trying to predict the probabilities and possibilities of maritime incidents. Although much can be done to avoid a maritime incident, there will always be unfortunate circumstances and situations that can lead to accidental oil pollution (Official Homepage of Global Marine Oil Pollution Information Gateway). Although there have been efforts by Arctic States to delineate maritime zones in order to control, prevent and respond to accidental oil pollution, the Arctic Ocean does not recognize the artificial boundaries set by international law for areas within and beyond national jurisdiction. The hypothesis is that activities or incidents taking place in the EEZ could substantially and adversely affect the Arctic high seas.

In retrospect, the high Seas have often been an area where maritime collisions occur and hence, result in a certain amount of discharge in the high seawaters. So, accidental oil spills in the high seas add to the burden of operational discharges by the increasing trans-Arctic shipping. From a broad perspective, the Global Ocean Commission highlights that contaminants can reach the high seas through deliberate or accidental discharges at sea from ships (Global Ocean Commission 2013). The Global Ocean Commission has also focused on the alarming number of accidents on oil and gas offshore platforms in recent years including grave accidents on offshore installations that occur every year (Montara, Australia 2009; Deep Water Horizon, US 2010; Penglai, China, 2011; Kulluk, Singapore, 2012). With the growing number of oil and gas offshore platforms and generic development plans and strategies, it is presumed that the Arctic high seas are exposed to greater risks from accidents than the average operational discharge of oil, which can be controlled by strict compliance measures prevailing in the national legislation of individual Arctic States.

4 Arctic Intervention Regime for Accidental Oil Pollution in the High Seas

As oil and gas prospects will be explored in the Arctic, chances of an accidental oil spill in remote ice-affected waters are presumed to increase at an alarming rate (The US Arctic Research Commission and the US Army Corps of Engineers 2012). The impact of oil on ice-waters is stated as being “long-term” since oil persists longer in Arctic conditions because it evaporates at a slow rate or may be trapped in or under ice and is, hence, less accessible to bacterial degradation (World Wildlife Fund 2007). Examples of the harmful impacts of oil on ice waters are evident from maritime incidents in the Antarctic areas, which are said to resemble the environmental features of the Arctic. One such incident is the grounding of the double purpose passenger and supply ship Bahia Paraiso, which grounded in January 1989 in the Antarctic, and exemplified the dangerous effects of pollution as a result of increased shipping traffic in ice-areas (Johansson and Donner 2014). From another instance, the tanker Exxon Valdez spilled 11 million gallons of crude oil into the Pacific Gulf of Alaska on 24 March 1989. Over 1200 miles of coastline of the Alaska Peninsula were contaminated with oil, which caused irreparable damage to the sensitive marine environment (Ford et al. 2006).

To address the potential for a major accidental spill, there exists a range of spill prevention, contingency planning and response readiness at the national level of the Arctic states (World Wildlife Fund 2007). Although the oil spill response systems are dependent on a consolidation of in-situ burning and dispersant application, the response options may be limited by the harsh environmental conditions of the Arctic operating environment or even be limited to areas within national jurisdiction. The areas beyond national jurisdiction may well be covered by international law or bi-lateral agreements under the International Convention of Oil Pollution Preparedness, Response and Co-operation of 1990 (OPRC) and the Protocol on the Preparedness, Response and Co-operation on Pollution Incidents by Hazardous and Noxious Substances of 2000 (OPRC-HNS Protocol). Parties to the OPRC Convention are also required to provide assistance to others in the event of a pollution emergency and a provision has been incorporated for the reimbursement of any assistance so provided. With reference to the Arctic high seas, the coastal states rely on Memorandum of Understandings (MOUs) and regional or bilateral arrangements are already in place that provides a framework for co-operation among Arctic States under the OPRC (Governance of Arctic Shipping 2009; Johansson and Donner 2014). Offshore units, i.e. fixed and floating offshore installation or structure have been included in the OPRC (OPRC 1990). Although the oil pollution emergency plans are to be coordinated with the national system in accordance with procedures established by the competent national authority, a unique and significant feature of the OPRC lies in its effort to promote international co-operation in combating oil pollution via bi-lateral or multilateral agreements (OPRC 1990). OPRC is thus seen as a framework for international cooperation in combating accidental oil pollution for the main five Arctic States, plus the other three States that are located in the Arctic-circle (Johansson and Donner 2014). The OPRC convention provides a concrete foundation for the Arctic states to jointly develop comprehensive strategies to respond to maritime incidents, which have environmental repercussions (Brubaker 2000). Arctic States have already initiated joint contingency planning arrangements and they include, among others, the Canada-US Joint Marine Pollution Contingency Plan for the Beaufort Sea area, the Russian-US Joint Marine Pollution Contingency Plan, the joint Russian-Norwegian Plan for the Combating of Oil Pollution in the Barents Sea and the Canada-Denmark Agreement for Marine Environmental Cooperation, which includes annexes for responding to shipping and offshore hydrocarbon spills (Governance of Arctic Shipping 2009; Johansson and Donner 2014).

Although the existing joint contingency plans exemplify the current trend of co-operation and joint initiatives compatible with measures promulgated in two individual states, the plans, unfortunately, do not extend to the Arctic high seas.

Although other contingency plans in their operating areas include "offshore" and "open ocean", the time standard for response in those areas is considerably higher (36–60 h) (Canada-US (Salish Sea) Response Organizations 2014). Based on the given harmful affect of oil on ice and ice-covered waters, the time standard for the Arctic areas and the high seas needs to be shorter than the average "open sea" response and prevention time standard.

An Arctic contingency plan that has emanated from strategic thinking is the Canada-US Joint Marine Pollution Contingency Plan for the Beaufort Sea area through which the coast guards of Canada and the US have entered into a Joint Marine Contingency Plan for dealing with the release of pollutants or harmful substances (Canada-US Joint Marine Pollution Contingency Plan 2003). The geographic scope of the contingency plan includes the Beaufort Sea, comprising those waters off the Arctic Coast of Canada and US (Canada-US Joint Marine Pollution Contingency Plan 2003). Although the initiatives involve "waters off the Arctic Coast", there are noteworthy joint initiatives in the past following the 1988 oil spill of the barge Nestucca and the Exxon Valdez. After these oil spills, the Governors of the states of Alaska, Washington, Oregon and California and the premier of the province of British Columbia established the Pacific States-British Columbia Oil Spill Taskforce (John 2006). For the Atlantic coast, large scale joint simulation exercises referred to as CANUSLANT exercises are routinely held to practice bilateral response and preparedness to a pollution event by respective authorities (Bellefontaine 2007). This exemplifies the current good governance practice of oil spill response for those areas of the Arctic led by respective agencies at the regional operational level (Bellefontaine 2007). More recently, under the Economic Action Plan 2015, the Canadian Coast Guard (CCG) is in the process of reviewing the Arctic maritime Search and Rescue (SAR) service requirements in order to enhance the CCGs auxiliary capacity to keep pace with the rising marine traffic levels in the given region (Official Homepage of the Government of Canada). It is also suggested that the CCG will immediately enhance emergency response and SAR capacity in the Arctic by enhancing the current coast guard auxiliary capacity presence in remote locations (Official Homepage of the Government of Canada), which demonstrates the Government of Canada's ongoing commitment to strengthen marine safety by improving the charting of Arctic waterways and developing options to improve immediate response in the Arctic (Official Homepage of the Government of Canada). For Canada, there also exists a bilateral agreement with Denmark covering Baffin Bay, Davis Strait and other joint seas around Denmark, which calls for co-operation between the Danish authorities and the CCG when responding to incidents in contiguous waters (John 2006). Although it does not extend to the high seas, the bi-lateral agreement serves as a common platform for joint decisions and initiatives to combat oil spills for those areas included in the agreement.

Although trans-Arctic shipping in areas beyond the national jurisdiction, i.e. beyond the EEZ and across the high seas of the Arctic raises governance issues, the International Maritime Organization (IMO) has implemented a convention that deals with instant measures on the high seas as may be necessary to prevent, mitigate or eliminate danger following upon a maritime casualty (Johansson and Donner

2014). The International Convention Relating to Intervention on the High Seas in Cases of Oil Pollution Casualties of 1969 (Intervention Convention) is one of a kind that deals with accidental oil pollution in the high seas and calls for "extreme urgency requiring measures to be taken immediately" (Intervention Convention, 1969). Although the Intervention Convention provides no explicit reference to the Arctic, the parties to the convention have an option to implement intervention policies under domestic contingency plans. Intervention in this context is directly connected to the preparedness and response actions and is introduced as a defence mechanism to respond rapidly to accidental pollution (Intervention Convention, 1969). "Oil spill intervention" is the new legal jargon when it comes to dealing with accidental pollution emerging on the high seas empowering the coastal states to take measures rendered necessary by the urgency of the situation, without prior notification or consultation or without continuing consultations which have already begun (Intervention Convention 1969; Johansson and Donner 2014). Article III (d) of the Intervention Convention states;

> in cases of extreme urgency requiring measures to be taken immediately, the coastal State may take measures rendered necessary by the urgency of the situation, without prior notification or consultation or without continuing consultations already begun; (Intervention Convention, 1969)

Although Article III (d) of the Intervention Convention bypasses the notion of "consultation with neighbouring states" and empowers the coastal state to proceed with unilateral actions, this approach can be considered pertinent, convenient and fitting for the Arctic high seas. The rationale of Article III (d) would certainly operate against all joint contingency plans currently in place, but the inevitable question is why not authorise individual Arctic States to take intervention measures in the high seas, which has been repeatedly cited as a sensitive area that can be damaged from operational discharge or accidental spill of oil and generic pollutants. On the other hand, if the Arctic states were parties to the Intervention Convention and decided to act upon Article III (d) following unilateral actions in dealing with extreme urgencies originating from a maritime incident, it would open the doorway and increase the chances of solving complicated issues. The problem for oil spill intervention in the Arctic can be termed as complex because first and foremost, there still remains the question of geographical delimitation issues. The concept of "places of refuge", as a part of the intervention solution, might act as a problem-catalyst to this status quo maritime boundary complexity where Arctic coastal states might be in a situation to select a place bordering another Arctic coastal state as a part of an oil spill response (Johansson and Donner 2014). This might invoke unnecessary conflicts since the Intervention Convention has given unlimited authority to state parties involved in immediate intervention that seek to control pollution of the Arctic environment via rapid response (Johansson and Donner 2014). If any of the Arctic Coastal states refrain from co-operating and refuse to provide a place of refuge to the acting authority—then it might complicate and interfere with the high sea intervention in question (Johansson and Donner 2014). Moreover, the exceptions to the provisions to act without "consultation", as embedded in article III (d), might not be acceptable to Canada if for any reason the US coast guard decides to act unilaterally without considering the negative impacts that the wrecked vessel from a maritime casualty may have on the Arctic waters of Canada (Johansson and Donner 2014). This observation is a major drawback in the context of oil spill intervention in the Arctic high seas and needs to be clarified by IMO (Johansson and Donner 2014).

5 Towards an Integrated Intervention Plan for the Arctic High Seas

Although the Intervention Convention calls for "such measures" on the high seas as may be necessary to address "imminent" danger, there are many dimensions to the word oil spill "intervention" (Intervention Convention, 1969) which need to be taken into account. When it comes to "intervention" in Norway, it is the duty of the Norwegian Coastal Directorate (NCD) to identify and list "places of refuge" and places of grounding in the Norwegian Coastal Administrations (NCA) Emergency Response Plan (Johansson and Donner 2014). These "places of refuge" are a haven for damaged ships and are utilized in cases where there is a danger of severe pollution as a result of accidents at sea (John 2006). In short, they are integral to the oil spill intervention plan for Norway.

To date, there have been numerous interventions that have taken place after various incidents, but they differ from each other in terms of ship-damage, authority-action and how individual states have defined intervention in their domestic law. In some instances, the immediate response was prolonged because of unavoidable circumstances and for some it was not possible because of late-response. For the Arctic areas and the high seas, if there is the need for a response, it has to be timely. Otherwise it defeats the very purpose of the word "intervention". "Response" would refer to a response action after an incident, but "intervention" would be different in so far as it requires an urgent action to contain the pollution before a single drop of oil touches the icy waters. This hypothesis takes into account the elements that comprise the word "intervention" in the Intervention Convention. The various quantitative research explanations relating to its vulnerability leading to its label as a "sensitive area" and the need to maintain its pristine features have always been a justification to take stringent measures in the Arctic. More recently, the five Arctic states that surround the central Arctic Ocean have also signed a declaration to prevent unregulated fishing in the central Arctic Ocean (Official Homepage of US Department of State 2015; Munir 2015). To that end, the declaration further acknowledges the interests of other states in preventing unregulated fishing in the high seas within the Arctic areas, and recommends the initiation of a broader process to develop measures consistent with the said declaration (Official Homepage of US Department of State 2015) (Fig. 1).

The above map shows the 2.8-million-square-kilometre area in the central Arctic Ocean that lies beyond the exclusive economic zone of the five Arctic coastal states:

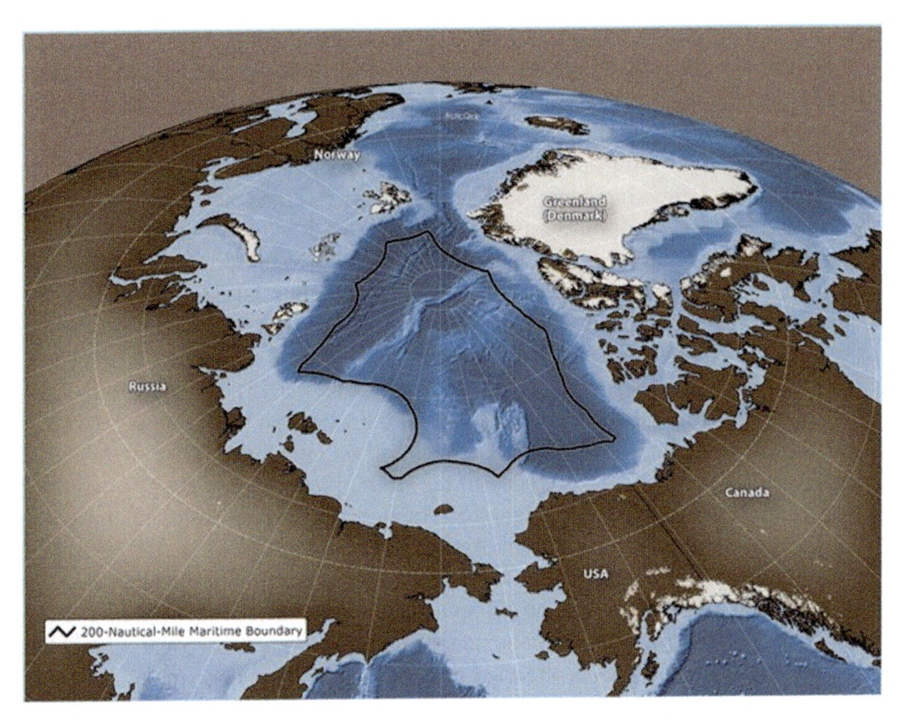

Fig. 1 Central Arctic Ocean. Source: retrieved from the article "Five Arctic countries, with Inuit support, sign moratorium on commercial fishing for the Central Arctic Ocean pending sustainable management regime that incorporates Inuit traditional knowledge July 22, 2015" by Magdalena A.K. Muir, Climate editor (posted by EUCC Editor in Fisheries & Aquaculture)

Canada, Russia, Norway, the U.S. (Alaska) and Denmark (Greenland). If the states that surround the central Arctic Ocean can take steps to regulate commercial fishing in the Arctic high seas, then the Arctic states could extend mandates to regulate accidental pollution in the high seas by defining the notion of "oil spill intervention" more stringently. The Arctic states could extend the agenda of the coast guards and maritime administrations of individual states, which can oversee a joint and integrated intervention plan in the Arctic high seas and North Pole, which is located roughly 400 miles to the north of any land. Maritime accidents cannot be predicted and the central Arctic Ocean is said to be a "common heritage of mankind", which will be a highly traversable area in the coming years bringing with it the risk of oil spill disasters that the world has witnessed in other waters of the globe.

At the 2013 Ministerial meeting in Sweden, the Arctic Council signed a second legal instrument known as the Agreement on Cooperation on Marine Oil Pollution Preparedness and Response in the Arctic of 2013 (ACMOPPRA). The agreement is centered on Arctic marine oil pollution and the prevention and response regime. The all-encompassing definition of "oil pollution" incident in Article 2 of the agreement includes "emergency response" and "immediate response" elements that help define "oil spill intervention" as explained previously (Agreement on Cooperation on

Marine Oil Pollution Preparedness and Response in the Arctic 2013). As such, the agreement was concluded between governments of the Arctic states and provided specific reference to the Intervention Convention of 1969. This is a clear indication and an indirect declaration to address the regulatory gaps that prevail in the Arctic high seas. This is also transparent from the way the preamble of the agreement has been structured expressing consciousness of "the threat from marine oil pollution to the vulnerable Arctic marine environment and to the livelihoods of local and indigenous communities", "that in the event of an oil pollution incident, prompt and effective action and cooperation among the Parties is essential in order to minimize damage that may result from such an incident" and "the Parties' obligation to protect the Arctic marine environment" (Agreement on Cooperation on Marine Oil Pollution Preparedness and Response in the Arctic 2013). Although the objectives highlight strengthened cooperation, coordination and mutual assistance in order to protect the marine environment from oil pollution, the scope of the agreement does not include any reference to the development of unilateral action with co-operation and consultation to intervene in an accidental oil spill. Again, Article 6 of the agreement is in conflict with Article III (d) of the Intervention Convention, which encourages unilateral action for which "consultation" with the other state can be avoided in order to satisfy the oil spill intervention objective. However, many provisions within the agreement concern contact points and exchange of information between parties, which gives the idea that the intention to develop an integrated Arctic Ocean management is now an implicit agenda.

An integrated intervention approach can also refer to a regional approach and from a regional perspective, the Arctic Council and its unique structure is considered to be a significant framework for the continuation of the development of such an approach. The Arctic Council has been termed as a high-level forum established to promote co-operation and coordination among the eight Arctic states (Arctic Governance in an Evolving Arctic Region 2012). One of the working groups of the Arctic Council is the Emergency Prevention, Preparedness and Response (EPPR) and the goal of EPPR is to contribute to the protection of the Arctic marine environment from the threat or impact that may result from an accidental release (Official Homepage of the Arctic Council EPPR Working Group). A proposal forwarded by the Standing committee of Parliamentarians of the Arctic Region is that the Arctic Council should become a fully-fledged international organization and in such event, the agreements and co-operation between and among the Arctic states could be made legally binding (Official Homepage of the Arctic Council EPPR Working Group). But prior to realizing this idea and reaping the benefits from the outcome, there is always the need to implement a regional intervention plan or a guideline, which embodies an integrated approach for the high seas and one that could have the force of "hard law". At the regional level, there already exists a legal instrument on dumping of wastes, also known as the Convention for the protection of the Marine Environment of the North-East Atlantic (OSPAR). OSPAR aims to amalgamate the Convention for the Prevention of Marine Pollution by Dumping from Ships and Aircraft of 1972 (OSLO Convention) and the Convention for the Prevention of Marine Pollution from Land-based Sources, 1974 as amended by the

Protocol of 26 March 1986 (Paris Convention) whereby Annex III deals with prevention and elimination of pollution from offshore sources. Although OSPAR addresses two primary concerns, the ocean dumping of radioactive wastes and the dumping of dredged material, the significant drawback of the convention is that it covers only a part of the Arctic marine environment. Then again, OSPAR has a membership different from the Arctic region and non-Arctic states with no national interest in the Arctic region may influence the decision-making process not unique to specific Arctic intervention concerns. Gaps will always remain, as Denmark and Norway are the only Arctic States parties to this convention and it will be impossible to enforce the provisions or policies of OSPAR on non-contracting states.

6 Conclusion

Despite the concern for the oil spill aftermath and disasters in ice-affected waters, there is a vacuum of law and a global instrument or a good governance body that can comprehensively regulate oil and gas activities in the Arctic high seas. The international regulations that brush on the idea of oil spill discharge, prevention and response have not taken the Arctic areas into account, let alone supplement a detailed guideline for oil spill intervention in the high seas. The harsh reality that no real global regulations exist apart from the Intervention Convention makes the reference to international standards rather insipid. OSPAR as a regional instrument does not necessarily relate to the Arctic. Even though there are many instances of Regional Seas Programmes in other parts of the world, which have expanded their remits beyond pollution prevention and enveloped a wider array of issues and responsibilities, there has been no attempt by respective authorities to address any form of integrated approach for the Arctic high seas. Examples are ripe at the international level and one such example is the UNEP Regional Seas Programme, which has set goals beyond the mere protection of the environment. It is not that the UNEP Regional Seas Programme has developed an integrated intervention strategy, rather the expansion and scope of ambitions of the UNEP Regional Seas Programme to take broader ecosystems into consideration should be observed as an example. Even if the oil pollution does not originate from colliding ships or a maritime incident, there are growing interests towards seabed mining in the Arctic, which could pose a greater risk of oil pollution in the Arctic high seas. Although the International Seabed Authority (ISA) has implemented its own codes and regulations with the aim to reduce environmental impacts from seabed mining in the high seas, it does not cover the EEZ and represents a problem in its own right (Global Ocean Commission 2013). In the near future, seabed mining in the arctic will pose an additional challenge as the offshore industries are trying to reach deeper and more distant waters in search of oil and gas (Global Ocean Commission 2013).

Even though there is a general lack of a marine infrastructure in the Arctic, several of the Arctic states are familiar with the term "integrated approach" that reflects regional co-operation for other parts of their sea-region that do not fall within the

Arctic. The European Union (EU) "Macro-Regional" strategy for the Baltic-Sea Region is a unique example of regional co-operation and alludes to an area including territories from a number of different countries or regions associated with one or more common features (Bellefontaine and Johansson 2015). The main aim of the "Macro-Regional" strategy is to address common challenges by amalgamating individual regional potentials and the integrated approach as envisioned in the "draft Council conclusions on the governance of macro-regional strategies" is commendable (Bellefontaine and Johansson 2015). The "Macro-Regional" strategy can serve as an analogous example for the Arctic States and the Arctic Council can combine the efforts of the inter-governmental institutions and bi-lateral or multi-lateral agreements on oil spill contingency plans and develop a more integrated oil spill intervention strategy. The Arctic maritime zones leading up to the North Pole is a heterogeneous area in economic, environmental and cultural terms and the Arctic states share the pristine area enriched with common natural resources. Thus, the rationale for developing a type of "Macro-Regional" Strategy for the Arctic is justified in the sense that the Arctic Council, among others, is guided by the objectives of sustaining the pristine nature of the Arctic Ocean, connecting the Arctic states and increasing its prosperity. Once established, the cohesive strategy would automatically set in place an integrated approach and then an integrated intervention strategy for the Arctic high seas.

Global warming has a direct impact on the oceans and the seas (Bellefontaine and Linden 2009). Rapid climate change caused by global warming will over the next few decades transform the Arctic Ocean from an impassable area into a seasonally navigable sea. Current trends in shipping indicate that the paradigm shift of taking advantage of the shorter Arctic sea-routes has already begun. The augmentation of intra- and trans-Arctic shipping in this pristine and remote area poses a threat of significant damage to the marine environment and its living resources. With a view to diminishing these risks counteractive measures have been simultaneously adopted under international law and its follow-up in private law, which corresponds to ascertaining safety of navigation and preventing accidental pollution from vessels in the Arctic. Although the limelight has been cast on accidental vessel-pollution and navigational safety, deliberate dumping and operational discharge as a source of vessel-pollution is considered to be more detrimental in this part of the world (Brubaker 1993). It seems that the international, regional and national instruments already exist, which label some ocean areas as "protected' or "sensitive" and this is done to safeguard the pristine environment from voluntary vessel-pollution. The Arctic is deprived of any such official label or indication and in terms of oil spill prevention, response and preparedness suggesting the Arctic states apparently suffer from bureaucratic fatigue from the 'soft law' approach. However, the ever-expanding international shipping through the Arctic may soon test the adequacy of existing regimes. Hence, the Arctic region remains vulnerable due to the absence of legally binding law on an oil spill integrated intervention model, the aim of which should be to strike a balance between safeguarding the marine ecosystem and commercial exploitation. Faced with useful natural resources and increasing user conflicts over sectoral and political boundaries in the Arctic high seas, there is an urgency to initiate a more detailed, holistic and integrated approach to intervention management policies by interested Arctic states. The intervention regimes already exist and an integrated refinement of those regimes will surely add to the strength of the Arctic states' challenge to protect the Arctic marine environment from detrimental exploitation, which continues today and against an increased exploitation that will commence tomorrow.

References

Agreement on Cooperation on Marine Oil Pollution Preparedness and Response in the Arctic. (2013). Retrieved July 3, 2015, from https://oaarchive.arctic-council.org/bitstream/handle/11374/529/EDOCS-2068-v1-ACMMSE08_KIRUNA_2013_agreement_on_oil_pollution_preparedness_and_response_signedAppendices_Original_130510.pdf

Arctic Governance in an Evolving Arctic Region. (2012). *10th Conference of parliamentarians of the Arctic region: Conference Statement.* Retrieved July 2, 2015, from http://www.arcticgovernance.org/10th-conference-of-parliamentarians-of-the-arctic-region-conference-statement.5091603-142902.html

Bellefontaine, N. (2007). *Oil spill preparedness and response – The Canadian Experience.* Shanghai, China: Shanghai International Maritime Forum.

Bellefontaine, N., & Johansson, T. (2015) *Ocean governance and the EU "Macro-Regional" strategy for the Baltic-Sea Region.* Baltic Rim Economies, University of Turku.

Bellefontaine, N., & Linden, O. (2009). Impacts of climate change on the maritime industry. In *Proceedings of the Conference on Impacts of Climate Change on the Maritime Industry.* Lund, Sweden, 2009.

Brubaker, D. (1993). *Marine pollution and international law: Principles and practice.* London: Belhaven Press.

Brubaker, D. (2000). Regulation of navigation and vessel source pollution in the Northern Sea Route: Article 234 and state practice. In D. Vidas (Ed.), *Protecting the polar marine environment: Law and policy for pollution prevention.* Cambridge University Press.

Canada-US Joint Marine Pollution Contingency Plan. (2003). Retrieved July 3, 2015, from http://www.ccg-gcc.gc.ca/folios/00025/docs/canadaus_pub-eng.pdf

Canada-US (Salish Sea) Response Organizations. (2014). International Oil Spill Conference Proceedings, 2014(1).

Cinelli, C. (2011). The law of the sea and the Arctic Ocean. *Arctic Review on Law and Politics, 2*(1).

Ford, J. D., Smith, B., & Wandel, J. (2006). Vulnerability to climate change in the Arctic: A case study from Arctic Bay, Canada. *Global Environmental Change, 16*(2), 145–160.

Global Ocean Commission. (2013). *Elimination of pollution that affects the high seas.* Policy Option Paper no. 3, A Series of Papers on Policy Options Prepared for the Third Meeting of the Global ocean Commission. Retrieved July 2, 2015, from http://www.globaloceancommission.org/wp-content/uploads/POP-3_Marine-Pollution-FINAL_no-options.pdf

Gold, E. (2006). *Gard handbook on protection of the marine environment* (3rd ed.). Arendal: Gard AS.

"Governance of Arctic Shipping". (2009, April 29). *Arctic Marine Shipping Assessment.* Retrieved July 2, 2015, from http://www.arctic.gov/publications/AMSA/governance.pdf

Guidelines for Ships Operating in Polar Waters. (2009, December 2). Adopted by IMO Assembly Resolution A.1024 (26). Retrieved July 31, 2015, from http://www.imo.org/Publications/Documents/Attachments/Pages%20from%20E190E.pdf

International Convention for the Prevention of Pollution from Ships, 1973 as modified by the Protocol of 1978, entered into force on 2 October 1983. International Maritime Organization.

Retrieved July 31, 2015, from http://www.mar.ist.utl.pt/mventura/Projecto-Navios-I/IMO-Conventions%20(copies)/MARPOL.pdf

International Convention on Oil Pollution Preparedness, Response and Co-operation. (1990). *International Maritime Organization*. Retrieved July 2, 2015, from http://www.ifrc.org/docs/idrl/I245EN.pdf

International Convention Relating to Intervention on the High Seas in Cases if Oil Pollution Casualties, 1969 (with annex official Russian and Spanish translations and Final Act of the International Legal Conference on marine pollution damage, 1969), Concluded at Brussels on 29 November 1969. Retrieved August 5, 2014, from https://treaties.un.org/doc/Publication/UNTS/Volume%20970/volume-970-I-14049-English.pdf

Jensen, O. (2007). *The IMO Guidelines for ships operating in Arctic ice-covered areas: From voluntary to mandatory tool for navigation safety and environmental protection* (FNI Report 2/2007). The Fridt JOF Nansen Institute.

Johansson, T., & Donner, P. (2014). *The shipping industry, ocean governance and environmental law in the paradigm shift: In Search of a pragmatic balance for the Arctic*. Springer.

John, P. (2006). Places of Refuge: Considerations for determining a Canadian approach. In A. Chircop & O. Linden (Eds.), *Places of refuge for ships: Emerging environmental concerns of a maritime custom*. Brill: Nijhoff publishers.

Kimball, L. A. (2005). *The International Legal Regime of the High Seas and the Seabed Beyond the Limits of National Jurisdiction and Options for Cooperation for the establishment of Marine Protected Areas (MPAs) in Marine Areas Beyond the Limits of National Jurisdiction* (Technical Series No. 19). Montreal: Secretariat of the Convention on Biological Diversity.

Koivurova, T. (2010). Environmental protection in the Arctic and the Antarctic. In N. Loukacheva (Ed.), *Polar law textbook*. Copenhagen: Nordic Council of Ministers.

Lag (1980:424) om åtgärder mot förorening från fartyg, Consolidated version of Act No. 424 of 1980 as amended last by SFS 2006:1318.

Marchenko, N. (2014). Northern Sea Route: Modern State and Challenges, Draft OMAE 2014-23626. In *Proceedings of the 33rd International Conference on Ocean*. Offshore and Arctic Engineering.

Munir, M. A. K. (2015, July 22). *Five Arctic countries, with Inuit support, sign moratorium on commercial fishing for the Central Arctic Ocean pending sustainable management regime that incorporates Inuit traditional knowledge*. Retrieved September 5, 2015, from https://euccnews.wordpress.com/2015/07/22/five-arctic-countries-sign-moratorium-on-commercial-fishing-for-the-central-arctic-ocean-pending-sustainable-management-regime-that-incorporates-inuit-traditional-knowledge-with-support-of-inuit-peop/

Official Homepage of Global Marine Oil Pollution Information Gateway. Retrieved July 2, 2015, from http://oils.gpa.unep.org/facts/operational.htm

Official Homepage of the Arctic Council EPPR Working Group. Retrieved July 3, 2015, from http://www.arctic-council.org/eppr/

Official Homepage of the Government of Canada, Enhancing the Arctic Search and Rescue Capacity. Retrieved August 7, 2015, from http://news.gc.ca/web/article-en.do?nid=1009409

Official Homepage of US Department of State. (2015). *Arctic Nations Sign Declaration to Prevent Unregulated Fishing in the Central Arctic Ocean*. Retrieved July 2, 2015, from http://www.state.gov/r/pa/prs/ps/2015/07/244969.htm

Rothwell, D. R., & Joyner, C. C. (2000). Domestic perspectives and regulations in protecting the polar marine environment: Australia, Canada and the United States. In D. Vidas (Ed.), *Protecting the polar marine environment: Law and policy for pollution prevention*. Cambridge University Press.

The US Arctic Research Commission and the US Army Corps of Engineers. (2012). *Oil spills in Arctic Waters: An introduction and Inventory of Research Activities and USARC Recommendations*. Retrieved July 2, 2015, from http://www.arctic.gov/publications/white%20papers/oil_spills_2012.pdf

United Nations Convention on the Law of the Sea. (1982). Retrieved July 6, 2015, from http://www.un.org/depts/los/convention_agreements/texts/unclos/unclos_e.pdf

World Wildlife Fund. (2007). *Oil spill: Response challenges in Arctic waters*. Norway: WWF International Arctic Programme.

Arctic Vessel Traffic and Indigenous Communities in the Bering Strait Region of Alaska

Julie Raymond-Yakoubian

Contents

Abstract The Bering Strait region of Alaska is home to three different groups of indigenous people and 20 federally-recognized Tribes. Indigenous communities in the Bering Strait have both a right and a strong desire to be included in discussions about the future of vessel traffic in the region, to have their Traditional Knowledge and expertise about the marine environment considered and utilized, and to have meaningful involvement in decision making about activities taking place in their homeland and with the potential to impact their lives. This chapter outlines some of the concerns that Tribes and Tribal organizations have regarding current and projected vessel traffic in the region. It also discusses recent research conducted by Kawerak and Tribes that can contribute to discussions about the future of arctic shipping, including GIS mapping, Traditional Knowledge documentation projects, and regional meetings that have focused on shipping.

Keywords Bering Strait · Vessel traffic · Indigenous · Subsistence · Traditional Knowledge · Food security

J. Raymond-Yakoubian (✉)
Kawerak, Inc., Nome, AK, USA
e-mail: juliery@kawerak.org

L. P. Hildebrand et al. (eds.), *Sustainable Shipping in a Changing Arctic*, WMU Studies in Maritime Affairs 7, https://doi.org/10.1007/978-3-319-78425-0_16

1 Introduction

The Bering Strait region of Alaska (Fig. 1) is the homeland of Inupiaq, Yup'ik and St. Lawrence Island Yupik people. The members of 20 federally recognized Tribes currently live in Nome, the "hub" city for the region, and 15 surrounding villages. Kawerak, Incorporated is the Alaska Native nonprofit tribal consortium for this region, based out of Nome. Kawerak provides services and programs to Tribes and region residents, which includes conducting social science research in the region through our Social Science Program. Kawerak also has a Marine Program and administers the Eskimo Walrus Commission, both of which also focus on marine and vessel traffic-related issues.

The Social Science Program conducts collaborative, community-based research. Our methods are grounded in the tradition of anthropology and include interviews, focus groups, workshops, mapping, community meetings and participant observation. Much of the research we conduct addresses Tribal needs, or information gaps, many of which are relevant to vessel traffic. We work with Traditional Knowledge holders—individuals who are recognized as subject matter experts by their Tribal

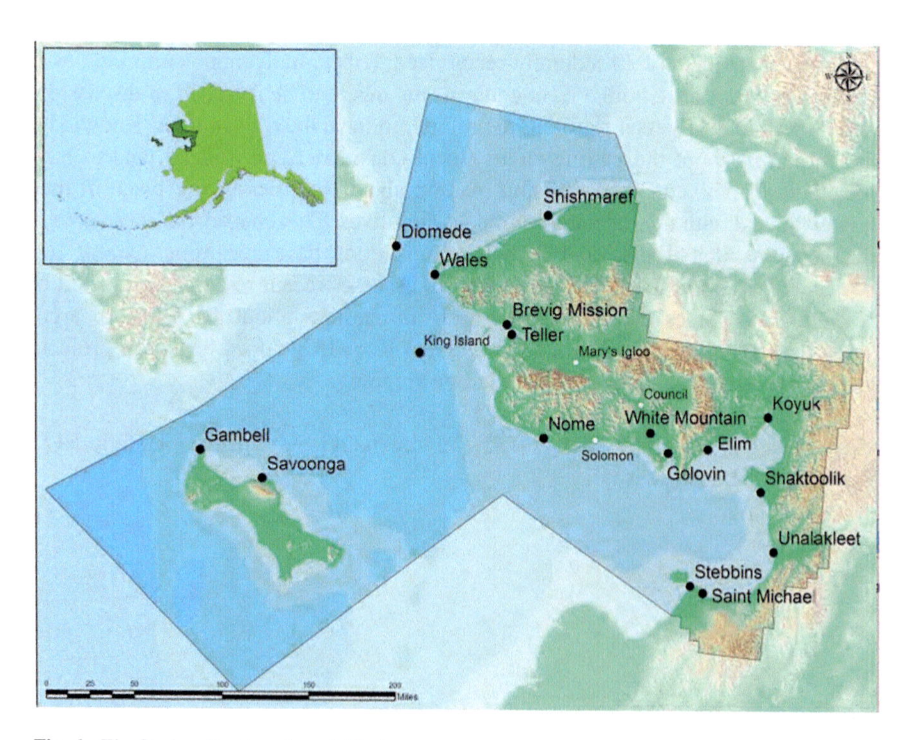

Fig. 1 The Bering Strait region of Alaska

leadership and peers—to document Traditional Knowledge[1] and community perspectives on a variety of topics. Our Social Science Program also partners with other indigenous organizations, with conservation groups, and other researchers. The results of our work are used by Kawerak, Tribes and others for a variety of purposes.

This chapter describes some of the concerns that Tribes and Tribal organizations have regarding current and projected vessel traffic in the Bering Strait region. It also discusses recent research conducted by Kawerak and Tribes that can contribute to discussions about the future of arctic shipping, including GIS mapping, Traditional Knowledge documentation projects, and regional meetings that have focused on shipping.

Bering Strait communities are remote and difficult to access. Most communities in the region are only accessible by small airplane, boat or snowmobile. The community of Diomede, for example, is only accessible by helicopter for most of the year. Region villages range in size from approximately 115 people to almost 700, and in most villages around 90% of residents are Alaska Native or American Indian (ADCCED 2017). The city of Nome, the largest community in the region, has a population of 3598 residents, of which approximately 55% are Alaska Native or American Indian (*ibid.*). The Nome Census Area has a population of approximately 9900 people, 75% of whom are Alaska Native or American Indian (US Census Bureau 2016).

Traditional cultural practices, including intensive use of the marine environment for subsistence activities and travel, remain critically important in region residents' lives today. Subsistence can be described as "hunting and gathering related activities which have a deep connection to history, culture, and tradition, and which are primarily understood to be separate from commercial activities" (Raymond-Yakoubian et al. 2017, p. 133). Communities in the region depend on marine species such as bowhead whales, beluga whales, walruses, ice seals, various birds and fish, and benthic fauna such as clams, sea peaches (Halocynthia aurantium) and other animals.

Marine subsistence activities take place primarily from small boats (e.g. 18 foot aluminum skiffs) with outboard motors, on the sea ice, and in the intertidal zone (e.g. collection of clams and driftwood). The safe and successful conduct of marine subsistence activities requires extensive knowledge of the local environment and

---

[1]Traditional Knowledge can be defined as: "a living body of knowledge which pertains to explaining and understanding the universe, and living and acting within it. It is acquired and utilized by indigenous communities and individuals in and through long-term sociocultural, spiritual and environmental engagement. TK is an integral part of the broader knowledge system of indigenous communities, is transmitted intergenerationally, is practically and widely applicable, and integrates personal experience with oral traditions. It provides perspectives applicable to an array of human and non-human phenomena. It is deeply rooted in history, time, and place, while also being rich, adaptable, and dynamic, all of which keep it relevant and useful in contemporary life. This knowledge is part of, and used in, everyday life, and is inextricably intertwined with peoples' identity, cosmology, values, and way of life. Tradition – and TK – does not preclude change, nor does it equal only 'the past'; in fact, it inherently entails change" (Raymond-Yakoubian et al. 2017, p. 133).

broader ecosystem, including information from both personal experiences and that which has been passed down through generations. The indigenous residents of the Bering Strait region rely upon the marine environment for their cultural, nutritional, economic and spiritual needs, and take very seriously their role as caretakers of their marine and terrestrial surroundings.

2 Recent Kawerak Work Related to Vessel Traffic

Much of the recent work of the Kawerak Social Science Program has involved mapping activities, such as the mapping of subsistence use areas and animal habitat. Through a variety of projects we have documented spatial information regarding fish, ice seals, walruses, indigenous place names and ocean currents. These projects have recorded information relevant in many ways to the issue of vessel traffic. In addition to and in conjunction with the mapping, we also document Traditional Knowledge related to harvest and processing of subsistence foods, climate changes, social relationships, cultural values and practices, and other information relevant to each particular project foci. Below we review some of this recent work and their connections to vessel traffic and vessel traffic related concerns of indigenous communities in the region.

For example, we collaborated with eleven Tribes in the region to document fish habitat and harvest areas spatially, in map format (see Fig. 2), as well as through interviews and workshops (Raymond-Yakoubian 2013; Raymond-Yakoubian and Raymond-Yakoubian 2015). Two of the communities involved in that work, Brevig Mission and Teller, sit on the north and south shores of Port Clarence, respectively. Port Clarence has also been under consideration as the location for the development of a deep-water port facility and has been used as a 'port of refuge' for over a century for vessels in need of safe harbor (USACE 2013). Brevig Mission and Teller residents, as well as people from other communities, are concerned about vessel traffic and pollution because they harvest fish and seals in the waters of Port Clarence, as well as various berries and plants along the shores. They also travel out into the waters of the northern Bering Sea and Bering Strait to hunt marine mammals.

Another recent project has involved documenting (spatially and through narratives) ice seal and walrus habitat and subsistence harvest areas in collaboration with nine Tribes (Figs. 3 and 4) (Gadamus and Raymond-Yakoubian 2015a, b; Gadamus et al. 2015). Figures 3 and 4 are part of a larger map atlas that includes information about ice seal and walrus subsistence use areas and habitat and Traditional Knowledge (Kawerak 2013a).

Figure 3 illustrates areas where King Island hunters may travel to harvest seals and walrus during the spring time (i.e. March through May). Proposed vessel traffic routes travel directly through these harvest areas, including close to King Island itself, which King Island Tribal members consider to be a particularly significant and sensitive area. Kawerak has recommended that the vessel route proposed in the Port Access Route Study be moved further to the west of King Island (Kawerak 2015b).

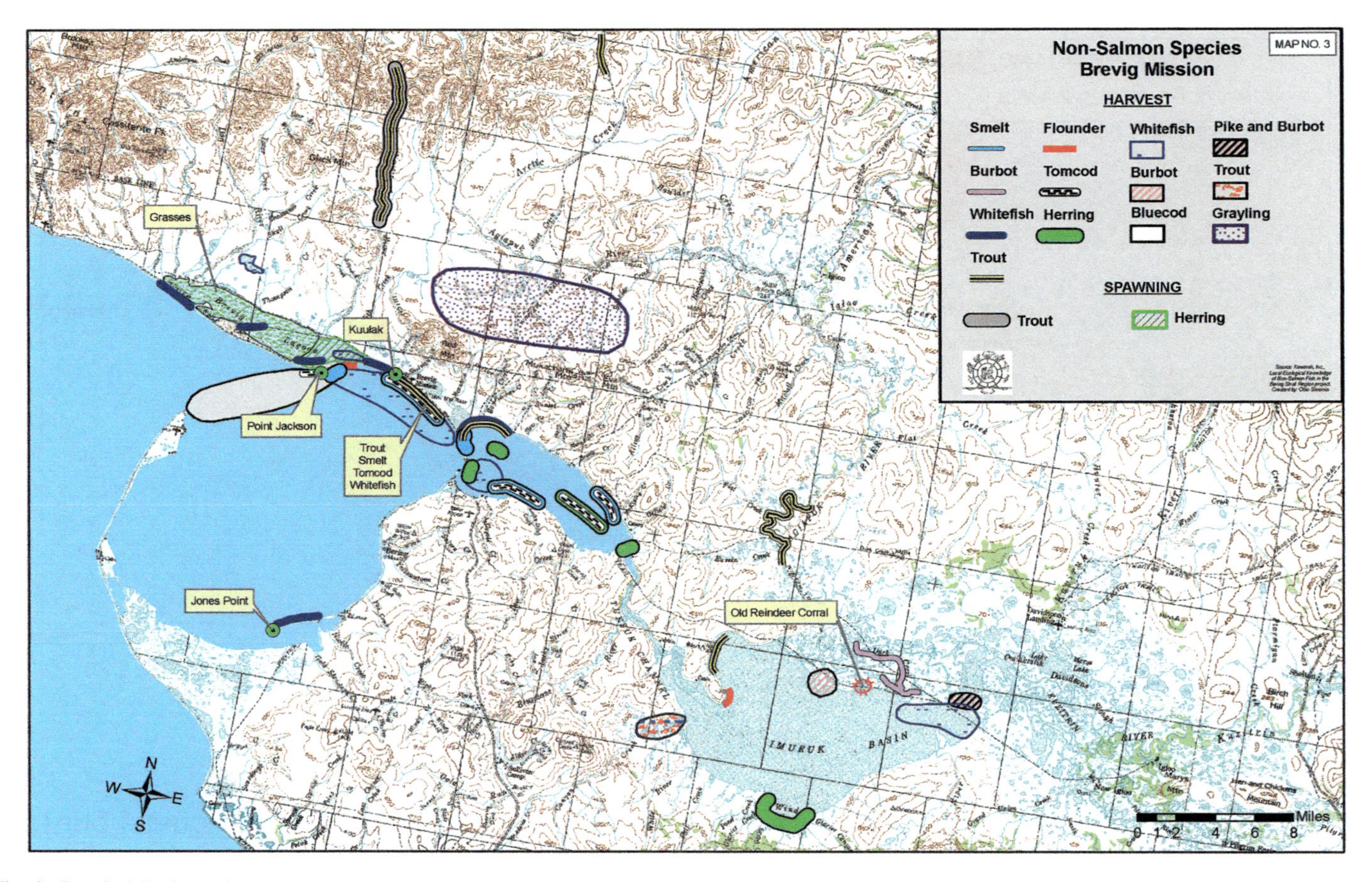

Fig. 2 Brevig Mission subsistence harvest areas for various fish species (originally published by Raymond-Yakoubian 2013; published with permission of Kawerak, Inc. All Rights Reserved)

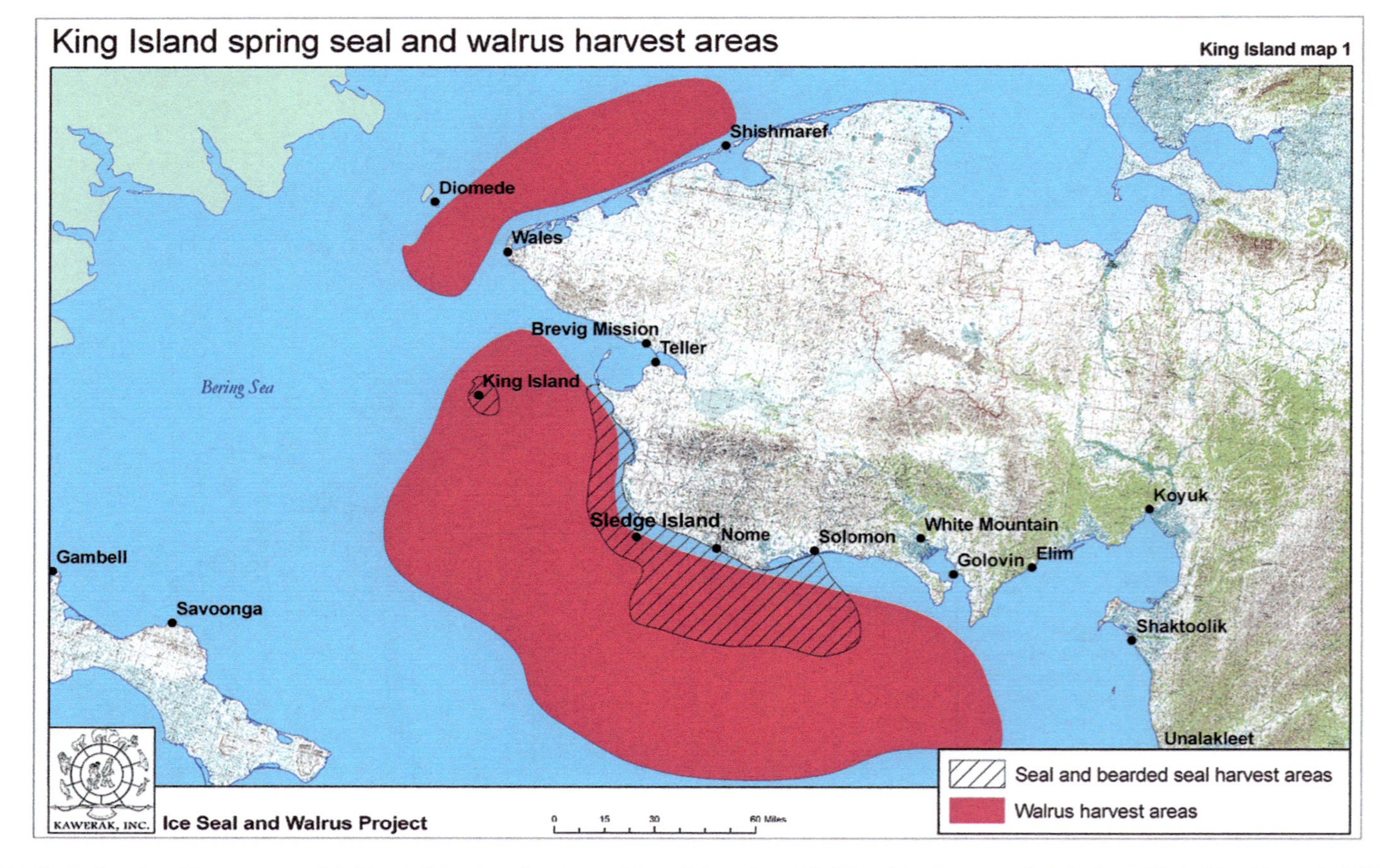

Fig. 3 Seal and walrus harvest and habitat information documented in collaboration with King Island experts (originally published by Kawerak 2013a; published with permission of Kawerak, Inc. All Rights Reserved)

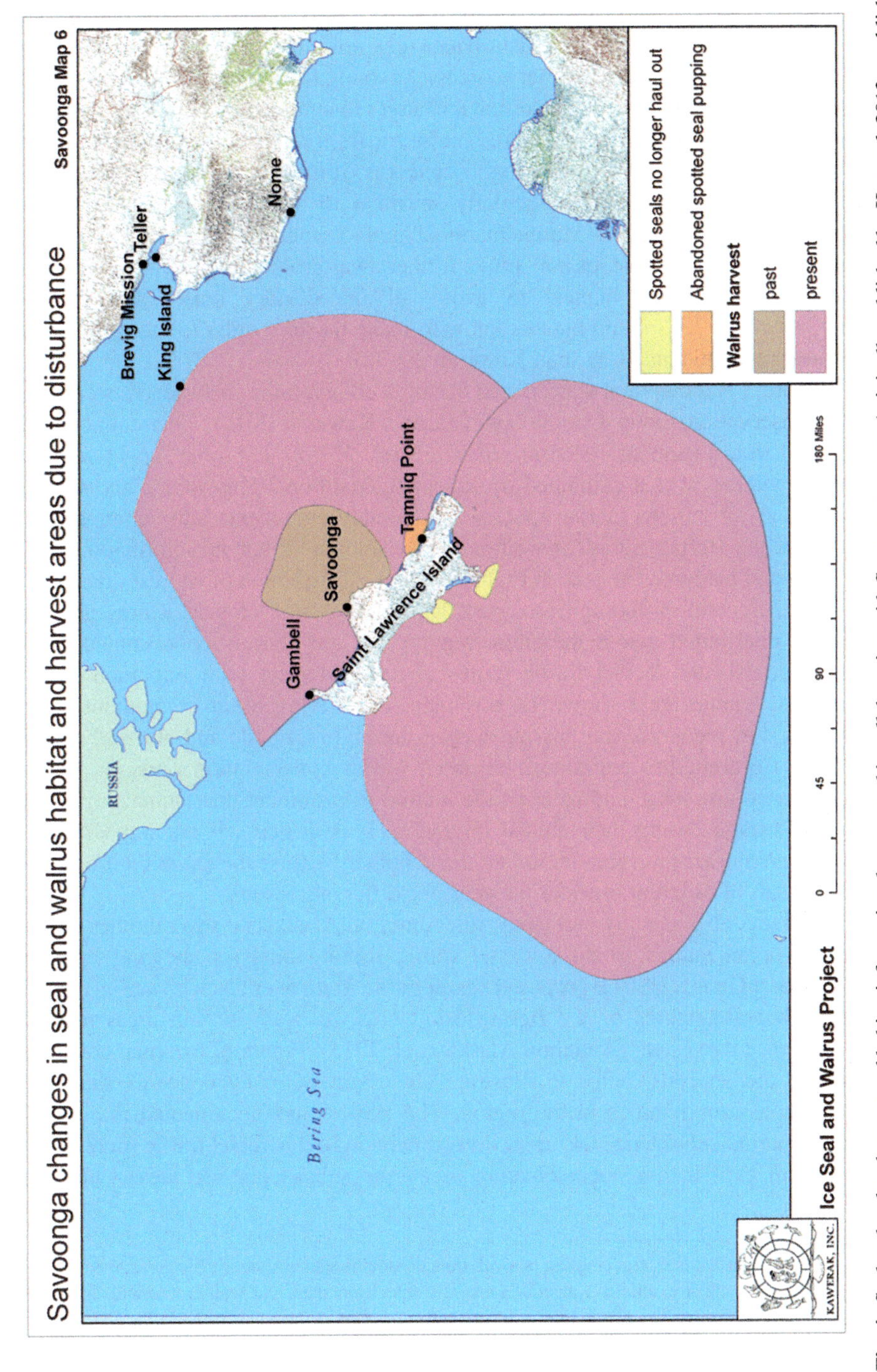

Fig. 4 Seal and walrus harvest and habitat information documented in collaboration with Savoonga experts (originally published by Kawerak 2013a; published with permission of Kawerak, Inc. All Rights Reserved)

Figure 4 shows both habitat and harvest information documented in collaboration with the Tribe in Savoonga, on St. Lawrence Island. This map shows changes in habitat and harvest areas due to disturbance (e.g. noise from human activities such as air traffic, vessel traffic, and other sources). Savoonga and other region communities are concerned about potential additional changes to animal distribution and behavior that may result from increased traffic, noise, or from any spills or accidents. Noise from vessel traffic is one of the major concerns of indigenous communities in the region. Some animals are particularly sensitive to noise and may change their behavior because of noise. Marine mammal hunters worry that animals may become more difficult to harvest, or may move further away from communities. Traditional Knowledge instructs hunters to avoid all unnecessary noise or movement (e.g. Kawerak 2013c) and hunters know to make noise in order to scare dangerous or unwanted animals away (e.g. Kawerak 2013d).

Figure 5 is a map that was created during a collaboration between Kawerak and the conservation group Oceana (Oceana and Kawerak 2014). This collaboration resulted in a document called the *Bering Strait Marine Life and Subsistence Use Data Synthesis* which combined documented Traditional Knowledge and western science about various marine species and habitat components into comprchensive maps of the Bering Strait (as well as a vast amount of textual information about species and habitat). The map in Fig. 5 shows the distribution of walruses during the fall (i.e. September through October). During this time of year walrus are very widely distributed across the entire region, and are densely concentrated in the Diomede Islands. Vessel traffic routes directly intersect with important marine mammal habitat areas shown on this map. For Bering Strait communities, it is important to recognize that marine species are highly mobile and utilize the entire region. Communities are concerned about walrus concentration areas, as well as subsistence use areas, and areas of the marine environment that animals may only travel through during their annual migration. It is difficult, if not impossible, to designate one area as more important than another because the entire region is used by animals at different times of the year, for different reasons.

Kawerak also recently partnered with Audubon Alaska and several other conservation organizations to do a vessel traffic routing analysis showing how the U.S. Coast Guard (USCG) proposed vessel traffic route overlaps with various habitat and physical features of the Bering Strait area, as well as with areas used by subsistence harvesters (Audubon Alaska et al. 2015). Figure 6, one map created as part of the analysis, helps to illustrate the origin of many concerns shared by indigenous communities in the region. The map shows documented information about marine subsistence use areas, in comparison to the vessel traffic route. While there are gaps in our documentation of marine subsistence use areas,[2] this map

[2]While there has been a large amount of work done to document subsistence use areas, there are still significant gaps. For example, Kawerak's recent research on seals and walrus was with 9 tribes of the 20 in the region. So while that work was comprehensive in each of the nine participating communities, it was not comprehensive for the region. This particular map also does not include any information from the Russian Federation, and does not illustrate habitat use by animals—only subsistence use areas.

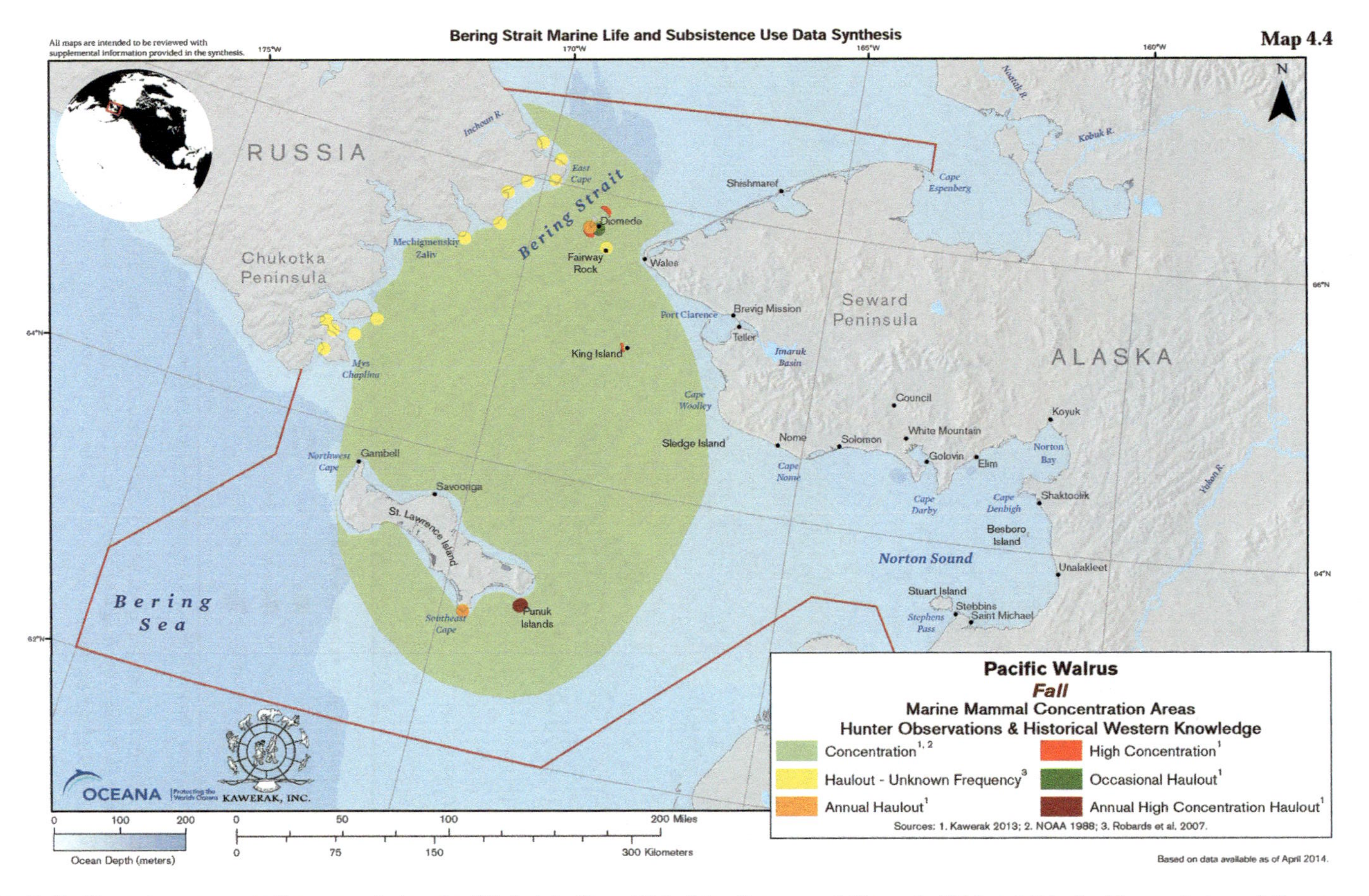

Fig. 5 Pacific walrus concentration areas during the fall (originally published by Oceana and Kawerak 2014; published with permission of Oceana and Kawerak. All Rights Reserved)

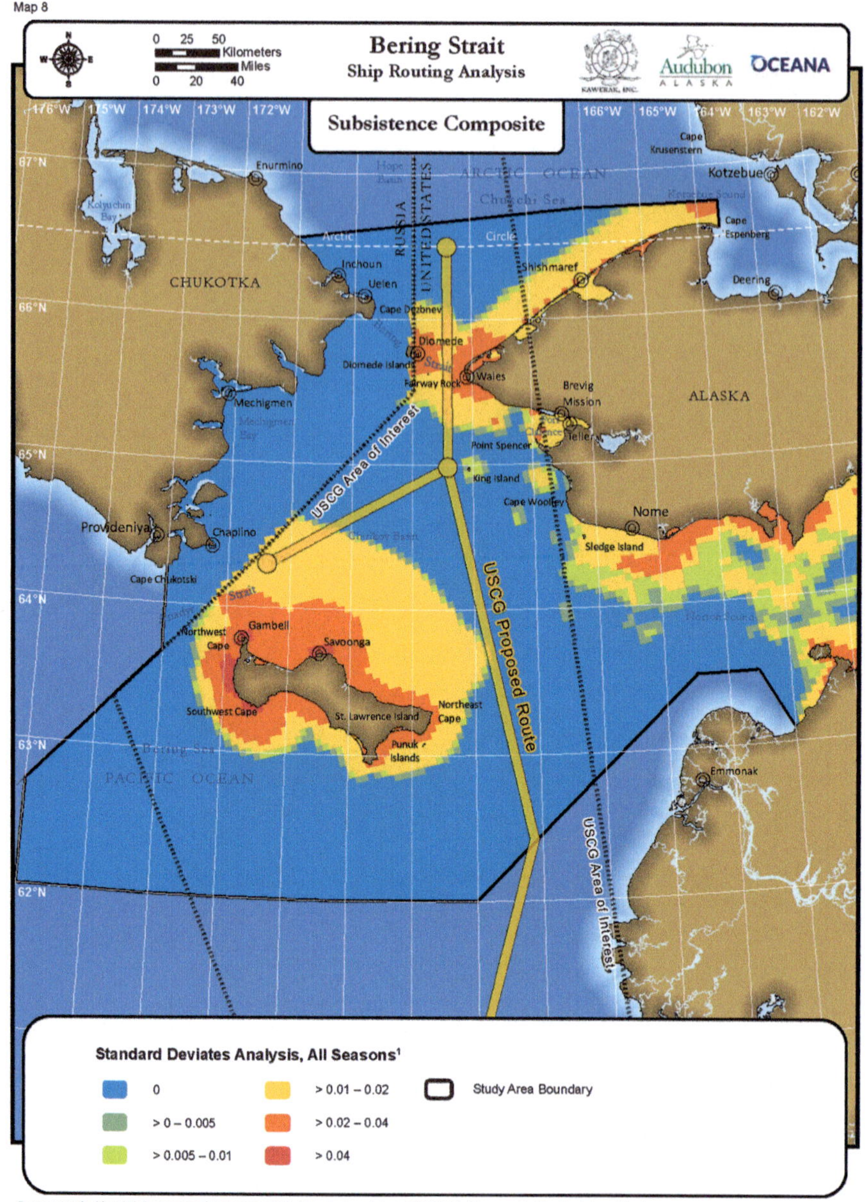

Fig. 6 Composite map of documented subsistence activity areas in the Bering Strait region (originally published by Audubon Alaska et al. 2015; courtesy of Melanie Smith of Audubon Alaska. All Rights Reserved)

illustrates how vessel traffic is and likely will continue to take place in highly sensitive marine areas. These areas are used by hunters in small boats (typically small, open, aluminum skiffs around 18 feet in length) carrying out subsistence activities and are also the main migration routes used by various marine mammals in their annual migrations. Any increases in traffic or associated noise, pollution or other disturbance in the region is of great concern, as is discussed further, spatial information from the region, such as was documented by Kawerak and in Kawerak and Oceana's Synthesis, has also recently been reviewed and updated when Kawerak and region Tribes cooperated with Audubon Alaska (Smith et al. 2017; Oceana and Kawerak 2014; Kawerak 2013a).

One final map that illustrates some of our recent research relevant to vessel traffic is of Bering Strait ocean currents (Fig. 7). This map includes information from experts from three Tribes and illustrates Bering Strait area ocean currents, as well as other marine features, such as areas where pressure ridges typically form. Marine animals also follow these currents throughout the year. There is a guide that goes along with this map that explains each of the features on it (see Raymond-Yakoubian et al. 2014). The project that this map is derived from highlighted Tribal concerns about hazardous materials spills, ship discharges, ships not under control and other situations where vessels or materials may be transported by currents to particularly sensitive places where people do not want them to go. Tribes wanted their knowledge about currents documented for accident and spill planning and response purposes.

Kawerak partners and collaborates with various organizations, agencies, bodies and researchers when we believe such relationships will benefit our Tribes. The determination of who to partner with and when is a decision made by Kawerak Administration and staff and is carefully considered and discussed. Inappropriate use or misuse of data (be it in spatial, numerical, narrative, photographic or other formats) is always Kawerak's major concern. Decisions are made based on a variety of considerations including availability of funding, shared values and goals, Tribal interest, Kawerak's own institutional capacity to partner, and other factors. Relationship building can be complicated and difficult at times, and it is crucial to have the terms of the collaboration clearly laid out ahead of time, to revisit them periodically, and to have an open dialogue throughout.

3 Indigenous Community Concerns

Many indigenous community concerns regarding arctic vessel traffic were noted above in relation to work recently completed by Kawerak and illustrated in the maps from that work. Those concerns are discussed below in additional detail.

Indigenous concerns primarily, but not exclusively, relate to the health of the marine environment and the health of marine resources that people harvest for subsistence. Marine resources are critical to Bering Strait region indigenous

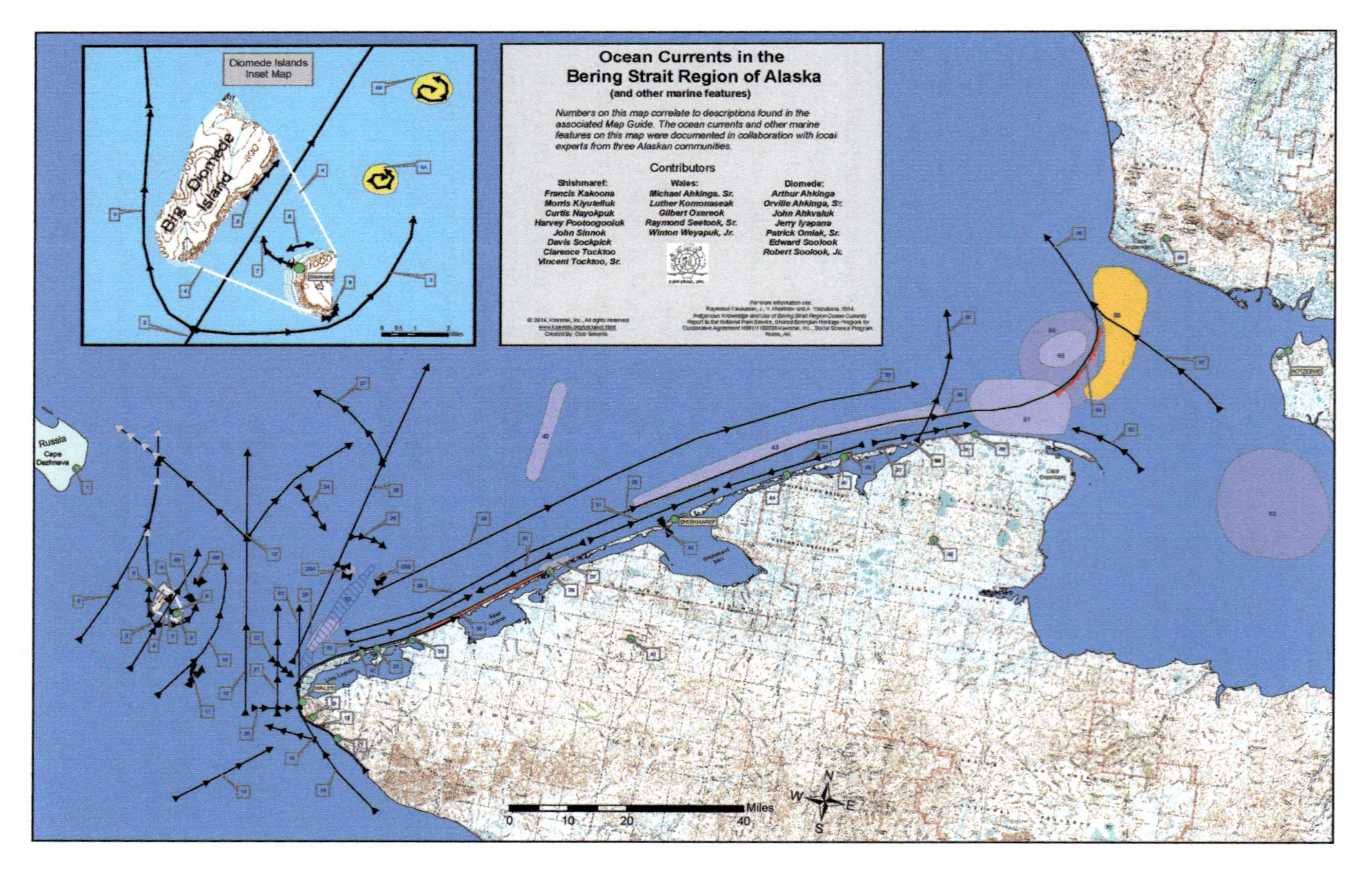

Fig. 7 Map of Bering Strait ocean currents (originally published by Raymond-Yakoubian et al. 2014; published with permission of Kawerak, Inc. All Rights Reserved)

communities for a variety of reasons,[3] as outlined below (see also, e.g. Gadamus 2013; Raymond-Yakoubian et al. 2014; Raymond-Yakoubian and Raymond-Yakoubian 2015; Raymond-Yakoubian 2013, 2015).

Cultural Marine-based subsistence activities encompass important cultural traditions related to language, dance, spirituality, eating, food preservation, cultural values, and individual and group senses of identity. This includes indigenous languages used during the harvest and processing of foods, the practice of cultural values such as sharing and not wasting, and a sense of identity and belonging derived from participating in traditions that have been passed down for millennia.

Social Marine foods play an important role in intergenerational relationship-building, knowledge transfer, and in the maintenance of ties between communities and within communities. Foods are often processed by groups of people of varying ages. For example, elders may supervise the butchering and distribution of a whale by younger adult community members. The distributed meat and blubber may then be processed by adults with young children observing or participating. These types of interactions not only pass on Traditional Knowledge and skills about food processing, preparation and other topics, but also foster relationships between generations. The sharing and exchange of foods also supports the maintenance of relationships between households within a community and between communities by promoting social interactions and communication.

Nutrition and Well-Being Marine foods harvested by Bering Strait indigenous residents are healthy and nutritious, and are also culturally preferred foods. The activities associated with harvesting and processing marine foods are important for both physical and mental health. Being physically active in the conduct of subsistence activities promotes physical health and many individuals report that such activities also increase positive feelings of mental well-being.

Economics Marine foods play a large role in household and community economies, as well as regional economies. Marine subsistence foods are consumed in large amounts by some individuals and households and are widely exchanged between households (e.g. Ahmasuk et al. 2008; Magdanz et al. 2007). Additionally, the more marine or other subsistence foods a community has, the less store-bought food is needed. Store-bought food is very expensive in rural Alaska and can be a significant household expense.

Kawerak's Marine Program has convened several gatherings with representatives from region Tribes and city governments to discuss vessel traffic-related issues. These gatherings and their associated reports (Kawerak 2015a, 2016, 2017) identified a variety of concerns shared by regional representatives. Additional concerns have been identified through further discussions with Tribes regarding specific

[3]Some of the below was presented at the Bering Strait Voices on Arctic Shipping workshop held in Nome, Alaska September 14–15, 2014 and was included in the workshop report (Kawerak 2015a).

actions or development activities or actions, such as the creation of a vessel traffic routing system by the USCG. Some of these concerns are outlined below:

- Spills and pollution (air pollution and water pollution, including the discharge of ballast water or other materials): because of their impact on the health of the marine environment and species harvested for subsistence.
- Vessel groundings: because of their potential for spilling hazardous materials which would then impact the health of the marine environment and species harvested for subsistence.
- A lack of true accident/spill response capability located in the Bering Strait region: because of the possibility of emergencies escalating due to the current long response times and the lack of response materials (e.g. boom, etc.) located along the northern Bering and southern Chukchi Sea coastlines.
- Vessel interactions with marine mammals (ship strikes or other interactions): because of the possibility of negative impacts to individual marine mammals or their broader populations.
- Increase in noise: because it may cause changes to animal behavior, including migration patterns, which may make it more difficult for communities to harvest animals.
- The possibility of large vessel interactions with small boats (boats with hunters, fishers, or travelers): because of safety concerns for the small boats, including vessel collisions and disturbances to subsistence activities (e.g. interference with hunts).
- The proximity of the USCG proposed vessel traffic route to King Island: because of the possibility of spills or other accidents that may negatively impact the environment around the Island, the animals that use that area, and the subsistence and cultural activities that take place there.
- The proximity of the USCG proposed vessel traffic route to Northeast Cape (on St. Lawrence Island): because of possible interference with bowhead whale migration routes and bowhead whale hunts.
- Communications between vessels and coastal communities: communities are concerned that they will have difficulty contacting large, transiting vessels if needed (e.g. to alert them to the locations of small boats).
- Communications between vessel regulatory or monitoring bodies and communities: communities would like a system to be in place for two-way communication to ensure that they are aware of what is happening in waters adjacent to them.
- The capacity for region residents to participate meaningfully in important venues related to vessel traffic management, monitoring and regulation: Tribes and communities feel as though it is difficult to have their voices and concerns heard and would like more mechanisms through which they can meaningfully engage with the various entities involved with vessel traffic management, monitoring and regulation.
- Cumulative threats and pressures (from vessel traffic and other sources): The cumulative impacts of the above noted concerns, and others, have the potential to greatly impact Tribes and the marine environment.

Fig. 8 The village of Wales, Alaska, located at Cape Prince of Wales on the Bering Strait, approximately 50 miles from Russia (copyright Vernae Angnaboogok, published with permission. All Rights Reserved)

These concerns are serious and of great importance to indigenous communities in the Bering Strait. There are, however, many other developments related to increasing vessel traffic that are currently impacting or have the potential to impact indigenous communities in the region. These other, cumulative impacts include: offshore oil and gas exploration and development; the potential for large commercial fisheries to move northward; salmon bycatch by the Pollock fishing industry; seal and walrus Endangered Species Act issues; offshore gold mining; research activities; climate change and its various impacts (see also, Kawerak 2013b; Raymond-Yakoubian and Raymond-Yakoubian 2017; Raymond-Yakoubian 2015). The cumulative impacts of all of these pressures and threats can be difficult to deal with and strain the capacity of Tribes, Tribal organizations, and communities as a whole. Taken in conjunction with increasing vessel traffic, and risks that such traffic poses, communities are faced with potentially serious challenges to their food security and cultural traditions (Fig. 8).

One of the main reasons these activities are all considered threats and pressures is because they are, or have the potential to, impact the food security of Bering Strait indigenous communities. Food security encompasses all aspects of Inuit culture and tradition and is synonymous with "environmental health" (ICC-Alaska 2014, 2015). Any one of the concerns noted above, or combinations of such concerns, have the potential to impact food security by limiting access to healthy, uncontaminated,

Fig. 9 The village of Diomede, located on Little Diomede Island, as seen from the helicopter landing pad during winter (copyright Meghan Topkok, published with permission. All Rights Reserved)

traditional foods and the distinct cultural traditions and practices associated with them.

For the residents of the region, even a small accident could have large-scale, long-term, intergenerational consequences. For example, even a small oil spill near a seal haul out could lead to the death of adult seals and pups, and could lead to the abandonment of the haul out, which could mean a community no longer has access to that food source. The cumulative effects of pressures on communities must be taken into account, and when they are it will become very clear that a precautionary approach to the Bering Strait region is needed (Fig. 9).

4 Measures to Address Indigenous Concerns

Many of the concerns noted above can be addressed by various means. On a broad scale, the Polar Code is, of course, a very positive step towards addressing these concerns, though additional measures have been proposed by Bering Strait region Tribes (and others). Other measures that could be taken include but are not limited to the following (see also Audubon Alaska et al. 2015; Kawerak 2015a, b).

Consultation: Consultation with indigenous peoples needs to be increased and formalized across the Arctic. Bering Strait Tribes and other Tribes in the U.S. Arctic have both a right and a strong desire to be consulted on a government-to-government basis (as provided by federal regulation, e.g. Federal Register 2000) about vessel traffic as well as other activities (e.g. Raymond-Yakoubian 2012; Raymond-Yakoubian et al. 2017), in addition to being consulted and having decision-making power at an international level. Consultation is just the first step in creating truly productive relationships and effective decision-making; we must go beyond consultation to have true equity.

Discharge: Indigenous communities in the Bering Strait want the entire Bering Strait region (which has not been formally defined by Tribes or internationally) to be a zero discharge zone. This would apply to ballast water, organic materials, oily water discharges, black and grey water, trash, and any other materials. Eliminating discharge in the region would help address concerns about pollution; specifically about marine animals becoming contaminated and people consuming them, but also in general for the health of the environment. This would also include strict standards on black carbon emissions.

Vessel Routing: Based on information from region Tribes, Kawerak and others have recommended that the USCG proposed vessel traffic route be moved further west and away from King Island because of its importance as a marine mammal hunting and haul out area and its cultural significance for King Island Tribal members. We have also recommended that the proposed route be moved slightly east in the vicinity of Northeast Cape on St. Lawrence Island because of that area's importance for subsistence activities.

Response capacity: Indigenous communities are likely to be first responders to any accident in the Bering Strait. The lack of response-oriented facilities and the small amount of response equipment currently in the region is inadequate for any major spill response. The nearest U.S. Coast Guard base is also located thousands of miles from the Bering Strait. Communities would like additional response equipment to be placed throughout the region and they would also like to receive training in the use of such equipment. The lack of infrastructure to address vessel-related concerns, such as waste disposal facilities, also needs to be addressed.

Speed limits: Indigenous communities are also concerned about marine mammal interactions with vessels, which could be catastrophic for animals. Speed limits should be created to offer some protection against vessel strikes for whales and walruses with calves, in particular. Speed limits may also reduce the amount of noise produced by vessels, creating less disturbance for marine animals.

Areas to be avoided (ATBA) and Protected Areas: Kawerak and Tribes have requested ATBAs in the waters around King, St. Lawrence Island, and in the Bering Strait proper. ATBAs should be located such that they do not impair the passage of transiting ships, but do protect the important characteristics of each area. Some of these ATBAs have been designated and others are likely to be in future.

Bering Strait Tribes have recognized the unique and globally important characteristics of the region. Protected areas have been discussed and proposed by region residents and others for many decades (e.g. Raymond-Yakoubian 2015). Some 'protected areas' have existed for some time, such as the Arctic Management Area (which currently prohibits commercial fisheries within the management area), for example (NMFS 2009; NPFMC 2009). In December 2016 President Obama, via Executive Order 13754, established the Northern Bering Sea Climate Resilience

Area (NBSCRA) (Federal Register 2016). The NBSCRA was supported by over 70 federally recognized Tribes (including Kawerak region Tribes), created a Federal Task Force and Tribal Advisory Council, and specifically directed Federal agencies to "consider traditional knowledge in decisions affecting the Northern Bering Sea Climate Resilience Area" (ibid.). This action was revoked by President Trump in April 2017 (Federal Register 2017). Re-implementing a similar Area and process, through consultation and equitable collaboration with Tribes, would be an important step towards giving Tribes authority and control over their maritime homelands.

Communications: A communication system should be established under which traffic in the Bering Strait is monitored in real-time. This system should also allow for weather and other information to be transmitted easily between ships, communities, and the shore. A communication system which includes vessel monitoring would help in preventing ship groundings, responding to vessels not under command, and in response to accidents, spills or other events. Communities also need free and unlimited access to Automatic Identification System information. This will allow hunters, fishers and travelers to better plan ocean travel and avoid large vessels that may be transiting offshore from their communities.

Traditional Knowledge: The use of Bering Strait communities' Traditional Knowledge (such as that documented by Kawerak and other entities) to address their concerns, as well as concerns of the shipping community is vital. Indigenous residents of the Bering Strait region have built (and continue to build) a body of detailed knowledge relating to the marine environment, based on first-hand experience, for many generations. If indigenous communities are consulted and included in decision-making this vast body of Traditional Knowledge will be accessible and can be used to formulate effective monitoring and management of vessel traffic and other activities in the region. Indigenous communities must be meaningfully and equitably involved in order for this to be successful.

5 Conclusion

This chapter reviewed some of the recent work carried out by Kawerak in collaboration with Bering Strait region Tribal communities and non-tribal partners. Much of this work, through its documentation of Traditional Knowledge, Tribal concerns, and suggested paths forward, is of direct relevance to vessel traffic activities. Indigenous residents of the region have millennia long connections to the marine environment, and a vast and important body of knowledge and experiences from those connections. This experience and knowledge can directly contribute to discussions about vessel traffic.

While Tribal communities in the region have many and serious concerns related to shipping activities, some of which are shared by others outside the region, they also have many valuable recommendations for addressing these matters.

Indigenous residents of the Bering Strait region are at ground zero for vessel traffic activity, climate changes, and a myriad of other developments. These communities bear the burden of risk (Kawerak 2015a) associated with these changes, developments and activities, and are in jeopardy of having important cultural traditions and practices negatively impacted or destroyed if the global community does not tread carefully and thoughtfully. Indigenous concerns and solutions must be considered at all points along our collective effort towards safe Arctic shipping.

Acknowledgements The work discussed in this chapter was made possible by funding from the following: Kawerak Incorporated, the National Fish and Wildlife Foundation, the National Park Service Shared Beringian Heritage Program, the National Science Foundation, Oak Foundation, Pew Charitable Trusts and the U.S. Fish and Wildlife Service. Dozens of local experts from the Bering Strait region Tribes of Brevig Mission, Diomede, Elim, Golovin, Koyuk, King Island, Nome Eskimo Community, Savoonga, Shaktoolik, Shishmaref, Stebbins, St. Michael, Teller, Wales, White Mountain and Unalakleet contributed their time and expertise to the projects outlined here. Please consult the references for more information about the individuals and Tribes that shared knowledge and expertise; that knowledge remains the property of those that contributed it. Other partners on projects discussed here include Audubon Alaska, the Eskimo Walrus Commission, the Ice Seal Committee, Kawerak's Marine Program, and Oceana. Thank you to the two anonymous reviewers for their comments on an earlier version of this chapter. For more information on Kawerak's Social Science Program visit: www.kawerak.org/socialsci.

References

Ahmasuk, A., Trigg, E., Magdanz, J., & Robbins, B. (2008). *Bering Strait region local and traditional knowledge pilot project: A comprehensive subsistence use study of the Bering Strait Region* (North Pacific Research Board Project Final Report, Project #643). Nome: Kawerak, Inc.

Alaska Department of Commerce, Community, and Economic Development (ADCCED). (2017). *US Census records from the community and regional affairs community database online*. Retrieved January 13, 2017, from https://www.commerce.alaska.gov/dcra/DCRAExternal/Community.

Audubon Alaska, Friends of the Earth, Oceana, Ocean Conservancy, Pacific Environment, Pew Charitable Trusts, Wildlife Conservancy, World Wildlife Fund. (2015). *Recommendations on the port access route study: In the Chukchi Sea, Bering Strait and Bering Sea* (Docket ID: USCG-2014-0941). Letter to the U.S. Coast Guard dated June 3, 2015. Anchorage, Alaska.

Federal Register. (2000). Executive order 13175. 1-9-2000. *Consultation and Coordination With Indian Tribal Governments, 65*(218), 67249–67252.

Federal Register. (2016). Executive order 13754. 12-9-2016. *Northern Bering Sea Climate Resilience, 81*(240), 90669–90674.

Federal Register. (2017). Executive order 13795. 28-4-2017. Implementing an America-first offshore energy strategy, 82(84), 20815–20818.

Gadamus, L. (2013). Linkages between human health and ocean health: a participatory climate change vulnerability assessment for marine mammal harvesters. *International Journal of Circumpolar Health, 72*. https://doi.org/10.3402/ijch.v72i0.20715.

Gadamus, L., & Raymond-Yakoubian, J. (2015a). Qualitative participatory mapping of seal and walrus harvest and habitat areas: Documenting indigenous knowledge, preserving local values, and discouraging map misuse. *International Journal of Applied Geospatial Research, 6*(1), 76–93.

Gadamus, L., & Raymond-Yakoubian, J. (2015b). A Bering Strait indigenous framework for resource management: Respectful seal and walrus hunting. *Arctic Anthropology, 52*(2), 87–101.
Gadamus, L., Raymond-Yakoubian, J., Ashenfelter, R., Ahmasuk, A., Metcalf, V., & Noongwook, G. (2015). Building an indigenous evidence-base for tribally-led habitat conservation policies. *Marine Policy, 62)*, 116–124.
ICC-Alaska. (2014). *Bering Strait Regional Food Security Workshop: How to assess food security from an Inuit perspective: Building a Conceptual Framework on how to Assess Food Security in the Alaskan Arctic*. Anchorage, AK: Inuit Circumpolar Council-Alaska.
ICC-Alaska. (2015). *Alaskan Inuit Food Security Conceptual Framework: How to assess the Arctic from an Inuit perspective*. Anchorage, AK: Inuit Circumpolar Council-Alaska.
Kawerak Inc. (2013a). *Seal and walrus harvest and habitat area for nine Bering Strait region communities*. Nome, AK: Kawerak Social Science Program.
Kawerak Inc. (2013b). *Policy-based recommendations from Kawerak's Ice Seal and Walrus project*. Nome, AK: Kawerak Social Science Program.
Kawerak Inc. (2013c). *Traditions of respect: Traditional knowledge from Kawerak's Ice Seal and Walrus project*. Nome, AK: Kawerak Social Science Program.
Kawerak Inc. (2013d). *Seal and Walrus hunting and safety: Traditional knowledge from Kawerak's Ice Seal and Walrus project*. Nome, AK: Kawerak Social Science Program.
Kawerak Inc. (2015a). *Bering Strait voices on Arctic shipping: Workshop report*. Nome, AK: Kawerak Marine Program.
Kawerak Inc. (2015b). *Port access route study: In the Chukchi Sea, Bering Strait, and Bering Sea PARS -USCG-2014-0941*. Letter to the U.S. Coast Guard dated June 1, 2015. Anchorage, Alaska.
Kawerak Inc. (2016). *Bering Strait voices on Arctic shipping: Moving forward to protect Alaska Native ways of life and the natural resources we rely on*. Nome, AK: Kawerak Marine Program.
Kawerak Inc. (2017). *Bering Strait voices on Arctic shipping: Vision for action summit*. Nome, AK: Kawerak Marine Program.
Magdanz, J., Tahbone, S., Ahmasuk, A., Koester, D., & Lewis, B. (2007). *Customary Trade and Barter in Fish in the Seward Peninsula Area, Alaska* (Technical Paper 328). Juneau, AK: Alaska Department of Fish and Game, Division of Subsistence.
NMFS (National Marine Fisheries Service). (2009). *Environmental Assessment/Regulatory Impact Review/Final Regulatory Flexibility Analysis for the Arctic Fishery Management Plan and Amendment 29 to the Fishery Management Plan for Bering Sea/Aleutian Islands King and Tanner Crabs*. Juneau, AK:National Marine Fisheries Service.
NPFMC (North Pacific Fishery Management Council). (2009). *Fishery Management Plan for Fish Resources of the Arctic Management Area*. Anchorage, AK: North Pacific Fishery Management Council.
Oceana and Kawerak. (2014). *Bering Strait Marine life and subsistence use data synthesis*. Juneau, AK: Oceana and Kawerak.
Raymond-Yakoubian, B., Kaplan, L., Topkok, M., & Raymond-Yakoubian, J. (2014). *"The World has Changed": Iŋalit Traditional Knowledge of Walrus in the Bering Strait* (Final Report to the North Pacific Research Board for Project 1013). Nome, AK: Kawerak Social Science Program.
Raymond-Yakoubian, B., & Raymond-Yakoubian, J. (2015). *"Always taught not to waste": Traditional Knowledge and Norton Sound/Bering Strait Salmon Populations* (Final report for Arctic-Yukon-Kuskokwim Sustainable Salmon Initiative Project 1333). Nome, AK: Kawerak Social Science Program.
Raymond-Yakoubian, B., & Raymond-Yakoubian, J. (2017). *Research processes and indigenous communities in Western Alaska: Workshop report*. Nome, AK: Kawerak Social Science Program.
Raymond-Yakoubian, J. (2012). Participation and resistance: Tribal involvement in Bering Sea fisheries management and policy. In C. Carothers, K. Criddle, C. Chambers, P. Cullenberg, J. Fall, A. Himes-Cornell, J. Johnsen, & E. Springer (Eds.), *Fishing people of the North: cultures, economies, and management responding to change* (pp. 117–130). Fairbanks, AK: Alaska Sea Grant and University of Alaska.

Raymond-Yakoubian, J. (2013). *When the Fish Come, We Go Fishing: Local ecological knowledge of Non-salmon fish used for subsistence in the Bering Strait Region*. Nome, AK: Kawerak Social Science Program.

Raymond-Yakoubian, J. (2015). Conceptual and institutional frameworks for protected areas and the status of indigenous involvement: Considerations for the Bering Strait Region of Alaska. In T. M. Herrmann & T. Martin (Eds.), *Indigenous peoples' governance of land and protected territories in the circumpolar Arctic* (pp. 83–103). Switzerland: Springer.

Raymond-Yakoubian, J., Khokhlov, Y., & Yarzutkina, A. (2014). *Indigenous knowledge and use of ocean currents in the Bering Strait Region* (Report to the National Park Service, Shared Beringian Heritage Program for Cooperative Agreement H99111100026). Nome, AK: Kawerak, Inc., Social Science Program.

Raymond-Yakoubian, J., Raymond-Yakoubian, B., & Moncrieff, C. (2017). The incorporation of traditional knowledge into Alaska federal fisheries management. *Marine Policy, 78*, 132–142.

Smith, M., Goldman, M., Knight, E., & Warrenchuk, J. (2017). *Ecological atlas of the Bering, Chukchi and Beaufort seas* (2nd ed.). Anchorage, AK: Audubon Alaska.

US Census Bureau. (2016). *State and country quick facts, Nome Census Area, Alaska*. Retrieved February 2, 2017. http://www.census.gov/quickfacts/table/PST045216/02180,00

USACE. (2013). *Alaska deep draft Arctic port system study*. Alaska District: U.S. Army Corps of Engineers.

Challenges for the Establishment of Marine Protected Areas in Response to Arctic Marine Operations and Shipping

Millicent McCreath and Lawson W. Brigham

Contents

The research and editing of this chapter and volume was supported by U.S. National Science Foundation grant award 1263678 to L. W. Brigham and the University of Alaska Fairbanks.

M. McCreath (✉)
Centre for International Law, National University of Singapore, Singapore, Singapore
e-mail: cilmmj@nus.edu.sg

L. W. Brigham
International Arctic Research Center, University of Alaska Fairbanks, Fairbanks, AK, USA
e-mail: lwbrigham@alaska.edu

L. P. Hildebrand et al. (eds.), *Sustainable Shipping in a Changing Arctic*, WMU Studies in Maritime Affairs 7, https://doi.org/10.1007/978-3-319-78425-0_17

Abstract Increasing Arctic marine use is driven primarily by natural resource development and greater marine access throughout the Arctic Ocean created by profound sea ice retreat. Significant management measures to enhance protection of Arctic people and the marine environment are emerging, including the development of marine protected areas (MPAs) which may be effective and valuable tools. MPAs have been established by individual Arctic coastal states within their respective national jurisdictions; however, a pan-Arctic network of MPAs has yet to be established despite Arctic Council deliberations. This overview focuses on those MPAs that can be designated by the International Maritime Organization and by international instrument or treaty to respond to increasing Arctic marine operations and shipping. Key challenges remain in the Arctic to the introduction of select MPAs and development of a circumpolar network of MPAs in response to greater marine use: the variability of sea ice; the rights and concerns of indigenous people; a lack of marine infrastructure; application to the Central Arctic Ocean; establishing effective monitoring; and, compliance and enforcement in remote polar seas. Robust bilateral and multilateral cooperation will be necessary not only to establish effective MPAs but also to sustain them for the long term. Reducing the large Arctic marine infrastructure gap will be a key requirement to achieve effective MPA management and attain critical conservation goals.

Keywords Marine protected area · Polar Code · UNCLOS · Arctic marine operations and shipping · Protective measures

1 Introduction

Natural resource development and profound sea ice changes are creating potential opportunities for increased Arctic marine operations and shipping (Brigham 2017). The Arctic marine environment will remain in winter dominated by ice, darkness, severe climatic conditions and limited infrastructure and facilities. However, the global interest in Arctic shipping brings with it an opportunity to implement measures to protect the environment from the impacts of shipping before it is substantially damaged. The Arctic is not only home to varied ecosystems and a wealth of biodiversity but also indigenous peoples, many of whom rely on the marine environment for their survival and way of life (ACIA 2004). Marine protected areas (MPAs) are increasingly recognised as an effective means of protecting the marine environment (Lalonde 2013). A number of MPAs of different forms have been established in the Arctic marine environment, however few if any have been implemented so as to affect international rights of navigation. This chapter is confined to MPAs designated under an international instrument or the auspices of the International Maritime Organization (IMO) for the protection of a specified area from the impacts of shipping.

The term "marine protected area" is notoriously ambiguous, and the concept attracts corresponding criticism. However, when used within clearly defined parameters, it serves as a useful umbrella for a variety of different tools used for different purposes. The Arctic marine environment is sensitive and vulnerable to human activity, including shipping. In order to achieve meaningful conservation outcomes, the designation of a particular Arctic marine area for protection must directly flow from the identification of the particular aspect of the environment needing protection. Once the target of protection and the corresponding conservation outcome are identified, appropriate and potentially effective protective measures can be designed and employed.

The Arctic marine environment presents unique challenges to the designation, implementation and enforcement of MPAs. The challenges include the lack of infrastructure and reception facilities, significant knowledge gaps, difficulties ensuring compliance with regulations in remote areas, designing appropriate measures in a changing and unpredictable environment, addressing the needs and rights of indigenous peoples and local inhabitants, managing competing coastal State priorities, and designating protected areas across boundaries and beyond national jurisdiction. This chapter explores the need for, and mechanisms to establish, Arctic MPAs both within and without national jurisdiction in the context of an Arctic Ocean under the Polar Code regime.

This first section of this chapter deals with the task of setting the parameters for the discussion on MPAs. In Sect. 2 the Arctic Marine Shipping Assessment (AMSA) and overall Arctic shipping are considered so as to identify the threats that warrant the establishment of MPAs in the Arctic. The Polar Code, its provisions and gaps, and its relevance to the development of MPAs, is examined in Sect. 3. Section 4 outlines in detail the main instruments or bodies under which MPAs may be established, as well as briefly touching on the multitude of other mechanisms. Section 5 examines the challenges that the Arctic poses to the establishment and implementation of MPAs.

1.1 Definitions and Parameters

For the purposes of this chapter, the marine Arctic is taken to mean the Arctic area covered by the Polar Code, as defined in MARPOL Annex I Ch XI Reg 46:

> Arctic waters means those waters which are located north of a line from the latitude 58°00′.0 N and longitude 042°00′.0 W to latitude 64°37′.0 N, longitude 035°27′.0 W and thence by a rhumb line to latitude 67°03′.9 N, longitude 026°33′.4 W and thence by a rhumb line to the latitude 70°49′.56 N and longitude 008°59′.61 W (Sørkapp, Jan Mayen) and by the southern shore of Jan Mayen to 73°31′.6 N and 019°01′.0 E by the Island of Bjørnøya, and thence by a great circle line to the latitude 68°38′.29 N and longitude 043°23′.08 E (Cap Kanin Nos) and hence by the northern shore of the Asian Continent eastward to the Bering Strait and thence from the Bering Strait westward to latitude 60° N as far as Il'pyrskiy and following the 60th North parallel eastward as far as and including Etolin Strait and thence by the northern shore of the North American continent as far south as latitude 60° N and thence eastward along parallel of latitude 60° N, to longitude 056°37′.1 W and thence to the latitude 58°00′.0 N, longitude 042°00′.0 W.

The term "Arctic shipping" also deserves discussion as it has many different meanings among the Arctic community and a global audience. Many see "Arctic shipping" only in the context of trans-Arctic navigation, envisioning a potential for large ships sailing between the Pacific and Atlantic across the top of the world. For the purposes of this chapter a more holistic approach is taken to include all ships (100 tons or more) that are operating in Arctic waters and can discharge into regional waters and release stack emissions into the surrounding atmosphere. Such an approach includes all ships on destinational and trans-Arctic voyages, vessels supporting offshore oil and gas operations, and those engaged in fishing, tourism, research, and other marine operations. A more inclusive term used in the Arctic Council is "Arctic marine operations and shipping" (Brigham 2017).

1.2 Conceptualizing MPAs Broadly

The term "marine protected area" is ambiguous, and therefore any meaningful discussion of MPAs must clearly set out the parameters for the use of the term in the particular context. An early starting point for an understanding of the meaning of "marine protected area" is the basic definition formulated by the General Assembly of the International Union for the Conservation of Nature (IUCN) as follows:

> Any area of intertidal or subtidal terrain, together with its overlying water and associated flora, fauna, historical and cultural features, which has been reserved by law or other effective means to protect part or all of the enclosed environment (Kelleher 1999).

This broad definition encompasses a wide range of areas designated for different purposes and targeting different activities. Under article 8(a) of the Convention on Biological Diversity (CBD), States parties are under a duty to "[e]stablish a system of protected areas or areas where special measures need to be taken to conserve biological diversity" (CBD 1992). The CBD does not define "marine protected area" but does define "protected area" in article 2 as "a geographically defined area which is designated or regulated and managed to achieve specific conservation objectives". The Ad Hoc Technical Expert Group to the CBD Secretariat has developed the "protected area" definition contained in the text of the CBD itself and applied it to the marine context. Its formulation of "marine and coastal protected area" rather than MPA is intended to clarify that it includes coastal areas as well as the sea:

> "Marine and Coastal Protected Area" means any defined area within or adjacent to the marine environment, together with its overlying waters and associated flora, fauna, and historical and cultural features, which has been reserved by legislation or other effective means, including custom, with the effect that its marine and/or coastal biodiversity enjoys a higher level of protection than its surroundings (CBD Secretariat 2004).

The above definitions are suitably broad to include an MPA that is focused solely on one activity, including for the purposes of this chapter, shipping. However the 2008 IUCN definition, which has been described as "the most universally accepted definition of a protected area" (Nicoll and Day 2017), and adopted by PAME[1] for its work on Arctic MPAs (PAME 2015), defines a protected area (not limited to marine protected areas) as:

> A clearly defined geographical space recognized, dedicated, and managed, through legal or other effective means, to achieve the long-term conservation of nature with associated ecosystem services and cultural values (IUCN 2008).

This definition has a more holistic and goal-based focus, requiring that the legal means for protecting the area are designed "to achieve the long-term conservation of nature with associated ecosystem services and cultural values" (PAME 2015). Arguably an area designated for single-sector protection, such as from shipping, is not capable of achieving such a comprehensive conservation outcome. It is widely suggested that a single sector designation does not meet the accepted IUCN MPA definition (Roberts et al. 2010; Nicoll and Day 2017; Jakobsen and Johansen 2017). Instead they are considered to be "area-based management tools" (ABMTs), of which MPAs are a sub-category of tool (Roberts et al. 2010).

The IUCN has adopted categories to which areas that meet its definition for MPAs (the 2008 definition) can be assigned. The categories "provide a framework for data collection, a set of international standards that allows comparison across countries, and a means of promoting international understanding" (UNEP-WCMC 2008). A summary of these categories obtained from the 2008 report of the United Nations Environment Programme (UNEP) World Conservation Monitoring Centre (WCMC) on "National and Regional Networks of Marine Protected Areas: A Review of Progress" (UNEP-WCMC 2008) is included in Table 1 below.

The internationally legally binding instrument currently being developed by the Preparatory Committee established by the United Nations General Assembly in resolution 69/292 (2015) under UNCLOS on conservation and sustainable use of marine biological diversity in areas beyond national jurisdiction will address the definitions of MPAs and ABMTs. The Chair's "streamlined non-paper on elements

Table 1 Summary of IUCN protected area management categories

Category	Definition—area mainly managed for
I	a. Science or as a Strict Nature Reserve
	b. Wilderness protection
II	Ecosystem protection and recreation; often called a National Park
III	Conservation of specific natural features; often called a National Monument
IV	Conservation through management intervention (e.g. habitat/species conservation areas)
V	Land/Seascape conservation and recreation
VI	Sustainable use of natural ecosystems (e.g. multiple-use protected area)

[1] Arctic Council Working Group, Protection of the Arctic Marine Environment.

of a draft text" of that legally binding instrument issued prior to the fourth session of the Preparatory Committee held from 10 to 21 July 2017, outlines various definitions put forward by participants for MPAs and ABMTs (PrepCom 2017). The definitional proposals have not yet been synthesised into single accepted formulations however there seems to be preliminary acceptance that ABMTs are a broad category, of which MPAs are a more specific subset. It remains to be seen how any future intergovernmental conference will finally define these terms.

Whilst the highest levels of protection in an MPA will be achieved through the regulation and control of all harmful uses of and activities in the MPA, this does not affect the legitimacy from a conservation perspective of an MPA solely regulating shipping. Shipping has the potential to have significant adverse effects on the marine environment, from, for example, operational and accidental discharges, damage to reefs from groundings, noise pollution, and disturbance to sensitive species and their habitats. The implementation of MPAs to regulate international shipping can reduce the risks of adverse environmental impacts and therefore achieve meaningful conservation outcomes.

1.3 Current Arctic MPAs

The Arctic marine environment includes areas under national jurisdiction and areas of high seas, beyond national jurisdiction. To date the only MPAs established in the Arctic have been in areas under national jurisdiction. The PAME report on a "Framework for a Pan-Arctic network of marine protected areas" (PAME 2015) contains an overview of the status of MPAs, "other measures" and MPA networks in the Arctic. It is beyond the scope of this chapter to provide a comprehensive analysis of the existing Arctic MPAs. It is sufficient for present purposes to state that a great many have been established in different forms for different objectives and by all the Arctic States, however none place restrictions on international navigation.

1.4 MPAs Targeting Shipping

Shipping is not the only or the greatest threat to the global marine environment, which also risks adverse impacts from other pollution sources and resource exploitation (Churchill 2013). However, the existence of greater threats does not render the regulation of shipping for environmental protection futile. Rather, protecting the marine environment from the impacts of shipping, particularly sensitive environments such as the Arctic, can have positive conservation outcomes. International navigational rights are firmly protected in international law, and therefore unilateral efforts to restrict them beyond the territorial sea are generally without legal basis or opposability to the ships of third States. Despite its protection, regulation of international navigation on a global level is achievable due to the universally accepted framework provided by the IMO. The IMO serves as a forum for States and certain industry and non-governmental organisations to reach agreement on regulations applicable internationally and capable of affecting the navigational rights of ships of third States in all maritime zones (Roberts et al. 2010).

1.5 Identifying Marine Areas for Protection from Shipping

Following AMSA recommendation IIC, Arctic Council working groups AMAP (Arctic Monitoring and Assessment Programme), CAFF (Conservation of Arctic Flora and Fauna) and SDWG (Sustainable Development Working Group) prepared an extensive report on the "Identification of Arctic marine areas of heightened ecological and cultural significance" (hereinafter the AMSA Recommendation IIC Report) (AMAP/CAFF/SDWG 2013). The drafters of the report took areas of heightened ecological significance to be areas that are "ecologically important", meaning that their ecological functions for the animals, plants and microbes in the area are more important than in other areas (AMAP/CAFF/SDWG 2013). The criteria from the IMO's Revised Guidelines for the Identification and Designation of Particularly Sensitive Sea Areas (PSSAs) were used to "evaluate the importance" of the identified areas, namely uniqueness or rarity; critical habitat; dependency; representativeness; diversity; productivity; spawning or breeding grounds; naturalness; integrity; fragility and bio-geographic importance (IMO 2005). The AMSA Recommendation IIC Report identified 97 areas of heightened ecological significance, constituting more than half of the total area of the marine Arctic (AMAP/CAFF/SDWG 2013). The areas of heightened ecological significance were distributed across the 14 large marine ecosystems (LME) in the Arctic area.[2] They were classed into four geographical groups: areas along mainland coasts, areas around Arctic archipelagos, areas on seasonally ice-covered Arctic shelves, and areas with drifting pack ice in the central Arctic (AMAP/CAFF/SDWG 2013).

In addition to work done under the auspices of the Arctic Council, the Conference of the Parties (COP) to the CBD is active in identifying "ecologically or biologically significant marine areas in need of protection in open-ocean waters and deep-sea habitats" through regional workshops (CBD COP 10 2010). These "ecologically or biologically significant marine areas" are known as "EBSAs". At its ninth meeting in 2008, the COP adopted Decision IX/20 adopting scientific criteria for identifying EBSAs (CBD COP 9 2008). These criteria are: uniqueness or rarity; special importance for life-history stages of species; importance for threatened, endangered or declining species and/or habitats; vulnerability, fragility, sensitivity, or slow

[2]The report defined the "Arctic area" as including "sub-Arctic, open-water areas south of the ice-covered areas. In the Pacific sector it extends south to include the Aleutian Islands and the east coast of Kamchatka. In the Atlantic the area extends south to the northern coast of Labrador in the west and to the Faroe Isles and the boundary to the North Sea at 62° N in the east".

recovery, biological productivity; biological diversity and naturalness (CBD COP 9 2008). Whilst these criteria differ from the IMO PSSA criteria used by the AMSA Recommendation IIC Report, there is substantial overlap and similarity. The Arctic regional workshop to facilitate the identification of EBSAs was held in Helsinki in 2014, at which a representative from CAFF highlighted the results of the AMSA Recommendation IIC Report (CBD 2014). The geographical scope of this study was significantly restricted as opposed to the AMSA Recommendation IIC Report, as excluded from consideration were areas within the national jurisdiction of Canada, Denmark, Norway and the United States. It did not consider areas further south than the Bering Strait. The workshop plenary agreed the following 11 areas met the EBSA criteria within the study area: the marginal ice zone and the seasonal ice-cover over the deep Arctic Ocean; multi-year ice of the Central Arctic Ocean; Murman Coast and Varanger Fjord; White Sea; the south-eastern Barents Sea (Pechora Sea); the coast of Western and Northern Novaya Zemlya; north-eastern Barents-Kara Sea; Ob-Enisei River Mouth Area; Great Siberian Polynya; Wrangel and Gerald Shallows and Ratmanov Gyre; and the coastal waters of western and northern Chukotka (CBD 2014). A comparison of the maps of the AMSA Recommendation IIC Report areas of heightened ecological significance and the Arctic CBD EBSAs shows significant concurrence in the overlapping areas of study (Figs. 1 and 2).

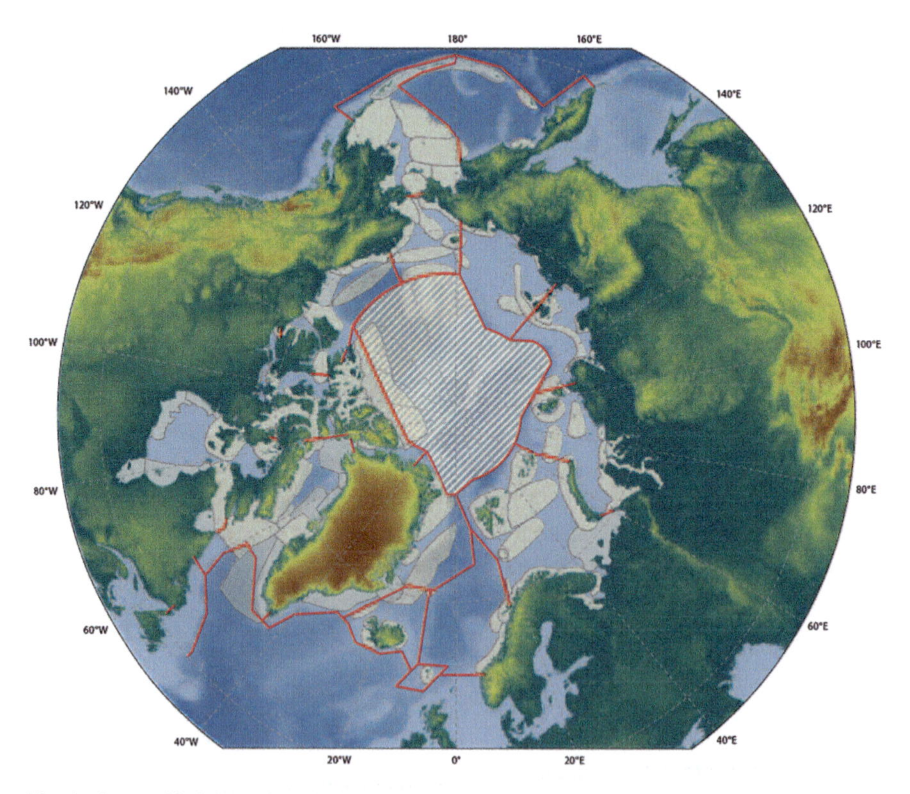

Fig. 1 Areas of heightened ecological significance as identified by the AMSA Recommendation IIC Report, divided into the sixteen Arctic LMEs (AMAP/CAFF/SDWG 2013)

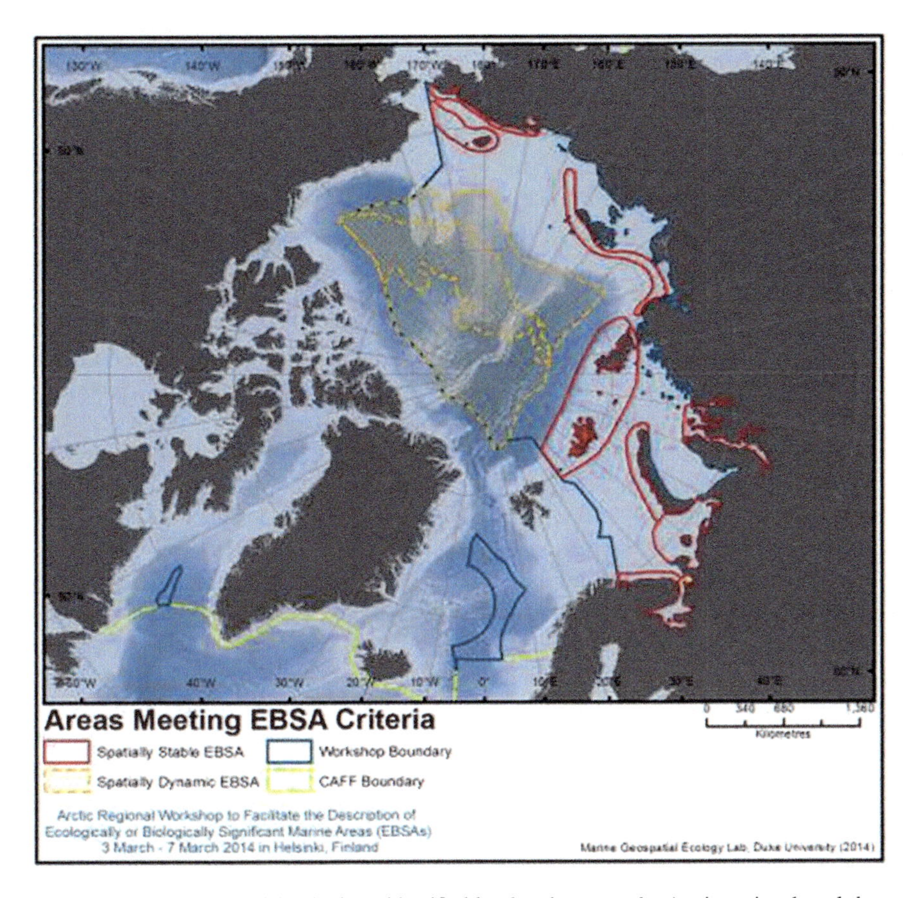

Fig. 2 Areas meeting EBSA criteria as identified by the plenary at the Arctic regional workshop (CBD 2014)

2 Arctic Marine Shipping Assessment

The 2009 Arctic Marine Shipping Assessment (AMSA) conducted by the Arctic Council's Protection of the Arctic Marine Environment (PAME) working group made 17 negotiated recommendations for future action by the Arctic Council, Arctic States, the IMO and the maritime industry (AMSA 2009). AMSA was a complex assessment of current and future marine operations and shipping and focused on measures or actions to enhance Arctic marine safety and environmental protection. The AMSA recommendations are grouped in three broad themes: I. Enhancing Arctic Marine Safety; II. Protecting Arctic People and the Environment; and, III. Building the Arctic Marine Infrastructure (see Appendix B). It is important to note that the study team considered the gap in marine infrastructure to be of the same level of requirements as safety and protection challenges. The significance of AMSA can be viewed in three interrelated perspectives: as a baseline of Arctic marine activity (an historic survey and snapshot of Arctic marine operations and shipping); as a strategic guide for the Arctic States, Permanent Participants (indigenous peoples' groups), and a host of stakeholders; and, as a policy document of the Council's eight State members who negotiated and reached consensus on its recommendations.

AMSA's Recommendation IIC called for the identification of areas of heightened ecological and cultural significance, with a view towards implementation of protective measures to protect those areas from shipping. Accordingly, three Arctic Council working groups published a report in response to the AMSA entitled "Identification of Arctic marine areas of heightened ecological and cultural significance: Arctic Marine Shipping Assessment (AMSA) IIC" (AMAP/CAFF/SDWG 2013). This report, a comprehensive "atlas" of these critical areas, will be used to shape a future circumpolar network of MPAs and also influence the creation of individual MPAs that may be transboundary in space.

AMSA's recommendation IID called for the Arctic States to explore the need for internationally designated areas for the purpose of environmental protection through the use of tools such as "Special Areas" or through PSSA designation through the IMO. Following that recommendation, Det Norske Veritas (DNV) was engaged by the Norwegian Environment Agency on behalf of PAME to produce a report (DNV Report) on specially designated marine areas restricted to the Arctic high seas areas (DNV 2014). The DNV Report identified the threats to Arctic high seas areas from shipping and examined the measures available under the IMO to protect the environment. It concluded that designation as a Special Area under MARPOL would not lead to a discernible increase in protection, but that all or part of the Arctic high seas should be designated as a PSSA. A preferred option was to designate the entire Arctic high seas as a PSSA with a ship reporting system to monitor traffic and dynamic areas to be avoided to reflect the moving sea ice edge. The second option was similar to the preferred option however absent the dynamic areas to be avoided which DNV considered logistically and politically challenging to establish and implement. The final discussed option was to designate one or more "core sea ice areas" as PSSAs and areas to be avoided (DNV 2014). However, the report's recommendations were not adopted by the Arctic States (PAME 2015).

3 IMO Polar Code

On 1 January 2017 the International Code for Ships Operating in Polar Waters (Polar Code) entered into force after a long period of development since 1993 (Brigham 2017). Through amendments to MARPOL (1973/78), SOLAS (1974) and the STCW Convention (1978) the Polar Code imposes on all commercial ships and passenger vessels (500 tons or higher) mandatory requirements relating to environmental protection and maritime safety. Negotiated with the objective of creating mandatory uniform standards for polar shipping, the Polar Code is a significant development in the protection of the polar marine environment. It is a new governance regime for polar waters that exhibit extreme environmental conditions and where marine infrastructure is very limited or non-existent. In summary the IMO Polar Code creates a set of new requirements for ships in Arctic and Antarctic waters including:

- Ship structural standards for Polar Class ships;
- Marine safety equipment designed for operation in polar environments;
- Training and experience standards for the ship's officers and crew;
- A Polar Ship Certificate issued by the flag State administrator or an authorized representative, normally a ship classification society;
- An onboard Polar Water Operational Manual that is unique to a given ship and includes operational capabilities and limitations; and
- Environmental rules regarding the discharge of oil, noxious liquids, sewage and garbage.

One of the critical features of the Polar Code is the "Arctic area" referred to in Sect. 1.1 above. The Polar Code boundary in the Arctic is adjusted (northward) due to the warmer waters in the North Atlantic and the general location of the seasonal sea ice extent in the region. In the Bering Sea, the Polar Code boundary is 60° North, corresponding closely to the seasonal maximum extent of winter sea ice in this seasonally ice-covered sea. This boundary was also chosen to provide additional measures of the protection to the Bering Sea's world class fishery. Broadly speaking, the new Polar Code Arctic area defines a large protected area although it is focuses on a single sector (shipping); this region can also be considered an "area-based management tool" that is under a set of international (IMO) standards.

4 Mechanisms to Establish MPAs in Response to Marine Operations and Shipping

Several mechanisms exist under the law of the sea and international environmental law for the establishment of MPAs. The IMO is recognised by the international community as the competent international organisation for the regulation of shipping. Although not mentioned by name in UNCLOS, it is understood that with regard to navigation, references in UNCLOS to "the competent international organization" are to the IMO (Roberts et al. 2010). Accordingly, States wishing to establish MPAs that will impact on navigation do so under the auspices of the IMO in order to ensure that the protective measures are agreeable to other States. This section focusses on mechanisms of most relevance to the protection of the marine environment from shipping in the Arctic, including measures under UNCLOS, MARPOL Special Areas and PSSAs. Protective measures available under the IMO mechanisms either separately or under PSSA designation such as SOLAS routeing measures are considered.

4.1 Protective Measures in UNCLOS

In order to preserve the freedom of navigation, it is vital to ensure that international shipping rules and standards are globally uniform (Chircop 2016). However, in light of the interest coastal States have in protecting their coastline and maritime zones, international regulations attempt to strike a balance between the competing rights and national interests (Molenaar 2009). The regime for the regulation of vessel source pollution in UNCLOS operates so that the coastal State's right of protection recedes as the flag State's right of freedom of navigation advances further from the coast (Ringbom 2015). All States have a duty under article 194(1) to take, individually or jointly, all measures consistent with UNCLOS to prevent, reduce and control marine pollution, using "the best practicable means at their disposal" (UNCLOS 1982). Considering the recognised effectiveness of appropriately designated and effectively implemented MPAs, the duty in article 194(1) could be read as encouraging the establishment of MPAs on a local or regional basis. Provided that they do not seek to enforce the regulations against third States, collective exercise of coastal State jurisdiction to establish an MPA regulating shipping is not inconsistent with UNCLOS and the role of the IMO (Molenaar 2014).

The following section will briefly outline the prescriptive jurisdiction provided for in UNCLOS of coastal States in their maritime zones to implement MPAs, as well as the jurisdiction of all States to establish high seas MPAs.

4.1.1 Territorial Sea (UNCLOS Part II)

The coastal State is sovereign in the territorial sea and consequently coastal State jurisdiction is extensive. However article 24(1)(a) of UNCLOS provides that the coastal State cannot enact or enforce any law which is discriminatory or which would have the practical effect of denying or impairing the right of innocent passage. Article 211(4) provides that within the territorial sea the coastal State may in exercising its sovereignty adopt laws and regulations for the prevention, reduction and control of marine pollution from vessels, so long as the right of innocent passage is not hampered. Under article 21(1) the coastal State has prescriptive jurisdiction to adopt laws and regulations affecting ships in innocent passage relating to matters including the safety of navigation and the regulation of maritime traffic, the conservation of the living resources of the sea, and the preservation of the environment of the coastal State and the prevention, reduction and control of pollution thereof. Where these laws regulate the design, construction, manning of equipment of foreign ships, article 21(2) states that they may only give effect to generally accepted international rules or standards. Under article 22, the coastal State has jurisdiction to require foreign ships in innocent passage to use sea lanes and traffic separation schemes "where necessary having regard to the safety of navigation". In doing so article 22(3)(a) provides that the coastal State must take into account the recommendations of the IMO. Despite the wording in article 22 referring only to maritime safety, considering the textual and practical inter-linkages between maritime safety and pollution, it is widely accepted that routeing measures may be designated in the territorial sea for environmental protection (Molenaar 1998; Jakobsen 2016; Henriksen 2013b; Roberts and Tsamenyi 2007). The coastal State therefore has considerable discretion to implement its prescriptive jurisdiction for environmental protection and maritime safety so as to create MPAs in its territorial sea, so long as the innocent passage regime is preserved.

4.1.2 Straits Used for International Navigation

In straits used for international navigation as defined in Part III of UNCLOS the coastal State has significantly restricted jurisdiction to regulate for ships in transit passage. It is the position of certain States that the Northwest Passage and the Northern Sea Route are straits used for international navigation and therefore that all ships enjoy a right of transit passage. Other States take the view that these routes do not meet the criteria for classification as straits used for international navigation under Pt III and therefore the transit passage regime does not apply (Chircop 2016). Canada and Russia claim parts of these routes as internal waters, to which legally there is no right of access by ships of third States (Boone 2013; Chircop 2016). Considering the zone-based jurisdiction in the law of the sea, the legal status of the Arctic waters and the applicable passage regime is relevant to determining the extent to which the coastal State can regulate international navigation for environmental protection. Also debated is whether the expanded jurisdiction under article 234 (to be discussed below in Sect. 4.1.3.3) "overrides" the regime of straits used for international navigation (Boone 2013).

4.1.3 Exclusive Economic Zone (UNCLOS Part V)

4.1.3.1 General

In the exclusive economic zone (EEZ), the coastal State has jurisdiction under article 56(1)(b) for the protection and preservation of the marine environment. Article 211 (5) provides that the coastal State may adopt laws and regulations for the prevention, reduction and control of pollution, "conforming to and giving effect to generally accepted international rules and standards established through the competent international organisation or diplomatic conference", particularly MARPOL, but also others such as SOLAS and COLREG (1972). This link to generally accepted international rules and standards ensures that interference with the freedom of navigation (which ships of third States enjoy in the EEZ) is avoided. Except for certain special circumstances discussed in the following sections, in the EEZ the coastal State cannot impose on ships flagged to other States environmental regulations, including MPAs, affecting their international navigation rights (Henriksen 2013a).

4.1.3.2 Article 211(6)

Despite the importance given to globally uniform shipping standards, there is provision in UNCLOS for the adoption of special measures where due to the special circumstances of the particular area the global rules and standards are not adequate. Article 211(6) articulates the several procedural requirements for the imposition of such special mandatory measures, which may essentially allow the creation of an MPA. The State applying for the special measures must identify a particular, clearly defined area within its EEZ that it has reasonable grounds to believe requires special mandatory measures for the prevention of pollution from vessels. The need for protection from vessel source pollution must be due to recognised technical reasons regarding its oceanographic and ecological conditions, and its utilisation or protection of its resources, and the particular character of its traffic. If these circumstances are met the coastal State, after consultations with other concerned States, may direct a communication to the IMO submitting scientific and technical evidence in support and information on necessary reception facilities.

If the IMO determines that the conditions in the particular area meet the requirements in article 211(6), the coastal State may adopt laws and regulations for the prevention, reduction and control of pollution from vessels. These laws and regulations must either implement IMO rules and standards or navigation practices for special areas or additional rules as accepted by the IMO. Any additional rules must not require foreign vessels to observe design, construction, manning or equipment standards beyond the generally accepted international rules and standards. To date the mechanisms under art 211(6) to designate an area with special pollution control measures have not been used. Rather, it has been used as the legal basis for associated protective measures proposed for PSSAs (IMO 2005; Chircop 2009b).

4.1.3.3 Article 234

The second special UNCLOS regime for environmental protection in the EEZ is provided for in art 234. Article 234 grants coastal States the right to adopt and enforce unilateral laws and regulations for the prevention, reduction and control of vessel-source pollution in ice-covered areas "within the limits of the exclusive economic zone". A number of cumulative requirements must be met in order for jurisdiction under article 234 to be enlivened. "[P]articularly severe climatic conditions and the presence of ice covering such areas for most of the year" must "create obstructions or exceptional hazards to navigation" and "pollution of the marine environment could cause major harm to or irreversible disturbance of the ecological balance". Laws and regulations adopted under article 234 must have "due regard to navigation and the protection and preservation of the marine environment based on the best available scientific evidence". The geographical scope of article 234 has been debated due to the use of the words "within the limits of the exclusive economic zone". On one interpretation this would mean that the special coastal State jurisdiction could only be exercised in the EEZ, not in the territorial sea. However, the more "logical" interpretation is that article 234 includes the territorial sea, as the opposing view runs contrary to the general system of coastal State jurisdiction receding with each maritime zone away from the coastline (Molenaar 1998).

An important feature of the jurisdiction afforded by article 234 is the absence of the requirement for IMO oversight. The right of the Arctic coastal States to adopt and enforce unilateral environmental regulations out to 200 nautical miles from their coastline is an exception to the general objective of the international law of the sea to ensure the uniformity of global shipping standards. On its face, art 234 only permits laws and regulations "for the prevention, reduction and control of marine pollution from vessels". However, it is suggested by commentators such as Molenaar (2014) and Chircop (2009a) that this subject matter may reasonably be extended to include some provision for the safety of navigation. Provided that the environmental protection purpose is equally or more important than the maritime safety purpose, the regulation would be included within the scope of article 234. In any case, considering the interconnectedness of the two issues, especially in an environment as fragile as the Arctic, it may be difficult to distinguish between the purposes (Chircop 2009a). This is the inverse scenario to that in straits used for international navigation discussed above in Sect. 4.1.2.

During the negotiation period of the Polar Code, some commentators raised the question of whether the introduction of mandatory regulations for Arctic shipping would have implications for the coastal States' continued exercise of their expanded jurisdiction under article 234 (Fauchald 2011; McDorman 2015). However considering that article 234 is absent reference to generally accepted international rules and standards and does not include an advisory or approval rule for the IMO, it would appear that the coastal States can continue to exercise their article 234 jurisdiction (Henriksen 2015). The savings provisions in MARPOL and SOLAS which specify that their provisions are without prejudice to the rights or obligations of States under international law (SOLAS 1974: Ch XIV Reg 2) or without prejudice to the present or future claims and legal views of any State concerning the law of the sea and the nature of coastal and flag state jurisdiction (MARPOL 1973: art 9), would seem to support the continued operation of article 234 coastal state expanded jurisdiction. It is beyond the scope of this chapter to consider the matter further here, however it has been suggested that although article 234 jurisdiction remains available to the coastal States, its application may be affected to some extent by the introduction of the Polar code (Henriksen 2015).

4.1.4 High Seas

On the high seas the flag State has exclusive jurisdiction regarding the protection of the marine environment from vessels. An indirect legal basis for the establishment of a high seas MPA can be found in the environmental duties contained in arts 192, 194 and 197 of UNCLOS. Article 237 of UNCLOS specifically permits the conclusion of other agreements between States in furtherance of UNCLOS' general principles. States are free to assume additional environmental obligations, however they must be carried out in a manner consistent with UNCLOS' general principles and objectives. Therefore, States parties to UNCLOS are able to conclude an agreement on the creation of a high seas MPA, however it would not be opposable to ships flagged to States not party to the MPA agreement.

4.2 MARPOL Special Areas

MARPOL is the primary international agreement on vessel-source pollution (Rothwell 2000). The main operative provisions of MARPOL are contained in its annexes which each deal with a different polluting substance. Participation in Annexes I (oil pollution) and II (noxious liquid substances in bulk) is compulsory for all parties, with the other four annexes optional. MARPOL has a very high level of participation, representing 99.14% of world tonnage as of 7 February 2017 (IMO 2017b). The optional annexes count as members 98.55%, 91.44%, 98.72% and 96.13% of world tonnage respectively. Of the States with maritime zones within the Polar Code boundary, there is complete membership of all MARPOL annexes, except for the United States which is not a party to Annex IV (IMO 2017a). Of the Arctic Council States, Finland and Sweden are parties to all annexes, whilst Iceland is not a party to Annex IV or VI.

Under Annexes I, II, IV and V, certain sea areas may be defined as "Special Areas" "where for recognised technical reasons in relation to its oceanographical and ecological condition and to the particular character of its traffic the adoption of special mandatory methods for the prevention of sea pollution" is required (MARPOL 73/78). Under Annex VI "Emission Control Areas" may be designated in which sulphur dioxide and particulate matter emissions from ships are further restricted. The IMO Guidelines for the Designation of Special Areas under MARPOL 73/78 adopted in 2001 (Special Area Guidelines) specify the three categories of criteria to be satisfied for an area to be given special area status: oceanographic conditions, ecological conditions and vessel traffic conditions (IMO 2001). The Arctic Ocean would most likely satisfy the oceanographic criteria—"the area possesses oceanographic conditions which may cause the concentration or retention of harmful substances in the waters or sediments of the area", and the ecological criteria (Henriksen 2013b). However, the vessel traffic criterion may be less clearly met. Regulation 2.6 of the Special Area Guidelines provides:

> The sea area is used by ships to an extent that the discharge of harmful substances by ships when operating in accordance with the requirements of MARPOL 73/78 for areas other than Special Areas would be unacceptable in the light of the existing oceanographic and ecological conditions in the area (IMO 2001).

Arctic shipping remains at a very low level by international standards. However, considering the environmental conditions in the Arctic and its ecological sensitivity, the impacts of the discharge of harmful substances from only a small number of ships may be sufficient to satisfy the vessel traffic criterion (Henriksen 2013b).

Designation as a special area requires amendment of MARPOL, as the text of the relevant annex lists each of the special areas (IMO 2001). The Polar Code includes a ban on the discharge of oil (Annex I), and restrictions on the discharge of sewage (Annex IV) and garbage (Annex V). Designation as a Special Area under the other applicable annexes either for the whole of the Polar Code area or for smaller areas could potentially be a protective mechanism for the Arctic. The Special Area Guidelines provide that in order for a Special Area designation to become effective, adequate reception facilities must be provided for ships in accordance with the provisions of MARPOL (IMO 2001). Considering the lack of port infrastructure in the Arctic, this requirement would present a challenge to Special Area designation, and is discussed below in Sect. 5.4. Further, the DNV Report concluded that in light of the discharge restrictions in the Polar Code (that DNV anticipated at that time), the existing discharge restrictions in the Arctic and the international shipping standards for Arctic operation, there was little benefit in special area designation under MARPOL for the Arctic high seas (DNV 2014). It did however consider that emissions restrictions corresponding to emission control area designation under Annex VI of MARPOL would be a beneficial associated protective measure in an Arctic PSSA. Coastal areas frequented by cruise ships may be better candidates for designation under Annexes IV (sewage—to combat eutrophication) and V (garbage) (DNV 2014).

4.3 PSSAs

Unlike measures under article 211(6) or MARPOL Special Areas, particularly sensitive sea areas (PSSAs) are not founded in an international instrument, but rather emerged from IMO practice. The IMO derives its competence from its founding 1948 convention (IMO 1948) which specifies the functions of its Assembly as including in article 15(j) to recommend to members for adoption regulations and guidelines concerning the prevention and control of marine pollution from ships and other matters concerning the effect of shipping on the marine environment. In December 2005 the IMO Assembly adopted resolution A.982(24) on Revised Guidelines for the Identification and Designation of Particularly Sensitive Sea Areas (hereinafter PSSA Guidelines) (IMO 2005). A PSSA is defined in the PSSA Guidelines as:

> An area that needs special protection through action by IMO because of its significance for recognised ecological, socio-economic, or scientific attributes where such attributes may be vulnerable to damage by international shipping activities.

PSSAs may be designated within or beyond the limits of the territorial sea (IMO 2005). A PSSA proposal is initiated by a member State or a group of States where those States have a common interest in the proposed PSSA area. The application must contain evidence to satisfy the three primary requirements for PSSA designation. First, the area must meet at least one of the specified ecological, socio-economic or scientific criteria; second, the area must be at risk from international shipping; and third, the proposed associated measures must have a clear legal basis and be within the competence of the IMO (IMO 2005).

The ecological criteria listed in the PSSA Guidelines are: uniqueness or rarity; critical habitat; dependency; representativeness; diversity; productivity; spawning or breeding grounds; naturalness; integrity; fragility and bio-geographic importance. The social, cultural and economic criteria are: social or economic dependency; human dependency and cultural heritage. The scientific and educational criteria are: research; baseline for monitoring studies and education (IMO 2005). As found in the AMSA Recommendation IIC Report, which assessed sensitive areas against the PSSA criteria, there are numerous marine areas in the Arctic that would satisfy all or many of the ecological criteria (AMAP/CAFF/SDWG 2013). Although that report was not able to complete its work in assessing the entire Arctic for cultural significance and sensitivity, it concluded that many areas would meet the three social, cultural and economic criteria (AMAP/CAFF/SDWG 2013). It is also highly likely that much of the Arctic would satisfy the scientific and educational criteria (DNV 2014).

Once at least one of the ecological, socio-economic or scientific criteria is shown to exist in the PSSA area, the application must also show that the "recognised attributes" are at risk from international shipping. The risk from international shipping is determined by reference to a number of factors (IMO 2005). These factors are divided into vessel traffic characteristics and natural factors, and include the types of vessels, volume or concentration of traffic, harmful substances carried, and the hydrographical, meteorological and oceanographic characteristics of the area that may increase the risk of harm from shipping. Although shipping remains at very low levels, as the AMSA concluded, the Arctic is vulnerable to an accidental oil spill incident (AMSA 2009). The traffic is low, however its increased vulnerability to the impacts of shipping compared to most other marine areas means that the Arctic would prima facie meet the requirement of being at risk from international shipping activities (DNV 2014).

4.3.1 Associated Protective Measures (APMs)

An application for PSSA designation must propose at least one associated protective measure (APM) (IMO 2005). APMs are limited to measures that are to be or have been approved or adopted by the IMO. The application must justify why the proposed APMs are appropriate for the protection of the area. As PSSAs do not have any legal basis in and of themselves, is not the declaration as a PSSA that has legal force, but rather the associated protective measures implemented within it. APMs may include such measures as routeing measures under SOLAS or discharge standards under MARPOL. PSSA designation is not always necessary for a coastal State to implement measures that may be APMs, as they can generally be approved separately through the relevant IMO committee. Once approved by the IMO, all associated protective measures are to be identified on charts in accordance with the symbols and methods of the International Hydrographic Organisation. The challenges of implementing appropriate APMs in the Arctic marine environment are discussed below in Sect. 5.

4.4 Others

Many of the protective measures available to be implemented as APMs in a PSSA are also able to be adopted separately of PSSA designation. Instead, the coastal State or a group of Arctic States may apply to the relevant IMO committee for the adoption of measures such as routeing or reporting systems. As stated above, the coastal State's jurisdiction in the EEZ for environmental protection is limited, and therefore rather than adopt such measures unilaterally, the Arctic coastal States may do so through the IMO. Under SOLAS available measures include routeing systems, vessel traffic services and ship reporting systems. Within the broader category of routeing systems, particular measures include traffic separation schemes, two-way routes, precautionary areas and areas to be avoided (IMO 1985). A measure such as an area to be avoided, which could be designated to direct traffic around an area of identified sensitivity, can be an effective tool to protect the environment from the impacts of shipping. The following section addresses the challenges of the Arctic marine environment that may render geographically fixed routeing systems impractical. Traffic separation schemes and two-way routes are more likely to be adopted in areas where shipping traffic and the corresponding risk of collision is high.

Vessel traffic services (VTS) and ship reporting systems (SRS) may also be adopted in the Arctic. A VTS is a service by which the competent authority of the coastal State can interact with vessels and respond to traffic situations in order to improve safety and efficiency of traffic and to protect the environment (IMO 1997a). An SRS operates from ship to land only, which notifies the coastal or port authorities of the presence of a vessel and certain basic information (IMO 1997b). In the Canadian Arctic the NORDREG vessel traffic system is in operation, which requires a vessel to provide reports at specified stages of its voyage to the Canadian Coast Guard (Canada Shipping Act 2001). The Northern Sea Route Administration regulates shipping through Russian Arctic waters, but no VTS has been established. In addition to rules on icebreaker assistance and pilotage, the Northern Sea Route Administration operates a system for communication between ships, icebreakers and the land authorities, as well as the provision of information relating to hazards for navigation (NSRA 2013).

5 Arctic Challenges

5.1 Environmental Challenges for MPA Designation

It is apparent that either the marine Arctic in its entirety, or multiple areas of heightened ecological importance, satisfy the eligibility requirements for PSSA designation or other IMO protective measures. However, identifying protective measures appropriate for the Arctic environment may be challenging primarily due to the presence of ice and further the fact of the ice's dynamic and unpredictable nature (DNV 2014). Common APMs in existing PSSAs such as routeing measures and areas to be avoided may be impractical in the Arctic where it is safer and more economically viable to navigate through open water rather than breaking ice. It is difficult to predict the location and thickness of ice from one year to the next. Requiring a ship to navigate through ice along a designated sea lane as opposed to an available open water route is not likely to be a good outcome from either an environmental or a safety perspective. Reflecting these practicalities, the IMO General Provisions on Ships' Routeing specify that routeing systems are intended for use "in ice-free waters or under light ice conditions where no extraordinary manoeuvres or ice-breaker assistance are required" (IMO 1985, par 8.2).

In light of the environmental challenges in the Arctic, dynamic areas to be avoided have been suggested as APMs appropriate for the Arctic reality (Jakobsen and Johansen 2017; DNV 2014). Although IMO rules require that areas to be avoided have defined limits, it has been suggested that defining an area to be avoided by reference to the ice edge could meet this requirement (Jakobsen and Johansen 2017; DNV 2014). However, although the DNV Report considered that a dynamic area to be avoided would be the most effective APM from an environmental perspective, it stated that it would be difficult to design and implement, would place "too big a burden" on shipping, and would be a strong violation of international navigation rights (DNV 2014).

5.2 Rights and Concerns of Indigenous Peoples and Local Inhabitants

Marine operations and shipping can directly impact the lives of indigenous people and their lifestyles, particularly affecting food security by disruptions to subsistence hunting and fishing. Any potential designation of an MPA must involve early in the process indigenous people and coastal communities. AMSA created a specific recommendation (IIB), "Engagement with Arctic Communities", that addressed the importance of effective communication mechanisms for the engagement of coastal communities with national and regional authorities as well as the maritime industry (AMSA 2009). AMSA also called for the conduct of surveys of indigenous use (Recommendation IIA) so that baseline data would be available to access the impacts of marine operations and shipping (AMSA 2009). The construct of any future MPA or network of MPAs should consider all marine users in a holistic manner with a top priority given to indigenous marine users.

5.3 MPAs in the Central Arctic Ocean

The high seas are outside the jurisdiction of any State, and ships navigating on the high seas are, except in certain limited circumstances, subject to the exclusive jurisdiction of the flag State. Therefore the protective measures established under any MPA in the high seas of the Central Arctic Ocean would only be binding on ships registered to a State that has agreed to be bound. This is a significant challenge to the effectiveness of an MPA in the Central Arctic Ocean. In order to be effective, not only the Arctic coastal States, but also the flag States of the ships navigating through the Central Arctic Ocean must be party to the agreement establishing the MPA.

The recent developments toward a Central Arctic Ocean fisheries agreement show that the Arctic States are interested in taking a precautionary approach towards the protection of the marine environment of the Arctic high seas (Pew 2017). Also involved in the discussions are non-Arctic States or organisations with major fishing interests, namely the European Union, the People's Republic of China, Japan and the Republic of Korea. These developments may be a positive indicator of future precautionary action regarding the protection of the Arctic high seas from shipping, as well as the involvement of non-Arctic States with shipping interests in the Central Arctic Ocean.

A polar high seas MPA would not be unprecedented. An MPA was recently adopted on the high seas in the Ross Sea under the auspices of the Commission on the Conservation of Antarctic Flora and Fauna (CCAMLR), which will enter into force on 1 December 2017 for a period of 35 years (CCAMLR 2016a). The Ross Sea MPA follows an earlier CCAMLR high seas MPA, the South Orkney Islands southern shelf MPA, which was the world's first high seas MPA (CCAMLR 2016b). However due to CCAMLR's mandate which is focused on the protection of flora and fauna, and does not extend to shipping regulation, CCAMLR MPAs do not affect international navigation.

5.4 Marine Infrastructure

The lack of marine infrastructure was identified in AMSA as a major limitation to enhancing marine safety and environmental protection in the region (AMSA 2009). AMSA called for in its recommendations: addressing the infrastructure deficit by the Arctic States; an Arctic marine traffic awareness system; continued development of circumpolar environmental response capacity; and, importantly, significant investment and improvement in hydrographic, meteorological, and oceanographic data (AMSA 2009). Infrastructure that is highly relevant to the establishment and functioning of MPAs (for shipping) include: hydrography and charting; monitoring and surveillance of marine traffic; aids to navigation; search and rescue capacity; environmental response capacity; salvage; environmental observing (sea ice, oceans, atmosphere and terrestrial (permafrost); communications; port services, and, more. Without these fundamental elements of infrastructure MPAs cannot become effective marine management tools that can enhance safety and protection.

5.5 Compliance and Enforcement

Ensuring compliance with any Arctic MPA will be a major challenge, as will enforcement of any violations of its regulations. Under UNCLOS, flag States generally have primary jurisdiction with respect to compliance and enforcement of shipping regulations. Within the territorial sea, the coastal State may only arrest a ship in the limited circumstances in which a ship's passage is no longer considered "innocent" (UNCLOS 1982, art 25). Within the EEZ, the coastal State's enforcement jurisdiction is essentially limited to the enforcement of laws related to its sovereign rights to the living resources of the EEZ (UNCLOS 1982, art 73). Otherwise, ships navigating in the Arctic are subject to the exclusive jurisdiction of their flag State.

The flag State has the duty to ensure compliance by its vessels with applicable international rules and standards, such as SOLAS and MARPOL regulations, as well as with their domestic laws adopted in accordance with UNCLOS for the prevention, reduction and control of vessel-source pollution (UNCLOS 1982, article 217). The flag State must also provide for the effective enforcement of such laws and regulations regardless of where the violation occurs, as well as all rules and standards established through the IMO (UNCLOS 1982, art 217). Although flag States are under legal duties to ensure compliance and effectively enforce international shipping regulations, the size, remoteness and environmental conditions in the Arctic render the exercise of this jurisdiction challenging. For non-Arctic flag States, and even Arctic flag States, to ensure that their ships navigating in the Arctic are complying with all relevant regulations under an MPA would require considerable financial investment as well as the infrastructure and data to do so. Port State control can play a vital role in ensuring compliance with shipping regulations, as due to its territorial sovereignty the port State can inspect and take action against ships voluntarily in port. However, the lack of port infrastructure in the Arctic would significantly limit the utility of port State control in the region.

6 Conclusions

MPAs can be effective tools or management strategies to enhance protection of the Arctic marine environment in the face of increases in marine operations and shipping. There are a host of unique challenges to how MPAs can be applied to a specific sector such as marine operations and shipping under changing climatic conditions and where there is a significant lack of maritime infrastructure. However, significant progress has been made in recent years. The Arctic Council's AMSA called for the identification of areas of heightened ecological and cultural significance, and for mandatory IMO measures to enhance Arctic marine safety and environmental protection. A comprehensive "atlas" of these key sites was produced by two of the Council's working groups for use as a guide to establishing protected areas. The IMO Polar Code came into force on 1 January 2017 and its designated region of application in the Arctic can be considered a large 'marine protected area' with regards to shipping. The Arctic Council continues to work on establishing a network of MPAs. How the Arctic's indigenous peoples, especially those in coastal communities, can have a shared vision regarding the use of local MPAs and even a transboundary network of MPAs is a critical issue to address by the Arctic States and the maritime industry. If established within a robust legal framework, two of the serious challenges to any Arctic MPA (addressing shipping) will be its implementation and subsequent enforcement. Observing and regular monitoring will be central to an MPA's overall effectiveness. These are clear challenges also to the new IMO Polar Code which is interwoven with international efforts to define protected areas at all levels.

References

ACIA. (2004). *Impacts of a warming Arctic: Arctic climate impact assessment*. Cambridge: Cambridge University Press.

AMAP/CAFF/SDWG. (2013). *Identification of Arctic marine areas of heightened ecological and cultural significance: Arctic Marine Shipping Assessment (AMSA) IIc*. Oslo: Arctic Monitoring and Assessment Programme (AMAP).

AMSA. (2009). Arctic Marine Shipping Assessment (AMSA). Arctic Council, April 2009, second printing.

Boone, L. (2013). International regulation of polar shipping. In E. J. Molenaar, A. G. Oude Elferink, & D. R. Rothwell (Eds.), *The law of the sea and the polar regions: Interactions between global and regional regimes* (p. 193). Leiden: Brill.

Brigham, L. W. (2017). The changing maritime Arctic and new marine operations. In R. C. Beckman et al. (Eds.), *Governance of Arctic shipping* (p. 3). Leiden: Brill.

Canada Shipping Act. (2001). Canada Shipping Act, 2001, Northern Canada Vessel Traffic Services Zone Regulations, SOR/2010-127.

CBD. (1992). Convention on Biological Diversity, opened for signature 5 June 1992, 1760 UNTS 79 (entered into force 29 December 1993).

CBD. (2014). CBD Doc. UNEP/CBD/EBSA/WS/2014/1/5 20 May 2014, Report of the Arctic Regional Workshop to Facilitate the Description of Ecologically or Biologically Significant Marine Areas.

CBD COP 10. (2010). Decision X/29 Marine and coastal biodiversity, adopted 29 October 2010.

CBD COP 9. (2008). Decision IX/20 Marine and coastal biodiversity, adopted 9 October 2008.

CBD Secretariat. (2004). Technical advice on the establishment and management of a national system of marine and coastal protected areas. CBD Technical Series No 13.

CCAMLR. (2016a). Conservation Measure 91-05 (2016), Ross Sea region marine protected area.

CCAMLR. (2016b). CCAMLR to create world's largest Marine Protected Area, 28 October 2016, https://www.ccamlr.org/en/organisation/ccamlr-create-worlds-largest-marine-protected-area.

Chircop, A. (2009a). The growth of international shipping in the Arctic: Is a regulatory review timely? *The International Journal of Marine and Coastal Law, 24*, 355.

Chircop, A. (2009b). The designation of particularly sensitive sea areas: A new layer in the regime for marine environmental protection from international shipping. In A. Chircop, T. McDorman, & S. Rolston (Eds.), *The future of ocean-regime building: Essays in tribute to Douglas M Johnston* (p. 573). Leiden: Brill.

Chircop, A. (2016). Sustainable Arctic shipping: Are current international rules for polar shipping sufficient? *The Journal of Ocean Technology, 11*(3), 39.

Churchill, R. (2013). The growing establishment of high seas marine protected areas: Implications for shipping. In R. Caddell & D. R. Thomas (Eds.), *Shipping, law and the marine environment in the 21st century: Emerging challenges for the law of the sea – Legal implications and liabilities* (p. 53). Oxon: Lawtext Publishing.

COLREG. (1972). Convention on the International Regulations for Preventing Collisions at Sea, opened for signature 20 October 1972, 1050 UNTS 16 (entered into force 15 July 1977).

DNV. (2014). Specially Designated Marine Areas in the Arctic High Seas. Det Norske Veritas. Oslo (Revision No. 1).

Fauchald, O. K. (2011). Regulatory frameworks for maritime transport in the Arctic: Will a Polar Code contribute to resolve conflicting interests? In J. Grue & R. Gabrielsen (Eds.), *Maritime transport in the High North* (p. 73). Oslo: Norwegian Academy of Science and Letters and Norwegian Academy of Technological Studies.

Henriksen, T. (2013a). Conservation of marine biodiversity and the international maritime organization. In C. Voigt (Ed.), *Rule of law for nature: New dimensions and ideas in environmental law* (p. 331). Cambridge: Cambridge University Press.

Henriksen, T. (2013b). The future of navigation in ice-covered areas: A view from the Arctic. In R. Caddell & D. R. Thomas (Eds.), *Shipping, law and the marine environment in the 21st century: Emerging challenges for the law of the sea – Legal implications and liabilities* (p. 8). Oxon: Lawtext Publishing.

Henriksen, T. (2015). Protecting polar environments: Coherency in regulating Arctic shipping. In R. Rayfuse (Ed.), *Research handbook on international marine environmental law* (p. 363). Cheltenham: Edward Elgar.

IMO. (1948). Convention on the International Maritime Organization, opened for signature 6 March 1948, 289 UNTS 3, 1520 UNTS 297 (entered into force 17 March 1958).

IMO. (1985). IMO Res.A.572(14), General Provisions on Ships' Routeing, adopted 20 November 1985, as amended.

IMO. (1997a). IMO Res. A.857(2), Guidelines for Vessel Traffic Services, adopted 27 November 1997.

IMO. (1997b). IMO Res. A.851(20), General Principles for Ship Reporting Systems and Ship Reporting Requirements, including Guidelines for Reporting Incidents involving Dangerous Goods, Harmful Substances and/or Marine Pollutants, adopted 27 November 1997.

IMO. (2001). IMO Res. A.927(22), Guidelines for the Designation of Special Are as under MARPOL 73/78 and Guidelines for the Identification and Designation of Particularly Sensitive Sea Areas, adopted 29 November 2001.

IMO. (2005). IMO Res. A.982(24), Revised Guidelines for the Identification and Designation of Particularly Sensitive Sea Areas, adopted 1 December 2005.

IMO. (2017a). Status of Conventions. 7 February 2017. <http://www.imo.org/en/About/Conventions/StatusOfConventions/Pages/Default.aspx>.

IMO. (2017b). Summary of Status of Conventions. 7 February 2017. <http://www.imo.org/en/About/Conventions/StatusOfConventions/Documents/Status%20of%20Treaties.pdf>.

IUCN. (2008). In N. Dudley (Ed.), *Guidelines for applying protected area management guidelines*. IUCN.

Jakobsen, I. U. (2016). *Marine protected areas in international law: An arctic perspective*. Leiden: Brill Nijhoff.

Jakobsen, I. U., & Johansen, E. (2017). Efforts of the Arctic Council to protect sensitive arctic high sea areas from the impact of shipping. *MarIus, 471*, 1–40.

Kelleher, G. (Ed.). (1999). *Guidelines for marine protected areas*. IUCN.

Lalonde, S. (2013). Marine protected areas in the Arctic. In E. J. Molenaar et al. (Eds.), *The law of the sea and the polar regions: Interactions between global and regional regimes* (p. 85). Leiden: Martinus Nijhoff Publishers.

LOSC. (1982). United Nations Convention on the Law of the Sea, opened for signature 10 December 1982, 1833 UNTS 396 (entered into force 16 November 1994).

MARPOL. (1973/78). International Convention for the Prevention of Pollution from Ships (as Modified by the Protocol of 1978 Relating Thereto), opened for signature 2 November 1973, 1340 UNTS 184 (entered into force 2 October 1983).

McDorman, T. (2015). A note on the potential conflicting treaty rights and obligations between the IMO's Polar Code and Article 234 of the Law of the Sea Convention. In S. Lalonde & T. L. McDorman (Eds.), *International law and politics of the Arctic Ocean: Essays in honor of Donat Pharand* (p. 141). Leiden: Brill Nijhoff.

Molenaar, E. J. (1998). *Coastal state jurisdiction over vessel-source pollution*. The Hague: Kluwer Law International.

Molenaar, E. J. (2009). Arctice marine shipping: Overview of the international legal framework, gaps and options. *Journal of Transnational Law and Policy, 18*(2), 289.

Molenaar, E. J. (2014). Status and reform of international arctic shipping law. In E. Tedsen et al. (Eds.), *Arctic marine governance* (p. 127). Berlin: Springer-Verlag.

Nicoll, R., & Day, J. C. (2017). Correct application of the IUCN protected area management categories to the CCAMLR Convention Area. *Marine Policy, 77*, 9.

NSRA. (2013). Ministry of Transport of Russia, Rules of Navigation in the Water Area of the Northern Sea Route, January 17, 2013, No 7.

PAME. (2015). Framework for a Pan-Arctic Network of Marine Protected Areas. Arctic Council, April 2015.

Pew. (2017). International Officials Close to Agreement to Protect Central Arctic Ocean Fisheries, Steve Ganey, April 10, 2017, available at http://www.pewtrusts.org/en/research-and-analysis/blogs/compass-points/2017/04/10/international-officials-close-to-agreement-to-protect-central-arctic-ocean-fisheries.

PrepCom. (2017). Chair's Streamlined non-paper on Elements of a Draft Text of an International Legally-Binding Instrument under the United Nations Convention on the Law of the Sea on the Conservation and Sustainable use of Marine Biological Diversity of Areas Beyond National Jurisdiction, available at http://www.un.org/depts/los/biodiversity/prepcom_files/Chairs_streamlined_non-paper_to_delegations.pdf.

Ringbom, H. (2015). Vessel-source pollution. In R. Rayfuse (Ed.), *Research handbook on international marine environmental law* (p. 105). Cheltenham: Edward Elgar.

Roberts, J., Chircop, A., & Prior, S. (2010). Area-based management on the high seas: Possible application of the IMO's particularly sensitive sea area concept. *The International Journal of Marine and Coastal Law, 25*, 483–522.

Roberts, J., & Tsamenyi, M. (2007). The regulation of navigation under international law: A tool for protecting sensitive marine environments. In T. Ndiaye & R. Wolfrum (Eds.), *Law of the sea, environmental law and settlement of disputes* (p. 787). Leiden: Koninklijke Brill.

Rothwell, D. (2000). Global environmental protection instruments and the polar marine environment. In D. Vidas (Ed.), *Protecting the polar marine environment: Law and policy for pollution prevention* (p. 57). Cambridge: Cambridge University Press.

SOLAS. (1974). International Convention for the Safety of Life at Sea, opened for signature 1 November 1974, UNTS 1184 (entered into force 25 May 1980).

STCW. (1978). International Convention on Standards of Training, Certification and Watchkeeping for Seafarers, opened for signature 7 July 1978, 1361 UNTS 190 (entered into force 28 April 1984).

UNEP-WCMC. (2008). *National and regional networks of marine protected areas: A review of progress*. Cambridge: UNEP-WCMC.

Part V
Training and Capacity Building

The Effects Toward Maritime Higher-Education in China After the Entry into Force of the Polar Code

Haibo Xie and Xiaori Gao

Contents

Abstract The International Code for Ships Operating in Polar Waters (Polar Code) is expected to enter into force on 1 January 2017. As a major exporter of seafarers, the maritime higher-education academy in China should take positive and appropriate measures to comply with the requirements of the Polar Code. Human errors dominated the causes of maritime accidents, so education, training and certification are the most important parts of the Polar Code. However, the education and training for seafarers involving the Polar waters are not covered in current maritime higher-education programmes based on the STCW Convention and codes. Consequently, additional courses and trainings should be developed to make the seafarers

H. Xie (✉) · X. Gao
Dalian Maritime University, Dalian, Republic of China
e-mail: halbertsport@126.com; s15031@alumni.wmu.se; gxrdlmu@126.com; s16902@wmu.alumni.se

L. P. Hildebrand et al. (eds.), *Sustainable Shipping in a Changing Arctic*, WMU Studies in Maritime Affairs 7, https://doi.org/10.1007/978-3-319-78425-0_18

competent with the operations in polar waters. These issues will be discussed from the following aspects: the courses and textbooks, training and assessment, selection of instructors and the development of simulators for polar waters. This chapter provides guidance to improve compulsory competence of seafarers required by the Polar Code.

Keywords Higher-education · Polar Code · Maritime education and training · Compliance

1 Introduction

With the shrinking of ice coverage in polar waters, more and more countries are paying increasingly close attention to Arctic routes. Due to the presence of sea ice, Arctic routes have their own peculiarities and greater danger. That means higher professional skills and psychology are required. Effective and complete education and training is an important prerequisite for improvement of safety of navigation in polar waters. Considering peculiarities and greater danger, the additional education and training demands, beyond the existing requirements, are adopted in the International Code for Ships Operating in Polar Waters (Polar Code), which is expected to enter into force on 1 January 2017 (International Maritime Organization 2015).

As a major exporter of seafarers, maritime higher-education academy in China should take positive and appropriate measures to comply with the requirements of the Polar Code. Although some limited knowledge for shipping in polar waters had been introduced in navigation courses and textbooks, these are still not enough to meet the newest requirements of the Polar Code (Wang 2010). Consequently, additional courses and trainings should be developed to make seafarers competent with operations in polar waters. These issues will be discussed from the following aspects: the courses and textbooks, training and assessment, selection of instructors and the development of simulators in polar waters.

2 Requirements of Polar Code Involving Education, Training and Certification

The Polar Code includes mandatory measures covering safety (part I-A) and pollution prevention (part II-A). Requirements of the Polar Code involving education, training and certification shall be fully considered to ensure the safety of the ship and protect the environment of polar waters (International Maritime Organization 2014). Besides the amendments to the Safety of Life at Sea (SOLAS) and Prevention of Pollution from Ships (MARPOL) conventions, another amendment to chapter V of the International Convention on Standards of Training, Certification and Watchkeeping for Seafarers (STCW) relating to training requirements for officers and crew on board in the Polar Code was issued in February of 2016 (DNV 2015).

Member States were invited to develop a draft model course after the aforesaid amendments to chapter V of the STCW Convention and the STCW Code have been prepared (Lloyd's Register Marine 2015). The Polar Code is far more restrictive than existing rules in other parts of the open ocean. These include new requirements in ship design and equipment, crew training and search-and-rescue operations (COSTAS PARIS 2015).

After the enforcement of Polar Code, it will pose some challenges to maritime higher-education that will play an important role in the process of implementation and enforcement involving education and training in the Polar Code. For example, according to the requirements of safety (part I-A) of the Polar Code, the training requirements are Masters, Chief mates and officers in charge of a navigational watch must have completed appropriate basic training and advanced training for other waters including ice-covered waters (International Maritime Organization 2014). It is the responsibility of higher-education agencies to build competency of personnel in compliance with the Code.

3 Current Maritime Higher-Education in China

Maritime higher-education universities in China have already cultivated thousands of international seafarers and laid a solid groundwork for the growth of navigation students with overall quality. There are four main maritime higher-education universities: Dalian Maritime University, Shanghai Maritime University, Jimei University and Wuhan University of Technology.

Compared with general higher-education, maritime higher-education has the dual nature of academic education and vocational education. Furthermore, it should be in compliance with international requirements together with the training requirements for seafarers.

Due to global warming and ice melting in polar waters, many ships had navigated the Arctic waters successfully. M.V. Yong Sheng of China has sailed its second Arctic voyage to Europe from China since July 2015, which can save two weeks compared with the normal Suez Canal route. Trends and forecasts indicate that polar shipping will grow in volume and diversify in nature over the coming years (International Maritime Organization 2015). However, the current education and training system is mainly developed based on the STCW convention and some national laws (Fig. 1) (Wang 2010). In current navigation courses and textbooks, just some basic knowledge for shipping in polar waters had been introduced, which still cannot follow the newest requirements of the Polar Code. Therefore, China has responsibilities to carry out appropriate updating of courses and training to comply with the requirements of the Polar Code.

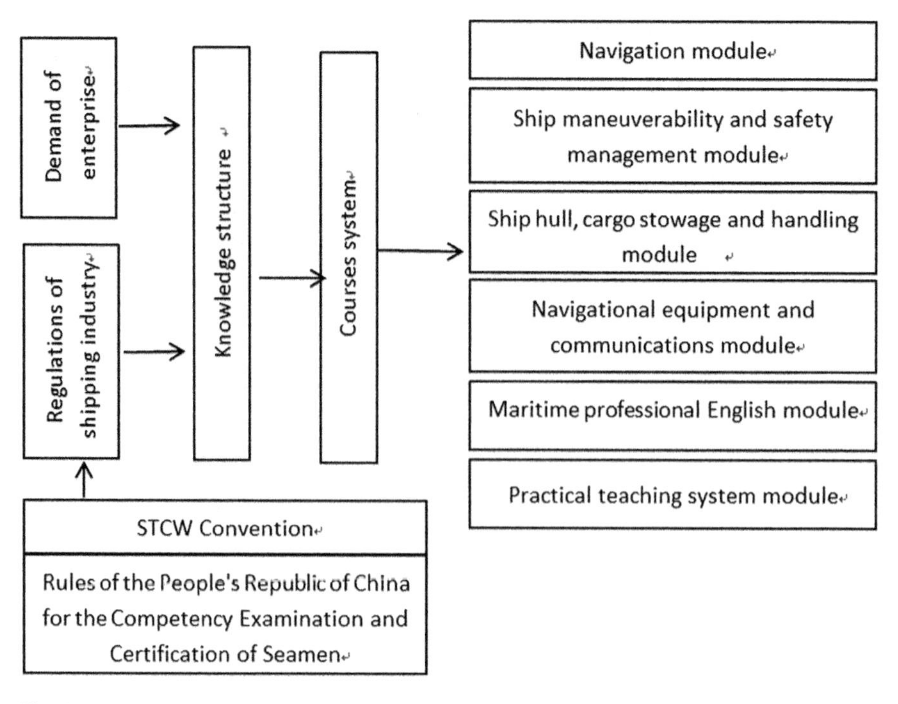

Fig. 1 The current course system in China

4 The Effects Toward Maritime Higher-Education

4.1 The Courses and Textbooks

The courses and textbooks provide compulsory knowledge of shipping in polar waters in theoretical teaching. These knowledge and skills should cover all relevant regulations, geographical environment, voyage plan and communication.

4.1.1 Arctic Navigation Routes and Geographical Demarcation

The current courses and textbooks do not cover knowledge involving the geographical demarcation and Arctic navigation routes. It is very important for seafarers to be familiar with the navigation environment. So the extent of Arctic waters and Antarctic Waters should be clarified to make the students' awareness of the range of the polar waters. The Arctic navigation routes including the 'Northwest Passage' and the 'Northern Sea Route' (NSR) should be fully understood by students (Norwegian Mapping Authority 2011).

Table 1 Main conventions and guidelines in polar waters

Types	Main convention and guidelines
International regulations	UNCLOS—United Nations Convention on the Law of the Sea
	Polar Code—the International Code for Ships Operating in Polar Water
	SOLAS—Safety of Life at Sea
	MARPOL—Prevention of Pollution from Ships
	BWM—Ballast Water Management (Not yet in force)
	STCW—Standards of Training, Certification and Watchkeeping
	COLREG Preventing Collisions at Sea
Various guidelines	Guidelines for ships operating in Polar waters
	Guidelines on voyage planning for passenger ships operating in remote areas
	Enhanced contingency planning guidance for passenger ships operating in areas remote from SAR facilities

4.1.2 Legal Issues in Polar Waters

Due to the great interests of the Arctic route, many countries have begun to pay attention to the Arctic navigation. For example, European Union Member States collectively own the world's largest merchant fleet; the EU has a number of significant interests in Arctic shipping while promoting stricter safety and environmental standards. It is necessary to learn the legal issues concerning shipping in Polar waters and comply with the international and national regulations (European Commission 2010). The main conventions and guidelines in polar waters are listed in Table 1 (Adviser 2012).

In this regard, the content of courses and textbooks should provide an overview of the relevant international legal regime that applies to the Arctic waters in terms of shipping and related activities. In addition to the international regulations and various guidelines, Russia and Canada also introduced a number of regulations on Arctic navigation, which also relates to crew training, qualification and experience requirements (Jensen 2007).

4.1.3 Voyage Planning

The requirements of voyage planning are listed in chapter 11 of the Polar Code (Table 2) (International Maritime Organization 2014), which shall take into account the potential hazards of the intended voyage to ensure that the Company, master and crew are provided with sufficient information (International Maritime Organization 2014).

Table 2 Requirements of Voyage planning

No.	Requirements of voyage planning
1	Limitations of the hydrographic information and aids to navigation available
2	Information on the extent and type of ice and icebergs in the vicinity of the intended route
3	Statistical information on ice and temperatures from former years
4	Places of refuge
5	National and international designated protected areas along the route
6	Operation in areas remote from search and rescue (SAR) capabilities
7	Information on relevant ships' routing systems, speed recommendations and vessel traffic services
8	Information and measures to be taken when marine mammals are encountered relating to known areas with densities of marine mammals, including seasonal migration areas

Source: IMO (2014)

4.1.4 Communication in Polar Waters

Communications including ship-to-ship or ship-to-survival craft and rescue boat may be limited because of the special environment in polar waters. The specific requirements are regulated in the Polar Code, which should also be incorporated into the courses and textbooks to ensure effective communication during ship operation or in an emergency (International Maritime Organization 2014).

4.2 Training and Assessment

Ice navigation has a major impact on structures of the ship and stricter demand for skills of seafears in polar waters. So ice navigation training and assessment are necessary supplements to enhance the practical ability in polar waters after academic teaching.

The training program can be carried out referring to the requirements of the Polar Code and recommendations of guidelines navigating in Polar Codes. Such a training program should provide knowledge, understanding and proficiency required for operating a ship in polar ice-covered waters (Østreng 2012), including recognition of ice formation and characteristics; laws and regulations; safety of navigation; impact on ships; effects of extreme low temperatures; impact on device performance; emergency response; and environmental Protection (Table 3) (Xie 2011). However, the training program should be revised and updated once the Model course on ice navigation is adopted.

Table 3 Relevant content in training program

Content	Training program
1. Ice regime	1.1 Ice physics: formation, growth, ageing and stages of melt
	1.2 Motion feature of Sea ice: movement characteristics under the wind and current
	1.3 Ice types and concentration
	1.4 Recognition of sea ice: Signs of approaching sea ice (including ice and icebergs)
	1.5 Ice reporting, coding and terminology
2. Laws and regulations	2.1 International laws and regulations: Manila amendment to the 2.2 STCW convention and Polar Code
	2.3 National laws: Rules adopted by Russia, Canada and other Coastal States
3. Safety of navigation	3.1 Navigation in ice-covered waters
	3.2 Selection of voyage routes
	3.3 Preparations before entering the ice
	3.4 Maneuvering in ice-covered waters
4. Impact on ships	4.1 Influence to hull strength and structure
	4.2 Impact on the stability of the goods
5. Effects of extreme low temperatures	5.1 Effects of extreme low temperatures
	5.2 Brittleness of ships components
	5.3 Methods and precautions in de-icing
6. Impact on device performance	6.1 Adjustment of Radar, Identification of sea ice by using Radar
	6.2 Positioning device: Limitations of electronic positioning device positioned at high latitudes
	6.3 Compass: Magnetic compass accuracy in high latitudes
	6.4 Limitations of the communication system
	6.5 Limitations of charts and nautical publications
7. Emergency response	7.1 Drill of emergency response
	7.2 Emergency contact
	7.3 Operation rescuing ship and personnel
8. Environmental Protection	8.1 Particularity of Arctic ecological environment
	8.2 Effect to the Arctic ecological environment from Atmospheric emissions of ships
	8.3 Emission and Control of garbage, bilge water, sewage and other substances

4.3 *Selection of Instructors*

Due to the shortage of competent seafarers and the large demand for training in polar waters in the near future, it is necessary to train and select more qualified instructors. Three criteria for selection may be referred to in order to import the high qualified instructors.

First of all, some experienced personnel who would like to contribute their knowledge in polar waters can be employed as a priority. Secondly, qualified

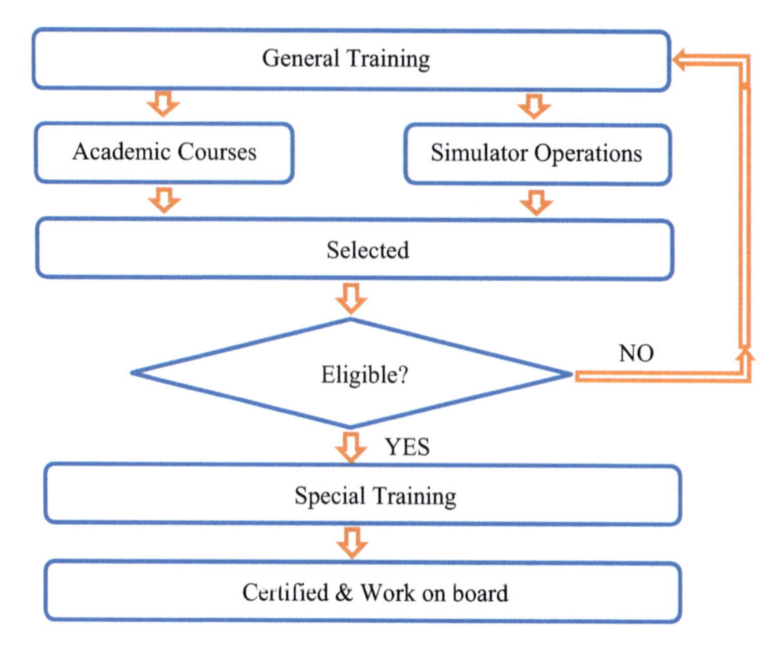

Fig. 2 The state flow to train the young teachers

instructors without sea experiences but who have guidance experiences for polar waters shipping in training centers can also be employed as instructors. Finally, taking into account shortage of practical instructor, it is vital to train the young teachers to be competent instructors. Figure 2 demonstrates the state flow to train our young teachers. Two steps shall be completed by young teachers. They will obtain both theoretical knowledge and practical ability by following this flow to meet the requirements of instructors.

4.4 Development of Simulators in Polar Waters

New technologies have been widely applied in the development of ship simulators. Ship simulator training is an effective way to train competent seafarers. Therefore, development of full mission simulators in polar waters is another major priority for maritime higher-education in China. The key technology lies in the development of ship mathematical motion models in polar waters, real world vision display by virtual reality technology, electronic charts in polar waters and convenient function setting module.

4.4.1 Ship Mathematical Motion Models in Polar Waters

The development of the new ship motion math model is the core technology in polar waters. The key point is the calculation of the hydrodynamic and other external forces which acted on the ship in the special navigational environment.

4.4.2 Real World Vision Virtual Display

Virtual reality technology has been widely used in daily life by computer science. Training results depend on the processing capabilities of the visual display server. In order to develop high fidelity navigation simulation environments, many photos or videos should be taken to collect the vision datum in the ship in Arctic and Antarctic water areas.

4.4.3 Electronic Chart

Different from mainstream electronic charts, electronic chart systems in polar waters shall be three dimensions, particularly because of the depth of icebergs. Therefore, how to collect the terrain datum in the ship Arctic and Antarctic water areas and develop the three dimensions for electronically chart becomes a challenging task.

4.4.4 Convenient Function Setting Module

The system shall be easy to understand and convenient for the user. The functions of the ship simulator are developed by model units. The employee can train the learner in certain situations repeatedly until the trainee fully understands the whole context. The ship simulator can be designed to set the certain environment in terms of the weakness of seafarers.

5 Conclusion

With the increasing number of ships navigating through the polar waters, the unique environment and eco-systems of the Polar Regions have been greatly affected. In this situation, the Polar Code was adopted by IMO, which is expected to enter into force in 2017. Training and education of seafarers can be regarded as an effective way to improve the competence of seafarers shipping in polar waters. As a major exporter of seafarers, China has responsibilities to prepare and update the current education system in advance. The effects toward the maritime higher-education are discussed from the courses, training, instructors and simulators. In order to minimize the gap between maritime higher-education in China and the requirements of the Polar Code, the preparation work of training and education shipping in polar waters has been carried out in China. The efforts of this chapter may contribute to improved compulsory competence of seafarers required by the Polar Code.

References

Adviser, S. (2012). *IMO Polar Code*. Available at: https://nsidc.org/noaa/iicwg/presentations/IICWG-2012/Stemre_IMO_Polar_Code.pdf

COSTAS PARIS. (2015). *U.N. agency adopts Polar Code to prevent sea pollution*. Available at: http://www.wsj.com/articles/u-n-agency-adopts-polar-code-to-prevent-sea-pollution-1431711578

DNV. (2015). *STCW revisions to include Polar Code requirements*. Available at: http://www.dnv.nl/nieuws_events/nieuws/2015/stcw_revisions_to_include_polar_code_requirements.asp?Edit=1

European Commission. (2010). *Legal aspects of Arctic shipping: Summary report*. Luxembourg: Publications Office of the European Union. https://doi.org/10.2771/51132.

International Maritime Organization. (2014). *International Code for Ships Operating in Polar Waters (POLAR CODE)*. London.

International Maritime Organization. (2015). *Shipping in polar waters*. Available at: http://www.imo.org/en/MediaCentre/HotTopics/polar/Pages/default.aspx

Jensen, Ø. (2007). *The IMO guidelines for ships operating in Arctic ice-covered waters*. FRIDTJOF NANSENS INSTITUTT(FNI) Report.

Lloyd's Register Marine. (2015). *Human Element, Training and Watchkeeping (HTW 2)*. Available at: http://www.lr.org/en/_images/213-51166_HTW_2_LR_Summary_Reportl.pdf

Norwegian Mapping Authority. (2011). *Marine traffic in the Arctic*. Available at: http://www.iho.int/mtg_docs/rhc/ArHC/ArHC2/ARHC204C_Marine_Traffic_in_the_Arctic_2011.pdf

Østreng, W. (2012). *Shipping in Arctic waters: A comparison of the Northeast, Northwest and Trans Polar passages* (pp. 35–38). Heidelberg: Springer.

Wang, Y. (2010). *Influence of cultivating model for higher maritime education on seafarers* (pp. 22–24). Dalian: World Shipping.

Xie, J. (2011). *Maritime training for seafarers navigating in ice-covered water*. Paper presented at the International Conference IMLA 19 Opatija 2011.

Continuing to Improve Oil Spill Response in the Arctic: A Joint Industry Programme

Joseph V. Mullin

Contents

J. V. Mullin (✉)
Arctic Oil Spill Response Technology, Joint Industry Programme, International Association of Oil and Gas Producers, London, UK
e-mail: josephmullinconsulting@comcast.net

L. P. Hildebrand et al. (eds.), *Sustainable Shipping in a Changing Arctic*, WMU Studies in Maritime Affairs 7, https://doi.org/10.1007/978-3-319-78425-0_19

Abstract For more than 50 years, the oil and gas industry has funded and conducted research to improve oil spill response technologies and methodologies with industry, government, academia, and stakeholders jointly involved. This research has included hundreds of studies, laboratory and basin experiments and field trials, specifically in the United States, Canada and Scandinavia. Recent examples include the SINTEF Oil in Ice JIP (2006–2009) and research conducted at Ohmsett—The National Oil Spill Response Research and Renewable Energy Test Facility. This sustained and frequently collaborative effort is not commonly known and recognised by those outside the field of oil spill response.

To build on this existing research and continue improving the technologies and methodologies for Arctic oil spill response, nine international oil and gas companies (BP, Chevron, ConocoPhillips, Eni, ExxonMobil, North Caspian Operating Company (NCOC), Shell, Statoil, and Total) are working collaboratively in the Arctic Oil Spill Response Technology—Joint Industry Programme (JIP). The goal is to advance Arctic oil spill response strategies and equipment as well as to increase understanding of potential impacts of oil on the marine environment. The $21.5M (USD) programme is coordinated by an Executive Steering Committee comprising representatives from each company under the auspices of the International Association of Oil and Gas Producers. The world's foremost experts on oil spill response, development, and operations from across industry, academia, and independent scientific institutions are being engaged to perform the scientific research.

The JIP has completed phase one of the programme which included technical assessments and state of knowledge reviews in the following six areas: dispersants, environmental effects, trajectory modelling, remote sensing, mechanical recovery, and in situ burning (ISB). Sixteen research reports that identify and summarise the state-of- knowledge and regulatory status for using dispersants, remote sensing and ISB in the Arctic are available on the JIP website (www.arcticresponsetechnology.org).

Phase two activities are now underway which include laboratory, small and medium scale tank tests, and field research. Eleven projects are in progress ranging from dispersant effectiveness testing; modelling the fate of dispersed oil in ice; assessing the environmental effects of an Arctic oil spill; advancing oil spill modelling trajectory capabilities in ice; extending the capability to detect and map oil in darkness, low visibility, in and under ice; herder application, fate and effects; and expanding the 'window of opportunity' for ISB response operations. This chapter presents recent JIP progress and key learnings from results.

Keywords Arctic · Oil spill · Dispersants · In situ burn · Mechanical recovery · Environmental effects

1 Introduction

The key characteristic that distinguishes the Arctic from other oil and gas production areas is the presence of ice. The ice environment varies substantially throughout the Arctic, depending on the season and location. As such, response options in the Arctic also vary depending on the time of year and location. Operational challenges for Arctic oil spill response are: remoteness, low temperatures, seasonal darkness, and the presence of seasonal sea ice. The selection of one or more strategies to deal with a spill in an Arctic environment will depend upon a variety of factors, including the size and type of spill, local weather and sea conditions, and the presence, concentration and characteristics of ice. Rapid and effective response can be achieved with the aid of a Net Environmental Benefit Analysis (NEBA), to determine the most effective strategies to minimise the impact of a spill for any given scenario. Decisions on response options depend on the spill conditions at the time, relative risks to response personnel and local environmental sensitivities. The flexibility to use a broad range of response options, as conditions change, is essential to mounting the most effective response possible.

2 Arctic Oil Spill Response Technology: Joint Industry Programme

The oil and gas industry has made significant advances in being able to detect, contain, and clean up oil spills in Arctic environments (API 2012). Ongoing research continues to build upon more than 50 years of examining all aspects of oil spill preparedness, oil spill behaviour, and field experiments to further advance oil spill response in the Arctic marine environment. To build on existing research and improve the technologies and methodologies for Arctic oil spill response, members from the IPIECA-Oil Spill Working Group, Industry Technical Advisory Committee (ITAC) and the American Petroleum Institute-Emergency Preparedness and Response Program Group formed a joint committee in 2009. The committee's task was to review the oil and gas industry's prior and future work scope on prevention and response to oil spills in ice, to identify technology advances and research needs in industry preparedness, and prioritise identified issues. One outcome was the recommendation to establish the Arctic Oil Spill Response Technology—Joint

Industry Programme (JIP) that would undertake specifically targeted research projects identified to improve industry capabilities and coordination in the area of Arctic oil spill response.

The JIP research is focusing on expanding industry knowledge of and capabilities in Arctic oil spill response in Dispersants, Environmental Effects, Trajectory Modelling, Remote Sensing, Mechanical Recovery, and ISB. Recognised subject matter experts with years of experience in oil spill response research and operations lead Technical Working Groups (TWGs), manage each research area. All research projects are being conducted using established protocols and proven scientific technologies, utilising the best researchers, consultants, and laboratories.

3 Project 1 Fate of Dispersed Oil Under Ice

3.1 Context

One of the requirements for efficient dispersion is adequate mixing in the water column, allowing for a cloud of dispersed oil to rapidly dilute to very low concentrations. A key parameter for stable dispersion is the level of turbulence to keep dispersed oil entrained in the water column. Ice cover dampens energy input from the wind into the ocean (SL Ross Environmental Research Ltd. 2010). This dampening may cause turbulence under the ice to be lower compared to an open ocean environment. Existing numerical models can determine how quickly dispersed oil plumes will rise on the basis of information on ambient turbulence conditions, dispersed oil droplet size distributions, and dispersed oil densities. For these models to predict dispersed oil behaviour under ice, improved understanding of the natural turbulence under a range of ice roughness conditions is required.

3.2 Project Goal

The overall goal of this research project is to provide critical information in support of dispersants use in ice-covered marine environments and develop a tool to support contingency planning decisions with respect to dispersant use. The aim is to provide additional evidence to support dispersant use and decision making in ice-covered waters and to determine optimal operational dispersion criteria. The primary research objective is to develop a numerical model that predicts the potential for a dispersed oil plume to resurface and reform a new slick under the ice. The model will then be run with varying ice concentrations, release types, environmental conditions, oil types, and levels of turbulence. The first phase determined what data already exists to support model development. The second phase is focused on developing the model, gathering the additional data required to run and validate the model, and then modelling surface and subsurface dispersant use scenarios. The model is being designed to evaluate whether or not dispersed oil droplets formed under continuous or concentrated ice could resurface under the ice to form a significant accumulation within 2 days.

3.3 Progress and Achievements

The JIP has already delivered a phase one report entitled: "Fate of Dispersed Oil Under Ice – Literature Review" (Beegle-Keause et al. 2013), which provides a summary of background information on the state of knowledge concerning under-ice turbulence and is available on the JIP website. The literature review supports the view that sufficient knowledge exists to develop an under-ice turbulence closure model, but that existing observations may not be sufficient to provide both calibration and verification data. The report provides a summary of methods for obtaining additional data as necessary, to allow the development of a reliable model to predict whether oil droplets could surface within a 2 day period, based upon an initial oil droplet size distribution.

Phase two research is now in progress, including flume tank turbulence experiments, under ice turbulence and dye study field experiments, modelling studies, and development of oil droplet rise tables. SINTEF mounted a field campaign at Svea, Norway, between 18th and 24th March 2015, to collect under ice turbulence measurements. The equipment (an Acoustic Current Doppler Profiler (ACDP), remote operated vehicle (ROV), and a conductivity, temperature, depth sensor) were deployed. The ROV was looking at under ice roughness to assist in placement of the under ice sensors. Field work was safely and successfully completed, all planned measurements collected, and data analysis is underway. Flume tank experiments were conducted at Plymouth University, UK, in January 2016, and a dye study field experiment was conducted at Svea, Norway, in April 2016.

4 Project 2 Dispersant Testing Under Realistic Conditions

4.1 Context

Researchers have examined dispersant effectiveness in cold waters with sea ice in laboratory scale, wave basin tests, and with at sea experiments using a variety of oils. Research has also demonstrated that inorganic mineral fines in turbid coastal waters function naturally to form oil mineral aggregates (OMA's) that can remove oil from contaminated shorelines. The Department of Fisheries and Oceans, Canada, and the Canadian Coast Guard have also demonstrated through field research that streams of mineral fines slurry combined with mixing energy from vessel propeller wash promoted rapid OMA formation and dispersion of oil slicks in ice (SL Ross Environmental Research Ltd. 2010). Use of chemical dispersants and/or mineral fines provides a response option with high encounter rates, high effectiveness, lower manpower requirements, and greater responder safety than mechanical recovery. Furthermore, mineral fine treatment may be suitable for use on spills in freshwater.

4.2 Project Goal

The overall goal of this research project is to provide additional information to support dispersant use in ice-covered waters. The primary objective is to define the operational limits of dispersant and mineral fines in Arctic marine waters with respect to oil type, oil viscosity, ice cover (type and concentration), air temperatures, and mixing energy (natural, water jet and propeller wash). A second objective is to summarize the regulatory requirements and permitting process for dispersant and mineral fines use for each Arctic nation/region. This project is being conducted by the JIP in a phased approach.

4.3 Progress and Achievements

The JIP has completed three tasks, with the reports available on the JIP website. The first report "Dispersant Testing Under Realistic Conditions: State of the Knowledge Review", summarizes the scientific literature and identifies previous research on dispersant effectiveness under Arctic conditions. Important parameters assessed were; oil type (naphthenic, asphaltenic, paraffinic, waxy crude or fuel oil); oil viscosity, oil weathering degree, dispersant type, dispersant to oil ratio, salinity, ice coverage, mixing energy and temperature (Lewis 2013). The second report "Dispersant Use in Ice Affected Waters: Status of Regulations and Outreach Opportunities" identifies and summarises the regulatory requirements and permitting process for use of dispersants and mineral fines for each Arctic nation (SEA 2013). The third report "Test Tank Inter-Calibration for Dispersant Efficiency" describes the energy conditions and test protocols for planned meso scale flume to test dispersant effectiveness under varying release and deployment conditions (Faksness et al. 2013). The main findings from these three reports are:

- Dispersants can work in the Arctic and will, under certain conditions, be more effective in the presence of ice than in open water.
- In addition to increasing effectiveness, the presence of ice can increase the time window within which dispersants can be used effectively.
- Except for the UK and the US, there is generally an absence of national policies and procedures to approve the use of dispersants during an incident.
- Some countries have good regulatory models established for dispersant use.

Mesoscale flume tank experiments were conducted at CEDRE, SINTEF and SL Ross to establish boundaries to define dispersant effectiveness and dispersant effectiveness experiments with natural mixing energy and propeller wash. More than 75 dispersant experiments were conducted. Five crude oils weathered for 18 h, four dispersants at two DOR's, three energy levels (low, high, propeller wash), two levels of ice coverage, and three salinity levels (5, 15, and 35 ppt) were used in the test matrix. Research experiments on the use of polyethylene blocks to improve dispersant effectiveness test repeatability were also being conducted. Upcoming tasks include laboratory and mesoscale flume tank experiments to evaluate dispersant effectiveness in open water conditions and after oil or oil-dispersant mixtures have been frozen in or on ice.

5 Project 3 Environmental Impacts from Arctic Oil Spills and Oil Spill Response Technologies

5.1 Project Goal

The overall goal of this research project is to improve the knowledge base and stakeholder acceptance for using "Net Environmental Benefit Analysis" (NEBA) in response decision making, and ultimately gain stakeholder acceptance of the role of environmental impact assessment in oil spill response plans and operations. Due to the fundamental role of comparing the effectiveness and impacts of different response options in NEBA, the information base needs to address both the acute and chronic effects of spilled oil as well as the impacts of various response options (e.g., natural attenuation, surface/subsea applied dispersants, in situ burning, etc.) on Arctic ecosystems. Review and tabulation of published measured effects (e.g., toxicity thresholds and recovery times) is anticipated to be an important part of this project.

The initial phase of this project was to perform a comprehensive review of the environmental impacts of Arctic oil spills and the technologies used to respond to such spills, and identify research activities to improve the knowledge base for using NEBA in the Arctic. This phase is complete. The second phase is to conducting the most crucial research activities identified in phase one. Four projects were identified that include laboratory and field research as well as modelling studies. The final phase will demonstrate how the information base resulting from the review and the data from new studies are used in optimising the NEBA process.

5.2 Progress and Achievements

The phase one review culminated in the publication online, by the JIP, of a report based on over 960 literature references from investigations into spilled oil and oil spill response technologies in the Arctic marine environment. The report is the first time the significant body of research on this area has been compiled and reviewed in one place, indicating that there is a large amount of literature already available on oil spill response decision-making in the Arctic. With nine chapters that cover technical topic areas in detail, the report covers a range of research from biodiversity assessments, to laboratory experiments and large-scale field studies focussing on oil spill response in Arctic environments.

The report was compiled by a consortium of international expert investigators in the field of Arctic biology, the physical environment, oil fate and biodegradation, oil spill response, toxicology, population modelling and recovery, and NEBA. For each category, the report provides priority research recommendations for enhancing Arctic NEBAs, as well as other research considerations not directly related to NEBA. Key findings are:

- That there is an extensive existing science base for Arctic NEBAs. Many baseline ecosystem and biodiversity assessments have been performed to better understand and protect the marine Arctic environment. In addition, field and laboratory studies on the fate of oil, oil spill response techniques and potential environmental effects under the different seasonal conditions in the Arctic have produced extensive data sets on oil fate and effects.
- There is also evidence that Arctic species are not more sensitive to dispersed oil than non-Arctic species and that they react to dispersed oil exposure in the same way as temperate species do. To fully understand how species populations are impacted and recover, the review has recommended follow-up work to study population resilience.
- Furthermore, data has been reviewed that shows that certified dispersants and oils treated with dispersants are not more toxic than the oil itself. Another important finding is that biodegradation of oil in the Arctic does occur and that certified dispersants do not reduce the ability of microbes to degrade oil.
- Biology tends to aggregate at interfaces like the water/ice interface, which is one of the unique features of the Arctic ecosystem. Undispersed oil might collect at this interface potentially interfering with unique Arctic resources. The review recommended that information on the potential effects of oil on these Arctic communities be developed in order to better address these in NEBA.
- The behavior of oil in ice can actually mitigate the environmental impact, as the presence of ice results in reduced evaporation, dispersion and emulsification and can form a barrier so that vulnerable resources like coastlines cannot be reached.

The report and referenced databases are housed within a 'NEBA tool' hosted on a dedicated microsite, accessible from the Arctic Response Technology JIP website and openly available to all visitors http://neba.arcticresponsetechnology.org. This education and resource tool for NEBA practitioners, stakeholders and the public links technical chapters with the literature database and the supporting references. The NEBA tool can be used as a one stop shop for NEBA practitioners and decision makers, to identify information and available literature relevant to Arctic oil spill response, including information on Arctic ecosystems, fate and effects of oil and the NEBA process itself. Researchers can also use the tool to look for inspiration for new research. The report is fully searchable and can be downloaded as a whole as well as by chapter.

Phase 2: Recommendations from the phase one review have led to four research projects initiated by the JIP. RAMBOLL/ENVIRONS, Emeryville, California, USA is the contractor for Project 1 developed the Analysis Tool for Evaluating the Ecological Consequences of Oil Spill Response (ARCAT) that summarize the important ecological and environmental attributes necessary for NEBA and support OSR decision making. The ARCAT project collected and organised nearly 3500 documents from the scientific, research and government literature describing the consequences of spilled oil in the Arctic environment. These documents, along with several other related data sources, form the basis of the current searchable Arctic NEBA website. The final version of the online NEBA information portal will include an updated version of the literature database, the data navigators, and summary tables. Each cell within the Arctic NEBA information and support tool will provide a "hot-link" to the relevant literature, providing the citation and abstract when available.

Project 2a is performing field studies using in situ mesocosms to measure the exposure potential, sensitivity and resiliency of sea ice communities. Eight mesocosms for sea ice were built in France and installed in Van Mijenfjorden, Svea, Norway in February 2015 with the following set up: two mesocosms were treated with crude oil mixed with dispersant, two with residuals of burnt crude oil, two controls, and an additional two controls outside of the immediate area. KOBBE crude oil from the Goliat field of the Barents Sea crude oil was used in these experiments. The mesocosms were periodically sampled in March, April, and May to gather data on the effects of naturally and chemically dispersed oil and in-situ burn residue on the composition and density of the neuston and ice ecology communities during the entire winter and spring including peak of biological activity. For assessing the sensitivity and resiliency of the sea surface micro layer, 12 smaller mesocosms for open water were installed in Van Mijenfjorden, Svea, Norway in May 2015.

Project 2B is characterizing oil weathering in sea ice, sediment and rocky bottom and the biodegradation processes by identifying microbial communities. Natural rock tiles and sediment samples were exposed to KOBBE crude oil and situated at the same location as the mesocosms in February 2015 to study the fate of the oil, natural attenuation and biodegradation on solid substrate and investigate the role of sediment microorganisms and rock surface biofilms in the oil biodegradation process. There were 3 sampling times (March, April, and May) to examine:

- The weathering processes
- The biodegradation rate from the residual oil composition
- The oil behavior and migration into the ice from the freezing period to the melting one
- The effect of oil on the natural microbial community (Fig. 1)

Project 3 is a modelling study examining the effects of oil components on the keystone ice edge fish and calanus species (Polar cod and Arctic calanus species).

Fig. 1 Mesocosms deployed in the Svea, Norway field experiments (source: IRIS (International Research Institute of Stavanger))

- Acute effects on populations
- Combined acute and chronic effects on populations

Results from this research project will be used to establish in what direction NEBA based decision making has to be changed when one has to deal with chronic effects as well as acute effects.

6 Project 4 Oil Spill Trajectory Modelling in Ice

6.1 Project Goal

The overall goal of this research project is to conduct research investigations in ice modelling and integrate the results into established industry oil spill trajectory models to improve their accuracy. Current ice models have intrinsic limitations, such as the inconsistent assumption of viscous-elastic rheology of the ice, that render them inaccurate. The primary research objective is to advance and expand the oil and gas industry's oil spill trajectory modelling for oil spills in ice affected waters. This project will create or adapt an existing model for predicting ice movement in the marginal and pack ice zones under applied (forecast) wind and current forcing. The new model is expected to provide increased accuracy on the behavior and movement of ice and it is intended that the model will be implementable in any of the leading oil fate and effects models. The model may also be applicable beyond the Arctic, for example, in non-Arctic but ice-prone areas (e.g., Baltic and Caspian seas). The outcome of this project will be an improvement of the oil spill trajectory models accuracy in presence of sea ice, along with an estimation of the uncertainties in these trajectories.

6.2 *Progress and Achievements*

NERSC is developing a new sea ice model that will be tested/evaluated/validated at a regional scale as well as a new very-high resolution model to simulate sea ice dynamics in the Marginal Ice Zone (MIZ). The project is underway and progressing on schedule. The JIP conducted a workshop in December 2015 with industry and government scientists to present the NERSC ice modelling results and discuss capabilities and potential improvements in the near-future to their modelling efforts using NERSC findings. The NERSC ice algorithms will be integrated into established oil spill trajectory models (e.g., OilMap and OSCAR) and scenarios run to verify/demonstrate results. The outcome of this project will be an increase of the oil spill trajectory models accuracy in presence of sea ice along with an estimation of the uncertainties in these trajectories.

7 Project 5 Oil Spill Detection and Mapping in Low Visibility and Ice

7.1 *Context*

Remote sensing is an important element of an effective response to marine oil spills. Timely response requires rapid and sustained reconnaissance of the spill site to determine the exact location and extent of oil (particularly the thickest portion of the slick) and updated projections of oil slick's movement and fate at sea. Remote detection and mapping are essential to effectively directing spill countermeasures such as mechanical containment and recovery, dispersant application, in situ burning, and for the preparation of resources required for shoreline clean-up. Previous industry and government supported research and development has yielded technologies such as strengthened beacons designed to track the location of oiled ice, ground penetrating radar to detect oil in, on, and under ice, laser fluorosensors, and enhanced marine radar. In addition, recent tests have shown that Autonomous Underwater Vehicles (AUVs) can carry sensors capable of locating and tracking oil under ice.

7.2 *Project Goal*

The overall goal of this project is to expand industry's remote sensing and monitoring capabilities in darkness and low visibility, in pack ice, and under ice. This project is split into two elements: surface remote sensing (i.e. satellite-borne, airborne, shipborne and on-ice detection technologies) and subsea remote sensing (i.e. mobile-ROV or AUV based and fixed detection technologies) and will be performed in a phased approach. First, an assessment and evaluation of existing and emerging technologies was performed that includes an evaluation of further research and development needs, logistical support requirements, and operational considerations including testing opportunities. Based on this assessment, a test programme was developed to identify and qualify the most promising sensors and platforms capable of determining the presence of oil on, in, and under ice and mapping its extent.

7.3 Progress and Achievements

Two reports have now been produced, available on the JIP website. The first entitled:" Oil Spill Detection and Mapping in Low Visibility and Ice: Surface Remote Sensing" summarizes the state-of-knowledge for surface remote sensing technologies to monitor oil under varying conditions of ice and visibility (Puestow et al. 2013). The second "Oil Spill Detection and Mapping in Low Visibility and Ice: Subsea Remote Sensing" summarizes the state-of-the-art for subsea remote sensing technologies to monitor oil under varying conditions of ice and visibility (Wilkinson et al. 2013). The main findings from the reports are:

- The current state of technology in remote sensing, confirms that the industry has a range of airborne and surface imaging systems utilized from helicopters, fixed-wing aircraft, vessels and drilling platforms that have been developed and tested for the "oil on open water scenario" that can be used for ice conditions.
- There are several technologies that exist today capable of, or having the potential for, effective sensing in a broad range of ice and environmental conditions that would be experienced in the Arctic.
- Unmanned underwater vehicles (UUVs) have been successfully operating in ice-covered waters and are now a viable technology for under sea ice operations.
- UUVs, and especially autonomous underwater vehicles (AUVs), have the dual advantages of being deployable in a range of ice and weather conditions, and importantly their sensor payloads will have a direct view of oil trapped beneath the ice.
- For logistical considerations, flexibility of deployment and range, AUVs are likely the most promising underwater platform for oil spill detection.
- Detection of oil encapsulated within the ice may also be possible with some sensors mounted on UUVs, and possibly more efficiently than with surface and airborne remote sensing methods.

In late 2014, first-of-their kind phase two research experiments were conducted to test and evaluate the performance of various surface and subsea remote sensing technologies with crude oil on, encapsulated in and under ice, in conditions that include low visibility. The experiments used the climate controlled test basin (37 m long × 9 m wide and 2.4 m deep) at the U.S. Army Corps of Engineers-Cold Regions Research and Engineering Laboratory (CRREL) located in Hanover, New Hampshire, USA. The CRREL test programme was the first time that an array of above surface and subsea sensors were deployed under controlled conditions and simultaneous multi-sensor data was collected from initial growth of sea ice through its melt.

To prepare for the experiments, an underwater trolley system was installed on the bottom of the test basin to position the subsea sensors and cameras. Mounted on the underwater trolley were twenty underwater acoustic transducers, five visible cameras, three laser fluorosensors, and two light sensors. On the carriage boom above the ice there were three Ground Penetrating Radar (GPR) units, two Frequency Modulated Continuous Wave (FMCW) radar units, five visible cameras, three infra-red cameras, one fluorosensor, and one radiometer.

On November 3, 2014, the CRREL tank was cooled to grow sea ice. Once the ice layer was formed, the team injected various amounts of Alaska North Slope (ANS) crude oil over a 2-month period into a series of 16 containment hoops. This design provides tests ranging from frazil (new) ice mixed with oil at the very beginning of the growth process, to columnar ice 80 cm thick, at the end. The oil thickness varied from a few millimeters to 5 cm. Data collection continued through the end of February 2015.

Ice cores were periodically taken to measure ice growth, document temperature and salinity profile, crystal structure, as well as the incorporation of oil into the ice matrix on both macroscopic (i.e. brine channels) and microscopic scales. Oil chemistry is being conducted to examine the changes of oil properties and the oil migrated up through the ice. The experiments were successfully and safely completed, the test basin was cleaned and returned to operational standards by the end of March 2015. The test team then modelled and extrapolated from the sensor measurements taken under the test conditions in the basin, to predict the performance of the instruments and model their performance in a wide range of field conditions. The model results are key to understanding the future potential of the different sensors under real world conditions. The final project report, available on the JIP website contains recommendations to maximize detection performance considering individual and multiple sensors (Fig. 2).

8 Project 6 Mechanical Recovery of Oil in Ice

8.1 Context

The rapid recovery of oil at or near the source is provided by on-site spill response vessels. Mechanical skimmers can be used to remove oil from the water surface and transfer it to a storage vessel. Floating barriers, including oil booms, are used to collect and contain spilled oil into a thicker layer. In the Arctic offshore, ice itself could act as a boom where the oil is contained in thicker layers between ice floes. Skimmers work most efficiently on thick oil layers and a variety of skimmer designs have been optimized for Arctic sea conditions and several have been proven to work well. In most countries, mechanical recovery of oil is the first response option,

Fig. 2 CRREL facility with oiled containment hoops and boom mounted with surface sensors (source: CRREL (Cold Regions Research and Engineering Laboratory))

requiring no pre-approvals. Mechanical recovery in broken ice is limited by the ability of the skimmer to encounter and remove spilled oil and to function effectively under extremely low temperatures. Another issue related to mechanical recovery is storage, transfer and disposal of the recovered oil/ice/water mixture, which is a special challenge in remote Arctic areas with limited onshore supporting infrastructure.

8.2 *Project Goal*

Recognizing the limitations of mechanical recovery systems available today, the JIP Mechanical Recovery research project was initiated with the following objectives:

- Examine results obtained from previous research projects and identify further improvement opportunities for design of mechanical recovery equipment and response strategies for oil spill recovery in ice;
- Develop a selection process by which novel concepts can be rigorously examined; and
- Select and develop the most promising concepts.

8.3 *Progress and Achievements*

An innovation workshop was conducted in March 2012 to identify new leads that could dramatically improve mechanical recovery. Following the workshop four novel ideas were selected to be evaluated. The JIP commissioned internal feasibility evaluations to identify the most promising technologies or equipment designs that can improve recovery of oil in ice and recommend any concepts that can be taken to the 'proof of concept stage'. *The contractor's selected were:*

- *New Vessel Design—Aker Arctic*
- *Remote Recovery Systems—Aker Arctic*
- *On Board Oil/Water/Ice Separation—LAMOUR*
- *Onboard Oil Incineration—SL Ross Environmental Research Ltd.*

The JIP has commissioned a high level summary of the four feasibility reports, which examines the JIP's research undertaken to evaluate the feasibility of existing equipment, and look at potential future development of equipment, to improve effectiveness of Mechanical Recovery in the Arctic. It provides an overview of mechanical recovery of oil in ice-covered waters and the results of JIP-contracted feasibility evaluations of methods for mechanical recovery in the four areas. Alaska Clean Seas was selected as the contractor to develop the summary report. The final project report available on the JIP website.

9 Project 7 In Situ Burning of Oil in Ice-Affected Waters

9.1 *Context*

Oil on water or between ice floes can be disposed of quickly, efficiently and safely by controlled burning (API 2012). This technique works most efficiently on thick oil layers, as oil is contained by fire-resistant booms or ice. Through burning, an average of about 80–95% of oil volume is eliminated as gas, 1–15% as soot and 1–10% remains as a residue. Controlled burning has been proven to work well in the Arctic. The objective of this project is to ensure in situ burning (ISB) is available to industry as a response option. This requires ISB to be incorporated into contingency planning and that response organisations have the necessary resources and training. The overall goal is to prepare educational materials to raise the awareness of industry, regulators and external stakeholders of the significant body of knowledge that currently exists on all aspects of ISB. The materials are also intended to inform specialists and stakeholders interested in operational, environmental and technological details of the ISB response technique.

9.2 *Progress and Achievements*

This project is complete and three reports have been produced, available on the JIP website. The first report entitled: "In Situ Burning of Oil in Ice-Affected Waters: State of Knowledge" (Buist et al. 2013a) provides a detailed state of knowledge that summarizes the role, function, benefits and limitations of ISB as a response option in the Arctic offshore environment and covers planning and operational aspects of ISB, including the potential impacts on human health and the environment. The second "In Situ Burning of Oil in Ice-Affected Waters: Technology Summary and Lessons Learned from Key Experiments" provides a summary of relevant scientific studies and experiments as well as previous research efforts on the use of ISB in Arctic environments both offshore and onshore, highlighting key findings and conclusions (Buist et al. 2013b). The third report "In Situ Burning in Ice-Affected Waters: Status of Regulations in Arctic and Sub-Arctic Countries" identifies and summarizes the regulatory requirements to obtain approval for use of ISB in Arctic nations (Buist et al. 2013c). The main findings from the reports are:

- Confirmation that technology exists to conduct controlled ISB of oil spilled in a wide variety of ice conditions and that ISB is one of the response techniques with the highest potential for oil spill removal in Arctic conditions.
- There is a considerable body of scientific and engineering knowledge on ISB to ensure safe and effective response in open water, broken pack ice and complete ice cover, gleaned from over 40 years of research, including large-scale field experiments.
- Most of the risks associated with burning oil can be mitigated by following approved procedures, using trained personnel, and maintaining appropriate separation distances.

10 Project 8 Aerial Ignition Systems for In Situ Burning

10.1 Context

Experience with in situ burning (ISB) includes many terrestrial spills ignited by hand using simple tools (e.g., flares, drip torches, or breakable bottles of gelled gasoline) as well as numerous field experiments in open water and ice where ignition was accomplished with a mix of surface and aerial ignitors such as the Helitorch™. ISB was used with great success offshore during the 2010 Gulf of Mexico oil spill where an estimated 11,000,000 gallons of oil was safely ignited and burned with ignitors deployed from small boats (Federal Interagency Solutions Group, Oil Budget Calculator Science and Engineering Team 2010). This was the first large-scale application of burning in an operational setting. Spills in smaller water bodies which are easy to reach, can be similarly ignited with surface methods; however airborne alternatives are needed to ignite spilled oil in areas with difficult/restricted access.

10.2 Project Goal

The overall goal of this research project is to develop improved ignition systems to facilitate the use of ISB in offshore arctic environments by extending offshore reach and lowering response times. This project is split into two elements; development of an integrated herder delivery and ignition system for slicks in very open pack ice and open water and development of a long-range aerial ignition system applicable to the spring scenario of oiled melt pools naturally wind-herded on the ice surface.

10.3 Progress and Achievements

The JIP has developed and tested a prototype integrated herder delivery and ignition system that allow both functions to be employed in one flight without landing or hovering to pick up another load. Development of new igniters for the system is complete with United States regulatory approvals granted for transportation. The engineering study that was conducted produced a conceptual design of a palletized airborne ignition system capable of rapid installation in a suitable fixed wing airplane or helicopter. This development could enable access to remote offshore sites at higher speeds with much greater capacity and endurance than existing aerial ignition tools.

11 Project 9 Herders and In Situ Burning

11.1 Context

Herding agents were initially developed in the 1970s as a method of thickening oil slicks prior to mechanical recovery and can provide an additional tool to support oil spill response in ice and open water. Herders use surface active agents to thicken slicks without the need to collect the oil in a physical boom and do not require a physical boundary to work. Herding agents cause the oil slick to contract, the same way a drop of dish soap in a wet, greasy pan forces the grease to the edges. As oil spills shrink in surface area, they get thicker, growing from about a millimeter (0.04 inch) to 6 mm (0.24 inch). This contraction makes it possible to ignite the slick and achieve an efficient burn: the thicker a spill is before it's burned, the more oil gets removed and the higher the overall response effectiveness. ISB herding agents can be useful in thickening oil in the 30–70% ice concentration range so that in situ burning can be effective (SL Ross 2007). They are effective in the open sea, with or without the presence of ice in sea conditions up to Beaufort Force 4, where breaking waves are present. In 2008, two field experiments using chemical herders, conducted during the SINTEF Oil in Ice JIP, demonstrated that herders work in cold open water with ice nearby and that the burns were effective, with greater than 90% removal efficiencies observed (Buist et al. 2011).

11.2 Project Goal

The overall goals of this research project are to advance the knowledge of herder fate, environmental effects, and window of opportunity to expand the operational utility of ISB in open water and in ice-affected waters. Herding agents thicken oil slicks that have spread too thinly to support combustion. This is accomplished by spraying a small amount of surfactant around the perimeter of the slick. The surfactants are not applied to the oil but to the water surface immediately adjacent to the slick. Once applied, the surfactants will spread to ultimately form a monomolecular layer that significantly reduces the surface tension of the water. The reduced water surface tension reverses the oil spreading tendency and a thin slick will rapidly re-thicken. The surfactants do not need a boundary to "push" against and work equally well offshore.

Herders are effective in fresh and marine waters. As herders are low toxicity and used in extremely small quantities (the recommended field application rate for herders is 150 mg/m^2, more than 30 times less than the design application rate for dispersants, which have already been shown to be non-toxic), they represent very little risk to the environment. The use of herders on an oil slick does not detract from the effectiveness of subsequent or concurrent chemical dispersant application or mechanical recovery. Using herders to contract slicks on open water can improve the operational efficiency of dispersants applied by vessels by reducing the slick area needing treatment. Two herding agents (ThickSlick 6535 and SilTech OP-40) are now on the U.S. Environmental Protection Agency, National Oil and Hazardous Substances Pollution Contingency Plan (NCP) Product Schedule for consideration for use in U.S. waters and both are commercially available as of June, 2012.

11.3 Progress and Achievement

The JIP recently published a report entitled "Herding Surfactants to Contract and Thicken Oil Spills for In-Situ Burning in Arctic Waters", available on the JIP website. This report summarizes the results of a 10-year research program by SL Ross Environmental Research on the feasibility of using oil herding surfactant chemicals to contract oil slicks spilled on water among drift ice. The findings of this research indicate that oil spill responders should consider utilizing herders as a method of enhancing in situ burning in light to medium ice concentrations and in salt marshes, where spilled oil can rapidly spread and use of fire containment booms is impractical.

Experiments were conducted at Aarhus University, Danish Center for Environment and Energy (DCE), Technical University of Denmark (DTU), and at the U.S. Army Corps of Engineers-Cold Regions Research and Engineering Laboratory (CRREL) in Hanover, New Hampshire, USA. Four tasks were conducted:

- Environmental Effects of Using Herders for In Situ Burning: This task includes laboratory burning experiments for the investigation of the environmental effects of using chemical herders for in situ burning operations, experiments to determine acute and chronic toxicity and bioaccumulation of chemical herders on Arctic copepods, studying biodegradation of herders in Arctic conditions with water collected in the high Arctic, determining the physical fate of the herder during burning, and of the smoke plume generated during test burns to determine if the herder or herder combustion products are being emitted.
- Windows-of-Opportunity for Herders: Experiments are being conducted to determine the window-of-opportunity for the two commercially-available herders (ThickSlick 6535 and OP 40) to herd slicks of different oils to ignitable thicknesses. A 2-week test programme was conducted at CRREL in March 2015 at a larger scale using a test protocol developed in 2009 herder experiments.
- Impacts of Herder Monolayer on Birds: Tasks include laboratory experiments to investigate the potential impact and effects of fouling by the herder sheen (prior to and after the burning operation) to the feathers from Arctic seabirds. At DCE an ongoing study is investigating the effects on seabird feathers of fouling by burn residues and the protocol for this test is established.
- Development of Educational Materials: The project team is developing summary information and material (e.g., text, images, and video) to describe chemical herders, how they function, their fate and effects, and relevant research findings.

This project is complete and the final project report is available on the JIP website.

12 Project 10 Field Research Using Herders to Advance In Situ Burning

12.1 Context

In Situ Burning (ISB) is an oil spill response option particularly suited to remote, ice-covered waters. Thick oil slicks are the key to effective ISB and if ice concentrations are high, the ice can limit oil spreading and keep slicks thick enough to burn. However, in drift ice conditions and open water, oil spills can rapidly spread to become too thin to ignite. Fire-resistant booms can collect and keep slicks thick in open water; however, even light ice conditions make using booms challenging. The slick thickness produced by herders, 3–5 mm, provides favorable conditions for effective ignition and ISB without the need for containment booms (Buist et al. 2011).

A series of research projects was initiated in 2004 by ExxonMobil URC and funded by many industry and government organizations to study oil-herding surfactants as an alternative to booms for thickening slicks in light ice conditions for ISB. Successful test programmes were conducted in small and large test tanks and in field settings. The work continues in 2014 under the auspices of the JIP with aerial application of both the herding agent and ignition source (igniter), the herder/burn combination becomes an extremely rapid and effective new response tool, independent from vessel support. The slower weathering of oil slicks in ice and cold water can also extend the window of opportunity for this new tool.

12.2 Project Goal

The primary objective of the field research is to validate the application of chemical herders by helicopter to enhance offshore in-situ burning in ice conditions ranging from limited ice cover to ice-free waters. The goals of the research are twofold:

- Prove the operational feasibility of an aerial herder/burn response strategy using a manned helicopter to both spray herder and ignite slick
- Use a remote-controlled helicopter to perform the same activities

12.3 Progress and Achievements

A test basin was constructed at the Poker Flat Research Range (PFRR), 50 km NE of Fairbanks, Alaska, USA. This is an extensive land area (thousands of acres) managed by the University of Alaska Fairbanks (UAF). Selection criteria included climate, distance from populated areas, land area, and logistics access. The large bermed above-ground test basin (90 m × 90 m × 1 m deep) was constructed at the PFRR in September 2014 and is lined with an impermeable cold weather membrane resistant to hydrocarbons, and geotextile protection above and below the liner.

From April 22–28, 2015, five field tests were conducted using Bell 407 helicopter to determine if herding agents could be applied to Alaska North Slope crude oil slicks in simulated drift ice conditions and then ignite the herded oil slicks using a Heli-torch. Oil volumes released varied from 25 to 200 l. After some initial failed attempts, two successful in-situ burns were accomplished using, one burn was accomplished with OP-40 herder and one with ThickSlick 6535 herder. The burning of the free-floating slicks resulted in the removal of approximately 70–85% of the oil on the water surface. It was successfully demonstrated for the first time that applying herder around and subsequently igniting a free-floating oil slick using equipment mounted on a helicopter is feasible. Further refinement of the herder application system is required to simplify operations and allow more control over its operation (Fig. 3).

Fig. 3 Successful burn of free-floating ANS crude oil slick herded with ThickSlick 6535 (source: SL Ross Environmental Research Ltd.)

Several operational tests with the ING Robotic Aviation RESPONDER (UAV helicopter) were also undertaken. UAV helicopters show promise as a herder application/igniter deployment vehicle, but require additional R&D, system redundancies and regulatory easing to be operationally viable. A combined herder/igniter concept would be useful for both helicopters and UAVs to allow for a one-flight herd and ignite operation. This project is complete and the final project report is available in the JIP website. The JIP successfully demonstrated the use of herders in offshore open water conditions during the June 2016 Norwegian field trials. These field experiments provided a valuable opportunity to transfer herder and ISB technology to the Norwegian Coastal Administration and the Norwegian Clean Seas Association for Operating Companies (NOFO). The projects in Alaska and Norway provided regulators and responders from both countries the opportunity to see the value of the new herder/burn response tool first hand.

13 Conclusions

Collaborative projects, such as the current JIP have been the hallmark of industry's oil spill response research. Advancing oil spill response is a key area where the oil and gas industry works to cooperate and collaborate and the companies involved in the JIP believe that working together gives them access to a wider range of technical expertise and experience. Uniting efforts and knowledge in this JIP increases opportunities to develop and test oil spill response technologies and methodologies, conduct large scale field experiments, and raise awareness of existing industry oil spill response capabilities in the Arctic region.

The JIP's results demonstrate that industry now has a more robust range of operationally proven tools to suit specific regional Arctic environments, encompassing ice and open water seasons. The JIP was completed in August 2017 and the release of 32 reports adds to existing industry knowledge and continues to build a comprehensive picture of Arctic state-of-the-art oil spill response technologies. The industry has a role to play in helping countries in Arctic jurisdictions understand the benefits of having a regulatory process in place to approve the use of all of these response methods and technologies. As such, the results of studies will be published in peer-reviewed journals, and materials have been developed for the benefit of the wider audience interested in Arctic oil spill response, including NGOs, policymakers and members of the environmental community. All JIP research reports, including the programme summary and synthesis reports, are available for download free of charge at http://www.arcticresponsetechnology.org/reports.

References

American Petroleum Institute (API). (2012). *Spill response in the Arctic offshore* (144 pp). Washington, DC: Final report for the American Petroleum Institute and the Joint Industry Programme on Oil Spill Recovery in Ice.

Beegle-Krause, C. J., Simmons, H., McPhee, M., Lundmark Daae, R., & Reed, M. (2013). *Fate of dispersed oil under ice – Literature review, SINTEF materials and chemistry* (57 pp) (Final Report 1.4 for the Arctic Oil Spill Response Technology – Joint Industry Programme).

Buist, I., Potter, S., Nedwed, T., & Mullin, J. (2011). Herding surfactants to contract and thicken oil spills in pack ice for in situ burning. *Cold Regions Science and Technology, 67*, 3–23.

Buist, I. A., Potter, S. G., Trudel, B. K., Shelnutt, S. R., Walker, A. H., Scholz, D. K., et al. (2013a). *In situ burning in ice affected waters: State of knowledge report* (Final Report 7.1.1 for the Arctic Oil Spill Response Technology – Joint Industry Programme).

Buist, I. A., Potter, S. G., Trudel, B. K., Walker, A. H., Scholz, D. K., Brandvik, P. J. et al. (2013b). *In situ burning in ice affected waters: Status of regulations in Arctic and Sub-Arctic Countries* (Final Report 7.2.1 for the Arctic Oil Spill Response Technology – Joint Industry Programme).

Buist, I. A., Potter, S. G., Trudel, B. K., Walker, A. H., Scholz, D. K., Brandvik, P. J., et al. (2013c). *In situ burning in ice affected waters: A technology summary and lessons learned from key experiments* (Final Report 7.1.2 for the Arctic Oil Spill Response Technology – Joint Industry Programme).

Faksness, L., Belore, R., & Merlin, F, (2013) *Test tank inter-calibration for dispersant efficiency* (72 pp) (Final Report 2.2 for the Arctic Oil Spill Response Technology – Joint Industry Programme).

Federal Interagency Solutions Group, Oil Budget Calculator Science and Engineering Team. (2010). *Oil Budget Calculator: Deepwater Horizon*. Technical Documentation.

Lewis, A. (2013). *Dispersant testing in realistic conditions* (33 pp) (Final Report 2.1 for the Arctic Oil Spill Response Technology – Joint Industry Programme).

Puestow, L., Parsons, L., Zakharov, I., Cater, N., Bobby, P., Fuglem, M., et al. (2013). *Oil spill detection and mapping in low visibility and ice: Surface remote sensing* (102 pp) Final Report 5.1 for the Arctic Oil Spill Response Technology – Joint Industry Programme.

SEA Consulting Group. (2013). *Final Report 2.8 for the Arctic Oil Spill Response Technology – Joint Industry Programme* (105 pp).

SL Ross. (2007). *Mid-scale test tank research on using oil herding surfactants to thicken oil slicks in broken ice* (Report to the U.S. Department of the Interior, Minerals Management Service).

SL Ross Environmental Research Ltd, DF Dickins Associates LLC., Envision Planning Solutions Inc. (2010). *Beaufort sea oil spill state of knowledge review and identification of key issues* (126 p) (Environmental Studies Research Funds Report No. 177). Calgary.

Wilkinson, J., Maksym, T., & Singh, H. (2013). *Capabilities for detection of oil spills under sea ice from autonomous underwater vehicles* (103 pp) (Final Report 5.2 for the Arctic Oil Spill Response Technology – Joint Industry Programme).

Emergency Management in Maritime Mass Rescue Operations: The Case of the High Arctic

Natalia Andreassen, Odd Jarl Borch, Svetlana Kuznetsova, and Sergey Markov

Contents

Abstract Maritime activity in turbulent environments represents a challenge to the emergency preparedness system. In particular, the Arctic may be turbulent as to weather, especially in winter time. The consequences of accidents may be severe due to long distances, cold climate and limited local resources. In this chapter we look into large scale emergencies causing mass rescue from ships. We elaborate on the coordination of the broad range of search and rescue actors included in such an incident both in the air, at sea and ashore with several institutions and management

N. Andreassen (✉) · O. J. Borch
Nord University Business School, Nord University, Bodø, Norway
e-mail: natalia.andreassen@nord.no; odd.j.borch@nord.no

S. Kuznetsova · S. Markov
Northern (Arctic) Federal University, Arkhangelsk, Russia
e-mail: mailsvete@mail.ru

L. P. Hildebrand et al. (eds.), *Sustainable Shipping in a Changing Arctic*, WMU Studies in Maritime Affairs 7, https://doi.org/10.1007/978-3-319-78425-0_20

levels included. We also describe the incorporation of host nation support from neighboring countries. We build upon the experiences from the accident of the cruise ship "Maxim Gorkiy" in the ice South-West of Svalbard. We illustrate the organizational structure of mass rescue operations and the coordinative roles at different levels. Finally, we discuss the implications for emergency management in extreme environments like the Arctic region.

Keywords Arctic · Emergency management · Mass rescue · Turbulent environments · Search and rescue · Cross-border cooperation · Incident command systems

1 Introduction

Maritime activity in the High North is influenced by a cold climate, long distances and a general lack of infrastructure (Borch et al. 2016). The passenger and cargo transportation, fishing, tourism, research and offshore resource exploration imply a diversity of vessels, some of them with a large number of passengers and crew on board. The probability of accidents is low, but the consequences may be high. Also, there is a change in activity patterns, which now includes large cruise vessels and offshore oil and gas activity that may change the risk picture (Jardine-Smith 2014; Marchenko et al. 2015).

The International Maritime Organization (IMO) defines mass rescue as "an immediate response to a large number of persons in distress so that the capabilities normally available for search and rescue authorities are inadequate". Severe risk for people's lives is imminent if a ship has to be evacuated in Arctic waters. Low temperatures, possible low visibility and long distances for the rescue units create extra operational complications. In most countries a range of agencies and departments will be involved from the preparedness system including both civilian and military, voluntary and government resources. Large scale operations may be hampered by lack of coordination and communication challenges. In joint operations with two or more countries, the coordination may be even more complex (Borch and Andreassen 2015). The involved actors may have different types of command structures and employ different coordination mechanisms, but they still need to cooperate effectively and control a broader set of resources. Knowledge about possible problems regarding command and control can contribute to better dynamics of coordination in emergency response in the High North and better mechanisms for collaboration.

The purpose of this chapter is to investigate the coordination tasks in situations of mass rescue in the context of the Arctic region. We elaborate on how the institutional aspects of emergency response systems and environmental factors influence the coordination.

The chapter starts with a literature review on mass rescue, its main issues and challenges, and the cross-border regulation of emergency management, and includes an analytical model used in our study. The case of the "Maxim Gorkiy" collision with ice is described and analyzed. Finally, we discuss managerial implications of the findings on maritime mass rescue operations and cross-border cooperation.

2 Literature Overview

2.1 Mass Rescue and Emergency Management

The IMO defines rescue as the "operation to retrieve persons in distress, provide for their initial medical or other needs and deliver them to a place of safety," and a mass rescue operation (MRO) as "characterized by the need for immediate response to large numbers of persons in distress such that the capabilities normally available to the search and rescue (SAR) authorities are inadequate."

Damage to property or the environment may also trigger rescue if it poses a risk to the safety, health, and welfare of people. Effective response to such major incidents typically implies immediate, well-planned and closely coordinated large-scale actions and use of resources from multiple organizations (IMO 2003). The highest priority in major incidents that cause mass rescue operations (MRO) is the preservation of lives, the second generally being environmental protection, and third, the protection of property (EMERCOM 1995).

Mass rescue plans need to allow command, control and communication structures that can accommodate air, sea and land operations simultaneously. Poor preparation for mass rescue and a lack of communication and coordination may have disastrous consequences including the risk for the personnel involved in the SAR operations.

Situational awareness and analyzing the scale and complexity of an event is part of the challenge in planning an emergency response. A mass rescue at sea represents an additional challenge given the rarity of such incidents the high severity of consequences. The rarity and volatility of MROs mean that responders have limited experience dealing with them though major emergency organizations provide MRO exercises both for land and sea operations (Barents Rescue Exercise, etc.). To conduct a MRO a large number of emergency organizations and other actors are usually involved. Such responders include the captains of casualty vessels, aircrafts and offshore installations, and the commanders of additional facilities such as ships at or near the scene of the incident who are ready to help in accordance with their obligations under international regulations. The management is taken care of by the SAR mission coordinator who is responsible for organizing the SAR response to the incident, designated SAR unit commanders responsible for ensuring that their units are prepared for their efforts, and the on-scene coordinator (OSC) responsible for putting the SAR response action plan into effect (IMO 2016). The OSC may have an aircraft coordinator responsible for the safety and effective use of air units, which may be operating in unusual numbers and circumstances in such a case. Finally, there could be other emergency response authorities, who must be ready to receive those involved as they are brought ashore by the maritime responders. As local SAR resources are limited, they may need to be obtained from distant national or international sources (IMO 2003), along with mobilizing the so-called host nation support system.

Depending on the magnitude, nature and complexity of a mass rescue incident, the rescue efforts may involve several layers of coordination with a broad range of considerations to make. The considerations might include aspects such as extensive rescue support by other organizations than those primarily used for maritime SAR (e.g. land SAR units, civil organizations, etc.), the need for substantial international diplomatic support, and potential problems following loss of lives, such as environmental threats, destructive (terrorist) actions, or national security issues (IMO 2003).

An important part of the MRO is that the vessels and installations are equipped with the evacuation methods or systems necessary in order to ensure safe year-round operations and mitigate risks during all emergency rescue scenarios. The new IMO Polar code, which came into force in 2017, demands more rescue and survival capability aboard the vessels in case of distress and the capability to keep the passenger and crew alive for a longer period of time.

Also an important part of MRO is that different stakeholders partake and dialogue in the process. The geography of stakeholders should therefore be multiscalar from the very local with context-specific nature of local knowledge and community-based actions, through to national and international levels (Gaillard and Mercer 2012). Local communities are involved for tier local and expert knowledge. With these possibilities in mind, MRO plans may provide guidance for various degrees of response, along with criteria for determining which amount of response will be implemented. For example, as local SAR resources are exhausted, there may be a need to obtain distant national or international resources (IMO 2003).

Mass rescue from emergencies in the Arctic represent more uncertainty. Rescuers must contend with vast stretches of cold water, possible ice floes, icebergs or smaller growlers, cold weather, wind and possibly fog. The Arctic conditions and the long distances between the potential emergency sites and the support bases increase the risks and limit potential victims' chances of survival. Other challenges include:

- shortage of duly equipped support vessels that may be called on for assistance with regards to their maneuvering and station-keeping abilities in ice,
- the effect of cold temperatures on human physiology and psychology, equipment, materials and supplies,
- the possible flight limits of the rescue helicopters due to technical limitations or military regulations,
- lack of experienced personnel and training facilities for the specific evacuation systems that have been proposed for the Arctic areas,
- the effect of the polar night with extended periods of darkness,
- the possible lack of qualified medical help, the recovery and transportation of large numbers of survivors (and bodies, if necessary), accounting for survivors potentially having injuries and lack of training, age limitation, hypothermia, etc. (Barents 2020 2009). This issue can be addressed by coordinating with hospitals in neighbouring regions/countries.

UNCLOS states that every coastal state shall promote the establishment, operation and maintenance of an adequate and effective search and rescue service regarding safety on and over the sea and, when circumstances require, by way of mutual regional arrangements cooperate with neighboring states for this purpose.

The coordination of mass rescue operations must run effectively from the very beginning to avoid delays, but communication difficulties may arise due to magnetic conditions and high latitudes, lack of satellite coverage and language barriers. SAR capability can vary geographically and according to the conditions at the time an incident occurs. In large-scale mass rescue operations, there may be need for supplemental communication capabilities, possibly including the need for interpreters both at the coordination centers and at the emergency site when international rescue resources are involved.

2.2 Joint Operations, Cross Border Support and Emergency Management

All SAR operations are governed by several international laws and agreements, such as the 1979 International Convention on Maritime Search and Rescue at Sea (the SAR Convention) and the International Aeronautical and Maritime Search and Rescue Manual (the IAMSAR Manual) with the latest edition in 2016. Additionally, there are multilateral and bilateral SAR agreements such as the Agreement on Cooperation in Aeronautical and Maritime Search and Rescue in the Arctic (Arctic SAR Agreement 2011), and bilateral agreements on search and rescue, such as in the Barents Sea between Norway and Russia that was signed in 1995 (Takei 2013; Rottem 2014; Yan et al. 2014).

The most revealing joint rescue operation is considered to be the evacuation of passengers and crew of the cruise ship “Maxim Gorkiy” after the ice accident on 20 June 1989 (Kvamstad et al. 2009). There were many causes of the accident: erroneous control of the vessel, the discrepancy of conditions of navigation, the lack of ice-specific reinforcement of the hull and others. Due to joint actions of rescuers and crew, casualties were avoided. After the incident, the national governments and international organizations initiated processes to improve international relations in SAR. In 1995, the Agreement on Search and Rescue Cooperation in the Barents Sea was signed by the Government of the Russian Federation and the Government of the Kingdom of Norway, in November 2002—an agreement on cooperation on border issues (http://ps.fsb.ru). The bilateral Agreement between the Russian Government and the Norwegian Government on cooperation in search and rescue of people suffering distress in the Barents Sea defines that the parties shall provide assistance in search and rescue in the Barents Sea, outlines the competent national authorities responsible for the implementation and their tasks, clarifies how requests for help are forwarded and procedures for information exchange.

The Barents exercise is held on an annual basis in accordance with the 1995 SAR Agreement and the 1994 Agreement on Oil Spill Response in the Barents Sea. The exercise is directed on a rotational basis by representatives of Russian and Norwegian SAR and OSR services. The exercise scenarios include elements such as lifesaving and oil spill recovery.

Learning from exercises and other joint actions establish better SAR collaboration. The increased traffic in the high seas in the High North calls for that. The flow of tourists to the Northern areas and the frequency of cruise shipping in the Arctic will continue to increase (Borch et al. 2016).

There is a broad political consensus to maintain the high priority of an effective SAR service with high focus on cross border support through the Host Nation Support (HNS) system (DSB 2014). The basic idea of it in the civil sector is the consensus that all available resources from abroad may be mobilized in the government-coordinated rescue service in a situation when the responsible authority does not have the necessary resources to handle a large-scale accident. Major international non-governmental organizations, and recently both the International Federation of Red Cross and Red Crescent Societies (IFRC) and the European Union (EU) have developed HNS guidelines. The national framework for HNS in Norway is developed by the Directorate for Civil Protection and Emergency Planning. The guidelines provide information about existing international mechanisms for cooperation for assistance, facilitates the reception and dispatch of assistance and aims to make the reception of assistance from abroad as efficient and effective as possible (DSB 2014).

HNS uses the four main principles for interaction within crisis management in case of assistance from abroad:

- *The responsibility principle*: means that the organization responsible for an area, also has the responsibility for necessary emergency preparedness in this area and for the response in case of a crisis or a catastrophe.
- *The similarity principle* states that an organization in emergency should be as similar as possible to this organization during its day-to-day activities.
- *The proximity principle* states that crises should be managed at the lowest organizational level possible.
- *The cooperation principle* states that governments, businesses or agencies are responsible to ensure the best possible cooperation between relevant actors and organizations in the prevention, preparedness and crisis management (DSB 2014).

The role of HNS can be critical in situations of mass rescue. The institutional aspects of countries and organizations involved in the response may influence the way the emergency response coordination is provided.

However, the difficulties for the emergency management can increase with several nations participating (Mitroff 2004; Comfort and Kapucu 2006; Kvamstad et al. 2009). A number of studies have found evidence of poor information sharing and coordination in inter-agency emergency response (Bharosa et al. 2010; Pan et al. 2012; Jardine-Smith 2015). Requests for help and information are considered a primary asset within inter-agency emergency response. The information needs to be sent timely, in an appropriate format and to both operational and strategic levels. Organisational interaction among several countries is complex because of different institutional frameworks, policies and standard procedures (Quarantelli 1986; Stacey 2001; Bigley and Roberts 2001). Border crossings must be arranged with the border authorities, and collaboration with the national forces must be governed (Borch et al. 2016). On the scene, there will be a strong need for functional management and good communication at an operational level, with additional liaison capacities (Borch and Andreassen 2015). At the tactical level, strong teams with strict coordination and control must be established without further delay. Therefore, multi-agency disaster response is the most important objective of emergency management exercises and drills (Bharosa et al. 2010). If not well managed, the effects of support from other countries may be limited and the additional forces may cause more problems than they solve.

2.3 The Role of Coordination

The importance of coordination has generally been acknowledged as central in organizations (Adizes 1979; Ekvall and Arvonen 1994; Mintzberg 2009). In emergency organizations, coordination is essential at all levels of crisis management. Comfort (2007) suggests to define the concept of coordination as aligning one's actions with those of other relevant actors and organizations to achieve a shared goal. Various organizational elements or teams with own processes, information and technology interact with each other, and coordination is achieved through assigning responsibility at different layers and leaders taken on different roles and following standard operating procedures and routines. However, it may also imply significant degrees of improvisation, in reallocating roles and responsibility, and innovative solutions (Cunha et al. 1999). Between teams or organizations, this interaction meets some challenges. The joint emergency operations may include a great number of actors and different responsibilities aligning to others' actions and at the same time operating with a large degree of autonomy. The coordination levels include the police, rescue coordination centers, the coast guard, fire and rescue services, and the health institutions. These institutions may have different organizational structures, routines and emergency response plans, and this is the most common reason why achieving smooth interaction between them can be challenging. As the number of involved actors increases, the range of coordination challenges widens (Comfort and Kapucu 2006), and the need for joint exercises on coordination becomes more important.

Uncertainties about roles and authority and the need to work effectively with different organizations who may otherwise seldom work together has led to the development of standardized incident management systems for emergency response (Lutz and Lindell 2008). The standardized incident management systems have been designed to be consistent with the general principles of organizational management. They include specified roles for response personnel and facilitate coordination of resources at the site of a broad range of incidents (Bigley and Roberts 2001). One of the most prominent standardized incident control system, the ICS, was created for fire departments fighting wildfires in Southern California in the 1970s. Since then, the ICS approach has been developed and revised in order to become suitable for teams across different jurisdictions. It turned out to be suitable for a wider range of emergencies, including multi-casualty accidents of nearly any size (Bigley and Roberts 2001; Lindell et al. 2005; Buck et al. 2006). The International Aeronautical and Maritime Search and Rescue Manual (the IAMSAR Manual) highlights the crucial importance of coordination of SAR procedures. The International Maritime Rescue Federation recommends the ICS as one simple and effective means of meeting the need for overall crisis management (Jardine-Smith 2015). In general, the ICS is constructed around five major management roles: command, planning, operations, logistics and finance/administration (Lindell et al. 2005). The basic ICS includes a standard management hierarchy of managerial roles and managing procedures. Managerial roles refer to a set of certain types of actions, while coordination mechanisms refer to a set of rules and practices to guide the action procedures (Bigley and Roberts 2001; Buck et al. 2006).

Emergency management is characterized by a strict interplay between the operational levels, from the headquarter down to the on-scene incident command, which implies the focus on coordination roles and capabilities (Borch and Andreassen 2015). Mintzberg (1973) defines managerial roles as sets of actions and responsibilities that are assigned for each of the managers and claims that all managerial roles within an organization can be separated into three main groups: interpersonal, decisional and informational. The starting point for them is the formal authority that provides the status.

Interpersonal roles include the figurehead, leader, and liaison roles. The figurehead is the head of an organizational unit and his role involves both internal motivation and inspiration and representing the crisis organization externally to different stakeholders, for example media and interest groups. The leader role constitutes leadership duties towards subordinates, like hiring and training the staff and the indirect duty to motivate and encourage individuals within the goals of the organization (Mintzberg 2003). The leader is also responsible for a long-term vision and transforming management and operational processes accordingly, as well as embracing the entire process and focusing on performance across the entire economic chain (Drucker 2007; Nieswandt 2015). Establishing effective liaison roles and mechanisms is essential for interorganizational coordination and contact outside the command system (Paton and Flin 1999; Bigley and Roberts 2001).

Informational roles include the monitor, disseminator, and spokesman roles. Managers develop a powerful database of information, and constantly work with the incoming information. In emergency management perspective information needs are critical and may differ from those within routine operating environments (Paton and Flin 1999). The monitor scans the environment, interrogates his liaison contacts and subordinates and receives all kinds of information. The disseminator passes some information that is needed within the organization along to subordinates. The spokesman sends some information to people outside the unit (Mintzberg 2003).

Decisional roles include the entrepreneurial action, disturbance handling, resource allocation, and negotiator roles. Making effective use of information leads to the need and importance of decision-making, especially large-scale (Turoff et al. 2011). In the entrepreneurial role, managers seek to improve the unit to changing conditions of the environment. They respond to information from the monitor and initiate new development projects or actions. The disturbance handler role is devoted to responding to different pressures and handling ad hoc problems. The resource allocators are responsible for decisions and strategies for resource distribution, including responsibilities for everybody in an organization, design of its structure in order to better coordinate the work, and time management. The negotiations are duties and routines for all managers and are an integral part of the job. Cosgrave (1996) highlights the importance of these four of the Mintzberg's roles especially in disaster management and explains that decision making is not a function in itself but is a critical part of all management functions. Paton and Flin (1999) highlights the importance of decision-making and the need for adaptation of management roles to utilize different decision-making procedures.

In mass rescue operations, the role of coordination has been discussed from various aspects. The main four coordinating roles are identified and discussed in the IAMSAR Manual—the SAR Coordinator, the SAR Mission Coordinator, the On Scene Coordinator and the Aircraft Coordinator (IMO 2016). **The SAR Coordinator** is defined as a person (persons) with overall responsibility for providing SAR services, establishing, staffing, equipping, and managing the SAR system. They are the top-level managers and determine the actual resource capability, engage with the response organizations and coordinate the planning.

The SAR Mission Coordinator is the operational role of an official temporarily assigned to coordinate response to an actual or apparent distress situation', usually based at a Rescue Coordination Centre and nominated from among its staff.

The On Scene Coordinator coordinates search and rescue operations within a specified geographic area. His role is intended to be mostly a communicative role together with the role of implementing the directions from the SAR Mission Coordinator.

The Aircraft Coordinator in a mass rescue operation has the primary task is to coordinate the involvement of multiple aircrafts (IMO 2016).

Coordination and command systems are difficult to configure in large disasters, which often involve multiple hazards and much damage, with a range of agent-generated demands, various responding agencies and conflicting goals that cannot be anticipated and reconciled. Coordinative mechanisms in emergency management depend on the complexity of the disaster response, recovery and mitigation tasks (Buck et al. 2006). As Kapucu (2005, p. 35) explores, "in complex and turbulent environments, organizations frequently develop formal or informal relationships in order to work together to pursue shared goals, address common concerns, and/or attain mutually beneficial ends". This dilemma of coordination in such management settings is highlighted also by Owen et al. (2013). In particular, they point out a paradox in which on the one hand, there is a need for tight structuring, formal command, control and hierarchical decision making to ensure a clear division of responsibilities, but on the other hand, there is a need for improvised coordination mechanisms to address emerging problems (Owen et al. 2013, p. 5). Owen et al. (2013) assert that when a larger number of organizations is involved, there is a need for the formation of complex teams that engage in hierarchical interactions within the organization and lateral interactions with related organizations. As Borch and Andreassen (2015) suggest, in high complexity-high volatility environments, there is a need for additional coordination roles and mechanisms added to the standard ICS, most importantly to deal with contextual challenges and in order to allow for improvisation. While in the recent years there are a lot of successful experiences of in areas of high complexity, still the High North context remains to be in need for more research on coordination mechanisms and resources capability.

In line with the specific features associated with maritime mass rescue situations, the regional factors of the High North and the institutional complexity of cross-border operations are the starting point. The high complexity and high volatility of the region mean that a wide range of institutions are involved in emergency rescue situations and the uncertainty of the consequences because of fast changes and unpredictable outcomes (Turoff et al. 2011; Borch and Batalden 2014). The under-developed infrastructure of rescue resource facilities and harsh weather conditions may have an influence on the reduced resource capacity available in particular crises in the High North. Therefore, the regional context influences the possible frequency and consequences of the accidents (Marchenko et al. 2016). Information on regional context in a SAR operation should therefore contain information about natural conditions, distance to resources and actual capacity, and the involved emergency institutions, which should plan how to collaborate.

IMO recognizes that in mass rescue operations the information about scale, complexity and rarity of the event is crucial (Jardine-Smith 2015). In larger-scale SAR operations, it is important to assess the number of people on board, the size of the ship, emergency resources available and the rarity of the accident to establish how much experience the crew has with similar cases.

These specific factors influence the efficiency of the SAR operation, which includes several stages: the distress signal and allocating the information, planning, operation, and rescue of people. It can be expected that in various settings, coordination responsibilities will be assigned differently. It is necessary to be aware of who takes the overall responsibility for coordination of these stages and to have an overview of available communication facilities, both with the vessel in distress and with other emergency actors.

3 Method

3.1 Case Study

In this study we build upon an illustrative case study of a major incident in the Arctic, i.e. the mass rescue situation with the "Maxim Gorkiy" cruise ship in 1989.

The data for analysis includes analytical reports and articles, media releases from 1989 and a detailed description of the accident (Hovden 2012). The data is also verified with log data and interviews with key personnel at the Joint Rescue Coordination Center Northern Norway in Bodø.

4 The Case of the Joint Norwegian-Russian Rescue of MV Maxim Gorkiy

4.1 The Institutional Framework

The Norwegian SAR is administrated by the Ministry of Justice and Public Security (Politidirektoratet 2011). Norwegian rescue services are carried out through cooperation between government agencies under the Ministries of Defense, Health and Transportation, voluntary organizations and private companies that have resources appropriate for rescue services.

The SAR operations in the High North are coordinated by the Joint Rescue Coordination Centre Northern Norway (JRCC NN) in Bodø and 21 local Rescue Coordination Centres (LRCC) located at the police district headquarters (www.hovedredningssentralen.no).

The Russian maritime SAR operation system is based on cooperation between different ministries, agencies and services such as the Ministry of Transport, the Ministry of the Russian Federation for Civil Defense, Emergencies and Elimination of Consequences of Natural Disasters known as EMERCOM, the Ministry of Defense, the Federal Agency of Fishery, the Border Guard of the Federal Security Service (FSB), and regional fire and rescue services. Additionally, the Ministry of Public Health is involved in maritime SAR operations to ensure medical assistance and evacuation of the injured to hospitals. The Federal Hydrometeorology and Environment Monitoring Service provides hydrometeorology information.

The organization of maritime SAR operations in Russia is the responsibility of the State Maritime Rescue Service, reporting to the Ministry of Transport. The responsibility for deployment and coordination of SAR resources lies with the seven rescue coordination centers and six sub-centers within marine centers (MRCC/MRCS). Rescue fleets, other vessels of all types from different jurisdictions that may be stationed close to the location of the accident site and, most importnatly, helicopter capacity are vital for mass rescue operations in the High North.

In Norway, the rescue helicopters Sea King from the 330 Squadron of the Royal Norwegian Air Force are considered the most significant lifesaving resource of the SAR service. In the Finnmark region, there are two Sea King helicopters based in Banak. In Nordland, there are two Sea King helicopters in Bodø. The new AW101 helicopters will between 2018 and 2020 replace the Sea King helicopters' role as the main rescue helicopter. The Governor of Svalbard has from 2014 two Super Puma helicopters capable of conducting SAR in Arctic areas (Borch et al. 2016).

The regions in Russia have different experience concerning the helicopters' SAR response in the Barents region. In the Murmansk region, the helicopters of the Northern Fleet are used to conduct SAR at sea with military personnel on board. In the Arkhangelsk region, the regional rescue authorities have come to an agreement with a private aviation enterprise to perform SAR activities at sea with regional rescue service specialists to handle the injured people. The air ambulance or rescue service doctors can provide medical assistance aboard a helicopter (Marchenko et al. 2016; Borch et al. 2016).

The Coast Guard plays an important role in civil emergency preparedness in Norwegian waters and for international SAR cooperation (Rottem 2014). Their role include search and rescue, ongoing coordination of operations, the transporting of police and defense special forces, medical support and other resources. The Norwegian Coast Guard operates 15 vessels of various types, sizes and capabilities, and is authorized to use maritime helicopters, civil aircrafts and get support from the Defense maritime surveillance aircrafts (Borch et al. 2016; Forsvaret 2016).

4.2 The "Maxim Gorkiy" Accident

The "Maxim Gorkiy" cruise ship accident happened in the Arctic waters of Northern Norway, to the west of Svalbard, in June 1989. The ship "Maxim Gorkiy" owned by Sovcomflot, Russia, started its cruise from Hamburg in Germany via Kirkwall on Orkney Islands, Reykjavik and Akureyri in Iceland, passed the Norwegian island Jan Mayen and from there set a course for the west side of Svalbard. As the New York Times reported, the passengers, mainly Germans and some other Western Europeans, were on a cruise to view the continuous daylight of the midnight sun north of the Arctic Circle (Lohr 1989). Around midnight on 19 June 1989, the ship hit an ice floe at high speed, which ripped two large holes in the hull and the vessel took in water. The ship had no ice reinforcement due to being intended for use in summer cruises when there was a very low possibility of encountering floating ice. The crew had little experience with icy waters The "Maxim Gorkiy" had 953 people on board, and the accident was treated as serious for lives at once (Hovden 2012).

The captain instructed all passengers and some of the crew to abandon the ship. One hour later, due to difficulties in identifying their exact location, the distress signal "SOS" was registered by a coastal radio station on Svalbard and ships near Svalbard in the Barents Sea (Lohr 1989). The distress signal on a poor VHF channel was incomplete and did not contain information that is now obligatory by international agreements. The Norwegian Rescue Coordination Center got the first distress call at 12:28 AM. The Norwegian Coast Guard ship "Senja" received a message from the Svalbard radio station about a Russian vessel in distress (Kvamstad et al. 2009). The radio connection was so poor that the coordination from the mainland was impossible. Without all the necessary details, the Norwegian Coast Guard vessel "Senja" was dispatched to assist with approx. 4 h of sailing to reach the vessel in distress.

At 01:13 AM the dramatic information that the ship in distress was the cruise ship "Maxim Gorkiy" was reaffirmed by the Russian Consulate. The signal "sinking" was received with an updated position of the "Maxim Gorkiy". LRCC at Svalbard received the signal from the Svalbard Radio and sent the ice-reinforced search ship Polarsyssel, which was 10 h away from the location of the accident, and established the preparedness plan for receiving injured people in Longeyarbyen. JRCC NN in Bodø sent the surveillance aircraft "Orion" from Andøya to locate the vessel and serve as observation post. Also, a stroke of luck was that a Sea King helicopter was situated at the Bear Island for exercises. JRCC NN in Bodø was planning the resources capacity for this operation. The police chief as head of the SAR operation began to plan the overall resource capability, coordinate resources and report to LRCC Svalbard, to the Defense Command North Norway in Reitan, the Sea Rescue Center in Murmansk, the Ministry of Justice, the Ministry of Defense and the Ministry of Foreign Affairs, who would contact the Soviet ambassador in Oslo. Additionally, hospitals in Hammerfest, Tromsø, Harstad and Bodø were alerted. The JRCC in Murmansk reported to JRCC NN that an "Ilyushin Il-38" surveillance aircraft was sent to the location of the "Maxim Gorkiy" and a "Tupolev TU-142" passenger aircraft was sent to Longyearbyen airport.

Because of the lack of information, the emergency team on "Senja" had to improvise and prepare for all possible situations. Key issues included planning to bring nearly a thousand people on board by rearranging rooms, warming possibilities by increasing temperature on board, available food, planning the best rescue method to take the passengers aboard, help with possible injures and diseases, registration, interaction with external resources such as helicopters, boats and aircrafts that were sent to the scene of the accident, the coordination of aircrafts and helicopters, communication plans, situation reports and preparing information for media. Their onboard resources included 53 people, a medical treatment capacity of 110 persons, medical personnel, and various equipment such as cranes and smaller boats. When they arrived at the accident site, they encountered many issues they had not anticipated and more improvisation was needed (Kvamstad et al. 2009).

By the time "Senja" arrived on the scene and the operation started almost 4 h after the incident occurred, the "Maksim Gorkiy" was already partially submerged. All passengers and some of the crewmembers were waiting in life boats, in rafts and on ice floes. The captain and the crew of "Maxim Gorkiy" had taken measures to keep the ship afloat and prevent casualties among the passengers (IFSMA 2015). However, the leakages was not stopped before the vessel was on the fringe of sinking. Tass, a major news agency in Russia, said the accident occurred in heavy fog, leaving holes in the vessel about 7 and 20 feet long (Lohr 1989). Fortunately, due to the midnight sun period, there was daylight throughout the night.

"Senja" was the first on the scene at 04:35 and began to evacuate passengers. Passengers located on the ice floes were not secured and towing people aboard the moving vessel was difficult. The Orion aircraft arrived shortly afterwards and began surveying and mapping the situation. For fear of buffeting the ice floes, they did not lower any life rafts down. "Senja" took the overall responsibility to make decisions about the action plan. They also assigned a helicopter control officer on board that had a good radio connection between the helicopters, which was important for updates on the situation. The Sea King helicopters from Bodø and Banak arrived and began retrieving passengers 05:52 AM. Two Soviet "Hip-8" helicopters from the Kap Heer base near Barentsburg arrived at 7:25 AM, but could not join in the rescue operation smoothly because of language problems. Their English skills were not advanced enough to communicate with the helicopter control officer from "Senja", so the captain from "Maxim Gorkiy" negotiated with them and confirmed that the situation was under control and that they could return to their base. They left the area at 07:55 AM. Later, it was discovered that they had shipwreck materials and instruments that may have stopped the water intrusion and stabilized the "Maxim Gorkiy". The Russian surveillance aircraft observed, took pictures and reported to the Murmansk rescue center, but did not participate in the rescue operation. The helicopter control officer made the decision to close off the area for all air traffic that was not involved in the rescue operation (Hovden 2012).

Some crewmembers remained on board the sinking cruise ship and struggled to stabilize it (Inquirer Wire Services 1989). After the evacuation the crew of the Norwegian Coast Guard ship "Senja" contributed with pumps to stop the "Maksim Gorkiy" from sinking, her bow already deep in the water. All the ship's passengers were rescued by 8:35 AM but 230 of crewmembers remained on board. Helicopters then ferried passengers from the rescue vessel's deck to the Spitsbergen archipelago. Planes picked them up and took them to the Norwegian mainland. The "Mayday" signal was cancelled at 23:52 almost 12 h after the incident occurred.

The New York Times reported on 21 June, that Norwegian officials and the Soviet press agency Tass informed all those aboard were safe and uninjured (Lohr 1989). On 21 June the "Maksim Gorkiy" was towed to Svalbard for quick repairs and to Germany for greater repairs. The ship sailed to Lloyd Werft, Bremerhaven without assistance, and after further repairs was back in service on 17 August 1989. The losses from the accident amounted to about $ 10 million (IFSMA 2015).

5 Discussion

5.1 Managerial Roles and Emergency Management

In this section, we analyze the main coordination roles, paying special attention to the elements of the SAR operation—the stages of the operation, the different responsibilities, the assignment of tasks to different actors and the communication aspects between the incident site and the mainland.

In the described rescue operation, the coast guard vessel took on the overall responsibility for the operation because they were closest to the scene. The rescue leader on duty at the JRCC NN in Bodø took the SAR Mission Coordinator role, and the JRCC NN began to plan the resource capabilities for this operation although communication with the vessel on the scene was problematic. Despite of this, they managed to scramble the capacity needed. The liaison role was also taking care of reporting to the Ministers of Justice, Defense and Foreign Affairs. The Ministry of Foreign Affairs established communication with the embassies of the passengers.

The **On Scene Coordinator** in the "Maxim Gorkiy" accident—the captain and crew of the rescue boat "Senja"—had to plan extensively and prepare for all possible situations while heading to the emergency site: rearranging rooms, warming possibilities by increasing the onboard temperature, food availability, help with possible injures and diseases, registration, interaction with external resources such as helicopters, boats and aircrafts that were sent to the scene, the coordination of aircrafts and helicopters, situation reports and information to the media. The helicopters and aircrafts were coordinated by the helicopter control officer from "Senja". In this case, the **Aircraft Coordinator** was appointed by the On Scene Coordinator instead of the SAR Mission Coordinator, but it worked out well. The communication between "Senja" and on-scene actors was satisfactory, while communication with the mainland was weak. Therefore, the On Scene Coordinator developed the action plan rather than the SAR Mission Coordinator.

To demonstrate the different nature of the coordination roles during this case, Table 1 summarizes the roles according to the theory about managerial roles by Mintzberg (1973). We try to focus on each of the groups and discuss which managerial roles the coordinators had to take on.

The case shows that pre-planned managerial roles and responsibilities can be completely different when facing a real accident. Because of the quickly changing working conditions, including lack of information exchange between departments that are supposed to collaborate and weather conditions influence the current action plan, coordinators may have to rearrange their roles and procedures. Sometimes, that means assigning people new roles or assigning several people the same role. Especially in case of joint operations, all sides should manage issues like resource coordination effectively. In the case of the Maxim Gorkiy accident, there were several crucial challenges that caused changes in the coordination. The most important of them were the limited radio communication, language problems and the lack of common SAR procedures between Russia and Norway.

5.2 *Cross-Border Support and Coordination Roles*

Mobilizing all the available resources was crucial in the Maksim Gorkiy case. Passengers were moved to unstable ice floes, and if the vessel sank, there were still a large crew aboard the ship that had to be evacuated quite fast. There were few other ships in the area, the mainland was far off in all directions and the SAR helicopters had few places in the area to refuel. Luckily, one of the Norwegian SAR helicopters was located at the Bear Island at the time of the incident, only a few hours' flight from the scene, and the Coast Guard was close to the Spitsbergen islands. Today two Super Puma helicopters at Longyearbyen would have represented a significant capacity.

Table 1 Coordination roles in emergency management vs. managerial roles in the "Maxim Gorkiy" accident

Mass rescue coordination roles	Managerial roles		
	Interpersonal	Informational	Decisional
The SAR Coordinator	Only the liaison role towards outside the national preparedness system	–	–
The SAR Mission Coordinator	Took the figurehead role and presented the crisis situation towards different stakeholders within the system	Took the monitor role in order to establish resources and the spokesman role in order to pass on information. The disseminator function failed because of bad communication	The resource allocator role
The On Scene Coordinator	Took the role of leader motivating the crew within the goal of the mission	As spokesman, reported some information to the SAR mission Coordinator. Had a disseminator role towards the Aircraft coordinator	Took the entrepreneurial role considering all possible information. When the new changing conditions of ice and waves came up, they initiated new actions and decisions. As to disturbance handling, they solved the situation of communication with Russian helicopters by finding the Russian-speaking captain who reported to the Russian side
The Aircraft Coordinator	–	Took the monitor role on scene in order to establish a plan on how to rescue the passengers	As resource allocator on scene, ensured that all helicopters and aircrafts had sufficient fuel and coordinated them in order to avoid panic, extra traffic and to maximize efficiency

The collision occurred in a region with well-equipped local communities both in the Russian village Barentsburg and in Longyearbyen. Even though the cold war had hampered the cooperation between Norway and Soviets, there had been good relations between Norwegian and Soviet authorities at Svalbard. The formal agreements and joint practices were, however, limited. In the "Maksim Gorkiy" case, the absence of routines for cooperation and means of communication between the Russian and Norwegian units prevented the Russian side from participating actively in a coordinated SAR operation even though Russian helicopter resources and a surveillance aircraft were on site. The bilateral agreement between Norway and

Russia on search and rescue first came in 1995 and opened for annual exercises. The IAMSAR Manual came into force later on. Since the "Maxim Gorkiy" accident emergency exercises have been run every year for almost 20 years already. Therefore, nowadays, regulatory framework and improved cooperation skills and procedures would contribute to a smoother rescue operation.

Communication challenges were present due to limited radio communication capacities in the region, as were military and language problems. Much of the same communication challenges are still present in spite of increased political focus, R&D efforts and joint exercises (Kvamstad et al. 2009; RCC NN 2015; Borch et al. 2016). In 1989, through on-site improvisation, the Russian and Norwegian units managed to work together as the Russian-speaking captain of "Maksim Gorkiy" served both as interpreter and coordinator even though his time for such tasks was limited. Nowadays, these communication issues have undoubtedly improved. The satellite communication is better than radio channels, although is still challenging in the High North. The language requirements have been also improved since then, and simple English has become the common language during joint operations the High North. At least, rescue coordination centers from Bodø and Murmansk have seldom any language problems. However, when it comes to interaction with ship crew, aircraft and helicopters directly, the language skills have a potential for improvement with Norwegians learning Russian, and/or Russians learning English.

This case illustrates that the SAR coordinators at all levels must be skilled at communicating, improvising, choosing the optimal action pattern and at the same time, work at adding resources to the scene. The emergency management must be skilled in the language and culture of the neighboring countries, in understanding technological capabilities in the specific maritime context, and knowledge about the competence and limitations of the personnel involved. These capabilities have to be present for emergency managers within logistics, staffing, and information and liaison personnel at every functional level. The rescue coordination center and the SAR mission Coordinator is at the core of the operation and need all the support available from higher authorities to speed up the decision processes of neighboring country institutions. This calls for well-functioning information channels at national levels to the government of supporting countries, as well as negotiation skills and fast decision-making skills for coordinators at top directorate and ministry levels.

5.3 The High North Context

The natural conditions of the High North, especially the Svalbard region, greatly influence the safety of shipping in that area (Marchenko 2015). The incidents with larger ships such as cruise vessels or oil installations may put the entire national preparedness system to the test (Marchenko et al. 2016). In our case, the limited visibility, summer fog, dynamic water conditions, and ice floes made it challenging to rescue the passengers of the damaged cruise ship. As for visibility, the incident happened in June with midnight sun. Long distances, lack of suitable means of communication and poorly developed SAR facilities and services make emergencies more difficult to manage (Kvamstad et al. 2009). Borch et al. (2016) described the limited emergency resource capacity and the increased activity level in the region. In our case, limited capacity of resources, both in terms of the number of helicopters and aircrafts and their flying capability influenced the mass rescue situation. The distance to the resources in the High North plays a crucial role. It took 4 h for the rescue vessel and other actors to get to the emergency site. Actually, in other weather conditions and other actors' position the first help could come up to 12 h later. Nowadays, two new helicopters from Svalbard could start the evacuation phase earlier than those 4 h. They can fly 100 nautical miles per hour and have only 30 min preparedness time.

There are many emergency institutions involved in a SAR situation. Frequent communication between the parties is crucial. In 1989, the coast guard vessel did not have satellite communication facilities. Around Svalbard, there were established military communication lines with poor signal. In this operation, the Svalbard radio played a critical role in communication with the mainland. The strategic and operation coordination with the mainland was limited. Kvamstad et al. (2009) highlight that getting access to necessary information and dependable means of communication are crucial for any SAR operations in the High North. Currently, the challenge may be connected to the problem of intensive load on communication lines, and therefore the principle of information priority has become important in emergency management, especially in the High North where the satellite communication can be limited. Additionally, the language problems of different nationalities can cause challenges in SAR coordination on scene. It is not always easy to find translators when they are needed in the region. In our case, the Russian helicopters could not participate in the SAR rescue operation because of language problems and sometimes even cultural differences. In fact, the helicopters had enough equipment on board to stop the hull leakage but were unable to provide assistance due to the language barriers. The communication issues related to the distress signal was due to the limited information via poor VHF channel and the time because it came through much later than it could have done. That delay cost the rescue operation an hour, and could have cost even more if the distress signal had not been treated as a serious emergency. Today, even if the distress signal is incomplete, such a serious emergency would be handled faster with larger precision due to position tracking systems. Although the satellite communication system is much improved, the language difficulties with Russian ships, aircrafts and helicopters remain.

To summarize, regional and situational peculiarities may challenge mass rescue situations at all stages. A lack of experience in managing a rescue operation in extreme conditions challenges the planning stage. Limited resource capacities challenge the operation and rescue efforts. Available facilities and equipment should be optimized for the northern climate conditions. Communication problems could be devastating in SAR operations, especially in mass rescue situations. There is a need to focus on communication infrastructure, language skills and increased knowledge about the capacities and understanding of modi of operandi in the neighboring countries.

6 Conclusion

In this chapter we have shown how the increased ship traffic in Arctic waters may put a heavy strain on the emergency preparedness system of the region. This is especially in case of emergency incidents calling for mass rescue from large passenger ships or oil and gas installations. There is a general lack of capacity and infrastructure for emergency operations in this area and SAR operations may be severely hampered by harsh weather, sea ice or icing, and communication limitations. The management of such operations is complicated due to the multiple institutions and management levels involved. Due to lack of resources, assistance from neighboring countries and the employment of the Host Nation Support framework is often needed. International agreements on search and rescue have defined the roles of mass rescue coordination at operational levels. This chapter shows that the managerial roles of the emergency management system has to be adapted to the context. There is a need for preparedness systems with a broader range of managerial tools to face the challenges of complexity and volatility in polar waters. Both the interpersonal, information and decision-making capabilities must be scaled up as fast as possible with competent personnel that can improvise and find solutions with limited resources available. Extra focus is on the incident commander's responsibility for the overall operation and the formal authority. People coming from different cultures and language groups may have problems with understanding each other as well as trusting each other. This calls for additional cross-cultural liaison roles, and fast reconfiguration of the organization, involving the roles and procedures of the units.

6.1 Implications for Industry

The discussion in this chapter shows the strain on the emergency management system of the Arctic region in a major crisis. Even though there have been improvements in capacities since the Maxim Gorkiy incident, many of the same problems are present today when it comes to dealing with a major accident. There is a continuous need for evaluation of government emergency capacity as the regional pattern of vessel operation changes, and this has to be an on-going process. In addition, the capabilities of the vessels and installations are crucial both for avoiding severe accidents, for self-reliance, and for comrade ship support. Industries have launched special standards for polar waters like the cruise industry organization AECO's "Guidelines for expedition cruise operations in the Arctic" and the oil and gas industry's "Guidelines for offshore marine operations" (G-OMO). The Polar code provides additional demands regarding capacity and competence on ships travelling in the most challenging parts of the polar regions including rules concerning ship structure, the qualifications of the navigators and safety equipment. However, industry standards and guidelines as well as the Polar code may be seen as minimum standards for the region. Thus, the SAR authorities within the region should emphasize further development of roles, routines and procedures for all the actors in the region. There is also a need for increasing the degree of training and exercises not only between SAR institutions, but also between private companies and government authorities.

6.2 *Implications for Further Research*

The findings in this chapter provide some interesting avenues for further research. In-depth case studies on real incidents showing how the crisis management systems work out in major incidents is valuable. In particular, there is a need to look into the different roles and adjacent capabilities of the command systems of the countries involved in close collaborative emergency actions. There should be increased knowledge of competence needed and the best-practice of education, training and exercise. We also need to look closer into how international agreements and conventions will function at operational level both related to governments and companies. The implementation of the Polar code here represents a great opportunity for longitudinal and comparative research on improvements in the SAR value chain. There have been efforts to improve both capacities and competences in the Arctic regions in the professional SAR system, but further gap analyses on response time should be given priority. More frequent exercises have been performed. The annual Exercise Barents between Norway and Russia provides valuable knowledge on technology and capacities available, as well as trust and competence exchange. The effects of such exercises and the need for additional education and training in joint operations should be emphasized in future research.

The cruise operators are introducing several preventive measures to avoid accidents in the Arctic. However, there is a need to look closer into the sailing routes, the preparations, the equipment and the competence needed for all sea areas in the Arctic, not only for the regions covered by the Polar code. For the larger cruise ships there is reason to discuss what sea areas they should be allowed to enter. More in-depth risk assessments are needed in this respect.

Acknowledgement This chapter is written under the MARPART project "Maritime International Partnership in the High North" funded by the Norwegian Ministry of Foreign Affairs, the Nordland County Administration and partner universities. We are also grateful for the kind support of the staff at the Joint Rescue Coordination Center—Northern Norway in Bodø.

References

Adizes, I. (1979). *How to solve the mismanagement crisis: Diagnosis and treatment of management problems*. The Adizes Institute Publishing.

Arctic SAR Agreement. (2011). Agreement on Cooperation on Aeronautical and Maritime Search and Rescue in the Arctic, Nuuk, Greenland, 12 May 2011. http://www.arctic-council.org/index.php/en/environment-and-people/oceans/search-and-rescue/157-sar-agreement

Barents 2020. (2009). Assessment of international standards for safe exploration, production and transportation of oil and gas in the Barents Sea, Final Report, Russian-Norwegian Cooperation Project, Panel RN04 Escape, Evacuation and Rescue of People.

Bharosa, N., Lee, J. K., & Janssen, M. (2010). Challenges and obstacles in sharing and coordinating information during multi-agency disaster response: Propositions from field exercises. *Information Systems Frontiers, 12*, 49–65.

Bigley, G. A., & Roberts, K. H. (2001). The incident command system: High reliability organizing for complex and volatile environments. *Academy of Management Journal, 44*(6), 1281–1299.

Borch, O. J. (2015). Observation from Exercise Barents. Internal notes.

Borch, O. J., & Andreassen, N. (2015). Joint-task force management in cross-border emergency response. Managerial roles and structuring mechanisms in high complexity-high volatility environments. In A. Weintrit & T. Neumann (Eds.), *Information, communication and environment: Marine navigation and safety of sea transportation*. Boca Raton: CRC Press.

Borch, O. J., Andreassen, N., Marchenko, N., Ingimundarson, V., Gunnarsdóttir, H., Iudin, I., et al. (2016). *Maritime activity in the high North – Current and estimated level up to 2025* (MARPART project report 1). Nord University.

Borch, O. J., & Batalden, B. (2014). Business-process management in high-turbulence environments: The case of the offshore service vessel industry. *Maritime Policy & Management, 42*(5).

Buck, D. A., Trainor, J. E., & Aguirre, B. E. (2006). A critical evaluation of the incident command system and NIMS. *Journal of Homeland Security and Emergency Management, 3*(3), Article 1. http://www.bepress.com/jhsem/vol3/iss3/1

Comfort, L. (2007). Crisis management in hindsight: Cognition, communication, coordination, and control. *Public Administration Review*, Special Issue, 189–197.

Comfort, L. K., & Kapucu, N. (2006). Inter-organizational coordination in extreme events: The World Trade Center attacks, September 11, 2001. *Natural Hazards, 39*, 309–327.

Cosgrave, J. (1996). Decision making in emergencies. *Disaster Prevention and Management: An International Journal, 5*(4), 28–35.

Cunha, M. P., Cunha, J. V., & Kamoche, K. (1999). Organizational improvisation: what, when, how and why. *International Journal of Management Reviews, 1*(3), 299–341.

Drucker, P. (2007). *The practice of management*. London: Routledge.

DSB Norwegian Directorate for Civil Protection. (2014). *Veileder for vertsnasjonstøtte I Norge (Host Nation Support). General Guideline*. http://www.dsb.no/Global/Publikasjoner/2014/Tema/veileder_vertsnasjonstotte.pdf

Ekvall, G., & Arvonen, J. (1994). Leadership profiles, situation and effectiveness. *Creativity and Innovation Management, 3*, 139–161.

EMERCOM. (1995). The manual of population evacuation in emergency situations.

Forsvaret. (2016). *Kystvakta* [Online]. Retrieved January 31, 2016, from https://forsvaret.no/kystvakten

Gaillard, J. C., & Mercer, J. (2012). From knowledge to action: Bridging gaps in disaster risk reduction. *Progress in Human Geography*. http://journals.sagepub.com/doi/full/10.1177/0309132512446717

Hovden, S. T. (2012). *Redningsdåden – om Maxim Gorkiy-havariet utenfor Svalbard I 1989*. Sandnes: Commentum Forlag.

Norwegian Coast Guard website http://forsvaret.no/kystvakten

IFSMA. (2015). Accident of the edge of catastrophe. *Captains' blogs*, published 15 July 2015. http://amko.org.ua/index.php/blogi/entry/avariya-na-grani-katastrofy

Inquirer Wire Services. (1989). Hundreds rescued from Arctic after Ship Rams Iceberg, philly.com, June 21, 1989. http://articles.philly.com/1989-06-21/news/26108856_1_sigurd-kleiven-senja-lifeboats

International Maritime Organization (IMO). (2003). *Guidance for mass rescue operations*. http://imo.udhb.gov.tr/dosyam/EKLER/201381214504COMSAR1Circ31GuidanceForMassRescueOperations.pdf
International Maritime Organization (IMO). (2016). *International aeronautical and maritime search and rescue IAMSAR manual*. London: International Maritime Organization.
Jardine-Smith, D. (2014). *Mass rescue*. Seaways, January 2014.
Jardine-Smith, D. (2015). *The challenge: Acknowledging the problem, and mass rescue incident types*. IMRF (International Maritime Rescue Federation) Guidance paper. http://www.international-maritime-rescue.org/index.php/librarymropublic/categoriesmrolibrarypublic/philosophyandfocusphoca/file/926-1-1-the-challenge-acknowledging-the-problem-and-mass-rescue-incident-types
Kapucu, N. (2005). Interorganizational coordination in dynamic context: Networks in emergency response management. *Connections, 26*(2), 33–48.
Kvamstad, B., Fjortoft, K. E., Bekkadal, F., Marchenko, A. V., & Ervik, J. L. (2009). A case study from an emergency operation in the Arctic Seas. *TransNav, the International Journal on Marine Navigation and Safety of Sea Transportation, 3*(2), 153–159.
Lindell, M. K., Perry, R. W., & Prater, C. S. (2005). *Organizing response to disasters with the incident command system/incident management system (ICS/IMS)*. International Workshop on Emergency Response and Rescue Oct 31 ~ Nov 1, 2005.
Lohr, S. (1989). All Safe in Soviet Ship Drama. *The New York Times*, June 21, 1989. http://www.nytimes.com/1989/06/21/world/all-safe-in-soviet-ship-drama.html
Lutz, L. D., & Lindell, M. K. (2008). Incident command system as a response model within emergency operation centers during hurricane Rita. *Journal of Contingencies and Crisis Management, 16*(3), 122–134.
Marchenko, N., Borch, O. J., Markov, S. V., & Andreassen, N. (2016). Maritime safety in the high north - Risk and preparedness. In *ISOPE - International Offshore and Polar Engineering Conference. Proceedings 2016*; Volume 2016-January (pp. 1233–1240).
Marchenko, N. A. (2015). Ship traffic in the Svalbard area and safety issues. In *Proceedings of the 23rd International Conference on Port and Ocean Engineering under Arctic Conditions*.
Marchenko, N. A., Borch, O. J., Markov, S. V., & Andreassen, N. (2015). Maritime activity in the High North – The range of unwanted incidents and risk patterns, conference paper. In *Proceedings of the 23rd International Conference on Port and Ocean Engineering under Arctic Conditions*.
Mintzberg, H. (1973). *The nature of managerial work*. New York: Harper Row.
Mintzberg, H. (2003). The manager's job: folklore and fact. In J. Reynolds, J. Henderson, J. Seden, J. Charlesworth, & A. Bullman (Eds.), *The managing care reader*. London: Routledge.
Mintzberg, H. (2009). *Managing*. Williston, VT: Berrett-Koehler Publishers.
Mitroff, I. I. (2004). *Crisis leadership. Planning for the unthinkable*. Hoboken: John Wiley & Sons Inc..
Nieswandt, M. (2015). *Fast cultural change. The role and influence of middle management*. Springer.
Owen, C., Bearman, C., Brooks, B., Chapman, J., Paton, D., & Hossain, L. (2013). Developing a research framework for complex multi-team coordination in emergency management. *International Journal of Emergency Management, 9*(1), 1–17.
Pan, S. L., Pan, G., & Leidner, D. (2012). Crisis response information networks. *Journal of the Association for Information Systems, 13*(1), 31–56.
Paton, D., & Flin, R. (1999). Disaster stress: An emergency management perspective. *Disaster Management: An International Journal, 8*(4), 261–267.
Politidirektoratet. (2011). PBS 1. Politiets beredskapssystem del 1. Retningslinjer for politiets beredskap.
Quarantelli, E. L. (1986). *Research findings on organizational behavior in disasters and their applicability in developing countries* (Preliminary paper # 107). Newark, DE: Disaster Research Center, University of Delaware.

RCC Northern Norway. (2015). Evaluation of Exercise Barents 2015. Internal report (in Norwegian).

Rottem, S. V. (2014). The Arctic Council and the search and rescue agreement: The case of Norway. *Polar Record, 50*(254), 284–292.

SAR Convention. (1979). International Convention on Maritime Search and Rescue, Hamburg, 27 April 1979. http://www.imo.org/en/About/Conventions/ListOfConventions/Pages/International-Convention-on-Maritime-Search-and-Rescue-(SAR).aspx

Stacey, R. D. (2001). *Complex responsive processes in organisations: Learning and knowledge creation*. London: Routledge.

Takei, Y. (2013). Agreement on cooperation on aeronautical and maritime search and rescue in the Arctic: An assessment. *Aegean Review of the Law of the Sea and Maritime Law, 2*(1), 81–109.

Turoff, M., White, C., & Plotnick, L. (2011). Dynamic emergency response management for large scale decision making in extreme hazardous events. In F. Burstein, P. Brezillon, & A. Zaslavsky (Eds.), *Supporting real time decision-making* (Vol. 13). Boston: Springer.

The Maritime Rescue Service of the Russian Federation website www.gmssr.ru

Joint Rescue Coordination Center Norway website www.hovedredningssentralen.no

Yan, S.-Y., Jiang, H.-Y., & Wan, C.-P. (2014). Analysis of the trend of globalization of regional search and rescue cooperation. In W.-P. Sung, J. C. M. Kao, & R. Chen (Eds.), *Environment, energy and sustainable development*. Abingdon: Taylor & Francis Group.

Maritime Transport in the Arctic After the Introduction of the Polar Code: A Discussion of the New Training Needs

Dimitrios Dalaklis and Evi Baxevani

Contents

Abstract Considering that the Arctic's ice-coverage maintains a downward trend, maritime routes that were previously covered with ice-pacts are—slowly, but steadily—becoming more available for shipping. Additionally, great interest is now openly expressed for the extraction of the natural resources available in the wider region and especially its seabed, another possible task for maritime transport. The International Maritime Organization (IMO) has already taken a very significant step to ensure a safer and cleaner shipping industry in the region under discussion through the adoption of the International Code for Ships Operating in Polar Waters, which strongly promotes maritime safety in these challenging waters. Issues such as uncharted areas, ice that is drifting and harsh environmental conditions are just a few examples of challenges for Arctic shipping. Strengthening the necessary technical infrastructure in order to support the expected increase of maritime traffic in the Arctic routes, with emphasis on facilitating timely response to emergencies and search and rescue (SAR) activities should be added to the equation. Even though there is encouraging institutional progress when it comes to ship building standards and the STCW provisions are continuously improved, due to the current

D. Dalaklis (✉)
World Maritime University, Malmö, Sweden
e-mail: dd@wmu.se

E. Baxevani
Department of Shipping, Trade and Transport (STT), University of the Aegean, Chios, Greece
e-mail: e.baxevani@stt.aegean.gr

L. P. Hildebrand et al. (eds.), *Sustainable Shipping in a Changing Arctic*, WMU Studies in Maritime Affairs 7, https://doi.org/10.1007/978-3-319-78425-0_21

occasional-limited use of polar waters for seaborne trade, there is obviously a lack of crews with the necessary experience. New preparatory training courses, some type of "field" activities, improved simulator capabilities and a new more proactive emergency response procedure that involves cooperation of all Arctic countries are needed to mitigate the high risks.

Keywords Arctic shipping · Polar Code · Maritime safety · Training needs · STCW

1 Introduction

There is a very high level of agreement among the scientific community of climate studies that the two Poles of the Earth are paying a heavy toll when it comes to climate change. The repercussions of global warming are evident (Intergovernmental Panel on Climate Change-IPCC 2014), with the overall situation of ice retreat in the Arctic being breathtaking; the West Antarctic Ice Sheet is also estimated to have already surpassed the threshold of collapsing. On the flipside, the Poles' status of lesser ice has opened the way for shipping activities; nevertheless with conditions remaining harsh and perilous, the International Maritime Organization (hereinafter referred to as IMO) has already taken a significant step towards the enhancement of safety and reducing the impact of accidents that could be devastating for the regional ecosystem: the introduction of the International Code for Ships Operating in Polar Waters (hereinafter referred to as PCD). The analysis in hand will present the innovation of the PCD when it comes to training and ship certification. Additionally, safety incidents statistics will be provided in order to support the argument that the quick introduction of the PCD was an action of priority, even if today there is still room for improvement. After all, already since the year 2002 the IMO had published recommendatory provisions on Arctic Shipping, noting the unique challenges encountered in these waters (MSC 2002).

Furthermore, a discussion of the Code's and the International Convention on Standards of Training, Certification and Watch keeping for Seafarers (STCW) provisions is included, in order to pin-point how to improve associated training needs. Polar shipping activities—and more specifically those in relation to the Arctic—are very challenging endeavors, not only because of the difficult environmental conditions. Low/inadequate quality nautical charts still remain an important issue; strict shipbuilding criteria are only the first step that needs to be met. The lack of a large pool of adequately trained seafarers, with concrete experience of challenges in the wider region, is also necessary to be addressed in the years to come. Finally, limited infrastructure to support the monitoring of maritime traffic and how to improve the response times for Search and Rescue (hereinafter referred to as SAR) are also part of the urgently needed preparative actions towards safer polar shipping.

2 Arctic Routes

The Arctic is a new promising field for economic activities and an emerging market; the fact that various interested parties have already used the term "the industrial Mediterranean of the future", such as the Head of the Icelandic shipping company "Emskip" Gylfi Sigfusson (McGwin 2013) is indicative. Even though adversities persist and are in fact an intrinsic element of Arctic shipping, ice free waters are navigable. This evolution may be relatively new (the North West Passage-NWP was open for the first time in recorded history only in 2007 and the first complete transit through the Northern Sea Route (hereinafter referred to as NSR) was completed in 2009), yet the downward trend of the Arctic's ice-coverage dictates proactive action by various stakeholders in the maritime domain so as to safely accommodate the traffic that is expected to increase furthermore. When it comes to "Arctic Routes" there are four possible alternatives, namely the Northeastern Passage (hereinafter referred to as NEP) along the northern Russian and Norwegian coastline, the Northwest Passage (hereinafter referred to as NWP) which crosses the Canadian Arctic Archipelago, the Arctic Bridge (hereinafter referred to as AB) which is a seasonal sea (and air) connection between the Canadian port of Churchill and the Norwegian port of Narvik or the Russian port of Murmansk, and finally the Central Arctic Route (hereinafter referred to as CAR) or Transpolar Route (hereinafter referred to as TPR) across the Arctic Ocean (Fig. 1). The latter shall become a reality for commercial shipping should climate change proceed according to the current trend, noting nevertheless that icebreaker capability shall remain very essential (Dalaklis and Baxevani 2015).

Globally, only 15% of the container fleet is able to navigate through ice, while ice capable ships take up only 1% of the new orders (Baxevani et al. 2015). Main traffic in the Arctic is not expected to pertain to container shipments, but, bulk carriers. Arctic shipping is a promising field as the travel times are reduced almost by half compared to traditional routes, translating into less fuel cost and emissions for the voyage. Piracy prone areas in the Indian Ocean (Dalaklis 2012) could be avoided; at the same time pressure on the main (rather heavily congested) choke-points associated with the current dominant transcontinental routes like the ones through the Straits of Malacca and Bad El Mandeb will also be reduced. Apart from opportunities for maritime transport, the receding ice also enables access to more resources (oil, natural gas, minerals) and current traffic patterns will dramatically change should the extraction of the Arctic natural resources take place at a larger scale; fisheries and tourism could also impact positively on the number of vessels operating in the Arctic. A National Oceanic and Atmospheric Administration (hereinafter referred to as NOAA) study provides estimates for complete ice-free waters as early as the year 2020 (Overland and Wang 2013); for the moment however, transit through the above mentioned passages is possible only for a rather limited time period (a couple of months per year, around summer) meaning that the need for ice capable vessels and experienced crew is persistent.

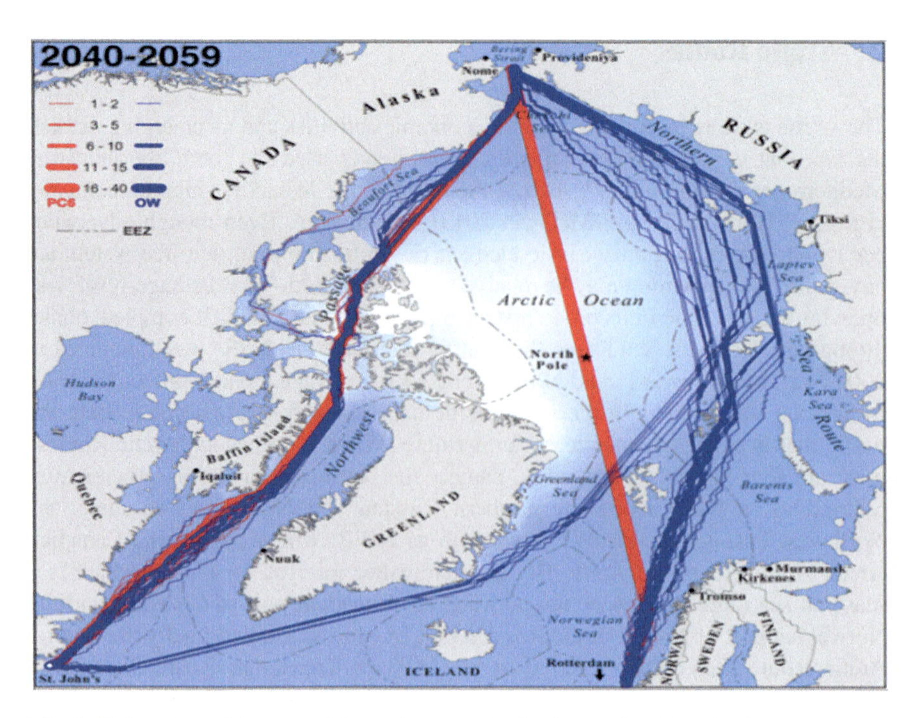

Fig. 1 Estimated maritime use of the arctic routes for the time period 2040–2059. Source: Smith and Stephenson (2013)

As more and more vessels will travel through the Arctic in the future, the possibility of a disastrous and costly in human lives accident can never be completely eliminated. This is why it is necessary to improve infrastructure supporting navigation and SAR response mechanisms; training people to tackle challenges and handle emergency situations is also needed. It is indicative that in 2014 there were 55 shipping safety incidents within the Arctic Circle—whereas only 3 took place a decade ago (Safety and Shipping Review 2015). The combination of a graph and table that follows is revealing an unexpectedly high numbers of incidents (Fig. 2 and Table 1). Increasing the level of safety was an action of priority: with already a significant number of accidents recorded during the last years, the introduction PCD with a very fast pace was a step towards the right direction. Needless to mention: further improvements of the PCD can still be implemented in the years to come.

3 Polar Code

The United Nations' Convention on the Law of the Sea (hereinafter referred to as UNCLOS), with its worldwide application, is also relevant to both Polar Regions. More specifically, Article 234 is acknowledging the navigational challenges and

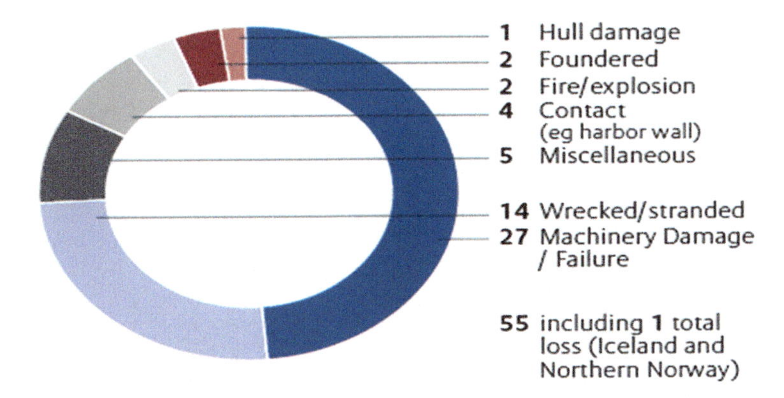

Fig. 2 Arctic Circle Waters: all casualties, including total losses for 2014. Source: Adaptation from Safety and Shipping Review (2015), p. 15

threats encountered in ice-covered waters and enables coastal states to introduce more rules (additional to the international rules and standards in place). In any case, the work of international organizations can support the aims put forth by codified law. Such is the case of the IMO, which as an agency specialized in shipping has the mandate to promote safety and security in the maritime industry and to safeguard the protection of the environment; the Maritime Safety Committee (hereinafter referred to as MSC) is responsible for both these issues. Codified texts such as International Convention for the Safety of Life at Sea (hereinafter referred to as SOLAS), International Convention for the Prevention of Pollution from Ships (hereinafter referred to as MARPOL) and International Convention on Standards of Training, Certification and Watch keeping for Seafarers (hereinafter referred to as STCW) further regulate maritime domain matters (Visvikis and Dalaklis 2014).

A very important safety evolution was the adoption of the Polar Code (International Code for Ships Operating in Polar Waters) by the IMO which applies to both Polar Regions (Arctic, Antarctic), albeit recognizing that "although the waters of the Arctic and the Antarctic are similar, they also have significant differences". Therefore, the PCD is covering both Poles, but the legal and geographic differences need to be weighed in nonetheless. In order for the new Code to become fully functional, there was a need for amendments in preexisting Conventions, such as SOLAS and MARPOL. These amendments have already been adopted in November 2014 and May 2015 respectively and the Code entered into force on January 1st 2017. PCD Part IA section on safety applies to new ships with SOLAS certificates (on or after January 1st 2017); for existing ships it applies from their first intermediate or renewal survey (whichever occurs first) on or after January 1st 2018. The new Code covers a wide array of issues, ranging from equipment, design and construction, operations and manning to invasive species, oil, chemicals and garbage.

Regarding its structure, the PCD is divided in two parts, one with mandatory measures on safety (Part IA) and one on pollution prevention (Part IIA), while recommendatory provisions on these areas are included in Parts IB and IIB

Table 1 Arctic Circle Waters: all safety incidents (including total losses) for 2005–2014

	2005	2006	2007	2008	2009	2010	2011	2012	2013	2014	Total
Machinery/damage/failure	2	3	5	13	14	16	12	13	20	27	125
Wrecked/stranded	1	4	10	11	14	9	9	8	10	14	90
Miscellaneous			5	1	4	4	2	6	5	5	32
Fire/explosion			3	1	2	6	6	1	4	2	25
Collision				1	4	10	4	4	2		25
Contact (e.g. harbor wall)			1	1	1	3	1	3	6	4	20
Hull damage		1	3	1	6	2	2	1	2	1	19
Foundered			1	1	2		3	1	1	2	11
Total	3	8	28	30	47	50	39	37	50	55	347

Source: Adaptation from Safety and Shipping Review (2015), p. 15

respectively. Ships are required to apply for a Polar Ship Certificate (hereinafter referred to as PSC) to be classified as Category A (operation in polar waters at least in medium first year ice which may include old ice inclusions), Category B (operation in polar waters in at least thin first year ice which may include old ice inclusions) or Category C (operation in open waters or in ice conditions less severe than Categories A and B). The Certificate will include an assessment of anticipated conditions and dangers, as well as information on identified operational limitations and procedures or additional equipment to enhance safety. Having a PSC is mandatory and confirms compliance with the (PCD) Code. The flag state (or classification society) will issue the certificate following a survey of the vessel that shall navigate through polar waters. This document (certificate) shall be supplemented by a Record of Equipment, where any additional equipment required by the PCD (beyond the minimum SOLAS requirements) shall be mentioned.

4 New Training Needs

When any one asks me how I can best describe my experiences of nearly forty years at sea I merely say uneventful. Of course, there have been winter gales and storms and fog and the like, but in all my experience I have never been in an accident of any sort worth speaking about – I never saw a wreck and have never been wrecked, nor was I ever in any predicament that threatened to end in disaster of any sort.—Source: Captain Edward J. Smith, RMS Titanic (1912) Disaster al last befalls Capt. Smith New York Times (ref. #3315, accessed Jan. 18th 2016)

"Because of the maneuvering exercises performed e.g. in a simulator, the bridge personnel was, however, able to react to this exceptional situation in a safe manner." Source: Investigation Report, M/S SILJA EUROPA (FIN), Breaking of the Starboard Rudder Shaft in the Aland Archipelago on Nov. 22nd 2009, p. 64 (accessed Jan. 18th 2016).—Source: Aboa Mare Maritime Training Academy and Training Centre, Simulation Training, p. 8.

All the tasks the Polar Code is called to regulate are challenging; a particular aspect needs however to be further highlighted: the subject of manning, qualification and training (Chapter 12). These provisions are mandatory, noting that no other guidance is included in Part IB. Given the prohibitive environmental conditions of these areas, experience in navigation has been extremely limited; therefore there is a justified yet severe lack of experienced crews and a thorough discussion of the new training needs is an action of priority.

As these issues were not sufficiently mitigated by SOLAS, STCW and MARPOL Conventions, the PCD introduced enhanced regulations for the "Polar Waters" (namely, south of 60°S for the Antarctic and north of 60°N for the Arctic—with a cut-out for Iceland and Norway). This section will focus on the discussion of the current STCW regime and its much needed provisions for facilitating safe and sustainable maritime operations in the wider region. The IMO Maritime Safety Committee (MSC) adopted two Resolutions, namely MSC.416(97) amending STCW Chapter V and MSC.417(97) amending STCW Code, Part A (Special Training Requirements for Personnel on Certain Types of Ships), in February

2015, during the IMO sub-committee on Human Element, Training and Watch keeping (HTW 2) meeting. These will come into effect on July 1st 2018, with a 6 month extension granted from the initial date of January 1st, so as to offer more time to maritime training centers to prepare courses and find qualified trainers and to masters, chief mates and watch officers to receive the necessary training. Ship owners, operators and masters have from January 2017 to July 2018 time to refer to IMO Circular STCW.6/Circ. 12, which the IMO issued to provide recommended provisions for applying the amendments that just entered into force (January 2017). The aim of these amendments (along with the PCD) is to establish a common basis so as to have trained and "ice-capable" seafarers (qualified, experienced and trained) available within the following years.

PCD Chapter 12 defines that every crew member onboard a vessel operating in Polar Waters needs to be familiar with the equipment of the ship and with procedures depending on his/her duties, in addition to any procedures mentioned in the Polar Water Operational Manual (hereinafter referred to as PWOM) which every ship is required to have. It also requires training and certification for masters, chief mates and officers in charge of a navigational watch on certain ships operating in Polar Waters. All crew members need to be familiar with the PWOM of the vessel, including risk management and emergency response operation in Polar Waters, for the specific vessel, complementing the STCW competency requirements. It is necessary to have enough officers who meet the above training requirements, so as to cover all watches and meet the minimum hours of rest requirements at all times. In concept, the PWOM is similar to safety management documentation already required on all SOLAS-certified ships by the IMO's ISM Code, with an indicative structure included in Appendix 1. It is also useful to note that the PWOM will not be subject to an approval by the flag state, although it is envisaged that a similar audit and verification scheme to ISM will apply.

The above mentioned requirements are dependent on ice coverage extend (ice free, open or other waters) and vessel category (passenger, tanker or other cargo). The PCD defines "ice free waters" as waters without any kind of ice and this is the lowest risk category and there are no "additional" training or certification requirements for vessels operating in such waters according to the PCD. No training or certification requirements are in place either for the "other" vessels (cargo) operating in "open waters" which are defined as a large area of freely navigable waters where sea ice is present, in concentrations less than 1/10 and no ice of land origin. Nevertheless, when it comes to passenger ships or tankers, the master, chief mate and officers in charge of a navigational watch need to hold a certificate in basic training and to have satisfactorily completed approved basic training and to meet the specified standard of competence.

When it comes to "other waters", namely waters with more than 1/10 ice cover or any ice of land origin, no distinction is made among vessel types, as masters and chief mates on all covered vessels (passenger, tanker, cargo) need to have an advanced training certificate and meet the requirements for certification in basic training and the specified standards of competence. The officers who are in charge of a navigational watch on such vessels must hold a basic training certificate and meet

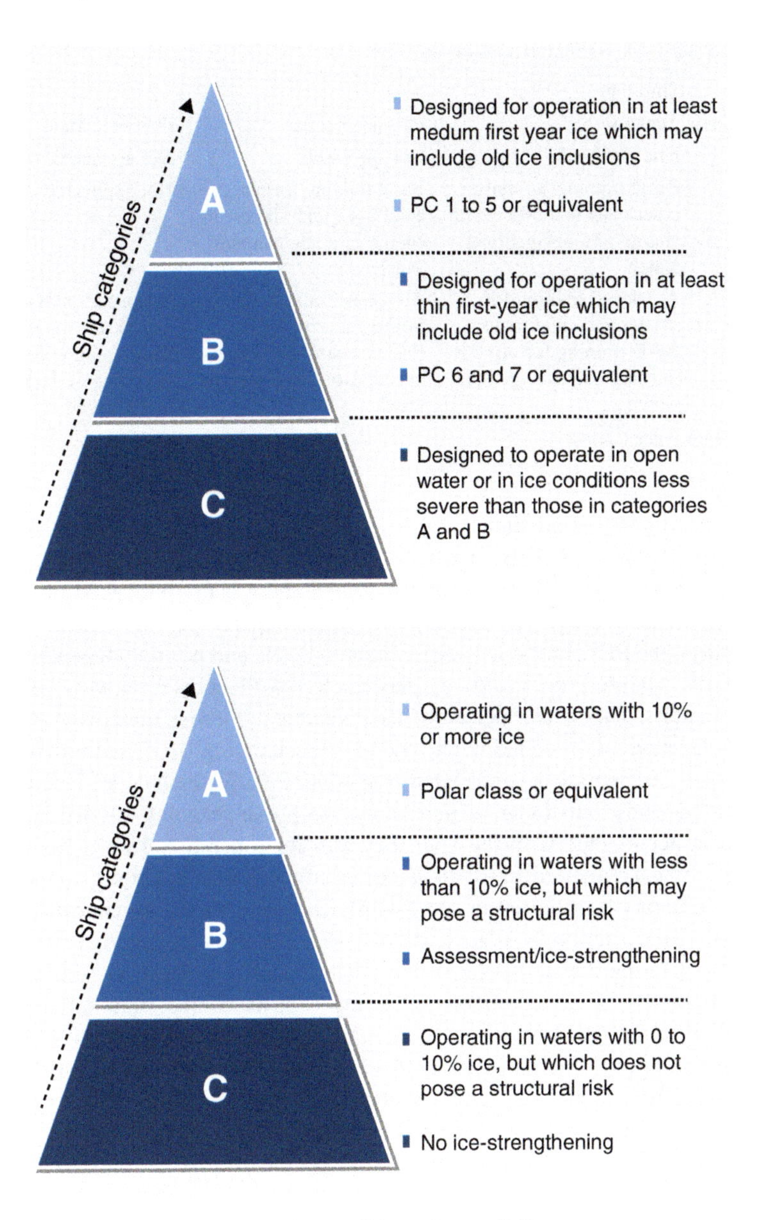

Fig. 3 Ship categories with respect to ice conditions. Source: IMO

the related competence standards, but they are not required to hold the advanced training certificate (Fig. 3 and Table 2).

The Certificate of Proficiency (hereinafter referred to as CoP) will complement the training courses and the necessary seagoing experience to sail through polar waters. It will be mandatory around 2020 (including the biennial transitional period)

Table 2 Polar Code training requirements

Ice conditions	Ship type		
	Tankers	Passenger ships	Others (Cargo)
Ice free	Not applicable	Not applicable	Not applicable
Open water	Basic training for master, chief mate and officers in charge of a navigational watch	Basic training for master, chief mate and officers in charge of a navigational watch	Not applicable
Other waters	Advanced training for master and chief mate. Basic training for officers in charge of a navigational watch	Advanced training for master and chief mate. Basic training for officers in charge of a navigational watch	Advanced training for master and chief mate. Basic training for officers in charge of a navigational watch

Source: IMO Polar Code

and will need to be revalidated every 5 years. The CoP was introduced following a discussion whether this is the optimum means so as to prove qualification. In any case, seagoing experience cannot be substituted and it is up to local administrations to define sea service equivalence to polar waters experience. Apart from the initial certification, the PCD introduces differentiated levels and training depending on the rank of the seafarer (deck officer, master etc.) with courses already offered by training centers. This makes perfect sense given the numerous unknown coefficients of the polar navigation equation and the always changing (real) sailing conditions. Transitional arrangements (until 2018, regarding STCW) include a grace period of 2 years after entry into force, seafarers need to prove seagoing service in relevant position in accordance with the CoP they aim for (operational level for the basic certificate, management level for advanced certificate) and depending on the level, it may be voluntary or required to successfully complete a training course in compliance with valid section B-V/g of the STCW Code (this is relevant for all deck officers navigating through polar waters). There will be a Basic Certificate, which will be obtained by deck officers following a basic course (familiarization) and satisfaction of STCW A-V/4 (regarding competence), though with no actual experience at sea. For the Advanced Polar Waters Certificate of Proficiency, officers will need to comply with all Basic prerequisites, along with seagoing service and STCW A-V/4 fulfillment. As mentioned, STCW will incorporate amendments so as to accommodate the new PCD demands. More specifically, this pertains to mandatory qualification and training prerequisites for all crew members, with amendments referring both to the Convention and parts A &B of the Code.

Given that for the time being on-the-job experience is not great due to practical reasons (i.e. office posts favor professional development/promotion), experienced personnel to deliver training is certainly a valuable asset. Instructors, usually have experience in ice navigation with cargo vessels and icebreakers in the Arctic and Baltic Area whereas the trainings aim at providing participants with knowledge on a wide array of ice navigation issues, namely navigational characteristics of sea ice, ship's hull interaction with ice, ice traffic restrictions, reading ice charts, use of radar and search light for ice edge detection, maneuvering in ice (ship handling, ice management), navigation of different types of ships/voyage planning/selecting route through ice, unassisted passage tactic, mooring with tug assistance, dangers pertaining to freezing of equipment, propulsion systems in ice operation, azimuthing propulsion systems (Azipod, Aquamaster), icebreaker operations (rules, signaling, convoy ops), escort towing, safe working practice, etc. Emergency scenarios are also applied, such as engine overheating or rudder jamming in order to provide training as close to reality as possible. The use of simulators must be extensive and valuable, as they offer the opportunity to experience conditions as close to real as possible. For instance, bridge simulators can be combined with engine room ones, so as to deliver near real conditions and enhance teamwork. Major shipping companies participate at trainings of the kind and most frequent scenarios are about near misses or accidents (as well as SAR duties).

In 2009, the IMO had proposed ice-related training topics, namely ice physics (formation, growth, aging, stage of melt), ice types and concentrations, ice pressure, friction from snow covered ice, implications of spray-icing (superstructure icing) and options during icing up, ice regimes in different regions; first year vs. multi-year; land (river) ice vs. sea ice; (glacier ice), use of (remote sensing) ice imagery to recognize consequences of rapid change in ice and weather conditions, knowledge of ice sky and water blink, knowledge of tides and currents in ice covered waters, marine mammal protection regulations etc. (IMO 2009). Training centers can use lectures, texts, photos, videos and simulator exercises to cover all these topics. Apart from the Polar Ship Certificate, ships shall be requested to have onboard an Operational Manual for the Polar Waters (PWOM). Its aim will be to provide information about the ship itself, namely its capabilities and limitations when it comes to such waters, which is turn aims at supporting decision-making (by owner, operator, master and crew). As already highlighted, this Manual includes risk based operational procedures assessed against the probability to occur and is similar to the documentation required by the ISM Code. Not every ship will include the same content for its PWOM, nor follow the same format. It is expected that every vessel shall hold a different PWOM according to its type, e.g. cruise ships may focus on passenger safety (ABS-IMO Polar Code Advisory 2016, p. 22).

This additional documentation must be aligned with the IMO's Guidelines and common practice, aiming at enhancing safety and ensuring avoidance of marine pollution. It should include information on engine operations in polar waters; it is noted that its approval by the Administration (or a Recognized Organization (hereinafter referred to as RO) on its behalf) is not mandatory. Furthermore, the POLAR Training Manual shall cover all the associated training requirements and must also be aligned with the above mentioned Guidelines. Training is generic, regarding the particular polar conditions (low temperatures and visibility, winds etc.), including the awareness issue; this document is also not subject to approval, although inspection may be held. Another important document that ships crossing polar waters must have onboard is the POLAR Contingency Plan, covering emergency action plan among others.

5 Concluding Thoughts

In case there is no dramatic shift regarding the retreat of ice, more traffic is clearly expected for the Arctic routes, despite the current conditions that it is obvious that they do not favor large scale activities. This development could radically influence certain current global shipping hubs (e.g. Singapore), gradually bringing the Arctic to the forefront of commercial shipping. Trade relations between Asia and Europe are increasing and so will competition and congestion. The interest of non-Arctic actors is explicit with China, India, South Korea, Japan and Singapore participating at discussions, scientific expeditions and debates about arctic maritime routes. On the downside, dangers and hindrances persist, given the uncertainty of the weather/ice conditions each of which may vary by year, despite the consistent tendency, the unrecorded ice floes, the inadequate infrastructure for traffic monitoring and timely response mechanisms. Hurdles persist and development depends on several directly and indirectly linked factors, namely settlement of the remaining legal disputes for the delimitation of the maritime zones, oil prices, fees (permission for transit, insurance), lack of infrastructure (e.g. deep water ports, hinterland connections, surveillance, charting) and ice capable vessels and/or availability of escorting ice-breakers. A "rush" of claims in the form of past acquirement of territories is not going to happen in the Arctic; UNCLOS offers the proper framework that ensures that. It is however very essential to have regulation in place when it comes to navigation in polar waters and train seafarers so that they are able to ensure the safety of operations. Up to now, there has been no major oil-spill in the Arctic; in such a disastrous event, repercussions will be not only regional, but also global.

As ice disappears from the Arctic for the first time in mankind's history, a new field opens up for economic profit. Shipping, being directly dependent on weather conditions is significantly affected by overturning the prevailing notion of an inaccessible Arctic. The sectors of transport, natural resource extraction, fishing, tourism, energy production and various others constitute the reasons the Arctic is going to receive increased maritime traffic. In order to reap only benefits and avoid a natural disaster greater than the already occurring one, there need to be regulations in place. Apart from the environmental aspect, it is pivotal to have properly trained personnel onboard: various scenarios in simulators can be used to cover this need. A new legal binding text has been adopted by the IMO; the Polar Code which is going to be supported by further amendments to specialized legal texts, namely STCW, SOLAS and MARPOL. The Polar Code is indeed a proactive action and provides common ground for further regulation. Keeping in mind that Arctic shipping is a new field, the Code as well as the other legal binding documents will have to adapt according to traffic and incidents that may occur, so as to maximize the protection of the environment and of course the seafarers. In addition to severe weather conditions that are threatening for survival at sea, there is rather limited infrastructure in place for SAR operations, while the current response times are also considerably long. The importance of escort by other vessels and strict compliance to the specific regulation can be easily understood. The Arctic Search and Rescue Agreement (Agreement on Cooperation on Aeronautical and Maritime Search and Rescue in the Arctic) is the first binding agreement of the Arctic Council (into force since 2013) and handles SAR response and coverage while it does not interfere with sovereignty or jurisdiction issues.

When it comes to training, realism is necessary as no simulator can duplicate actual conditions and alertness needed. Nonetheless, ice navigation simulators in the pre-deployment phase of any navigating officer in the Arctic would be extremely helpful in order to have a first contact with these extreme/strange conditions in a controlled and safe environment. Preparatory training courses would also provide vital information to crews and through them, to passengers, about how to use survival equipment and things to do when in distress in the Polar waters. In any case, for such dire conditions the value of cooperation among states is multiplied. In fact cooperation is vital for polar shipping due to the vast distances that need to be covered and the lack of knowledge at the same scale as for other geographical location. The Arctic Council provides an excellent forum to share opinions, knowledge and proceed to agreements promoting this scope. There are working groups which contribute to the same aims as IMO. There is for instance the Arctic Monitoring and Assessment Program (hereinafter referred to as AMAP) Working Group. In addition, the complementarity of the conventional texts is very important so as to have an all-embracing approach. For instance, the Polar Code regulates the "quantitative" characteristics of the crew (number, origins, experience) while the STW Code handles the "qualitative" aspect, namely the requirements when it comes to training and the competences of individuals.

On a final note, although not connected with the training and certification issue, the Arctic Council can play a major role when it comes to safety of navigation, as indicated by the progress achieved so far (e.g. binding Agreement on Cooperation on Aeronautical and Maritime Search and Rescue in the Arctic). As in every endeavor, the challenges of the Arctic call for a timely, coordinated response. Thankfully, positive steps have already been made which need to be followed by supplementary SAR training courses and even conduct of the respective exercises-live drills.

Appendix: Polar Water Operational Manual

1 - Operational Capabilities & Limitations	**1.1**	**Operations in ice**
	1.1.1	Operator guidance for safe operation
	1.1.2	Icebreaking capabilities
	1.1.3	Maneuvering in ice
	1.1.4	Special features
	1.2	**Operations in low air temperatures**
	1.2.1	System design
	1.2.2	Protection of personnel
	1.3	**Communication and navigation capabilities in high latitudes**
	1.4	**Voyage duration**
2 - Ship Operations	**2.1**	**Strategic planning**
	2.1.1	Avoidance of hazardous ice
	2.1.2	Avoidance of hazardous temperatures
	2.1.3	Voyage duration and endurance
	2.1.4	Manning
	2.2	**Arrangements for receiving forecasts of environmental conditions**
	2.2.1	Ice information
	2.2.2	Meteorological information
	2.3	**Verification of hydrographic, meteorological and navigational information**
	2.4	**Operation of special equipment**
	2.4.1	Navigation systems
	2.4.2	Communications systems
	2.5	**Procedures to maintain equipment and system functionality**
	2.5.1	Icing prevention and de-icing
	2.5.2	Operation of seawater systems
	2.5.3	Procedures for low temperature operations
3 - Risk Management	**3.1**	**Risk mitigation in limiting environmental condition**
	3.1.1	Measures to be considered in adverse ice conditions
	3.1.2	Measures to be considered in adverse temperature conditions
	3.2	**Emergency response**
	3.2.1	Damage control
	3.2.2	Firefighting
	3.2.3	Pollution response
	3.2.4	Escape and evacuation
	3.3	**Coordination with emergency response providers**
	3.3.1	Ship emergency response services
	3.3.2	Salvage
	3.3.3	Search and rescue
	3.3.4	Spill response
	3.4	**Procedures for prolonged entrapment by ice**
	3.4.1	System configuration
	3.4.2	System operation
4 - Joint Operations	**4.1**	**Escorted operations**
	4.2	**Convoy operations**

Source: American Bureau of Shipping, IMO Polar Code Advisory

References

Allianz Global Corporate & Specialty. (2015). *Safety and Shipping Review*. Retrieved June 20, 2015, from https://www.allianz.at/v_1427110816000/privatkunden/media-newsroom/news/aktuelle-news/pa-download/20150324studie-agcs-safety-and-shipping-2015.pdf
American Bureau of Shipping. (2016). *IMO Polar Code Advisory*. Retrieved January 30, 2017.
Arctic Council. (2011, April 21). *Arctic search and rescue agreement*. Retrieved February 8, 2013, from http://www.arctic-council.org/index.php/en/document-archive/category/20-main-documents-from-nuuk
ARCTIS. (2012). *Arctic Sea Routes*. Retrieved June 1, 2013, from http://www.arctis-search.com/Arctic+Sea+Routes
Baxevani, P., Kourounioti, I., Polydoropoulou, A., & Sioussiouras, P. (2015). Comparing container shipping and logistics costs between the Northern Sea Route and the Suez Canal. In *ECONSHIP Conference presentation*, 24th – 27th June, Chios, Greece.
Black, R. (2007, May 18). Earth-melting in the heat? *BBC News*. Retrieved April 12, 2013, from http://news.bbc.co.uk/2/hi/science/nature/4315968.stm#arctic
Borgerson, S. G. (2009). *The national interest and the Law of the Sea* (Council on Foreign Relations Special Report No. 46).
Climate Communication. (2011, September 7). *Current extreme weather and climate change*. https://www.climatecommunication.org/wp-content/uploads/2011/09/Extreme-Weather-and-Climate-Change.pdf
Dalaklis, D. (2012, December). Somali piracy: Some good news, but a lot more needs to be done. *Maritime Security Review* (9).
Dalaklis, D. (2015). *Maritime Security Energy Issues: The role of the Suez Canal*. Presentation in the 6th NMIOTC Conference: "Current and future challenges to Energy Security in the Maritime Environment", 2nd – 4th of June, Chania, Greece.
Dalaklis, D., & Baxevani, P. (2015, June). Arctic in the global warming phenomenon era: New maritime routes & geopolitical tensions. In *International Symposium: New Maritime Routes: origins, evolution and prospectives*, Nantes, France.
Davenport, C. (2015, May 11). *White house gives conditional approval for shell to drill in Arctic*. Retrieved May 29, 2015, from http://www.cnbc.com/id/102668322
Eger, K. M. (2011, August 15). Marine Traffic in the Arctic. A Report Commissioned by the Norwegian Mapping Authority. *Analyse and Strategy*.
Funkdec, M. (2014, December 30). The Wreck of the Kulluk. *New York Times*. Retrieved January 3, 2015, from http://www.nytimes.com/2015/01/04/magazine/the-wreck-of-the-kulluk.html?emc=edit_tnt_20141230&nlid=64813105&tntemail0=y&_r=0
Hulme, M., Mitchell, J., Ingram, W., Lowe, J., Johns, T., New, M., et al. (1999). Climate change scenarios for global impacts studies. *Global Environmental Change, 9*, S3–S19.
Humpert, M. (2014, October 31). *Arctic shipping: An analysis of the 2013 Northern Sea Route Season*. The Arctic Institute Center for Circumpolar Security Studies. Retrieved December 10, 2014, from http://www.thearcticinstitute.org/2014/10/NSR-Shipping-Report.html
Intergovernmental Panel on Climate Change. Retrieved June 2, 2015, from http://www.ipcc.ch/
International Association of Classification Societies. (2011). "Requirements concerning Polar Class" IACS, Req. Retrieved May 12, 2015, from http://www.iacs.org.uk/document/public/Publications/Unified_requirements/PDF/UR_I_pdf410.pdf
International Maritime Organization. (2009.) http://www.imo.org
Keil, K., & Raspotnik, A. (2014, October 22). *Commercial Arctic shipping through the Northeast Passage: Routes, Resources, Governance, Technology, and Infrastructure*. Retrieved March 3, 2015, from http://www.thearcticinstitute.org/2014/10/102214-Northeast-Passage-Commercial-Shipping.html
Loskutova, O. (2006). The Northern Sea Route expects active maritime traffic. *Maritime Market, 2* (16). Retrieved March 3, 2013, from http://www.maritimemarket.ru/article.phtml?id=723&lang=en

McGwin, K. (2013, October 13). *The Arctic Ocean is an industrial Mediterranean Sea for the future*. Retrieved June 11, 2014, from http://arcticjournal.com/business/shippers-plenty-potential-arctic-sea-route

Merk, O. (2015, February 3). *Industry viewpoint: the six challenges facing the next head of the IMO*. Retrieved April 18, 2015, from http://www.lloydslist.com/ll/sector/regulation/article456408.ece

MSC. (2002). *MEPC circular on guidelines for ships operating in Arctic ice covered waters*. In Annex 10 of the forty-fifth session of the Sub-Committee (DE 45/27)-reference Chapter 11 of the Circular. Draft Guidelines submitted to MSC76.

National Research Council. (2005). *Polar icebreaker roles and U.S. future needs: A preliminary assessment*. Washington, DC: The National Academies Press.

National Snow and Ice Data Center. (2014). *Arctic sea ice news*. Retrieved June 23, 2014, from nsidc.org/arcticseaicenews/

Nilsen, T. (2015, June 8). North Pole nuclear cruises season started. *Barents Observer*. Retrieved June 9, 2015, from http://barentsobserver.com/en/arctic/2015/06/north-pole-nuclear-cruises-season-started-08-06

NOAA Sea Ice Community Report. (2014). Issue 1.

Northern Sea Route Information Office, Transit data. Retrieved June 6, 2015, from http://www.arctic-lio.com/nsr_transits

Offshore Energy Today.com. (2015, March 12). *Viking scoops 'Spotlight on Arctic Technology' award*. Retrieved May 19, 2015, from http://www.offshoreenergytoday.com/viking-scoops-spotlight-on-arctic-technology-award/

Overland, J. E., & Wang, M. (2013). When will the summer Arctic be nearly sea iceee? *Geophysical Research Letters, 40*, 2097–2101, https://doi.org/10.1002/grl.50316

Pettersen, T. (2011, November 18). Russia to have ten Arctic rescue centers by 2015. *Barents Observer*. Retrieved August 8, 2014, from http://barentsobserver.com/en/topics/russia-have-ten-arctic-rescue-centers-2015

Ragner, C. L. (2000). *Northern Sea Route Cargo Flows and Infrastructure –Present State and Future Potential* (The Fridtjof Nansen Institute FNI Report 13/2000).

Shettar, G. (2015, June 4). *Polar Code "guideline" presents challenges*. Retrieved June 12, 2015, from http://www.ihsmaritime360.com/article/18173/polar-code-guideline-presents-challenges

Smith, L. C., & Stephenson, S. R. (2013). New Trans-Arctic shipping routes navigable by mid-century. *PNAS, 110*(13), E1191–E1195.

Snider, D. (2014, December 16). Review: Polar Shipping. *The Maritime Executive*. Retrieved December 17, 2014, from http://www.maritime-executive.com/features/2014-in-Review-Polar-Shipping

Suez Canal Authority, Brief Yearly Statistics. Retrieved June 6, 2015, from http://www.suezcanal.gov.eg/TRstat.aspx?reportId=4

Visvikis, I., & Dalaklis, D. (2014, September). Managers in Today's Competitive Maritime Industry: Staying Ahead of the Curve. *NAFS Magazine-World Shipping News, 100*.

Zhang, R. Polar Code and STCW Convention on Seafarers Training.

Part VI
Sustainable Arctic Business Development

The Offshore Oil and Gas Operations in Ice Infested Water: Resource Configuration and Operational Process Management

Odd Jarl Borch and Norvald Kjerstad

Contents

Abstract In this chapter, we emphasize the fleet configuration challenges of Arctic offshore oil and gas exploration. We highlight the role of offshore service vessels in achieving effective and safe oil and gas exploration activity in Arctic waters. We elaborate on the fleet resource configuration and operational management challenges. Data from case studies of operations in two High Arctic regions, the Disco Bay, Western Greenland and the Kara Sea in northwest Russia are revealed. The results show that the context of ice-infested waters, lack of infrastructure and risk related to weather and cold climate demands a more in-depth planning process including more companies and institutions, a more complex resource configuration with multi-functional vessels, and advanced Polar water competence as to logistics, managerial capacities, ice management and emergency preparedness. Implications for the industry and for further research are discussed.

O. J. Borch (✉)
Nord University, Bodø, Norway
e-mail: Odd.Jarl.Borch@uin.no; odd.j.borch@nord.no

N. Kjerstad
NTNU, Aalesund, Norway
e-mail: norvald.kjerstad@ntnu.no

L. P. Hildebrand et al. (eds.), *Sustainable Shipping in a Changing Arctic*, WMU Studies in Maritime Affairs 7, https://doi.org/10.1007/978-3-319-78425-0_22

Keywords Offshore service · Shipping · Oil and gas · Icy waters · Resource configuration · Operational management

1 Introduction

Strategic resource configuration in dynamic environments challenges routines and the decision-making process of companies (Borch et al. 2009; Schilke 2014). Arctic oil and gas related shipping has been regarded as extra challenging due to less predictability and higher complexity than offshore operations in more Southern regions as the North Sea (Borch and Batalden 2014). Oil companies have targeted offshore oil and gas resources in the High North including Alaska, Newfoundland, the Baffin Bay area, the Barents Sea and Arctic Russia. Limited experience and research efforts from both the government and the industry side cause uncertainty about the vessel configuration, the equipment and the managerial resources and capabilities in demand. For oil and gas operations, the stakes are high. An example is the reaction the oil major Shell experienced in Alaska after the rig Kulluk grounded in 2013 due to lack of towing vessel capability. There is a need to discuss operational concepts including rules and regulations beyond standard operating procedures for the sea areas in question (Borch and Batalden 2014). We are in need of increased knowledge on how operational challenges should be met with adequate vessel resources as well as operational management concepts.

In this chapter, we build upon the resource-based theory of the firm to emphasize the configuration of different types of resources and capabilities for safe operations in Arctic waters. The complexity of operations represents a cost driver. The companies operating in this region must consider a broad range of international, regional, and industry specific regulations, extra precautions to safeguard both people, environment and the local societies, and deal with a broad range of society stakeholders (Buixadé Farré et al. 2014). A nature that is both vulnerable and not easily predictable means increased preparedness emphasis. To be prepared for the unknown, the vessels may have a broader range of functions and advanced equipment. This enhanced technology level calls for added competence resources not the least among operational managers. This includes the operational level competence of operating vessels and equipment, as well the organizational capability of pooling efforts from several vessel units and companies into a smooth running operation (Jenssen and Randøy 2006).

In the first part of this chapter, we present the resource-based theory of the firm as an analytical platform for the study of maritime operations in complex and volatile environments. In the second part, we present the methodology for studying composite maritime operations emphasizing the importance of in-depth, longitudinal data collection and a broad range of data sources.

In part three, we present and analyze data from two field studies in Arctic environments, and conclude on the fleet composition and managerial concepts that may prove effective in icy waters.

2 Theory

2.1 *Operational Process Management*

Operational process management is about integrating resources and linking the basic activities within a value chain, to work as a whole towards highest possible value (Jeston and Nelis 2006). An operational process is an activity that includes physical and competence resources, patterns of action and dependency links to other activities. It also includes working patterns and written rules for making the resources work according to the objectives of the process. In complex environments, there will be a number of reciprocal links between different organizational parts and the environment (Thompson 1967). This calls for additional coordination and control within the organization. In a volatile or turbulent environment where the cause-effect relations are not well explored, the effects of others' and own actions may be not easily predicted (Eisenhardt and Martin 2000) and will require coordination in close interaction with top management and the other parts of the value chain (Trkman and McCormack 2009). In a maritime context, this calls for extra physical capacity to safeguard the operation and extra managerial efforts at the operational level due to distance from top management. The challenge of the operational process management is to tackle the increased dependency of other companies and at the same time make their own processes run smooth. This is in particular the case in operations where high technology resources are bundled into more complex resource configurations such as offshore oil and gas operations. The operations include managerial tasks related to logistics management coordinating the stream of goods, personnel and floating units both in—and outbound. There will be a need for safety management such as ice surveillance and defense to safeguard the drilling units, and emergency preparedness management to reduce risk and prepare for incidents threating life, environment and society. These mostly inter-business processes have to be sufficiently governed to achieve an effective interplay between the actors in the value chain.

2.2 *The Resource Configuration*

To achieve high performance, an operation has to be supported by appropriate resources and distinct competencies (Barney 2002). The resource-based view of the firm emphasizes the significance of an organization's unique or distinctive resources, and how they are pooled together to achieve superior performance (Amit and Schoemaker 1993; Barney 1991; Black and Boal 1994).

However, resources may not be easily accessible or it may take time to develop them within an organization. Special challenges are present if requiring resources that are rare, costly to buy or copy, and less mobile (Barney 1991, 2002). The lack of available technology and managerial competences means that organizations have to build resources on their own, or enter into alliances with other firms that may have some of this capability. Black and Boal (1994) claim that building a complex network of firm resources, competences, and relationships are critical elements in creating advantage for the firm and its customers.

At the operational level, the business process managers' task will be to decide upon the resources needed and how they should be bundled together. In this chapter, a distinction is made between basic resources as building blocks for more advanced structures and capabilities where different resources are mixed together into action preparedness. Basic resources refer to resources such as skilled workers, machines and financial capital. Capabilities are combinations of physical resources and individual competence bundled in the organization to provide a special output (Amit and Schoemaker 1993). Some capabilities may be history-dependent and tacit knowledge that build upon single person experience that may take significant practice to acquire. A third type of resources is the dynamic capability emphasizing the firms's ability to adopt or innovate, i.e. capabilities to develop new resources, to reconfigure new and existing resources and remove abundant resources to improve efficiency (Teece et al. 1997; Borch and Madsen 2007). Firms in some industries and not the least the mature shipping industry have to be careful about new resource investments as margins are low. At the same time, there is a need for redundant resources to meet unforeseeable situations and increase safety. If not the vessel may risk being "off hire" and stuck in the Arctic with significant income losses and extra costs. Thus, the firm has to be innovative in combining existing resources and increasing flexibility. The dynamic capability is crucial in stimulating the organization towards innovative action developing combinations of positioning tools not thought of among their competitors. By identifying, utilizing, and recombining its valuable assets, the organization should be able to meet unforeseen challenges, increase operational effectiveness and keep costs down. In complex and volatile environments, business process management has to be efficient on the specified tasks of an operation in coordinated action with other firms. At the same time, the management has to be agile to react proactively on changes in the external environment. Thus, the manager has to be efficient in both exploiting present resources, and in exploring new ones to meet new challenges, all performed in a different context of contract relations (Black and Boal 1994; March 1991; Borch 1994). Thus, operational efficiency is dependent on the configuration of vessel resources, human capabilities and managerial core competence to prosper in the volatile and highly complex environment of the High Arctic.

3 Methodology

3.1 Research Strategy

The focus on process studies in the High Arctic context with limited knowledge available calls for an in-depth longitudinal research strategy (Borch and Arthur 1995). We build upon data from two oil and gas offshore exploration cases. In depth, observation studies were conducted within companies in the offshore service

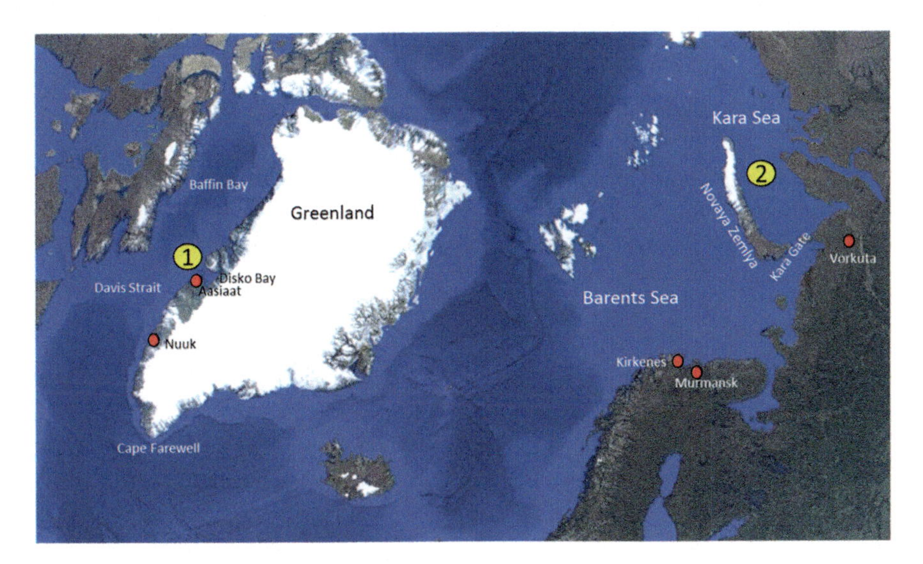

Fig. 1 Chart illustrating the two areas described in this chapter. West-Greenland (1) and the Kara Sea (2)

vessel (OSV) industry participating in large-scale offshore oil exploration expeditions. The empirical studies took place in two different environmental settings (1) the Disco Bay, West-Greenland and (2) the Kara Sea in Northwest Russia (Fig. 1).

3.2 Data Collection

For collection of primary data we used observation techniques and in-depth interviews onboard the vessels and within the shipping administration. The authors observed the two operations as part of the crew. We also performed interviews with the top managers of the owner companies, and with operational personnel within the oil and gas companies. Secondary studies of documents, reports, and minutes from meetings were combined with in-depth interviews and served to increase the validity of data through data-triangulation. The secondary data included vessel-shore communication and studies of the documentation distributed between the different parts of the organization. It also included company market reports, audit reports, laws, regulations, and descriptions of different types of vessels operating offshore.

Third, secondary information was collected from other operations in the Arctic. This included annual reports, company newsletters, investor reports and features in industry magazines.

3.3 Data Analysis

Observation and interview data were categorized according to the main variables in the research model. Data from the secondary sources was triangulated with the data from the interviews and observations in the field. We compared the two environmental settings along the dimensions shown in the research model. Patterns of both resource configuration and process management and how they related to changes in the environment were explored.

3.3.1 Data

3.3.1.1 Case 1 West-Greenland

Context The operation took place at the west coast of Greenland out of the town Aasiaat in Disco Bay. The area is known as the "iceberg alley" as a large number of icebergs break out of the mainland glaciers and travels North in the Baffin Bay West Greenland current.

Licenses The Disco Bay area was expected to hold significant amounts of oil and gas resources. US Geological Survey estimated in 2011 the amount of 50 bboe offshore Greenland, with 17 bboe in Baffin Bay, 7.3 bb of this was expected to be oil. In this area, oil is of highest interest because of the high development costs and better prices compared with gas. The seismic surveys for this area was performed in the summer 2009. The operator calculated a 7–14% chance of finding resources rich enough for exploitation and expected to spend approx. 400 million USD on 4 wells.

Government The National government Naalakkersuisut (landsstyret) of Greenland decides upon licenses after recommendation from the Minister of Industry and Natural resources. The Directorate of Raw Materials follows up license applications and operations. The Bureau of Minerals and Petroleum is responsible for the management of mineral resource activities. With incomes from field owner shares and government taxes on production income, Greenland expected to receive 60% of the net surplus from the production. The government demanded approx. 10 billion USD in guarantees for oil spill recovery. Before the operation started, the Minister of industry declared that the highest safety measures were taken according to best Arctic Standards: "The Norwegian North Sea standard is to be met, increasing the safety three times compared to the Mexican Gulf". As for public opinion, the majority of population was in favor of exploration. However, there were several organizations working against offshore oil and gas activity, such as the Inuit Circumpolar Council, World Wildlife Fund (WWF), Greenland environmentalist group AvataQ, and Greenpeace.

Nature The weather conditions in the region are challenging in both winter and summer:

- Mean air temperatures are below 10° C all year round. The coldest month is February and the warmest month in the coastal area is July.
- Fog or polar lows are common features near the South West and South Greenland shores.
- Frequency of fog increases during May and peaks in June/July, when the temperature contrast between the cool sea surface and the relatively warm atmosphere is at a maximum. It fades out in late August.
- Sea temperature is 5° C.

As for ice, the Baffin Bay basin with depth greater than 2000 m has ice cover 6–9 month of the year. Slush ice emerges in October out of Disco Bay. Sea ice normally covers most of the Davis Strait north of 65° N, except areas close to the Greenland coast, where a flaw lead (open water or thin ice) of varying width often appears between the more solid coastal ice and the drift ice offshore as far north as latitude 67° N. Normally, the warm northwesterly West Greenland current keeps the coastline waters ice free between 58° and 67° N during wintertime.

In the summer, icebergs from Disco Bay and Melville Bay drift in northwest direction. The region produces 10–15,000 icebergs per year. The "iceberg alley" out of Disco Bay creates medium-sized icebergs heading north and west with a speed up to 3 knots in storms. These icebergs have a length up to 200 m with the largest depth 250–300 m. An iceberg with a length of 50 m has a depth of approximately 40 m. Floating "growlers" and bergy bits comes from glacial ice out of Disco Bay and Isfjorden at Ilullissat.

Winter pack ice appears in small first year ice floes (20–100 m diameter) and is 70–150 cm thick. Icebergs and growlers originating from glaciers occur in the entire region, but the density of icebergs is normally low, increasing towards the Cape Farewell area to the south.

Fog There is much fog along the coast and out in the sea, with very low visibility. The frequency of fog in July is 20–30% of the total time over the coldest parts of the sea area.

Visibility is less than 0.5 nm 20–30% of the time in June and July, when fog is a significant problem. For helicopters, the fog makes it difficult to take off and land for a large part of the day, and provoke the need for intense observation and reduced speed of vessels.

Charts and Instruments The maps in the area are imprecise. Maps are not Electronic Navigational Charts (ENC) approved for electronic navigation. There is extra uncertainty in harbors as to depths. Further, up north there are problems with stable differential GPS (DGPS)-signals, where the correction signals for the DGPS signals are not strong enough. This is a challenge to the dynamic positioning system of the vessels. The closest Fugro correction signals are more than 2000 nm away with quality not guaranteed. It is difficult to notice sea ice on ordinary radars, especially in rough sea. Satellite communication (phone, internet) fallout happens quite often.

Infrastructure There are long distances between harbors and a lack of infrastructure for support. There was a limited oil recovery capacity locally, as well as SAR capacity. The closest harbor of some size is in Nuuk 2 days sail away from the operation site. Nuuk also has the nearest hospital. A small harbor and a short-range airport are situated in the small town of Aasiaat. Sailing time from the operation site to the Aasiaat harbor was approx. 15 h for the vessels involved.

Resource Configuration

The activity took place with the following units:

Drilling Units Two drilling units were employed among others to have a back up if something went wrong on one site. One 6th generation drilling ship, Stena Forth, and the 5th generation semi-submersible drilling rig Stena Don drilled two wells each. Drilling commenced June 30. None of these units had ice class.

ERRV-Vessels Two emergency response and rescue vessels Esvagt Don and Esvagt Connector were hired for stand by duties and preparedness.

Anchor Handling, Towing and Ice Management For anchor handling and ice management, ice class vessels were employed. Balder Viking and Vidar Viking were Icebreaker Ice 10/Polar 4 from Viking Supply Ships were hired for ice management, anchor handling and towing. They also performed supply duties. Loke Viking AHTV was no icebreaker and had Ice class 1A (operate in Summer/autumn ice up to 80 cm). Loke Viking had additional tasks within emergency response and oil recovery preparedness duties. The AHTS Fennica Ice 10/Polar 4 Icebreaker (thick 1st year ice) was also employed for a period.

Depot Ship The multi-purpose platform supply vessel Troms Vision was located outside Aasiaat harbor as a depot and hotel ship, with doctor and divers onboard. Two divers with specialized diving boat were included. It also included a 100t heavy lift crane and helicopter platform. A cabin cruiser was used for land transport.

Platform Supply Vessels Three platform supply vessels were involved in supply duties:

- Troms Pollux and Troms Artemis Ice 1C (Summer/Autumn 40 cm)
- Olympic Poseidon Ice 1C (Summer/Autumn 40 cm)

These vessels travelled between the drilling units and the bases at Aasiaat, Nuuk and Peterhead (Table 1).

Air Capacity The operator had three transport helicopters with a base in Aasiaat for transport to and from rigs. They also contributed to SAR operations.

Emergency Preparedness Resources The Danish coast guard was responsible for emergency preparedness at sea together with smaller police cutters. Two larger helicopter carrying coast guard vessels Knud Rasmussen (67 m) and its sister ship Einar Mikkelsen were available. Two national maritime radio stations were

Table 1 Vessels and roles in the West-Greenland operation

Vessel	Ice Class in DNV-GL (or equiv.)	Anchor handling	Ice Def. & recon.	Ice def. & close sup.	Rig Towing	FiFi & St. by	Oil spill response	Hospital	Crew change	Storage crit. spare	Hotel	Cargo	Drilling
Stena Forth	None												P
Stena Don	None												P
Esvagt Connector	Ice-1D	S				P	P	S	A			S	
Esvagt Don	–	S				P	P	S	A			S	
Balder Viking	Ice-10	P	P	A	S	A			A			S	
Loke Viking	Ice-1A	P	P	S	S	S			A			S	
Troms Pollux	Ice-1C					S	S		A	S		P	
Troms Artemis	Ice-1C					S	S		A	S		P	
Olympic Poseidon	Ice-1C					S	S		A	S		P	
Troms Vision	None							S	S	P	P	S	
Fennica	Ice-10												

P Primary function, *S* Secondary function, *A* Possible Additional duties communication centers- one was located in Aasiaat. Island Commander Greenland in cooperation with head of police in Nuuk and the local police stations coordinated the rescue operations. Greenland had no stand by helicopter as a part of its preparedness system. The government had to use passenger helicopters from Air Greenland. Smaller planes and air force Hercules-transport planes were available from Danish air forces base in Søndre Strømfjord. A special device tracked the vessels involved in the operation. In addition, they had to report position to Island Commander Greenland every 6 h. The vessels were equipped with ice searchlights to be used in dark periods.

Anti-terror resources were available from the Danish Navy Arctic Command to protect the rigs from destructive action. Among others navy vessels were deployed to follow the Greenpeace vessel Esperanza closing up on the rig. Seventeen environmentalists, however, managed to reach the rig by rib boats and entering the rig causing a stop in production.

Operational Process Management

The total operation was handled at the strategic level from the operator's headquarter (HQ) in Scotland. The domestic issues including contact with the government were taken care of by a site manager at Nuuk. The operational processes of the drilling expedition could be divided into three different sub-processes that were partly interlinked. The vessels involved in ice management were utilized in towing the drilling units and in cargo-run within the logistics process. They were also prepared for SAR and oil spill response emergencies. Some of the platform supply vessels were multifunctional and could work as stand by ERRV vessel, to support in search and rescue and oil spill recovery operations.

The Logistics Process The drilling took place with two units serving as back up for each other. The main base was in Scotland for larger supplies. There were local bases in the harbor in Nuuk and Aasiaat, plus at the MPSV Troms Vision. The distance from the Aasiaat base was 162 nm with a normal transit time of 15 h at 11–12 knots of economic speed, which may increase significantly in case of fog and ice where the speed would be reduced to 5–6 knots.

The platform supply vessels (PSV) were constantly on the move between the bases and the drilling sites in a given pattern, with many delays caused by inefficient base handling and reduced visibility. The logistics processes were coordinated in meetings among the rig's offshore installation manager (OIM), the drilling coordinator at the rig and the store master, and linked up to the logistics coordinators at the bases in Aasiaat, Nuuk and Peterhead. The logistics supervisor and the drilling supervisor in the operator's headquarter were governing the whole process, emphasizing the transport to and from the drilling sites, and the drilling process and needs, respectively. Due to weather and unforeseen situations there were frequent changes of plans and delays in transport, resulting in change of orders and many stand by hours waiting for new orders from the logistics coordinator.

The Ice Handling Process The area outside Disco Bay included a stream of icebergs moving north. To prepare for this situation anchor handling and towing vessels both with and without icebreaker capacity were chartered. There was also an ice management plan developed for the whole operation. Ice maps were bought from an external supplier and distributed to each ship frequently. The ice maps were quite accurate for the larger ice, but not for icy bits. At the platform there was an ice surveyor giving directions to the vessels doing ice management.

The ice management was based on actions in different zones by the AHTV reducing threats from icebergs that might be on collision course with the rigs. The following danger zones were used:

- Zone 1: 500 m or 1 h drift—Reaction: Ice alert Quick departure of rig
- Zone 2: 9 nm. T-zone Reaction—orderly departure of drilling well
- Zone 3: Outside 9 nm. Ice monitoring tracking, and management of the icebergs that created a threat to enter zone 2 towards a different course

The ice management was based on ringing in the icebergs that might enter the threat zones with ropes and nets and towing them into a different course. There was also an opportunity for using prop washing and water cannon on the smaller ice features.

Emergency Preparedness Process There were two vessels serving as stand by guard and emergency preparedness vessels. The operator's Emergency Response Group in Scotland was responsible for coordinating emergency operations together with the Offshore Installation Manager (OIM). The Emergency Coordination Center in operator HQ had continuous guards 24/7.

- Response levels as to oil spill recovery:
  - Stand by ships with first line oil recovery facilities
  - Depot at Aasiaat
  - Hercules with oil recovery equipment and dispergents out of South England

For SAR-operations, the helicopters at Aasiaat were available. There were also the local emergency services, particularly the fire service provide assistance in cases of fire, injury to persons and property as well as the environment from accidents and disasters. They assist in rescue missions and the search for missing persons. The police are responsible for directing the search and rescue operations in local marine waters and on land, pollution control outside the three-mile and assistance to other operators. The Danish Navy Arctic Command is responsible for the management of maritime rescue services—search and rescue of distressed ships at or below sea level, whether through the operation carried out at sea, by air or by land. Naviair is designated to provide infrastructure for aviation. They are responsible for providing Briefing Service from Flight Information Services as well as Flight information service from the Air Intelligence Unit. Technical operation and maintenance of CNS equipment in Greenland with regard to navigation and communication facilities in Greenland, Surveillance (ADS-B) in Greenland.

3.3.1.2 Case 2 The Kara Sea Expedition, Northern Russia

Context The Kara Sea operation took place in the Kara Sea east of Novaya Zemlya in Russia, in an area with winter ice and smaller icebergs in the summer time.

Licenses The drilling operation in the Kara Sea took place at the field called University 1 (later re-named Victory). Operators were a joint venture called Karmorneftegaz (KMNG) owned by Rosneft (51%) and ExxonMobil (49%). The field is located on the Russian continental shelf Northwest in the Kara Sea, approximately at N 74°30′ and E 064°, 50 nm from the coast of Novaya Zemlya.

Nature The area is covered by drift ice most of the year, and only 3–4 months from August to October can be expected to be ice-free. Most prominent is the first-year sea ice, but dependent on the severity of the previous winter, an influx of tougher multi-year ice, as well as tick river ice can be expected. In addition to the drift ice, a significant number of relatively small icebergs are mixed into the ice or open water along the northeastern coast of Novaya Zemlya. The nature and variability of the ice in the Kara Sea is one of the most demanding problems seen from an operational and logistic viewpoint. After a mild winter, it is possible to meet the decided start criteria for the operation in early July and work until mid-November (4 months). In other years, like in 1997 and 1998, the ice condition never met the start criteria, and consequently an operation like the one in 2014 could never have started. For obvious reasons, this variability has a major impact on when to start the operation, and what type of equipment to mobilize.

Infrastructure In addition to the ice, the remoteness and special regulatory status makes the Kara Sea a challenging area for offshore operations. Unavailability of civil airports within range, as well as military regulations, makes helicopter operations impossible for the time being. Lacking regional port facilities forces all supply services and crew changes to operate out of new offshore base in Murmansk, which is the closest alternative, 850 nm from the drill site. This results in a minimum 8 days roundtrip time for the service vessels. An alternative port could be the Norwegian town Kirkenes, where visa-free travel and easier custom procedures could contribute to logistics that are more effective. The sailing distance to Kirkenes is approximately equal to Murmansk, but due to the Russian demand for "local content", this Norwegian alternative has not been utilized so fare.

Government The Kara Sea has a special regulatory status. The entire northern Russian coast between Novaya Zemlya and the Bering Strait is subject to a special maritime regulatory regime under the Northern Sea Route Administration in Moscow. Further, most of Novaya Zemlya is highly restricted due to military regulations. Even potential use as a place of refuge for ships is prohibited—even if this could be disputed according to international laws of the sea. During the 2014 season, some of the service vessels experienced conflict with naval vessels, forcing them to change course and route between the drill site and Murmansk. This served as a limiting factor for preparedness planning and was considered as both a political risk and a safety issue. The 2014 expedition experienced this political risk as the showstopper, since the companies were forced out of the Russian Arctic after US and EU sanctions against the Russian role in the Ukraine conflict.

Resource Configuration
The operation in the Kara Sea in 2014 was the first drilling operation for KMNG after the agreement was signed in 2011. The semisubmersible drilling rig West Alpha, from North Atlantic Drilling was chartered for the operation, while C-Logistics should provide logistic support and base function in Murmansk. Viking Supply Ships (VSS) were given a major contract for the entire ice defense system, including ice and metocean data coordination. Due to the new area of operation and high safety margins, a massive fleet of vessels was mobilized for the operation. A total of 13 vessels (+ rig) took part on a regular basis (Table 2).

Rig Details West Alpha: This is a conventional 17193 GT semisub with Panama flag. The class notation in DNV-GL is: +1A1 Column Stabilized Unit, E0, HELDK, F-A, CRANE, POSMOOR ATA, Drill-(N). The rig have no ice class, and the criteria of operation was "no ice impact at all". In spite of this, some minor precautions were made during mobilization. Wire nets between the legs for ice protection was one of the measures. Special personnel and equipment for quick release of anchor lines were taken onboard.

In addition to the vessels involved, an ice reconnaissance airplane was hired to fly on a daily basis (or as often as needed) over the drill site and potential ice infested waters nearby. The airplane, which normally flew from Vorkuta airport (780km away), was equipped with Synthetic Aperture Radar (SAR) for ice detection under all weather conditions. The airplane was a fixed-wing Antonov 26 operated in a cooperation between Blom and Arctic and Antarctic Research Institute (AARI).

Helicopters could have been very helpful for ice reconnaissance and personnel transfer, but were excluded based on a safety evaluation (and Russian legal issues). In spite of this a drone was tested from the Russian icebreaker Kapitan Khlebnikov for reconnaissance and ice drift measurements.

Before the rig was towed to the drill site the nuclear icebreaker, Yamal had been on an ice and oceanographic research mission for Rosneft in the area.

Operational Process Management

The resource management at top level included a joint venture between the partnering oil companies. The joint venture was accumulating and bringing together all the resources needed for the operation.

At the operational level there were three different centers coordinating the operation:

(a) Shore operation center
(b) Field operation center
(c) Shore logistics base

Table 2 Vessels and roles in the Kara Sea operation

Vessel	Ice Class in DNV-GL (or equiv.)	Anchor handling	Ice Def. & recon.	Ice def. & close sup.	Rig towing	FiFi & St. by	Oil spill response	Hospital	Crew change	Storage crit. spare	Hotel	ROV	Cargo	Drilling
West Alpha rig	None													P
Balder Viking	Ice-10	P	P	A	S	A	–	–	A	–	A	–	A	
Brage Viking	Ice-1A	P	P	S	S	S	–	–	A	–	S	–	A	
Loke Viking	Ice-1A	P	P	S	S	S	–	–	A	–	S	–	A	
Magne Viking	Ice-1A	P	P	S	S	S	–	–	A	–	S	–	A	
Siem Amethyst	Ice-1C	S	S	P	P	P	S	–	A	S	S	–	A	
Siem Topaz	Ice-1C	S	S	P	P	P	A	–	A	S	S	–	A	
Siem Pilot	Ice-1C	–	–	S	–	A	–	S	A	–	–	–	P	
Rem Supporter	Ice-1C	–	–	A	–	–	–	S	A	–	–	–	P	
Rem Server	Ice-1C	–	–	A	–	–	–	S	A	–	–	–	P	
Botnica	Ice-10	–	A	P	–	–	–	P	(1)	P	P	P	–	
Island Crown	Ice-1C	–	–	–	–	–	–	P	P	–	–	–	A	
Kapitan Khlebnikov	Polar-10 (LL-3)	–	P	–	–	–	–	–	–	–	–	–	–	
Spasatel Kavdeykin	Ice-1A* (Arc-5)	–	A	S	–	–	P	–	–	–	A	–	–	

P Primary function, *S* Secondary function, *A* Possible Additional duties. (1) = Botnica could be used for transfer of crew between rig and Island Crown
Ice 1A* = able to break first year ice with a thickness of 1.0 m

Frequent contact was necessary between these centers.

The Logistics Process From an operational point of view, there can be two alternatives modes for offshore drilling operations in waters with ice: (a) With a significantly ice-strengthened rig and icebreakers, an extended season with ice management could be considered. (b) More relevant in the exploration phase is to operate in the ice-free window where ice strengthening can be limited to the vessels occupied with ice defense on a safe distance from the rig. The operation that took place in 2014 operated in this mode.

In an early stage of the planning process, it was made quite clear from KMNG that there was zero tolerance for accidents and oil spill. Lowering of risk was therefore thoroughly introduced in all planning processes as well as the execution of the operation. Since the rig had no ice-strengthening, ice-free water along the towing route and around the drill site was an absolute requirement. The convoy was not allowed to enter the inlet straits to the Kara Sea before ice-free water in a radius of 40 nm from the drill site and no potential hazardous ice (PHI) 10 nm on each side of the planned towing route, could be documented by air and satellite reconnaissance. Towing speed in the Kara Sea was reduced to maximum 5kts (from 9kts in the Barents Sea). In addition, a minimum weather window was required for leaving the staging area and entering the Kara Sea. With these types of entrance criteria, it is obvious that much effort had to be done in terms of Metocean services and ice reconnaissance.

The Ice Handling Process Before starting an offshore operation in an area where none of the contract partners had experience required special precautions. Therefore planning the ice defense system had very high priority. Viking Supply ships, which had the main contract, had to sign subcontracts with several service providers and strengthen their organization with extra ice expertise. Several individual consultants were hired. Many of these were maritime professionals with long experience from different types of operation in ice-infested waters.

Storm Geo was chosen as partner for metocean services. In addition to providing special forecasts, the company also took part with experts located in the shore operation center (SOC) in Moscow as well as on the rig. Storm Geo was further teamed up with expertise at the Arctic and Antarctic Research Institute (AARI) in St. Petersburg. All vessels were equipped for regularly metocean reporting to SOC.

Kongsberg Satellite Service was chosen as the provider of satellite imagery. These services was closely linked to the metocean service at SOC. Ice experts from Viking and Storm Geo planned in advance from where to order updated radar-based imagery data from. Since no SAR-satellite have continuous coverage, data from three different satellites were used. Proper planning is important since the price is related to priority and how far in advance the booking takes place. Short notice is highly priced.

Egersund Group was chosen as the supplier for ice towing equipment. This included tow-ropes, as well as special designed ice-nets. Since the expected ice was anticipated to have a different character compared to that experienced in

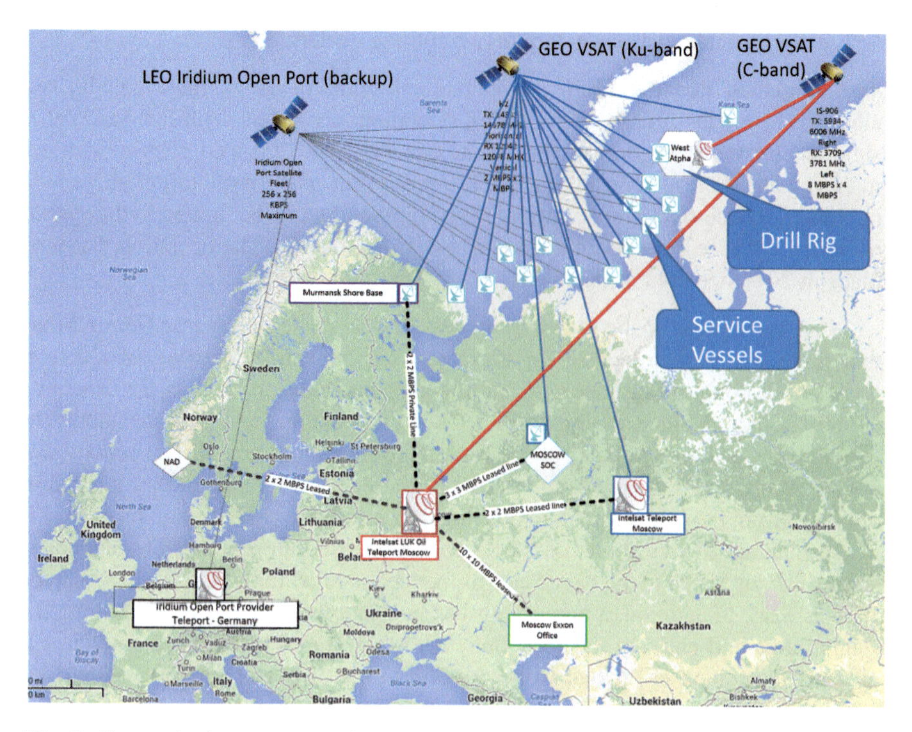

Fig. 2 Communication systems and organization integrated in the Common Operation Picture Display. Source: Marine Technologies

Greenland earlier, new and different nets were provided. The different nets and ropes were mobilized on the different AHTVs and tested before the expedition started (Table 2). In addition to the towing gear, most of the ice defense vessels had installed fire-fighting (fi-fi) equipment for potential ice handling close to the rig.

At an early planning stage, it was decided to design a special purpose information and communication system for the entire operation. This was called the Common Operation Picture Display (COPD). The system should be based on closed network communication and provide all units with metocean data, position and role of vessels, pictures, cargo documentation, etc. (Fig. 2). Marine Technologies (MT) was given the contract to develop the system, and install all the equipment onboard all units. The COPD should also serve as a closed telephone network for the operation. In the contract with MT there was an add-on made for the vessels involved in the ice defense. This was a dedicated ice-radar based on well-proven technology from Rutter. The rig and some vessels also had Infrared (IR) cameras installed for ice surveillance.

At an early stage, the planning team from VSS developed a comprehensive manual for the operation—the Ice Defense Operation Manual (IDOM). The manual was an important tool for how to organize and perform the operation as safe as possible. The IDOM described detailed procedures and responsibilities for all foreseen and unforeseen incidents related to the ice defense operation.

This central document was reviewed and updated, based on input from multi-disciplinary experts and experience gained through the training process of the personnel. Some of this training was based on simulator scenarios at the training center in Kalmar—a type of training that was found to be very helpful to team up key-personnel in ice defense and bridge cultural gaps. Since the drill site was located in a remote area at high latitude (N74°30′) serious challenges related to broadband satellite communication, as well as differential augmentation of GNSS systems. Redundant antenna systems were therefore installed to mitigate poor signal quality due to low-elevation satellites (Kjerstad 2006, 2015). At the end of season evaluation, it was concluded that both communication and DP-operation suffered from the limited GEO-satellite coverage, e.g. long downtime periods on the COPD. In the planning process, it was also identified to be a problem that the DP-operators on the Russian oil spill response vessel did not have the industry-required certificates for DP-operations.

Gathering Experience Data

In areas with large yearly variation of ice extent, such as the Kara Sea, one of the most difficult tasks is to predict the starting date for the operations. The planning group made therefore great efforts to study the ice variations back to 1972 in an attempt to investigate the ice-windows that met the "no-ice criteria". This study indicated 2 years with no ice-window at all, while start date varied from approximately 28th of June to 20th of September. Seen from a logistic and operational point of view, this is probably one of the most challenging and risky characteristics of the Kara Sea.

Based on different starting scenarios the movement and allocation of vessels were simulated from day to day throughout the operation. The experts in the planning group got feedback on the simulation from different users and put together dedicated training based on expected ice situation and available resources. The feedback from the training and simulator scenarios became valuable input to the final versions of the operation manuals.

The conclusion of the study was to have the rig ready at the Kara Gate on 1 August, but be prepared to anchor in a safe staging area close by if there was still ice along the towing path or in the drilling area. At the same time or in advance, ice strengthening AH-vessels should do reconnaissance and pre-lay the anchors, to reduce time for AH when the rig could be towed to the drill site. This proved to be a relatively good decision, since the rig could single up the anchor lines after a few days in the staging area and sail in a well-planned convoy configuration into the Kara Sea on 3 August. The rig was finally anchored, and the first drilling in the Kara Sea was ready to start on 6 August 2014.

Planning of the stop date was not as difficult as the start day, since the dynamics of the sea-ice freeze-up is more predictable than the melting. A potential extension of the season will require different approaches. If the expedition should start drilling earlier (June–July), it will require far more ice-strengthening and high ice classes on all involved units. In the end of the season (November–December) it will be more a matter of winterization to handle the low temperatures and darkness.

During the drilling operation in August and September, the geopolitics situation tensioned due to the Ukraine crisis. The EU and the US did not accept Russia's armed involvement in Ukraine and introduced a spectrum of sanctions. Among those was a ban on companies involved in Arctic or deep-water oil explorations in Russia. ExxonMobil was given a final date in the end of September to be out of Russian waters. Consequently, the rig, together with the 13 support vessels, had to close down the operation as fast as possible and move out of the Kara Sea. On 5 October, when the ice edge still was approximately 400 nm away from the drill site, the well was safely shut down and the rig was towed towards Norwegian waters and a very difficult market situation. A few days before, the director of Rosneft, Igor Sechin, could announce in public on Russian TV that the operation had been successful and more than 100 mill tons of oil and more than 300 billion cubic meter (bcm) of gas had been discovered. He also announced that the name of the field should be renamed to Pobeda (Victory). Since than, figures like 950 mill barrels of oil and 500bcm of gas have been published.

The ice defense organization was divided in three different segments (Fig. 3):

- Shore Operation Center (SOC) where ice advisors, ice analysts and meteorologists worked 24/7 with preparation of satellite imagery, aerial ice reconnaissance and weather forecasts. Together with the lead ice advisor on the rig, SOC proposed ice patrol strategy for the ice defense vessels. SOC had also a special editorial role with metocean data distributed on the COPD. The ice director headed SOC.
- Infield Operations Command Centre (IOCC) on the rig had the coordination function in the field and planned how ice defense should be executed in the different zones around the rig (Fig. 4). Lead ice advisors and meteorologists worked 24/7 to coordinate search patterns for potential hazardous ice (PHI) for each vessel in the ice defense. IOCC reported to SOC. During the tow OCC was strengthen with special trained radar observers for ice surveillance. The ice advisors took part in the daily meetings between marine- and drilling management on the rig, and were updated on different T-times during the different drilling stages.
- Each of the seven vessels dedicated for ice defense had an ice advisor on board. Based on discussion with the IOCC the ice advisors planned and reported on ice search and defense. Tracking the drift of PHI was an important activity—especially in the beginning of the expedition when there was limited knowledge of the ocean currents. The ice advisor operated the vessel's COPD and ice radar and had a special responsibility for metocean reports. On vessels where the crew had limited Arctic ice experience the ice advisors had a special educational role.

The entire ice defense organization had daily meetings via the communication network and the COPD. Most communication with voice was done by the integrated Iridium telephone system. Since the ice defense vessels often were out of range of

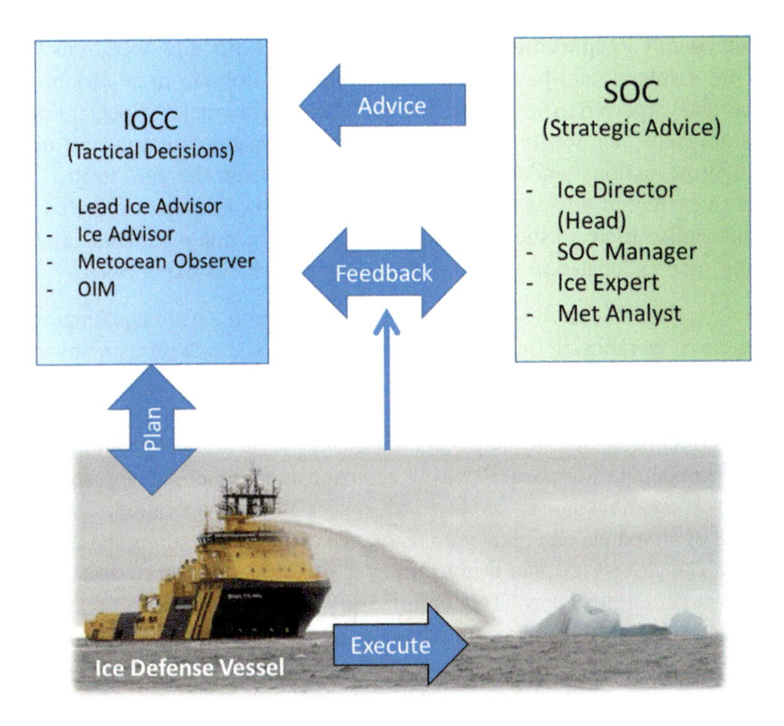

Fig. 3 The ice defense organization with roles and communication paths

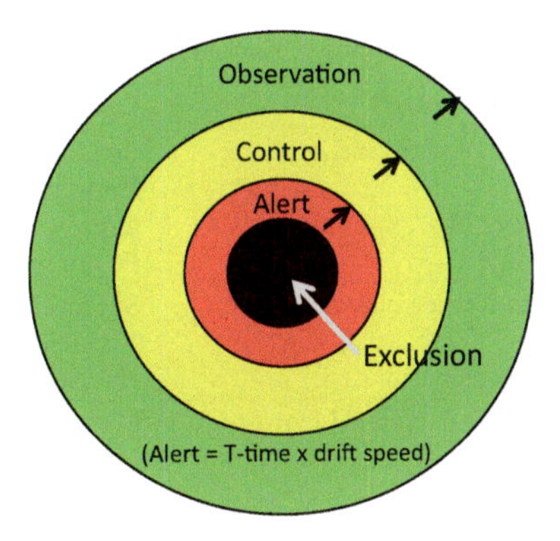

Fig. 4 Different levels of ice defense zones around the rig

VHF, the MF-radio was in some cases used between the different ships and IOCC. The ice defense group worked on a 5 weeks on and 5 weeks off schedule, like most of the vessel crews. Crew changes took place in Murmansk, and most of the homeward journeys were done by bus from Murmansk to Kirkenes (Norway) and flights from there.

The Emergency Preparedness Process Drilling and offshore operations in Arctic waters are disputed, and have often faced demonstrations from environmentalists. This was also expected to happen for the Kara Sea operation, and Special Forces and negotiators were therefor hired. The priority in this matter was to avoid unwanted and dangerous actions while the rig was under tow from the yard to the drill site. Different coastguard vessels from Norway and Russia therefore followed the tow along the entire route to the Kara Gate. Except for a minor Greenpeace-incident while the rig was located in the yard, no confrontation was experienced.

The very strict military requirements at Novaya Zemlya were challenging for the planning and operation in the Kara Sea. Even if this island, with its many sheltered bays and fjords, was situated only 50 nm away from the drill site, no preparedness role could be planned for here. Both the remoteness and the military regime led to the planning conclusion that no helicopter should be used in the operation. There are very few ports in the Kara Sea. Places like Dikson and Amderma, both situated close to 300 nm away, have no service facilities for an offshore operation.

In case of a medical emergency no resources are available along the coasts of the Kara Sea. The closest hospital is in Murmansk or Archangel, approximately 900 nm and 3 days of sailing away. To minimize the medical risks one of the stand-by vessels had a dedicated hospital function and doctor onboard. In addition dedicated paramedics were located on most vessels (Table 2).

Most offshore operations have a variety of regulations related to competence and health certificates—often related to the region of operation. For the Kara Sea

Table 3 Comparison of offshore operations in Kara Sea and West Greenland

	Case 1 West Greenland	Case 2 Kara Sea
Context	Large amount of large and small icebergs Floating bergy bits Foggy weather	Small icebergs Very limited government infrastructure Foggy weather
Resource configuration	2 PSVs with low ice class Depot ship Several helicopters	Several PSVs with low ice class No helicopters Hospital/Depot ship
The logistics process	Coordination through local logistics personnel and use of local distributor for base-to-base transport	Much planning to secure all resources available. Took care of everything within operation Challenging customs and immigration
The ice handling process	Challenges related to transport due to bergy bits in transport lanes Heavier icebergs a challenge for icebreakers	Challenging to plan operation due limited drift ice dynamics data (start/end of season) Exclusion from glazier fronts due to military regulations
The preparedness process	Limited capacity on oil spill preparedness. Limited government capacity	Limited local capacity on all levels. Limited government capacity

operation, special vaccination as well as special visas reflected this. Since the planning took consideration of a long time at sea and remoteness from hospitals, all crew members also had to document good dental health.

When operating in a remote and sensitive area there will be zero-tolerance for an oil spill. This was reflected both in the drilling procedures as well as the preparedness plan. The main resource in case of an oil spill was a dedicated response vessel located close to the rig during the entire operation. In addition, two of the AH-vessels had installed special equipment and trained operators for oil spill response. In case of a major spill, the closest base and depot for oil spill resources is Archangel, 900 nm away.

3.3.1.3 Analysis

The previous case descriptions show the context of operation, the fleet and the managerial capabilities in different functional areas. Table 3 below compares the two areas of operation.

The Context and Resource Configuration

The two cases illuminate different challenges related to the task environment of an Arctic operation. Case 1 shows that there may be a broad range of stakeholders influencing the operation that have to be considered with interest groups wanting to stop the operation due to environmental risks. This would call the government to introduce increased environment precautions such as using two rigs serving as back up for each other. In addition, additional oil spill response capacity was built up as a backup system and extra helicopter capacity was established. In case 2 we see that the demands for safety and environment precautions were not that high and only one rig was used and no helicopters mobilized. Increased complexity and volatility were, however, caused by military sensitivity in this area leading to lack of helicopter base capacity and long distance to base with extra platform service vessels included. Also, environmentalist activists threatened to stop the rig transport, causing additional capacity for safeguarding the rig in transit. Finally, there were unforeseen international political conflicts with sanctions that eventually led to a complete stop in operation and cancellation of all contracts with extra costs involved.

In both cases, the ice conditions with icebergs in the vicinity demanded additional towing capacity with large and costly anchor handling and towing vessels. Operating close up to the ice meant that the vessels should have icebreaking capacity or highest ice class. For the other vessels and the platform supply vessels in particular there were no vessels with higher ice class available in the market. All the vessels therefore had lowest ice class (1C). In case 1, this caused increased risk for the vessel and an extra burden for the crew with the combination of fog and bergy bits that could not be discovered on the radar. A collision may have caused the vessels to be transported to yards in Europe for repairs, as there were no yards large enough at Greenland.

In both cases there was drilling under summer conditions only. The timing of the drilling start related to ice melting was difficult especially in the Kara Sea case. If winter operations should be considered it would bring significant challenges with highest ice class on all vessels and rigs, and extended icebreaker capacity. In the case of the Kara Sea with long distances to base, several icebreakers would have to be mobilized not only for ice management but also for transport.

The Context, Resource Configuration and Operational Process Management

We see in both cases that the logistics challenges with long distance to base and lack of infrastructure had to be handled through the inclusion of more units that have to be coordinated. The transportation was hampered by fog and bergy bits in case 1 causing delays in vessel transport. This gave extra logistical management challenges and added costs. In case 2, the ice conditions with more winter ice called for in depth planning processes and involvement of more actors, including satellite imagery and research on site by icebreakers. Ice management became a very important issue.

Logistics Process Management The logistics process management was in case 1 challenged by lack of predictability as to weather conditions and iceberg threats. The drilling management including the offshore installation manager and the drill team had to be supplemented by an ice defense management and a marine management capacity. The operational phase meant coordination at several levels, where communication was essential. The logistics process management involved bringing together a broad range of resources from several companies. The bundling of the adequate resources had to take into consideration the limited predictability of the environment and access to personnel with necessary formal skills. Also, there was a need for training to prepare for the region.

Ice Handling Process Management The environment called for ice management operations and vessels with ice class and icebreaker capacity. There was a need for specialized and well-equipped vessels that made the operations much more costly. A special department had to be developed with an ice management director and ice advisors. This included the Shore Operation Center, the In-field Operations Command Centre (IOCC) and ice advisors at the different units. Even with some high-latitude communication challenges, this setup proved to be effective. There was a need for additional information services, and increased management capabilities. Ice pilot competence is rare and difficult to hire in the market. The Kara Sea operation was especially challenging due to more winter ice in the region and lack of units for surveillance of icebergs. The consequences of a collision with icebergs in this region would be large due to distance to base. Extra efforts had to be made by mobilizing both personnel with core competence and gaining data from satellites and other costly sources.

The Emergency Preparedness Management The preparedness issues were complicated due to the lack of government infrastructure and long distances. At West-Greenland, the operator had helicopters that could take SAR duties at a close by airport. The Kara Sea operation was severely handicapped in this respect by lack of landing sites for helicopters close by and government military regulations. In the Kara Sea, they had to add medical capacity at one of the vessels.

Oil spill recovery operation in icy waters is a challenging task. Adequate technology is available only to a certain degree. Both operations relied on technology and rules and regulations made for the North Sea. As the governments also had very limited capacity in these regions, the oil spill recovery capability was very limited.

As for police resources, long distances made this very challenging. However, military resources were available for anti-terror actions and other types of action against destructive behavior. At the West-Greenland operation, the authorities did not succeed in stopping Greenpeace from entering the rig and delaying the operation. In the Kara Sea, a more strict control regime was followed by the Russian anti-terror forces.

4 Conclusion

This chapter shows how firms faced with high complexity and volatility in an Arctic environment had to interact closely in a complicated bundle of different resources. This calls for an operational process management with good coordination skills at several levels. Operations in the ice-infested waters of the High Arctic demand a very broad resource base and the bundling of both high tech physical resources and core competence resources. The operation demands a tailor-made value chain and broad set of organizational adaptations within the organizations involved. Increased complexity due to a broad range of stakeholders, institutional arrangements and other factors call for a broader range of services including ice management, additional or different type of communication capacity and a number of units involved for emergency preparedness reasons. Dynamism or volatility was related to natural conditions like the icebergs, floes or bergy bits, fog, distances to base for spare parts and repair, and political and military sensitivity. This called for a broader range of physical resources including more and better equipped vessels with a broader range of functions needed if something unpredicted were to have happened. More costly vessels with ice class and icebreaker capacity had to be included even in summer operations. Winter operations would demand all vessels and rigs with the highest ice class and a much larger capacity of ice breaking vessels for both ice management and escort of platform service vessels, increasing the costs and the risk related to the operation significantly. Increased risk also calls for a significant upgrading of the maritime preparedness system, including both land bases, emergency rescue helicopters and oil recovery vessels. A broad range of competence resources at several levels were needed related to dealing with ice, marine environment challenges, and safety and security issues. For winter operations more specialized competence and cold climate experienced crew, including ice pilot competence would be necessary.

The implications of these findings is that offshore oil and gas operation in the High Arctic environment demands both redundant resources and a broader range of physical resources including a broad range of multi-functional vessels. These have to be matched by core competence resources that have to be added in due time before the operation. Much of this competence is experience- based and difficult to copy and imitate. This means that it will take time to build the necessary core competence, especially for winter operations. The new Polar Code will demand more specialized education and training infrastructure as well. The distances and resource scarcity of the operational area means that only to a limited extent will it be possible to add resources after the operation has started. The multi-functionality of vessels and multi-competence personnel have to be included and trained in realistic environments. A broader range of specialized functions and related competence resources bundled together in complex patterns mean resource management capabilities have to be developed at higher levels.

Implications for the Industry This study shows that offshore operation in icy waters is not "business as usual" and implies a broad range of physical and competence resources. Several of the physical resources may be included in the same vessels to keep the costs down. Multi-functionality and multi-competent personnel have to be mobilized. This means that the planning process has to start earlier and include both technology development and training. The largest problem is to gain the necessary resources for resource management as this demands a very special core competence. The resource management includes more specialized functions and management structures. Good communication between the different functions and between managerial levels is crucial. This means great effort for both physical communication equipment, as well as cross- cultural and cross-professional skills.

Implications for Further Studies This study shows that the bundling of resources in organizations has to take into consideration both the context of the operation, the resource types and number involved, and the number and type of organizations. There is a need for much more research both on vessel technology and equipment needed as to both specialization of ships and multi-functionality. There is a need for looking into the man-machine interface in the harsh environments, and to look at the need for redundant capacities and management structures, especially at operational levels. Research on flexible organizations, team management and improvisation is in order.

References

Amit, R., & Schoemaker, P. J. (1993). Strategic assets and organizational rent. *Strategic Management Journal, 14*(1), 3–46.

Barney, J. (1991). Firm resources and sustained competitive advantage. *Journal of Management, 17*, 99.

Barney, J. B. (2002). *Gaining and sustaining competitive advantage* (2nd ed.). Upper Saddle River, NJ: Prentice Hall.

Black, J. A., & Boal, K. B. (1994). Strategic resources: Traits, configuration and paths to sustainable competitive advantage. *Strategic Management Journal, 15*(Summer), 131–148.

Borch, O. J. (1994). The process of relational contracting. Developing trust-based strategic alliances among small business enterprises. In I. Paul Shrivastava, J. Dutton, & A. Huff (Eds.), *Advances in strategic management*. Greenwich, CT: JAI Press Inc.

Borch, O. J., & Arthur, M. B. (1995). Strategic network among small firms: Implications for strategy research methodology. *Journal of Management Studies, 32*(4), 419–441.

Borch, O. J., & Batalden, B. M. (2014). Business-process management in high-turbulence environments: The case of the offshore service vessel industry. *Maritime Policy & Management*, 1–18.

Borch, O. J., Huse, M., & Senneseth, K. (2009). Resource configuration, competitive strategies, and corporate entrepreneurship an empirical examination of small firms. *Entrepreneurship: Theory and Practice Fall, 1999*, 49.

Borch, O. J., & Madsen, E. L. (2007). Dynamic capabilities facilitating innovative strategies in SMEs. *International Journal of Technoentrepreneurship, 1*(1), 109–125.

Buixadé Farré, A., Stephenson, S. R., Chen, L., Czub, M., Dai, Y., & Demchev, D. (2014). Commercial Arctic shipping through the Northeast Passage: routes, resources, governance, technology, and infrastructure. *Polar Geography, 37*(4), 298–324.

Eisenhardt, K. M., & Martin, J. A. (2000). Dynamic capabilities: What are they? *Strategic Management Journal, 21*(10/11), 1105–1121.

Jenssen, J. I., & Randøy, T. (2006). The performance effect of innovation in shipping companies. *Maritime Policy and Management, 33*(4), 327–343.

Jeston, J., & Nelis, J. (2006). *Business process management: Practical guidelines to successful implementations*. Jordan Hill, GBR, Routledge.

Kjerstad, N. (2006). EGNOS - user experiences at high latitudes. *European Journal of Navigation, 4*(2).

Kjerstad, N. (2015). *Navigation and communication challenges at high latitudes – Experience from the Kara Sea*. Presentation on Nordic Institute of Navigation Workshop on eNav and the Arctic, Oslo, March 2015.

March, J. G. (1991). Exploration and exploitation in organizational learning. *Organization Science, 2*(1), 71–87.

Schilke, O. (2014). On the contingent value of dynamic capabilities for competitive advantage: The nonlinear moderating effect of environmental dynamism. *Strategic Management Journal, 35*(2), 179–203.

Teece, D. J., Pisano, G., & Shuen, A. (1997). Dynamic capabilities and strategic management. *Strategic Management Journal, 18*(7), 509–533.

Thompson, J. D. (1967). *Organizations in action*. New York: McGraw-Hill.

Trkman, P., & McCormack, K. (2009). Supply chain risk in turbulent environments—A conceptual model for managing supply chain network risk. *International Journal of Production Economics, 119*(2), 247–258.

Part VII
Conclusion

Navigating the Future: Towards Sustainable Arctic Marine Operations and Shipping in a Changing Arctic

Lawrence P. Hildebrand and Lawson W. Brigham

Contents

Abstract This volume is focused on a broad set of challenges and issues related to sustainable marine operations and shipping in a future Arctic, a region experiencing extraordinary change and increasingly intense attention. The numerous chapters in this volume highlight the key current and future issues in the Arctic, with a sharp focus on what remains to be done and how we must proceed.

1 Introduction

This volume is focused on a broad set of challenges and issues related to sustainable marine operations and shipping in a future Arctic, a region experiencing extraordinary change and increasingly intense attention. The numerous chapters in this

The research and editing of this chapter and volume was supported by U.S. National Science Foundation grant award 1263678 to L. W. Brigham and the University of Alaska Fairbanks.

L. P. Hildebrand (✉)
Ocean Sustainability, Governance & Management, World Maritime University, Malmö, Sweden
e-mail: lh@wmu.se

L. W. Brigham
International Arctic Research Center, University of Alaska Fairbanks, Fairbanks, AK, USA
e-mail: lwbrigham@alaska.edu

L. P. Hildebrand et al. (eds.), *Sustainable Shipping in a Changing Arctic*, WMU Studies in Maritime Affairs 7, https://doi.org/10.1007/978-3-319-78425-0_23

volume highlight the key current and future issues in the Arctic, with a sharp focus on what remains to be done and how we must proceed.

The Introductory chapter of this volume set the scene for the fundamental changes that are reshaping the maritime Arctic and compelling legal, regulatory, social and capacity-building considerations to nascent Arctic marine operations and shipping. The subsequent chapters build the picture that the Arctic is a new frontier, indeed one of the last on Earth, both in terms of its estimated economic potential and as a poorly understood and rapidly changing ecosystem. This concluding chapter highlights the main messages and themes that run throughout this volume and focuses our attention on how Arctic marine operations and shipping must proceed in a sustainable manner.

How, when and at what pace development of the Arctic takes place, is still subject to considerable speculation, uncertainty and constraints. Regardless, all proposed activities—destinational and trans-Arctic shipping; oil and gas development; fishing; mining; and, cruise and adventure tourism—must, we have learned throughout this volume—be pursued in a manner that is sustainable for the Arctic ecosystem and the economies and ways of life of indigenous communities.

Physically, the Arctic is warming twice as fast as the rest of the planet and the Arctic Ocean's sea ice cover is undergoing a profound transformation that has not been experienced in millennia. We are recording earlier sea-ice break-up and later freeze-up, leading to greater periods of open water, albeit mostly in the summer so far, that favours shipping, resource extraction and tourism. This greater marine access infers potentially longer seasons of navigation. However, as the region is one of the most poorly studied due to the harsh environment, remoteness and high costs involved with conducting assessments, these physical realities will dictate if, how and when we proceed.

With new and growing Arctic economic activities, increasingly integrated with the global economy, and the new marine transportation systems these require, we can anticipate increases in environmental risks. We are likely to see: increasing greenhouse gas emissions and air pollution and their associated risks to public health; disruption of marine life from anthropogenic noise and potentially costly marine pollution incidents; conflicts among traditional and new ocean users; safety of life at sea concerns; and, disruption of northern peoples' uses of their traditional territories. In addition to negative environmental impacts from each of these pressures, failure to plan for integrated and cross-sectoral management could potentially lead to user-user and user-environment disputes or conflicts.

From a governance perspective, states and increasingly non-state actors, are cooperating within the Arctic Council and within IMO, increasing governance complexity in Arctic affairs, but strengthening a unified, more global approach that will guide marine operations and shipping. While UNCLOS provides the over-arching legal framework for the Arctic Ocean and the IMO Polar Code a new governance regime for commercial ships in polar waters, bringing additional actors to the diplomatic table could increase the risk of creating an even more complex picture (UNCLOS; IMO Resolution MSC.385(94) 2014). How can all of these interests be accommodated and what impact will they have on an evolving regime which is already characterized by institutional overlaps and multi-layered governance?

Non-state actors too, especially Arctic indigenous peoples who live in coastal communities and use Arctic waters and ice as a critical part of their survival, are particularly concerned and will be most directly affected by increases in Arctic marine traffic. Encouragingly, their voices and rights are increasingly being heard and they are being engaged in the international forums such as the UN, IMO and Arctic Council. The new maritime Arctic compels governments and industry to foster greater communication and involvement of the Arctic indigenous people in decision-making, and to respond to their range of concerns and interests with regard to Arctic maritime affairs. One area of collaboration in which all can benefit is data and information sharing. This would be of mutual benefit to the research community, operational authorities and policy makers in the areas of regulatory compliance, law enforcement, search and rescue, and emergency response.

2 The Legislative and Regulatory Context

This volume begins with a chapter on the most recent and significant legislative initiative under the IMO—The mandatory International Code for Ships Operating in Polar Waters (Polar Code) (IMO Resolution MSC.385(94) 2014). IMO has addressed international concern about the protection of the polar environment and the safety of seafarers and passengers on ships operating in Arctic areas through the adoption of this internationally binding instrument, which became effective on 1 January 2017 under the SOLAS and MARPOL Conventions (SOLAS; MARPOL 73/78). The Polar Code's place in the existing global framework regulating international shipping, and the important requirements with regard to maritime safety and marine environmental protection, are clearly acknowledged. It is recognized as embodying positive changes for safe and environmentally friendly future shipping in the Arctic.

Important and timely as the Polar Code is, several authors emphasize that it should be viewed as only the beginning of a long process to further protect polar waters in an era of increasing marine operations and but one part of a broader regime to govern activities in the Arctic. It is important to note that resources and ecosystems in the Arctic extend across political boundaries and the Arctic Ocean contains a large area of high seas. IMO and external analysts writing in this volume agree that although the Polar Code is an important first step in protecting the Arctic socio-ecological system, there is a lot more work to be done to address those issues that go well beyond the Polar Code.

Against the hope and plans for economic development and increasing shipping operations in the Arctic, stands the region's lack of a host of infrastructure that is central to marine safety and environmental protection. These include: hydrographic data and adequate charting; environmental monitoring and forecasting (sea ice, weather and icebergs); SAR capacity; environmental response capacity; ship monitoring and tracking systems; salvage; deep water ports and port facilities; aids to navigation; adequate communications; and, more. This major gap in marine infrastructure was highlighted previously in the Arctic Council's Arctic Marine Shipping Assessment (see Appendix B). In particular, the mobilization of people to polar regions to assist in emergency response and SAR is a logistics bottleneck. While equipment and supplies can be prepositioned and cached in Arctic regions without the need for much attention, the same cannot be said for trained and qualified personnel who will be the lynchpin in the event of an emergency.

In time, IMO may also look at the introduction of completely new requirements and amendments, such as extending the Polar code to non-SOLAS ships, introducing additional performance and test standards and amendments to the survey guidelines, establishing emission control areas and port reception facilities, strengthening icebreaker support and routeing and reporting systems in straits, Port State control, designating PSSAs (Particular Sensitive Sea Areas), introducing measures to reduce underwater noise, and potentially more (IMO 2005, 2011, 2012a, b; AMAP/CAFF/SDWG 2013).

For the IMO, the marine insurance industry, the ship classification societies, and the global shipping enterprise, the Polar Code represents a new regulatory regime for polar waters and importantly, a set of uniform, non-discriminatory standards. However, the Polar Code presents a host of policy and practical challenges in its implementation as well as enforcement by the flag and port states. While the Polar Code is a seminal advance in governance of polar waters, the continued gap in maritime infrastructure hinders robust Arctic development. This is relevant to ship navigation in the Arctic requiring transit through shallow, draft-constrained coastal and archipelago waters that are relatively uncharted, lack aids to navigation and adequate search and rescue facilities, and teaming with surface and underwater hazards to navigation (MSC 2008).

3 Increasing Awareness

Awareness is a theme running throughout the chapters in this volume: the need for increased situational awareness in marine operations, awareness of the changing marine environment and what that implies for the sensitive and threatened ecosystems, and awareness of the unique needs and rights of the indigenous people of the Arctic (Raymond-Yakoubian et al. 2014; Raymond-Yakoubian 2015). And above all, awareness of the need for all concerned actors—state, non-state, indigenous and non-Arctic—to cooperate in the development, protection and sustainability of this emerging region.

The chapter authors cover the breadth of issues important to the changing and developing Arctic. They highlight and call for increased awareness and attention to a broad suite of central issues: legislative, maritime safety, governance, Indigenous rights and engagement, training, emergency preparedness and response, safe and environmentally sound Arctic navigation, vessel detection and tracking, Arctic strategies, marine insurance, biodiversity and marine protected areas. On this latter point, the Arctic marine environment presents unique challenges to the designation, implementation and enforcement of Marine Protected Areas (MPAs), including the lack of infrastructure and reception facilities, significant knowledge gaps, difficulties in ensuring compliance with regulations in remote areas, designing appropriate measures in a changing and unpredictable environment, recognizing the needs and rights of indigenous peoples and local inhabitants, managing competing coastal state priorities, and designating protected areas across boundaries and beyond national jurisdiction.

The chapters address these important issues and more by exploring and assessing the progress and implications of the Polar Code and the regulatory steps beyond the Code yet to be taken to ensure safe and sustainable shipping operations, including the human element in maritime safety and environmentally sound Arctic navigation [note that the human element is taken into account in new STCW requirements that are in force July 2018], ship monitoring and surveillance tracking, grounding avoidance, open-access ice data, cooperative governance, the role of non-state actors, marine insurance, high-seas biodiversity and protection, oil spill intervention and response, indigenous communities, and the need for training and capacity building (STCW Convention 1996; Hollnagel 2009, 2014).

4 Conclusion

The Arctic stands as one of the last places on Earth where human development and exploitation have been constrained by long-term climatic conditions. But this situation is changing rapidly. The central question is, then, how will we proceed? Will we rush in as fast as the ice is melting to tap these 'vital' natural resources as we have everywhere else in the world, reaping huge short-term profits, but significant losses, or will we take the little time we have left to think carefully and thoughtfully and develop the Arctic according to the central tenets of precaution and sustainable development?

New opportunities for Arctic marine operations and shipping are emerging, but significant challenges remain. These include: the effective implementation and enforcement of the IMO Polar Code; a huge gap in Arctic marine infrastructure; enhancing the monitoring and surveillance of Arctic waters; the challenge of developing a set of marine protected areas; additional Polar Code measures for the circumpolar region; and the need for large public and private investments, as well as potential public-private partnerships in the Arctic.

Cooperation among the Arctic states, the non-Arctic shipping states, and the global maritime enterprise, will be critical to effective protection of Arctic people and the marine environment. It is only through ongoing and sustained cooperation among all concerned stakeholders and actors, including the central role of the shipping industry and interests, that the sustainable development of the Arctic can be achieved.

To conclude on an encouraging note, in late 2017, officials from five Arctic countries and five major distant fishing powers, reached an unprecedented agreement on a legally binding international accord that will protect nearly three million square kilometres of the Central Arctic Ocean from unregulated fishing (Hoag 2017; Appendix H). Once signed, the agreement will prevent commercial fishing in the high seas of the world's smallest ocean for at least 16 years while scientific research is conducted and we gain a better understanding of the area's ecosystems, and appropriate conservation and management measures can be established.

Hope remains. If we can apply this same precautionary approach to Arctic oil and gas exploration and development, mining, tourism and, especially Arctic marine operations and shipping, we may indeed develop the Arctic in a sustainable way not seen in any of the world's other oceans.

References

AMAP/CAFF/SDWG. (2013). *Identification of Arctic marine areas of heightened ecological and cultural significance: Arctic Marine Shipping Assessment (AMSA) IIc*. Oslo: Arctic Monitoring and Assessment Programme (AMAP).

Hoag, H. (2017). *Nations agree to ban fishing in Arctic Ocean for at least 16 years*. Retrieved December 7, 2017, from http://www.sciencemag.org/news/2017/12/nations-agree-ban-fishing-arctic-ocean-least-16-years

Hollnagel, E. (2009). *ETTO principle: Efficiency – Thoroughness trade-off: Why things that go right sometimes go wrong*. Surrey: Ashgate Publishing Limited.

Hollnagel, E. (2014). *Safety-I and Safety-II: The past and future of safety management*. Surrey: Ashgate Publishing Limited.

IMO. (2005). IMO Res. A.982(24), Revised Guidelines for the Identification and Designation of Particularly Sensitive Sea Areas, adopted 1 December 2005.

IMO (Sub-Committee on Ship Design and Equipment). (2012a). DE Report to the MSC, DE 56/25 (28 February 2012), p. 22, para. 10.7.

IMO (Sub-Committee on Ship Design and Equipment). (2012b). Report of the Maritime Safety Committee on its 91st Session, MSC 91/22 (17 December 2012), p. 35, para. 8.5.

IMO Resolution MSC.385(94). (2014). *International Code for Ships Operating in Polar Water, 2014*. London: International Maritime Organization.

International Maritime Organization (Sub-Committee on Ship Design and Equipment). (2011). DE Report to the MSC, DE 55/22 (15 April 2011). p. 23, para. 12.7.1.

MARPOL 73/78, International Maritime Organization (IMO), International Convention for the Prevention of Pollution from Ships, 1340 UNTS 184, 1973 as modified by the Protocol 1978 relating to the International Convention for the Prevention of Pollution from Ships of 1973, 17 February 1978, 1340 UNTS 61. Retrieved December 7, 2017, from http://www.mar.ist.utl.pt/mventura/Projecto-Navios-I/IMO-Conventions%20(copies)/MARPOL.pdf

MSC. (2008). Casualty-Related Matters', Reports on Marine Casualties and Incidents, Ref. T1/12.01 International Maritime Organization. MSC-MEPC.Circ.3. 18 December 2008. Annex 2.

Raymond-Yakoubian, J. (2015). Conceptual and institutional frameworks for protected areas and the status of indigenous involvement: Considerations for the Bering Strait Region of Alaska. In T. M. Herrmann & T. Martin (Eds.), *Indigenous peoples' governance of land and protected territories in the circumpolar Arctic* (pp. 83–103). Switzerland: Springer.

Raymond-Yakoubian, J., Khokhlov, Y., & Yarzutkina, A. (2014). *Indigenous knowledge and use of ocean currents in the Bering Strait Region* (Report to the National Park Service, Shared Beringian Heritage Program for Cooperative Agreement H99111100026). Nome, AK: Kawerak, Inc., Social Science Program.

SOLAS, International Maritime Organization (IMO), International Convention for the Safety of Life At Sea, 1 November 1974, 1184 UNTS 3. Retrieved December 7, 2017, from http://www.refworld.org/docid/46920bf32.htm

STCW Convention. (1996). *International Convention on Strandards of Training, Certification, and Watchkeeping for Seafarers*. London: International Maritime Organization.

United Nations Convention on the Law of the Sea, 1982, 1833 U.N.T.S. 397. Retrieved December 7, 2017, from http://www.un.org/depts/los/convention_agreements/texts/unclos/unclos_e.pdf

Agreement on Enhancing International Arctic Scientific Cooperation

The Government of Canada, the Government of the Kingdom of Denmark, the Government of the Republic of Finland, the Government of Iceland, the Government of the Kingdom of Norway, the Government of the Russian Federation, the Government of the Kingdom of Sweden, and the Government of the United States of America (hereinafter referred to as the "Parties"),

Recognizing the importance of maintaining peace, stability, and constructive cooperation in the Arctic;

Recognizing the importance of the sustainable use of resources, economic development, human health, and environmental protection;

Reiterating the urgent need for increased actions to mitigate and adapt to climate change;

Emphasizing the importance of using the best available knowledge for decisionmaking;

Noting the importance of international scientific cooperation in that regard; Fully taking into account the relevant provisions of the 1982 United Nations Convention on the Law of the Sea, in particular the provisions in Part XIII on marine scientific research as they relate to promoting and facilitating the development and conduct of marine scientific research for peaceful purposes;

Recalling the Kiruna Declaration on the occasion of the Eighth Ministerial meeting of the Arctic Council held in May 2013 and the Iqaluit Declaration on the occasion of the Ninth Ministerial meeting of the Arctic Council held in April 2015;

Recognizing the ongoing development of the International Polar Partnership Initiative as determined by the Executive Council of the World Meteorological Organization;

Recognizing the significance of the research priorities as determined by the International Conference on Arctic Research Planning;

Recognizing the efforts of the Arctic Council and its subsidiary bodies;

Recognizing the significant scientific expertise and invaluable contributions to scientific activities being made by non-Parties and specifically by the Arctic Council Permanent Participants and Arctic Council Observers;

L. P. Hildebrand et al. (eds.), *Sustainable Shipping in a Changing Arctic*, WMU Studies in Maritime Affairs 7, https://doi.org/10.1007/978-3-319-78425-0

Recognizing the substantial benefit gained from the financial and other investments by the Arctic States and other nations in the International Polar Year and its outcomes, including in particular new scientific knowledge, infrastructure and technologies for observation and analysis;

Recognizing the excellent existing scientific cooperation already under way in many organizations and initiatives, such as the Sustaining Arctic Observing Networks, the International Arctic Science Committee, the University of the Arctic, the Forum of Arctic Research Operators, the International Network for Terrestrial Research and Monitoring in the Arctic, the World Meteorological Organization, the International Council for the Exploration of the Sea, the Pacific Arctic Group, the Association of Polar Early Career Scientists, indigenous knowledge institutions, the International Arctic Social Sciences Association, and many others; and

Desiring to contribute to and build upon existing cooperation and make efforts to develop and expand international Arctic scientific cooperation,

Have agreed as follows:

Article 1
Terms and definitions

For the purposes of this Agreement:

"Facilitate" means pursuing all necessary procedures, including giving timely consideration and making decisions as expeditiously as possible;

"Participant" means the Parties' scientific and technological departments and agencies, research centers, universities and colleges, and contractors, grantees and other partners acting with or on behalf of any Party or Parties, involved in Scientific Activities under this Agreement;

"Scientific Activities" means efforts to advance understanding of the Arctic through scientific research, monitoring and assessment. These activities may include, but are not limited to, planning and implementing scientific research projects and programs, expeditions, observations, monitoring initiatives, surveys, modelling, and assessments; training personnel; planning, organizing and executing scientific seminars, symposia, conferences, workshops, and meetings; collecting, processing, analyzing, and sharing scientific data, ideas, results, methods, experiences, and traditional and local knowledge; developing sampling methodologies and protocols; preparing publications; and developing, implementing, and using research support logistics and research infrastructure;

"Identified Geographic Areas" means those areas described in Annex 1.

Article 2
Purpose

The purpose of this Agreement is to enhance cooperation in Scientific Activities in order to increase effectiveness and efficiency in the development of scientific knowledge about the Arctic.

Article 3
Intellectual property and other matters

Where appropriate, cooperative activities under this Agreement shall take place pursuant to specific implementing agreements or arrangements concluded between the Parties or Participants pertaining to their activities, particularly the financing of such activities, the use of scientific and research results, facilities, and equipment, and dispute settlement. Through such specific agreements or arrangements, the Parties shall, where appropriate, ensure, either directly or through the Participants, adequate and effective protection and fair allocation of intellectual property rights, in accordance with the applicable laws, regulations, procedures, and policies as well as the international legal obligations of the Parties concerned, and address other matters that may result from activities under this Agreement.

Article 4
Entry and exit of persons, equipment, and material

Each Party shall use its best efforts to facilitate entry to, and exit from, its territory of persons, research platforms, material, samples, data, and equipment of the Participants as needed to advance the objectives of this Agreement.

Article 5
Access to research infrastructure and facilities

The Parties shall use their best efforts to facilitate access by the Participants to national civilian research infrastructure and facilities and logistical services such as transportation and storage of equipment and material for the purpose of conducting Scientific Activities in Identified Geographic Areas under this Agreement.

Article 6
Access to research areas

1. The Parties shall facilitate access by the Participants to terrestrial, coastal, atmospheric, and marine areas in the Identified Geographic Areas, consistent with international law, for the purpose of conducting Scientific Activities.
2. The Parties shall facilitate the processing of applications to conduct marine scientific research under this Agreement consistent with the 1982 United Nations Convention on the Law of the Sea.
3. The Parties also shall facilitate joint Scientific Activities that require airborne scientific data collection in the Identified Geographic Areas, and that are subject to specific implementing agreements or arrangements concluded between the Parties or Participants pertaining to those activities.

Article 7
Access to data

1. The Parties shall facilitate access to scientific information in connection with Scientific Activities under this Agreement.
2. The Parties shall support full and open access to scientific metadata and shall encourage open access to scientific data and data products and published results with minimum time delay, preferably online and free of charge or at no more than the cost of reproduction and delivery.
3. The Parties shall facilitate the distribution and sharing of scientific data and metadata by, as appropriate and to the extent practicable, adhering to commonly accepted standards, formats, protocols, and reporting.

Article 8
Education, career development and training opportunities

The Parties shall promote opportunities to include students at all levels of education, and early career scientists, in the Scientific Activities conducted under this Agreement to foster future generations of researchers and to build capacity and expertise to advance knowledge about the Arctic.

Article 9
Traditional and local knowledge

1. The Parties shall encourage Participants to utilize, as appropriate, traditional and local knowledge in the planning and conduct of Scientific Activities under this Agreement.

2. The Parties shall encourage communication, as appropriate, between holders of traditional and local knowledge and Participants conducting Scientific Activities under this Agreement.

3. The Parties shall encourage holders of traditional and local knowledge, as appropriate, to participate in Scientific Activities under this Agreement.

Article 10
Laws, regulations, procedures, and policies

Activities and obligations under this Agreement shall be conducted subject to applicable international law and the applicable laws, regulations, procedures, and policies of the Parties concerned. For those Parties that have subnational governments, the applicable laws, regulations, procedures, and policies include those of their subnational governments.

Article 11
Resources

1. Unless otherwise agreed, each Party shall bear its own costs deriving from its implementation of this Agreement.
2. Implementation of this Agreement shall be subject to the availability of relevant resources.

Article 12
Review of this Agreement

1. The Parties shall meet no later than one year after the entry into force of this Agreement, as convened by the depositary, and from then on as decided by the Parties. The Parties may elect to convene such meetings in conjunction with meetings of the Arctic Council including inviting Arctic Council Permanent Participants and Arctic Council Observers to observe and provide information. Scientific cooperation activities with non-Parties related to Arctic science may be taken into account when reviewing the implementation of this Agreement.
2. At such meetings the Parties shall consider the implementation of this Agreement, including successes achieved and obstacles to implementation, as well as ways to improve the effectiveness and implementation of this Agreement.

Article 13
Authorities and contact points

Each Party shall designate a competent national authority or authorities as the responsible point of contact for this Agreement. The names of and contact information for the designated points of contact are specified in Annex 2 to this Agreement. Each Party shall promptly inform the other Parties in writing through its competent national authority or authorities and through diplomatic channels of any changes to those designations.

Article 14
Annexes

1. Annex 1 referred to in Article 1 constitutes an integral part of this Agreement and is legally binding.
2. Annex 2 referred to in Article 13 does not constitute an integral part of this Agreement and is not legally binding.
3. At meetings of the Parties referred to in Article 12, the Parties may adopt additional legally non-binding Annexes. Annex 2 referred to in Article 13 may be modified as provided in that Article.

Article 15
Settlement of disputes

The Parties shall resolve any disputes concerning the application or interpretation of this Agreement through direct negotiations.

Article 16
Relationship with other international agreements

Nothing in this Agreement shall be construed as altering the rights or obligations of any Party under other relevant international agreements or international law.

Article 17
Cooperation with non-Parties

1. The Parties may continue to enhance and facilitate cooperation with non-Parties with regard to Arctic science.
2. Parties may in their discretion undertake with non-Parties cooperation described in this Agreement and apply measures consistent with those described in this Agreement in cooperation with non-Parties.
3. Nothing in this Agreement shall affect the rights and obligations of the Parties under agreements with non-Parties, nor preclude cooperation between the Parties and nonParties.

Article 18
Amendments to this Agreement

1. This Agreement may be amended by written agreement of all the Parties.
2. An amendment shall enter into force 30 days after the date on which the depositary has received the last written notification through diplomatic channels that the Parties have completed the internal procedures required for its entry into force.

Article 19
Provisional application, entry into force, and withdrawal

1. This Agreement may be applied provisionally by any signatory that provides a written statement to the depositary of its intention to do so. Any such signatory shall apply this Agreement provisionally in its relations with any other signatory having made the same notification from the date of its statement or from such other date as indicated in its statement.
2. This Agreement shall enter into force for a period of five years 30 days after the date of receipt by the depositary of the last written notification through diplomatic channels that the Parties have completed the internal procedures required for its entry into force.
3. This Agreement shall be automatically renewed for further periods of five years unless a Party notifies the other Parties in writing at least six months

prior to the expiration of the first period of five years or any succeeding period of five years of its intent to withdraw from this Agreement, in which event this Agreement shall continue between the remaining Parties.

4. Any Party may at any time withdraw from this Agreement by sending written notification thereof to the depositary through diplomatic channels at least six months in advance, specifying the effective date of its withdrawal. Withdrawal from this Agreement shall not affect its application among the remaining Parties.
5. Withdrawal from this Agreement by a Party shall not affect the obligations of that Party with regard to activities undertaken under this Agreement where those obligations have arisen prior to the effective date of withdrawal.

Article 20
Depositary

The Government of the Kingdom of Denmark shall be the depositary for this Agreement.

DONE at Fairbanks, Alaska, United States of America this 11th day of May, 2017. This Agreement is established in a single copy in the English, French, and Russian languages, all texts being equally authentic. The working language of this Agreement shall be English, the language in which this Agreement was negotiated. The Depositary shall transmit certified copies of this Agreement to the Parties.

Annex 1: Identified Geographic Areas

Identified Geographic Areas for purposes of this Agreement are described by each Party below and include areas over which a State whose government is a Party to this Agreement exercises sovereignty, sovereign rights or jurisdiction, including land and internal waters within those areas and the adjacent territorial sea, exclusive economic zone, and continental shelf, consistent with international law. Identified Geographic Areas also include areas beyond national jurisdiction in the -high seas north of 62 degrees north latitude.

The Parties agree that the Identified Geographic Areas are described solely for the purposes of this Agreement. Nothing in this Agreement shall affect the existence or delineation of any maritime entitlement or the delimitation of any boundary between States in accordance with international law.

CANADA – The territories of Yukon, Northwest Territories, and Nunavut and the adjacent marine areas of Canada.

KINGDOM OF DENMARK – The territory of the Kingdom of Denmark including Greenland and the Faroes and its marine areas above the southern limit of the Greenland exclusive economic zone and the Faroese fisheries zone.

FINLAND – The territory of Finland and its marine areas.

ICELAND – The territory of Iceland and its marine areas.

NORWAY – Marine areas north of 62 degrees north latitude, and land areas north of the Arctic Circle (66.6 degrees north latitude).

RUSSIAN FEDERATION

1. Territory of the Murmansk Region;
2. Territory of the Nenets Autonomous Area;
3. Territory of the Chukchi Autonomous Area;
4. Territory of the Yamalo-Nenets Autonomous Area;
5. Territory of the municipal entity "Vorkuta" (Komi Republic);
6. Territories of Allaikhov Ulus (District), Anabar National (Dolgano-Evenk) Ulus (District), Bulun Ulus (District), Nizhnekolymsk District, Ust-Yan Ulus (District) (Sakha Republic (Yakutia));
7. Territories of the Urban District of Norilsk, Taimyr Dolgan-Nenets Municipal District, Turukhan District (Krasnoyarsk Territory);
8. Territories of the municipal entities "The City of Arkhangelsk", "Mezen Municipal District", "Novaya Zemlya", "The City of Novodvinsk", "Onega Municipal District", "Primorsky Municipal District", "Severodvinsk" (Arkhangelsk region);
9. Lands and islands of the Arctic Ocean, identified in the Resolution of the Presidium of the Central Executive Committee of the USSR dated April 15, 1926 "On the announcement of lands and islands situated in the Arctic Ocean as a territory of the Union of SSR" and other legislative acts of the USSR; as well as adjacent marine areas.

Note: Territories of the municipal entities, listed in the abovementioned items 5–8, identified within the borders as of April 1, 2014.

SWEDEN – The territory of Sweden and its marine areas north of 60.5 degrees north latitude.

UNITED STATES OF AMERICA – All United States territory north of the Arctic Circle and north and west of the boundary formed by the Porcupine, Yukon, and Kuskokwim Rivers; the Aleutian chain; and adjacent marine areas .in the Arctic Ocean and the Beaufort, Bering, and Chukchi Seas.

Annex 2: Authorities and Contact Points

CANADA

Polar Knowledge Canada
170 Laurier Avenue West, 2ND Floor, Suite 200, Ottawa, Ontario KIP 5V5
Telephone:+ 1 613 943 8605
Email: info@polar.gc.ca

Point of contact for Marine Scientific Research requests:
Global Affairs Canada
Security and Defense Relations, 125 Sussex Drive, Ottawa, Ontario KIA 0G2

Telephone: + 1 343 203 3208
Email: chris.conway(@intcmational.gc.ca; EXTOTT-IGR@international.gc.ca

KINGDOM OF DENMARK

The Ministry of Foreign Affairs
Department for Northern America and the Arctic
Asiatisk Plads 2; 1448 Copenhagen K
Telephone: + 45 33 92 00 00
Email: ana@um.dk

Danish Agency for Science and Higher Education
Bredgade 40
DK-1260 Copenhagen K
Telephone: +45 3544 6200
Email: sfu@ufm.dk

Department of Foreign Affairs
Postboks 1340, 3900 Nuuk
Telephone: +299 34 50 00
Email: nap@nanoq.gl

Ministry of Education, Culture, Research and Church
Postboks 1029, 3900 Nuuk
Telephone: +299 34 50 00
Email: ikiin@nanog.gl

Ministry of Foreign Affairs and Trade
Gongin 7, Postbox 377, 110 T6rshavn
Telephone: +298 30 66 00
Email: uvmr@uvmr.fo

Ministry of Education, Research and Culture
Hoyviksvegur 72, Postbox 3279, 110 T6rshavn
Telephone: +298 30 65 00
Email: mmr@mmr.fo

FINLAND

Ministry of Education and Culture
P.O. Box 29, FI-00023 Government
(Visiting addresses: Meritullinkatu 10, Helsinki;
Meritullinkatu 1, Helsinki)
Telephone: +358 2953 30004 (Switchboard)
Email: kirjaamo@minedu.fi

ICELAND

Ministry of Education, Science and Culture
Solvh6lsgata 4, 150 Reykjavik
Tel: +354 545 9500
Email: postur@mmr.stjr.is

The Icelandic Center for Research
Borgartún 30, 105 Reykjavik
Tel: +354 515 5800
Email: rannis@rannis.is

NORWAY

Ministry of Education and Research
P.O. Box 8119 Dep, N-0032 Oslo
Visitor address: Kirkegata 18, Oslo
Telephone: +47 22 24 90 90
Email: postmottak@kd.dep.no

The Research Council of Norway
P.O Box 564 N-1327 Lysaker
Visitor address: Drammensveien 288, Oslo
Telephone: +47 22 03 70 00
Email: post@forskningsradet.no

RUSSIAN FEDERATION

Ministry of Education and Science
Department of Science and Technology
Tverskaya st., 11, Moscow 125993
Telephone: + 7 495 629 03 64
Email: D-14@mon.gov.ru

SWEDEN

Ministry of Education and Research
103 33 Stockholm
Telephone: +46 8 405 1000
Email: u.registrator@regeringskansliet.se

UNITED STATES OF AMERICA

US Arctic Research Commission
Executive Director, US Arctic Research Commission
4350 N. Fairfax Dr., Suite 510, Arlington, VA 22203
Telephone:+ 1 703 525 0113
Email: info@arctic.gov

Arctic Marine Shipping Assessment Recommendations

The focus of the AMSA is marine safety and marine environmental protection, which is consistent with the Arctic Council's mandates of environmental protection and sustainable development. Based on the findings of the AMSA, recommendations were developed to provide a guide for future action by the Arctic

Council, Arctic states and many others. The AMSA recommendations are presented under three broad, inter-related themes that are fundamental to understanding the AMSA: Enhancing Arctic Marine Safety, Protecting Arctic People and the Environment, and Building Arctic Marine Infrastructure. It is recognized that implementation of these recommendations could come from the Arctic states, industry and/or public-private partnerships.

Enhancing Arctic Marine Safety

A. Linking with International Organizations: That the Arctic states decide to, on a case by case basis, identify areas of common interest and develop unified positions and approaches with respect to international organizations such as: the International Maritime Organization (IMO), the International Hydrographic Organization (IHO), the World Meteorological Organization (WMO) and the International Maritime Satellite Organization (IMSO) to advance the safety of Arctic marine shipping; and encourage meetings, as appropriate, of member state national maritime safety organizations to coordinate, harmonize and enhance the implementation of the Arctic maritime regulatory framework.
B. IMO Measures for Arctic Shipping: That the Arctic states, in recognition of the unique environmental and navigational conditions in the Arctic, decide to cooperatively support efforts at the International Maritime Organization to strengthen, harmonize and regularly update international standards for vessels operating in the Arctic. These efforts include:

L. P. Hildebrand et al. (eds.), *Sustainable Shipping in a Changing Arctic*, WMU Studies in Maritime Affairs 7, https://doi.org/10.1007/978-3-319-78425-0

- Support the updating and the mandatory application of relevant parts of the *Guidelines for Ships Operating in Arctic Ice-covered* Waters (Arctic Guidelines); and,
- Drawing from IMO instruments, in particular the Arctic Guidelines, augment global IMO ship safety and pollution prevention conventions with specific mandatory requirements or other provisions for ship construction, design, equipment, crewing, training and operations, aimed at safety and protection of the Arctic environment.

C. Uniformity of Arctic Shipping Governance: That the Arctic states should explore the possible harmonization of Arctic marine shipping regulatory regimes within their own jurisdiction and uniform Arctic safety and environmental protection regulatory regimes, consistent with UNCLOS, that could provide a basis for protection measures in regions of the central Arctic Ocean beyond coastal state jurisdiction for consideration by the IMO.
D. Strengthening Passenger Ship Safety in Arctic Waters: That the Arctic states should support the application of the IMO's *Enhanced Contingency Planning Guidance for Passenger Ships Operating in Areas Remote from SAR Facilities*, given the extreme challenges associated with rescue operations in the remote and cold Arctic region; and strongly encourage cruise ship operators to develop, implement and share their own best practices for operating in such conditions, including consideration of measures such as timing voyages so that other ships are within rescue distance in case of emergency.
E. Arctic Search and Rescue (SAR) Instrument: That the Arctic states decide to support developing and implementing a comprehensive, multi-national Arctic Search and Rescue (SAR) instrument, including aeronautical and maritime SAR, among the eight Arctic nations and, if appropriate, with other interested parties in recognition of the remoteness and limited resources in the region.

Protecting Arctic People and the Environment

A. Survey of Arctic Indigenous Marine Use: That the Arctic states should consider conducting surveys on Arctic marine use by indigenous communities where gaps are identified to collect information for establishing up-to-date baseline data to assess the impacts from Arctic shipping activities.
B. Engagement with Arctic Communities: That the Arctic states decide to determine if effective communication mechanisms exist to ensure engagement of their Arctic coastal communities and, where there are none, to develop their own mechanisms to engage and coordinate with the shipping industry, relevant economic activities and Arctic communities (in particular during the planning phase of a new marine activity) to increase benefits and help reduce the impacts from shipping.

C. Areas of Heightened Ecological and Cultural Significance: That the Arctic states should identify areas of heightened ecological and cultural significance in light of changing climate conditions and increasing multiple marine use and, where appropriate, should encourage implementation of measures to protect these areas from the impacts of Arctic marine shipping, in coordination with all stakeholders and consistent with international law.
D. Specially Designated Arctic Marine Areas: That the Arctic states should, taking into account the special characteristics of the Arctic marine environment, explore the need for internationally designated areas for the purpose of environmental protection in regions of the Arctic Ocean. This could be done through the use of appropriate tools, such as "Special Areas" or Particularly Sensitive Sea Areas (PSSA) designation through the IMO and consistent with the existing international legal framework in the Arctic.
E. Protection from Invasive Species: That the Arctic states should consider ratification of the IMO *International Convention for the Control and Management of Ships Ballast Water and Sediments*, as soon as practical. Arctic states should also assess the risk of introducing invasive species through ballast water and other means so that adequate prevention measures can be implemented in waters under their jurisdiction.
F. Oil Spill Prevention: That the Arctic states decide to enhance the mutual cooperation in the field of oil spill prevention and, in collaboration with industry, support research and technology transfer to prevent release of oil into Arctic waters, since prevention of oil spills is the highest priority in the Arctic for environmental protection.
G. Addressing Impacts on Marine Mammals: That the Arctic states decide to engage with relevant international organizations to further assess the effects on marine mammals due to ship noise, disturbance and strikes in Arctic waters; and consider, where needed, to work with the IMO in developing and implementing mitigation strategies.
H. Reducing Air Emissions: That the Arctic states decide to support the development of improved practices and innovative technologies for ships in port and at sea to help reduce current and future emissions of greenhouse gases (GHGs), Nitrogen Oxides (NOx), Sulfur Oxides (SOx) and Particulate Matter (PM), taking into account the relevant IMO regulations.

Building the Arctic Marine Infrastructure

A. Addressing the Infrastructure Deficit: That the Arctic states should recognize that improvements in Arctic marine infrastructure are needed to enhance safety and environmental protection in support of sustainable development. Examples of infrastructure where critical improvements are needed include: ice navigation training; navigational charts; communications systems; port services, including reception facilities for ship-generated waste; accurate and timely ice information (ice centers); places of refuge; and icebreakers to assist in response.

B. Arctic Marine Traffic System: That the Arctic states should support continued development of a comprehensive Arctic marine traffic awareness system to improve monitoring and tracking of marine activity, to enhance data sharing in near real-time, and to augment vessel management service in order to reduce the risk of incidents, facilitate response and provide awareness of potential user conflict. The Arctic states should encourage shipping companies to cooperate in the improvement and development of national monitoring systems.

C. Circumpolar Environmental Response Capacity: That the Arctic states decide to continue to develop circumpolar environmental pollution response capabilities that are critical to protecting the unique Arctic ecosystem. This can be accomplished, for example, through circumpolar cooperation and agreement(s), as well as regional bilateral capacity agreements.

D. Investing in Hydrographic, Meteorological and Oceanographic Data: That the Arctic states should significantly improve, where appropriate, the level of and access to data and information in support of safe navigation and voyage planning in Arctic waters. This would entail increased efforts for: hydrographic surveys to bring Arctic navigation charts up to a level acceptable to support current and future safe navigation; and systems to support realtime acquisition, analysis and transfer of meteorological, oceanographic, sea ice and iceberg information.

Agreement on Cooperation on Marine Oil Pollution Preparedness and Response in the Arctic

The Government of Canada, the Government of the Kingdom of Denmark, the Government of the Republic of Finland, the Government of Iceland, the Government of the Kingdom of Norway, the Government of the Russian Federation, the Government of the Kingdom of Sweden, and the Government of the United States of America, hereinafter referred to as "the Parties",

Taking into account the relevant provisions of the 1982 United Nations Convention on the Law of the Sea,

Being Parties to the 1990 International Convention on Oil Pollution Preparedness, Response and Co-operation,

Taking also into account the 1969 International Convention Relating to Intervention on the High Seas in Cases of Oil Pollution Casualties,

Taking further into account the "polluter pays" principle as a general principle to be applied,

Recalling the 1996 Ottawa Declaration on the Establishment of the Arctic Council,

Highlighting that in the 2011 Nuuk Declaration on the occasion of the Seventh Ministerial Meeting of the Arctic Council, ministers representing the eight Arctic States decided to establish a Task Force to develop an international instrument on Arctic marine oil pollution preparedness and response,

Acknowledging the role of the International Maritime Organization, in particular in the development and adoption of additional rules and standards to address risks specific for operations in the Arctic environment,

Conscious of the threat from marine oil pollution to the vulnerable Arctic marine environment and to the livelihoods of local and indigenous communities,

Mindful that in the event of an oil pollution incident, prompt and effective action and cooperation among the Parties is essential in order to minimize damage that may result from such an incident,

Recognizing the challenges posed by harsh and remote Arctic conditions on oil pollution preparedness and response operations,

L. P. Hildebrand et al. (eds.), *Sustainable Shipping in a Changing Arctic*, WMU Studies in Maritime Affairs 7, https://doi.org/10.1007/978-3-319-78425-0

Mindful also of the increase in maritime traffic and other human activities in the Arctic region, including activity of Arctic residents and of people coming to the Arctic,

Mindful further that indigenous peoples, local communities, local and regional governments, and individual Arctic residents can provide valuable resources and knowledge regarding the Arctic marine environment in support of oil pollution preparedness and response,

Recognizing also the expertise and roles of various stakeholders relating to oil pollution preparedness and response,

Aware of the Parties' obligation to protect the Arctic marine environment and mindful of the importance of precautionary measures to avoid oil pollution in the first instance,

Recognizing further the importance of the Arctic marine ecosystem and of cooperation to promote and encourage the conservation and sustainable use of the marine and coastal environment and its natural resources,

Emphasizing the importance of exchanging information, data and experience in the field of marine oil pollution preparedness and response, especially regarding the Arctic environment, and on the effects of pollution on the environment, and of regularly conducting joint training and exercises, as well as joint research and development,

Have agreed as follows:

Article 1
Objective of this Agreement

The objective of this Agreement is to strengthen cooperation, coordination and mutual assistance among the Parties on oil pollution preparedness and response in the Arctic in order to protect the marine environment from pollution by oil.

Article 2
Terms and Definitions

For the purposes of this Agreement:

1. "Oil" means petroleum in any form including crude oil, fuel oil, sludge, oil refuse and refined products.
2. "Oil pollution incident" means an occurrence or series of occurrences having the same origin, which results or may result in a discharge of oil and which poses or may pose a threat to the marine environment, or to the coastline or related interests of one or more states, and which requires emergency action or other immediate response.
3. "Ship" means a vessel of any type whatsoever operating in the marine environment and includes hydrofoil boats, air-cushion vehicles, submersibles, and floating craft of any type.

Article 3
Scope of Application of this Agreement

1. This Agreement shall apply with respect to oil pollution incidents that occur in or may pose a threat to any marine area over which a State whose government is a Party to this Agreement exercises sovereignty, sovereign rights or jurisdiction, including its internal waters, territorial sea, exclusive economic zone and continental shelf, consistent with international law and above a southern limit as follows:

   Canada – marine areas above 60 degrees North;
   The Kingdom of Denmark, including Greenland and the Faroes – marine areas above the southern limit of the Greenland exclusive economic zone and the Faroese fisheries zone;
   Finland – marine areas above 63 degrees 30 minutes North;
   Iceland – marine areas above the southern limit of the exclusive economic zone of Iceland;
   Norway – marine areas above the Arctic Circle;
   The Russian Federation – marine areas above the coastlines of the White Sea, the Barents Sea, the Kara Sea, the Laptev Sea, the East Siberian Sea and the Chukchi Sea, and the mouths of the rivers flowing into these seas seaward of the baselines from which the breadth of the territorial sea is measured;
   Sweden – marine areas above 63 degrees 30 minutes North; and
   The United States of America – Marine areas seaward of the coastal baseline from the border between the United States and Canada at the Beaufort Sea along the north side of the mainland of Alaska to the Aleutian Islands, above 24 nautical miles south of the Aleutian Islands, and, in the Bering Sea, east of the limits of the exclusive economic zone of the United States.

2. Each Party shall also apply Articles 6, 7, 8, 10, and 15 and other provisions of this Agreement as appropriate to areas beyond the jurisdiction of any State, above the southern limit set forth in paragraph 1 of this Article, to the extent consistent with international law.
3. This Agreement shall not apply to any warship, naval auxiliary or other ship owned or operated by a State and used, for the time being, only on government non-commercial service. However, each Party shall ensure by the adoption of appropriate measures not impairing the operations or operational capabilities of such ships owned or operated by it, that such ships act in a manner consistent, so far as is reasonable and practicable, with this Agreement.

Article 4
Systems for Oil Pollution Preparedness and Response

1. Each Party shall maintain a national system for responding promptly and effectively to oil pollution incidents. This system shall take into account particular activities and locales most likely to give rise to or suffer an oil

pollution incident and anticipated risks to areas of special ecological significance, and shall include at a minimum a national contingency plan or plans for preparedness and response to oil pollution incidents. Such contingency plan or plans shall include the organizational relationship of the various bodies involved, whether public or private, taking into account guidelines developed pursuant to this Agreement and other relevant international agreements.

2. Each Party, as appropriate, in cooperation with other Parties and with the oil and shipping industries, port authorities and other relevant entities, shall establish:

   a. a minimum level of pre-positioned oil spill combating equipment, commensurate with the risk involved, and programs for its use;
   b. a program of exercises for oil pollution response organizations and training of relevant personnel;
   c. plans and communications capabilities for responding to an oil pollution incident; and
   d. a mechanism or arrangement to coordinate the response to an oil pollution incident with, if appropriate, the capabilities to mobilize the necessary resources.

Article 5
Authorities and Contact Points

1. Each Party's national system for responding promptly and effectively to oil pollution incidents shall include as a minimum the designation of:

   a. the competent national authority or authorities with responsibility for oil pollution preparedness and response;
   b. the national 24-hour operational contact point or points, which shall be responsible for the receipt and transmission of oil pollution reports; and
   c. an authority or authorities entitled to act on behalf of the Party to request assistance or to decide to render the assistance requested.

2. The entities designated by each Party pursuant to paragraph 1 of this Article are specified in Appendices to this Agreement. Each Party shall promptly inform the other Parties in writing through its competent national authority or authorities and through diplomatic channels of any changes to those designations. The Appendices to this Agreement shall be modified accordingly.

Article 6
Notification

1. Whenever a Party receives information on oil pollution, or possible oil pollution, it shall:

   a. assess the event to determine whether it is an oil pollution incident;
   b. assess the nature, extent and possible consequences of the oil pollution incident, including taking appropriate steps within available resources to identify possible sources; and

c. then, without delay, inform all States whose interests are affected or likely to be affected by such oil pollution incident, together with

   (i) details of its assessments and any action it has taken, or intends to take, to deal with the incident, including mitigation measures, and
   (ii) further information as appropriate, until the action taken to respond to the incident has been concluded or until joint action has been decided by such States.

2. When the severity of such oil pollution incident so justifies, the Party shall notify all the other Parties without unnecessary delay.

Article 7
Monitoring

1. Each Party shall endeavor to undertake appropriate monitoring activities in order to identify oil pollution incidents in areas under its jurisdiction and, to the extent feasible, in adjacent areas beyond the jurisdiction of any State.
2. In the event of an oil pollution incident, the Party or Parties affected shall, to the extent possible, monitor the incident to facilitate efficient and timely response operations and to minimize any adverse environmental impacts.
3. The Parties shall endeavor to cooperate in organizing and conducting monitoring, especially regarding transboundary oil pollution, inter alia, through conclusion of bilateral or multilateral agreements or arrangements.

Article 8
Requests for Assistance and Coordination and Cooperation in Response Operations

1. The Parties may request assistance from any other Party or Parties to respond to an oil pollution incident.
2. The Parties requesting assistance shall endeavor to specify the type and extent of assistance requested.
3. The Parties shall cooperate and provide assistance, which may include advisory services, technical support, equipment or personnel, for the purpose of responding to an oil pollution incident upon the request of any Party affected or likely to be affected.

Article 9
Movement and Removal of Resources across Borders

In accordance with applicable national and international law, each Party shall take the necessary legal or administrative measures to facilitate:

a. the arrival and utilization in, and departure from, its territory of ships, aircraft and other modes of transport engaged in responding to an oil pollution incident or transporting personnel, cargoes, materials and equipment required to deal with an oil pollution incident;

b. the expeditious movement into, through, and out of its territory of personnel, cargoes, materials, response supplies and other equipment referred to in subparagraph (a).

Article 10
Reimbursement of Costs of Assistance

1. Unless an agreement concerning the financial arrangements governing actions of the Parties to deal with oil pollution incidents has been concluded on a bilateral or multilateral basis prior to an oil pollution incident, the Parties shall bear the costs of their respective actions in dealing with pollution in accordance with subparagraph (a) or subparagraph (b). The principles laid down in this paragraph apply unless the Parties concerned otherwise agree in any individual case.
   a. If the action was taken by one Party at the express request of another Party, the requesting Party shall reimburse to the assisting Party the cost of its action. The requesting Party may cancel its request at any time, but in that case it shall bear the costs already incurred or committed by the assisting Party.
   b. If the action was taken by a Party on its own initiative, this Party shall bear the costs of its action.
2. Unless otherwise agreed, the costs of action taken by a Party at the request of another Party shall be fairly calculated according to the law and current practice of the assisting Party concerning the reimbursement of such costs.
3. The assisting Party shall be prepared to provide upon request documentation and information to the requesting Party on the assisting Party's estimated costs for the assistance and on the assisting Party's actual costs following the provision of any assistance. The Party requesting assistance and the assisting Party shall, where appropriate, cooperate in concluding any action in response to a compensation claim.
4. The provisions of this Agreement shall not be interpreted as in any way prejudicing the rights of Parties to recover from third parties the costs of actions to deal with pollution or the threat of pollution under other applicable rules of national and international law. Special attention shall be paid to international instruments and national law on liability and compensation for oil pollution damage.

Article 11
Joint Review of Oil Pollution Incident Response Operations

After a joint response operation, the Parties shall make best efforts to conduct a joint review of the operation, led by the Party or Parties that coordinated the operation. Where appropriate, and subject to relevant national law, Parties involved in a joint review should document their findings and conclusions and make the results of such joint review publicly available.

Article 12
Cooperation and Exchange of Information

1. The Parties shall promote cooperation and exchange of information that may serve to improve the effectiveness of oil pollution preparedness and response operations. Such cooperation and information exchange may include, inter alia, the topics identified in the Appendices to this Agreement.
2. Each Party, subject to its national law and international law, should endeavor to make information provided to other Parties under paragraph 1 of this Article publicly available.

Article 13
Joint Exercises and Training

1. The Parties shall promote cooperation and coordination by endeavoring to carry out joint exercises and training, including alerting or call-out exercises, table-top exercises, equipment deployment exercises, and other relevant activities.
2. Joint exercises and training should be designed to incorporate lessons learned.
3. Where appropriate, the Parties should include stakeholders in the planning and execution of joint exercises and training.
4. When conducting joint exercises and training, the Parties should apply the relevant provisions of this Agreement to the extent possible.

Article 14
Meetings of the Parties

1. The Parties shall meet no later than one year after the entry into force of this Agreement, as convened by the depositary, and from then on as decided by the Parties. At these meetings, the Parties shall review issues related to the implementation of this Agreement, adopt Appendices to this Agreement or modifications to the Appendices as provided in Article 20 of this Agreement, as appropriate, and consider any other issues as decided by the Parties. Parties may elect to convene such meetings in conjunction with meetings of the Arctic Council.
2. On a regular basis the Parties through their competent national authorities shall discuss and review operational issues related to the implementation of this Agreement, in cooperation, as appropriate, with relevant bodies including but not limited to the Arctic Council. Operational issues include, but are not limited to, cooperation and exchange of available information.

Article 15
Resources

1. Except as otherwise provided in Article 10 of this Agreement or otherwise agreed, each Party shall bear its own costs deriving from its implementation of this Agreement.
2. Implementation of this Agreement, except for Article 10, shall be subject to the capabilities of the Parties and the availability of relevant resources.

Article 16
Relationship with Other International Agreements

Nothing in this Agreement shall be construed as altering the rights or obligations of any Party under other relevant international agreements or customary international law as reflected in the 1982 United Nations Convention on the Law of the Sea.

Article 17
Non-Parties

Any Party may, where appropriate, seek cooperation with States not party to this Agreement that may be able to contribute to activities envisaged in this Agreement, consistent with international law.

Article 18
Settlement of Disputes

The Parties shall resolve any disputes concerning the application or interpretation of this Agreement through direct consultations.

Article 19
Amendments to this Agreement

1. This Agreement may be amended by written agreement of all the Parties.
2. An amendment shall enter into force 120 days after the date on which the depositary has received the last written notification through diplomatic channels that the Parties have completed the internal procedures required for its entry into force.

Article 20
Appendices

1. The Appendices to this Agreement do not constitute an integral part of this Agreement and are not legally binding.
2. At meetings of the Parties referred to in Article 14 of this Agreement, the Parties may adopt additional Appendices or modifications to existing Appendices, except

for those Appendices referred to in Article 5 of this Agreement, which may be modified as provided therein.

Article 21
Operational Guidelines

1. The Parties shall develop and maintain a set of Operational Guidelines to assist in the implementation of this Agreement. The Operational Guidelines will be included among the Appendices to this Agreement and be modified as appropriate.
2. The Operational Guidelines shall address, inter alia, the following topics:
   a. a system and formats for notification, requests for assistance, and other related information;
   b. provision of assistance, as well as coordination and cooperation in response operations involving more than one Party, including in areas beyond the jurisdiction of any State;
   c. movement and removal of resources across borders;
   d. procedures for conducting joint reviews of oil pollution incident response operations;
   e. procedures for conducting joint exercises and training; and
   f. reimbursement of costs of assistance.
3. In developing and modifying the Operational Guidelines, the Parties shall seek input from relevant stakeholders as appropriate.

Article 22
Provisional application, Entry into Force and Withdrawal

1. This Agreement may be applied provisionally by any signatory that provides a written statement to the depositary of its intention to do so. Any such signatory shall apply this Agreement provisionally from the date of its statement or from such other date as indicated in its statement.
2. This Agreement shall enter into force 30 days after the date of receipt by the depositary of the last written notification through diplomatic channels that the Parties have completed the internal procedures required for its entry into force.
3. Any Party may at any time withdraw from this Agreement by sending written notification thereof to the depositary through diplomatic channels at least six months in advance, specifying the effective date of its withdrawal. Withdrawal from this Agreement shall not affect its application among the remaining Parties.
4. Withdrawal from this Agreement by a Party shall not affect the obligations of that Party with regard to activities undertaken under this Agreement where those obligations have arisen prior to the effective date of withdrawal.

Article 23
Depositary

The Government of Norway shall be the depositary for this Agreement.

DONE at Kiruna this 15th day of May, 2013, in the English, French and Russian languages, all texts being equally authentic. The working language of this Agreement shall be English, the language in which this Agreement was negotiated.

Agreement of Cooperation on Aeronautical and Maritime Search and Rescue in the Arctic

The Government of Canada, the Government of the Kingdom of Denmark, the Government of the Republic of Finland, the Government of Iceland, the Government of the Kingdom of Norway, the Government of the Russian Federation, the Government of the Kingdom of Sweden, and the Government of the United States of America, hereinafter referred to as "the Parties",

Taking into account the relevant provisions of the 1982 United Nations Convention on the Law of the Sea,

Being Parties to the 1979 International Convention on Maritime Search and Rescue, hereinafter referred to as "the SAR Convention", and the 1944 Convention on International Civil Aviation, hereinafter referred to as "the Chicago Convention",

Noting the International Aeronautical and Maritime Search and Rescue Manual, hereinafter referred to as "the IAMSAR Manual",

Recalling the 1996 Ottawa Declaration on the Establishment of the Arctic Council,

Highlighting the 2009 Tromsø Declaration on the occasion of the Sixth Ministerial Meeting of the Arctic Council, which approved the establishment of a task force to develop and complete negotiation of an international instrument on cooperation on search and rescue operations in the Arctic,

Conscious of the challenges posed by harsh Arctic conditions on search and rescue operations and the vital importance of providing rapid assistance to persons in distress in such conditions,

Mindful of the increase in aeronautical and maritime traffic and other human activity in the Arctic, including activity of Arctic residents and of people coming to the Arctic,

Recognizing the great importance of cooperation among the Parties in conducting search and rescue operations,

Emphasizing the usefulness of exchanging information and experience in the field of search and rescue and of conducting joint training and exercises,

Have agreed as follows:

L. P. Hildebrand et al. (eds.), *Sustainable Shipping in a Changing Arctic*, WMU Studies in Maritime Affairs 7, https://doi.org/10.1007/978-3-319-78425-0

Article 1
Terms and Definitions

1. For purposes of this Agreement, the terms and definitions contained in Chapter 1 of the Annex to the SAR Convention and in Chapter 1 of Annex 12 to the Chicago Convention shall apply.
2. For purposes of this Agreement, "territory of a Party" shall mean the land area of a State, its internal waters and its territorial sea, including the airspace above those areas.

Article 2
Objective of this Agreement

The objective of this Agreement is to strengthen aeronautical and maritime search and rescue cooperation and coordination in the Arctic.

Article 3
Scope of Application of this Agreement

1. The delimitations of the aeronautical and maritime search and rescue regions relevant to this Agreement are specified in paragraph 1 of the Annex to this Agreement. The area in which each Party shall apply this Agreement is set forth in paragraph 2 of the Annex to this Agreement.
2. The delimitation of search and rescue regions is not related to and shall not prejudice the delimitation of any boundary between States or their sovereignty, sovereign rights or jurisdiction.
3. Each Party shall promote the establishment, operation and maintenance of an adequate and effective search and rescue capability within its area as set forth in paragraph 2 of the Annex to this Agreement.

Article 4
Competent Authorities of the Parties

1. The Competent Authorities of the Parties are specified in Appendix I to this Agreement.
2. Each Party shall promptly inform the other Parties in writing through diplomatic channels of any changes regarding its Competent Authorities.

Article 5
Agencies Responsible for Aeronautical and Maritime Search and Rescue

1. The agencies responsible for aeronautical and maritime search and rescue, hereinafter referred to as "search and rescue agencies", are specified in Appendix II to this Agreement.

2. Each Party, through its Competent Authorities, shall promptly inform the other Parties of any changes regarding its search and rescue agency or agencies.

Article 6
Rescue Coordination Centers

1. The list of aeronautical and/or maritime rescue coordination centers, hereinafter referred to as "RCCs", of the Parties for the purposes of this Agreement is contained in Appendix III to this Agreement.
2. Each Party, through its Competent Authorities, shall promptly inform the other Parties of any changes regarding its RCCs.

Article 7
Conduct of Aeronautical and Maritime Search and Rescue Operations

1. The SAR Convention and the Chicago Convention shall be used as the basis for conducting search and rescue operations under this Agreement.
2. The IAMSAR Manual provides additional guidelines for implementing this Agreement.
3. Without prejudice to paragraph 1 of this Article, the Parties shall conduct aeronautical and maritime search and rescue operations pursuant to this Agreement consistent with the following:

   (a) search and rescue operations conducted pursuant to this Agreement in the territory of a Party shall be carried out consistent with the laws and regulations of that Party;
   (b) if a search and rescue agency and/or RCC of a Party receives information that any person is, or appears to be, in distress, that Party shall take urgent steps to ensure that the necessary assistance is provided;
   (c) any Party having reason to believe that a person, a vessel or other craft or aircraft is in a state of emergency in the area of another Party as set forth in paragraph 2 of the Annex shall forward as soon as possible all available information to the Party or Parties concerned;
   (d) the search and rescue agency and/or RCC of a Party that has received information concerning a situation provided for in subparagraph (b) of this paragraph may request assistance from the other Parties;
   (e) the Party to whom a request for assistance is submitted shall promptly decide on and inform the requesting Party whether or not it is in a position to render the assistance requested and shall promptly indicate the scope and the terms of the assistance that can be rendered;
   (f) the Parties shall ensure that assistance be provided to any person in distress. They shall do so regardless of the nationality or status of such a person or the circumstances in which that person is found; and
   (g) a Party shall promptly provide all relevant information regarding the search and rescue of any person to the consular or diplomatic authorities concerned.

Article 8
Request to Enter the Territory of a Party

for Purposes of Search and Rescue Operations

1. A Party requesting permission to enter the territory of a Party or Parties for search and rescue purposes, including for refueling, shall send its request to a search and rescue agency and/or RCC of the relevant Party or Parties.
2. The Party receiving such a request shall immediately confirm such receipt. The receiving Party, through its RCCs, shall advise as soon as possible as to whether entry into its territory has been permitted and the conditions, if any, under which the mission may be undertaken.
3. The Party receiving such a request, as well as any Party through whose territory permission to transit is needed, shall apply, in accordance with its law and international obligations, the most expeditious border crossing procedure possible.

Article 9
Cooperation Among the Parties

1. The Parties shall enhance cooperation among themselves in matters relevant to this Agreement.
2. The Parties shall exchange information that may serve to improve the effectiveness of search and rescue operations. This may include, but is not limited to:

   (a) communication details;
   (b) information about search and rescue facilities;
   (c) lists of available airfields and ports and their refueling and resupply capabilities;
   (d) knowledge of fueling, supply and medical facilities; and
   (e) information useful for training search and rescue personnel.

3. The Parties shall promote mutual search and rescue cooperation by giving due consideration to collaborative efforts including, but not limited to:

   (a) exchange of experience;
   (b) sharing of real-time meteorological and oceanographic observations, analyses, forecasts, and warnings;
   (c) arranging exchanges of visits between search and rescue personnel;
   (d) carrying out joint search and rescue exercises and training;
   (e) using ship reporting systems for search and rescue purposes;
   (f) sharing information systems, search and rescue procedures, techniques, equipment, and facilities;
   (g) providing services in support of search and rescue operations;

(h) sharing national positions on search and rescue issues of mutual interest within the scope of this Agreement;
(i) supporting and implementing joint research and development initiatives aimed, inter alia, at reducing search time, improving rescue effectiveness, and minimizing risk to search and rescue personnel; and
(j) conducting regular communications checks and exercises, including the use of alternative means of communications for handling communication overloads during major search and rescue operations.

4. When conducting joint exercises, the Parties should apply the principles of this Agreement to the extent possible.

Article 10
Meetings of the Parties

The Parties shall meet on a regular basis in order to consider and resolve issues regarding practical cooperation. At these meetings they should consider issues including but not limited to:

(a) reciprocal visits by search and rescue experts;
(b) conducting joint search and rescue exercises and training;
(c) possible participation of search and rescue experts as observers at national search and rescue exercises of any other Party;
(d) preparation of proposals for the development of cooperation under this Agreement;
(e) planning, development, and use of communication systems;
(f) mechanisms to review and, where necessary, improve the application of international guidelines to issues concerning search and rescue in the Arctic; and
(g) review of relevant guidance on Arctic meteorological services.

Article 11
Joint Review of Search and Rescue Operations

After a major joint search and rescue operation, the search and rescue agencies of the Parties may conduct a joint review of the operation led by the Party that coordinated the operation.

Article 12
Funding

1. Unless otherwise agreed, each Party shall bear its own costs deriving from its implementation of this Agreement.
2. Implementation of this Agreement shall be subject to the availability of relevant resources.

Article 13
Annex

The Annex to this Agreement forms an integral part of this Agreement. All references to this Agreement are understood to include the Annex.

Article 14
Amendments

1. This Agreement may be amended by written agreement of all the Parties.
2. An amendment shall enter into force 120 days after the date on which the depositary has received the last written notification through diplomatic channels that the Parties have completed the internal procedures required for its entry into force.

Article 15
Amendment Procedure for the Annex

1. Notwithstanding Article 14 of this Agreement, any two Parties with adjacent search and rescue regions may by mutual agreement amend information contained in paragraph 1 of the Annex to this Agreement setting forth the delimitation between those regions. Such amendment shall enter into force 120 days after the date on which the depositary has received confirmation through diplomatic channels from both Parties that such mutual agreement has entered into force.
2. Notwithstanding Article 14 of this Agreement, any Party may amend that portion of paragraph 2 of the Annex to this Agreement that does not affect the area of any other Party and shall notify the depositary of any such amendment through diplomatic channels. Such amendment shall enter into force 120 days after the date of such notification.

Article 16
Relationship with Other Agreements

With the exception of paragraph 1 of the Annex to this Agreement, the provisions of this Agreement shall not affect the rights and obligations of Parties under agreements between them which are in force on the date of the entry into force of this Agreement.

Article 17
Settlement of Disputes

The Parties shall resolve any disputes concerning the application or interpretation of this Agreement through direct negotiations.

Article 18
Non-Parties

Any Party to this Agreement may, where appropriate, seek cooperation with States not party to this Agreement that may be able to contribute to the conduct of search and rescue operations, consistent with existing international agreements.

Article 19
Provisional Application, Entry into Force and Withdrawal

1. This Agreement may be applied provisionally by any signatory that provides a written statement to the depositary of its intention to do so. Any such signatory shall apply this Agreement provisionally from the date of its statement or from such other date as indicated in its statement.
2. This Agreement shall enter into force 30 days after the date of receipt by the depositary of the last written notification through diplomatic channels that the Parties have completed the internal procedures required for its entry into force.
3. Any Party may at any time withdraw from this Agreement by sending written notification thereof to the depositary through diplomatic channels at least six months in advance, specifying the effective date of its withdrawal. Withdrawal from this Agreement shall not affect its application among the remaining Parties.

Article 20
Depositary

The Government of Canada shall be the depositary for this Agreement.

DONE at this day of 2011, in the English, French and Russian languages, all texts being equally authentic. The working language of this Agreement shall be English, the language in which this Agreement was negotiated.

ANNEX
Scope of Application of this Agreement

1. The search and rescue regions relevant to this Agreement are delimited as follows:[1]

Canada – Denmark
The aeronautical and maritime search and rescue regions of Canada and Denmark shall be delimited by a continuous line connecting the following coordinates:

[1]The coordinates in this Annex use the World Geodetic System 1984 ("WGS 84"). All coordinates are connected by geodetic lines. The North Pole refers to the Geographic North Pole, located at 90 degrees North latitude, and the Arctic Circle refers to 66°33′44″N latitude.

- 58°30′00″N, 043°00′00″W;
- 58°30′00″N, 050°00′00″W;
- 63°00′00″N, 055°40′00″W;
- 65°00′00″N, 057°45′00″W;
- 76°00′00″N, 076°00′00″W;
- 78°00′00″N, 075°00′00″W;
- 82°00′00″N, 060°00′00″W; and
- Thence north to the North Pole.

Canada – United States of America

The aeronautical and maritime search and rescue regions of Canada and the United States of America shall be delimited by a continuous line connecting the following coordinates:

- 48°20′00″N, 145°00′00″W;
- 54°40′00″N, 140°00′00″W;
- 54°40′00″N, 136°00′00″W;
- 54°00′00″N, 136°00′00″W;
- 54°13′00″N, 134°57′00″W;
- 54°39′27″N, 132°41′00″W;
- 54°42′30″N, 130°36′30″W; and

Beaufort Sea and thence to the North Pole.

- North along the land border to the Beaufort Sea and thence to the North Pole.

Denmark – Iceland

The aeronautical and maritime search and rescue regions of Denmark and Iceland shall be delimited by a continuous line connecting the following coordinates:

- 58°30′00″N, 043°00′00″W;
- 63°30′00″N, 039°00′00″W;
- 70°00′00″N, 020°00′00″W;
- 73°00′00″N, 020°00′00″W; and
- 73°00′00″N, 000°00′00″E/W.

Denmark – Norway

The aeronautical and maritime search and rescue regions of Denmark and Norway shall be delimited by a continuous line connecting the following coordinates:

- 73°00′00″N, 000°00′00″E/W; and
- Thence north to the North Pole.

Finland – Norway

The land border between Finland and Norway shall be the limit of their respective search and rescue regions.

Finland – Sweden

The land border between Finland and Sweden shall be the limit of their respective search and rescue regions.

Finland – Russian Federation

The land border between Finland and the Russian Federation shall be the limit of their respective search and rescue regions.

Iceland – Norway

The aeronautical and maritime search and rescue regions of Iceland and Norway shall be delimited by a continuous line connecting the following coordinates:

- 66°33′44″N, 000°00′00″E/W; and
- 73°00′00″N, 000°00′00″E/W.

Norway – Sweden

The land border between Norway and Sweden shall be the limit of their respective search and rescue regions.

Norway – Russian Federation

The aeronautical and maritime search and rescue regions of Norway and the Russian Federation shall be delimited by the land border between Norway and the Russian Federation and then by a continuous line connecting the following coordinates:

- 69°47′41.42″N, 030°49′03.55″E;
- 69°58′45.49″N, 031°06′15.58″E;
- 70°05′58.84″N, 031°26′41.28″E;
- 70°07′15.20″N, 031°30′19.43″E;
- 70°11′51.68″N, 031°46′33.57″E;
- 70°16′28.95″N, 032°04′23.00″E;
- 72°27′51.00″N, 035°00′00.00″E; and
- Thence to the North Pole.

Russian Federation – United States of America

The aeronautical and maritime search and rescue regions of the Russian Federation and the United States of America shall be delimited by a continuous line connecting the following coordinates:

- 50°05′00″N, 159°00′00″E;
- 54°00′00″N, 169°00′00″E;
- 54°49′00″N, 170°12′00″E;
- 60°00′00″N, 180°00′00″E/W;
- 64°03′00″N, 172°12′00″W;
- 65°00′00″N, 168°58′24″W; and
- Thence north to the North Pole.

2. Each Party shall apply this Agreement in the following areas as encompassed by a continuous line, respectively:

Canada

- From the North Pole south to 82°00′00″N, 060°00′00″W;
- 78°00′00″N, 075°00′00″W;
- 76°00′00″N, 076°00′00″W;
- 65°00′00″N, 057°45′00″W;
- 63°00′00″N, 055°40′00″W;
- To the point where it intersects 60°00′00″N latitude;
- West along 60°00′00″N latitude until it intersects with Canada and the United States of America;
- North along the land border to the Beaufort
- Thence north to the North Pole.

Denmark

- From the North Pole south to 82°00′00″N, 060°00′00″W;
- 78°00′00″N, 075°00′00″W;
- 76°00′00″N, 076°00′00″W;
- 65°00′00″N, 057°45′00″W;
- 63°00′00″N, 055°40′00″W;
- 58°30′00″N, 050°00′00″W;
- 58°30′00″N, 043°00′00″W;
- 63°30′00″N, 039°00′00″W;
- 70°00′00″N, 020°00′00″W;
- 73°00′00″N, 020°00′00″W;
- 73°00′00″N, 000°00′00″E/W; and
- Thence north to the North Pole.

Finland

- From the tripoint at which the land borders of Finland, Norway and Sweden meet, south along the land border between Finland and Sweden to the point awhich that border intersects the Arctic Circle; East along the Arctic Circle to the point at which that border intersects the Arctic Circle;
- East along the Arctic Circle to the point at which the Arctic Circle intersects the land border between Finland and the Russian Federation;
- North along the land border between Finland and the Russian Federation to the tripoint at which the land borders of Finland, Norway and the Russian Federation meet; and
- Thence tot he tripoint at which the land borders of Finland, Norway and Sweden meet.

Iceland

- From 73°00′00″N, 000°00′00″E/W west to 73°00′00″N, 020°00′00″W;
- 70°00′00″N, 020°00′00″W;

- 63°30′00″N, 039°00′00″W;
- 58°30′00″N, 043°00′00″W;
- 58°30′00″N, 030°00′00″W;
- 61°00′00″N, 030°00′00″W;
- 61°00′00″N, 000°00′00″E/W; and
- Thence north to 73°00′00″N, 000°00′00″E/W;

Norway

- From the North Pole south to 66°33′44″N, 000°00′00″E/W;
- East along the Arctic Circle to the point at which the Arctic Circle intersects the land border between Sweden and Norway;
- North along the land border between Sweden and Norway to the tripoint at which the land borders of Sweden, Norway and Finland meet;
- East along the land border between Finland and Norway to the tripoint at which the land borders of Finland, Norway and the Russian Federation meet;
- North along the land border to the Barents Sea;
- 69°47′41.42″N, 030°49′03.55″E;
- 69°58′45.49″N, 031°06′15.58″E;
- 70°05′58.84″N, 031°26′41.28″E;
- 70°07′15.20″N, 031°30′19.43″E;
- 70°11′51.68″N, 031°46′33.57″E;
- 70°16′28.95″N, 032°04′23.00″E; and
- 72°27′51.00″N, 035°00′00.00″E;
- Thence north to the North Pole.

Russian Federation

- From the North Pole south to 72°27′51.00″N, 035°00′00.00″E;
- 70°16′28.95″N, 032°04′23.00″E;
- 70°11′51.68″N, 031°46′33.57″E;
- 70°07′15.20″N, 031°30′19.43″E;
- 70°05′58.84″N, 031°26′41.28″E;
- 69°58′45.49″N, 031°06′15.58″E;
- 69°47′41.42″N, 030°49′03.55″E south tot he point at which that meridian meets the land border between the RUSSIAN federation and Norway;
- South along the land border between the Russian Federation and Finland to the point at which that border intersects the Arctic Circle;
- East along the Arctic Circle until the Arctic circle intersects the 180°00′00″E/W meridian;
- 50°05′00″N, 159°00′00″E;
- 54°00′00″N, 169°00′00″E;
- 54°49′00″N, 170°12′00″E;
- 60°00′00″N, 180°00′00″E/W;
- 64°03′00″N, 172°12′00″W;
- 65°00′00″N, 168°58′24″W; and
- Thence north to the North Pole.

Sweden

- From the North Pole south to 65°00′00″N, 168°58′24″W;
- 64°03′00″N, 172°12′00″W;
- 60°00′00″N, 180°00′00″E/W;
- 54°49′00″N, 170°12′00″E;
- 54°00′00″N, 169°00′00″E;
- 50°05′00″N, 159°00′00″E;
- 50°05′00″N, 143°40′00″W;
- 54°40′00″N, 140°00′00″W;
- 54°40′00″N, 136°00′00″W;
- 54°00′00″N, 136°00′00″W;
- 54°13′00″N, 134°57′00″W;
- 54°39′27″N, 132°41′00″W;
- 54°42′30″N, 130°36′30″W;
- North along the land border tot he Beaufort Sea; and
- Thence north to the North Pole.

APPENDIX I
Competent Authorities

The Competent Authorities of the Parties are:

Canada – Minister of National Defence;
Denmark – Danish Maritime Authority;
Finland –Ministry of the Interior; Finnish Transport Safety Agency;
Iceland –Ministry of the Interior;
Norway –Ministry of Justice and the Police;
Russian Federation – Ministry of Transport of the Russian Federation; Ministry of the Russian Federation for Civil Defense, Emergency and Elimination of Consequences of Natural Disasters;
Sweden – Swedish Maritime Administration; and
United States of America – United States Coast Guard.

APPENDIX II
Search and Rescue Agencies

The search and rescue agencies of the Parties are:

Canada – Canadian Forces; Canadian Coast Guard;
Denmark – Danish Maritime Authority, Danish Transport Authority, Ministry of Fisheries – Faroe Islands;
Finland – Finnish Border Guard;
Iceland – Icelandic Coast Guard;
Norway – Joint Rescue Coordination Centre, Northern Norway (JRCC NN Bodø);
Russian Federation – Federal Air Transport Agency; Federal Agency for Marine and River Transport;
Sweden – Swedish Maritime Administration; and

United States of America – United States Coast Guard; United States Department of Defense.

APPENDIX III
Rescue Coordination Centers

The rescue coordination centers of the Parties are:

Canada – Joint Rescue Coordination Centre, Trenton;
Denmark – Maritime Rescue Coordination Center Grønnedal (MRCC Grønnedal); Rescue Coordination Center Søndrestrøm/Kangerlussuaq (RCC Søndrestrøm); Maritime Rescue and Coordination Center Torshavn (MRCC Torshavn);
Finland – Maritime Rescue Coordination Centre Turku (MRCC Turku); Aeronautical Rescue Coordination Centre Finland (ARCC Finland);
Iceland – Joint Rescue Coordination Center Iceland (JRCC Iceland);
Norway – Joint Rescue Coordination Centre, Northern Norway (JRCC NN Bodø);
Russian Federation – State Maritime Rescue Coordination Center (SMRCC); Main Aviation Coordination Center for Search and Rescue (MACC);
Sweden – Joint Rescue Coordination Center Gothenburg (JRCC Gothenburg); and
United States of America – Joint Rescue Coordination Center Juneau (JRCC Juneau); Aviation Rescue Coordination Center Elmendorf (ARCC Elmendorf).

Declaration Concerning the Prevention of Unregulated High Seas Fishing in the Central Arctic Ocean

Meeting in Oslo on 16 July 2015, Canada, the Kingdom of Denmark, the Kingdom of Norway, the Russian Federation and the United States of America continued discussions toward the implementation of interim measures to prevent unregulated fishing in the high seas portion of the central Arctic Ocean. They adopted the following Declaration:

We recognize that until recently ice has generally covered the high seas portion of the central Arctic Ocean on a year-round basis, which has made fishing in those waters impossible to conduct. We acknowledge that, due to climate change resulting in changes in ice distribution and related environmental phenomena, the marine ecosystems of the Arctic Ocean are evolving and that the effects of these changes are poorly understood. We note that the Arctic Ocean ecosystems until now have been relatively unexposed to human activities.

We recognize the crucial role of healthy marine ecosystems and sustainable fisheries for food and nutrition. We are aware that fish stocks in the Arctic Ocean may occur both within areas under the fisheries jurisdiction of the coastal States and in the high seas portion of the central Arctic Ocean, including straddling fish stocks. We note further that the ice cover in the Arctic Ocean has been diminishing in recent years, including over some of the high seas portion of the central Arctic Ocean.

We recognize that, based on available scientific information, commercial fishing in the high seas portion of the central Arctic Ocean is unlikely to occur in the near future and, therefore, that there is no need at present to establish any additional regional fisheries management organization for this area. Nevertheless, recalling the obligations of States under international law to cooperate with each other in the conservation and management of living marine resources in high seas areas, including the obligation to apply the precautionary approach, we share the view that it is desirable to implement appropriate interim measures to deter unregulated fishing in the future in the high seas portion of the central Arctic Ocean.

L. P. Hildebrand et al. (eds.), *Sustainable Shipping in a Changing Arctic*, WMU Studies in Maritime Affairs 7, https://doi.org/10.1007/978-3-319-78425-0

We recognize that subsistence harvesting of living marine resources is ongoing in some Arctic Ocean coastal States, and that traditional and local knowledge exists among the users of these resources. We desire to promote scientific research, and to integrate scientific knowledge with traditional and local knowledge, with the aim of improving the understanding of the living marine resources of the Arctic Ocean and the ecosystems in which they occur. We also recognize the interests of Arctic residents, particularly the Arctic indigenous peoples, in the proper management of living marine resources in the Arctic Ocean.

We therefore intend to implement, in the single high seas portion of the central Arctic Ocean that is entirely surrounded by waters under the fisheries jurisdiction of Canada, the Kingdom of Denmark in respect of Greenland, the Kingdom of Norway, the Russian Federation and the United States of America, the following interim measures:

- We will authorize our vessels to conduct commercial fishing in this high seas area only pursuant to one or more regional or subregional fisheries management organizations or arrangements that are or may be established to manage such fishing in accordance with recognized international standards.
- We will establish a joint program of scientific research with the aim of improving understanding of the ecosystems of this area and promote cooperation with relevant scientific bodies, including but not limited to the International Council for the Exploration of the Sea (ICES) and the North Pacific Marine Science Organization (PICES).
- We will promote compliance with these interim measures and with relevant international law, including by coordinating our monitoring, control and surveillance activities in this area.
- We will ensure that any non-commercial fishing in this area does not undermine the purpose of the interim measures, is based on scientific advice and is monitored, and that data obtained through any such fishing is shared.

We recall that an extensive international legal framework applies to the Arctic Ocean. These interim measures will neither undermine nor conflict with the role and mandate of any existing international mechanism relating to fisheries, including the North East Atlantic Fisheries Commission. Nor will these interim measures prejudice the rights, jurisdiction and duties of States under relevant provisions of international law as reflected in the 1982 United Nations Convention on the Law of the Sea, or the 1995 United Nations Agreement for the Implementation of the Provisions of the United Nations Convention on the Law of the Sea of 10 December 1982 relating to the Conservation and Management of Straddling Fish Stocks and Highly Migratory Fish Stocks, or alter the rights and obligations of States that arise from relevant international agreements.

In implementing these interim measures, we will continue to engage with Arctic residents, particularly the Arctic indigenous peoples, as appropriate.

We intend to continue to work together to encourage other States to take measures in respect of vessels entitled to fly their flags that are consistent with these interim measures.

We acknowledge the interest of other States in preventing unregulated high seas fisheries in the central Arctic Ocean and look forward to working with them in a broader process to develop measures consistent with this Declaration that would include commitments by all interested States.

Oslo, 16 July 2015

The Ilulissat Declaration Arctic Ocean Conference Ilulissat, Greenland, 27–29 May 2008

At the invitation of the Danish Minister for Foreign Affairs and the Premier of Greenland, representatives of the five coastal States bordering on the Arctic Ocean – Canada, Denmark, Norway, the Russian Federation and the United States of America – met at the political level on 28 May 2008 in Ilulissat, Greenland, to hold discussions. They adopted the following declaration:

The Arctic Ocean stands at the threshold of significant changes. Climate change and the melting of ice have a potential impact on vulnerable ecosystems, the livelihoods of local inhabitants and indigenous communities, and the potential exploitation of natural resources.

By virtue of their sovereignty, sovereign rights and jurisdiction in large areas of the Arctic Ocean the five coastal states are in a unique position to address these possibilities and challenges. In this regard, we recall that an extensive international legal framework applies to the Arctic Ocean as discussed between our representatives at the meeting in Oslo on 15 and 16 October 2007 at the level of senior officials. Notably, the law of the sea provides for important rights and obligations concerning the delineation of the outer limits of the continental shelf, the protection of the marine environment, including ice-covered areas, freedom of navigation, marine scientific research, and other uses of the sea. We remain committed to this legal framework and to the orderly settlement of any possible overlapping claims.

This framework provides a solid foundation for responsible management by the five coastal States and other users of this Ocean through national implementation and application of relevant provisions. We therefore see no need to develop a new comprehensive international legal regime to govern the Arctic Ocean. We will keep abreast of the developments in the Arctic Ocean and continue to implement appropriate measures.

The Arctic Ocean is a unique ecosystem, which the five coastal states have a stewardship role in protecting. Experience has shown how shipping disasters and subsequent pollution of the marine environment may cause irreversible disturbance of the ecological balance and major harm to the livelihoods of local inhabitants and indigenous communities. We will take steps in accordance with international law

L. P. Hildebrand et al. (eds.), *Sustainable Shipping in a Changing Arctic*, WMU Studies in Maritime Affairs 7, https://doi.org/10.1007/978-3-319-78425-0

both nationally and in cooperation among the five states and other interested parties to ensure the protection and preservation of the fragile marine environment of the Arctic Ocean. In this regard we intend to work together including through the International Maritime Organization to strengthen existing measures and develop new measures to improve the safety of maritime navigation and prevent or reduce the risk of ship-based pollution in the Arctic Ocean.

The increased use of Arctic waters for tourism, shipping, research and resource development also increases the risk of accidents and therefore the need to further strengthen search and rescue capabilities and capacity around the Arctic Ocean to ensure an appropriate response from states to any accident. Cooperation, including on the sharing of information, is a prerequisite for addressing these challenges. We will work to promote safety of life at sea in the Arctic Ocean, including through bilateral and multilateral arrangements between or among relevant states.

The five coastal states currently cooperate closely in the Arctic Ocean with each other and with other interested parties. This cooperation includes the collection of scientific data concerning the continental shelf, the protection of the marine environment and other scientific research. We will work to strengthen this cooperation, which is based on mutual trust and transparency, inter alia, through timely exchange of data and analyses.

The Arctic Council and other international fora, including the Barents Euro-Arctic Council, have already taken important steps on specific issues, for example with regard to safety of navigation, search and rescue, environmental monitoring and disaster response and scientific cooperation, which are relevant also to the Arctic Ocean. The five coastal states of the Arctic Ocean will continue to contribute actively to the work of the Arctic Council and other relevant international fora.

Ilulissat, 28 May 2008

Guidelines for Ships Operating in Polar Waters (2010 Edition), International Maritime Organization

N.B. only relevant parts of the Guidelines are incorporated herewith (footnote number modified)

Chapter 1
General

1.1 **Application**

1.1.1 Except where specifically stated otherwise, these Guidelines provide guidance for ships operating in Antarctic waters or while engaged in international voyages in Arctic waters.
1.1.2 Part A of the Guidelines provides guidance for new Polar Class ships.
1.1.3 Parts B, C and D of the Guidelines provide guidance for Polar Class and all other ships.

L. P. Hildebrand et al. (eds.), *Sustainable Shipping in a Changing Arctic*, WMU Studies in Maritime Affairs 7, https://doi.org/10.1007/978-3-319-78425-0

Table 1.1 – *Class descriptions*

POLAR CLASS	GENERAL DESCRIPTION
PC 1	Year-round operation in all ice-covered waters
PC 2	Year-round operation in moderate multi-year ice conditions
PC 3	Year-round operation in second-year ice which may include old ice inclusions
PC 4	Year-round operation in thick first-year ice which may include old ice inclusions
PC 5	Year-round operation in medium first-year ice which may include old ice inclusions
PC 6	Summer/autumn operation in medium first-year ice which may include old ice inclusions
PC 7	Summer/autumn operation in thin first-year ice which may include old ice inclusions

Note: Ice descriptions follow the WMO Sea-ice nomenclature.[2]

1.1.4 All Polar Class ships and the equipment to be carried in accordance with the Guidelines should be designed, constructed and maintained in compliance with applicable national standards of the Administration or the appropriate requirements of a recognized organization which provide an equivalent level of safety[3] for its intended service. Special attention should be drawn to the need for winterization aspects. Ships intending to operate as an icebreaker are to receive special consideration.

1.1.5 The structures, equipment and arrangements essential for the safety and operation of the ship should take account of the anticipated temperatures.

1.1.6 Special attention should be given to essential operating equipment and systems and safety equipment and systems. For example, the potential for ice building up inside the ballast tanks and sea chests should be considered. The life-saving and fire-extinguishing equipment specified in part B of the Guidelines, when stored or located in an exposed position, should be of a type that is rated to perform its design functions at the minimum anticipated air temperature. In particular, attention is drawn to the inflation of life saving equipment and the starting of engines in lifeboats and rescue boats.

1.1.7 Operations in polar waters should take due account of factors such as: ship class, environmental conditions, icebreaker escort, prepared tracks, short or local routes, crew experience, support technology and services such as ice-mapping, availability of hydrographic information, communications, safe ports, repair facilities and other ships in convoy.

[2]The WMO Sea-ice nomenclature is available at http://www.jcomm-services.org.

[3]Refer to SOLAS chapter II-1 and to the IACS Unified Requirements concerning Polar Class.

1.1.8 Equipment, fittings, materials, appliances and arrangements may deviate from the provisions of the Guidelines provided that their replacement is at least as effective as that specified in the Guidelines.

1.1.9 The provisions of the Guidelines do not apply to any warship, naval auxiliary, other vessels or aircraft owned or operated by a State and used, for the time being, only on government non -commercial service. However, each State should ensure, by the adoption of appropriate measures not impairing operations or operational capabilities of such vessels or aircraft owned or operated by it, that such vessels or aircraft act in a manner consistent, so far as is reasonable and practicable, with the Guidelines.

1.2 **Ice navigator**

1.2.1 All ships operating in polar ice-covered waters should carry at least one Ice Navigator qualified in accordance with chapter 14. Consideration should also be given to carrying an Ice Navigator when planning voyages into polar waters.

1.2.2 Continuous monitoring of ice conditions by an Ice Navigator should be available at all times while the ship is underway and making way in the presence of ice.[4]

Part A
Construction provisions

Chapter 2
Structures

2.1 **General**

2.1.1 All ships should have structural arrangements adequate to resist the global and local ice loads characteristic of their Polar Class.*

2.1.2 Each area of the hull and all appendages should be strengthened to resist design structure/ice interaction scenarios applicable to each case.

2.1.3 Structural arrangements should aim to limit damage resulting from accidental overloads to local areas.

2.1.4 Polar Class ships may experience in-service structural degradation at an accelerated rate. Structural surveys should, therefore, cover areas identified as being at high risk of accelerated degradation, and areas where physical evidence such as coating breakdown indicates a potential for high wastage rates.

[4]Refer to the Guidelines for voyage planning, as adopted by resolution A.893(21), and the Guidelines on voyage planning for passenger ships operating in remote areas, as adopted by resolution A.999(25).

2.2 **Materials**

2.2.1 Materials used in ice-strengthened and other areas of the hull should be suitable for operation in the environment that prevails at their location.
2.2.2 Materials used in ice-strengthened areas should have adequate ductility to match the selected structural design approach.
2.2.3 Abrasion and corrosion resistant coatings and claddings used in ice-strengthened areas should be matched to the anticipated loads and structural response.

Chapter 3
Subdivision and stability

3.1 **General**

3.1.1 Account should be taken of the effect of icing in the stability calculations in accordance with the 2008 IS Code.

3.2 Intact stability in ice

3.2.1 Suitable calculations should be carried out and/or tests conducted to demonstrate the following:

.1 the ship, when operated in ice within approved limitations, during a disturbance causing roll, pitch, heave or heel due to turning or any other cause, should maintain sufficient positive stability; and
.2 ships of Polar Classes 1 to 3 and icebreakers of all classes, when riding up in ice and remaining momentarily poised at the lowest stem extremity, should maintain sufficient positive stability.

3.2.2 "Sufficient positive stability" in paragraphs 3.2.1.1 and 3.2.1.2 means that the ship is in a state of equilibrium with a positive metacentric height of at least 150 mm, and a line 150 mm below the edge of the freeboard deck as defined in the applicable provisions of the ICLL is not submerged.
3.2.3 For performing stability calculations on ships that ride up onto the ice, the ship should be assumed to remain momentarily poised at the lowest stem extremity as follows:

.1 for a regular stem profile, at the point at which the stem contour is tangent to the keel line;
.2 for a stem fitted with a structurally defined skeg, at the point at which the stem contour meets the top of the skeg;
.3 for a stem profile where the skeg is defined by shape alone, at the point at which the stem contour tangent intersects the tangent of the skeg; or
.4 for a stem profile of novel design, the position should be specially considered.

Meeting on High Seas Fisheries in the Central Arctic Ocean, Reykjavik, Iceland, 15–18 March 2017

Chairman's Statement[5]

Introduction

Delegations from Canada, the People's Republic of China, the Kingdom of Denmark in respect of the Faroe Islands and Greenland, the European Union, Iceland, Japan, the Republic of Korea, the Kingdom of Norway, the Russian Federation and the United States of America met in Reykjavik, Iceland, from 15 – 18 March 2017, to continue discussions concerning the prevention of unregulated fishing in the high seas area of the central Arctic Ocean and related scientific matters.

The meeting followed previous talks that took place in Washington, D.C. from 1–3 December 2015 and from 19–21 April 2016, in Iqaluit, Canada from 6–8 July 2016, and in Tórshavn, The Faroe Islands, from 29 November to 1 December 2016.

All delegations reaffirmed their commitment to prevent unregulated high seas fishing in the central Arctic Ocean as well as a commitment to promote the conservation and sustainable use of living marine resources and to safeguard healthy marine ecosystems in the central Arctic Ocean. Most delegations view this as part of a "stepwise" process in advance of establishing in the future one or more additional regional or subregional fisheries management organizations or arrangements for this area.

Delegations worked on the basis of a Chairman's Text circulated in March 2017 that was in the format of a legally binding agreement. Delegations made considerable progress in resolving differences of view on many issues under discussion, such that only a small number of key provisions remained to be agreed. There was a general commitment to conclude the negotiations in the near future.

[5]This Chairman's Statement attempts to capture the basic elements of the meeting but does not necessarily reflect the views of any individual delegation.

L. P. Hildebrand et al. (eds.), *Sustainable Shipping in a Changing Arctic*, WMU Studies in Maritime Affairs 7, https://doi.org/10.1007/978-3-319-78425-0

Summary of Negotiations

On the basis of the Chairman's text and many proposals made during the meeting, delegations made considerable progress in resolving differences of view on many issues under negotiation.

Recognizing that "nothing is agreed until everything is agreed," delegations resolved language in the draft Agreement concerning the use of terms, its objective, many of the measures that would apply under the draft agreement, all provisions relating to a Joint Program of Scientific Research and Monitoring, the value of incorporating indigenous peoples' knowledge, all provisions relating to exploratory fishing, dispute settlement and most provisions concerning signature, accession, entry into force, withdrawal, and relation to other agreements. To the extent that some issues remain to be resolved among these provisions, there was a belief that this would be accomplished easily in the near future. Delegations welcomed with appreciation Canada's offer to serve as the Depositary.

Delegations also made progress on other elements of the draft Agreement that nevertheless remain under discussion for the near future. These include:

- a description or definition of the Agreement Area;
- the conditions under which a decision might be made to commence negotiations on an agreement to establish one or more additional regional or subregional fisheries management organizations or arrangements for the high seas portion of the central Arctic Ocean;
- the possibility to adopt other conservation and management measures that could apply after such negotiations have commenced; and
- decision-making procedures.

The Way Forward

The Chairman circulated an updated text immediately following the end of the meeting. The Chairman also offered to circulate no later than 24 March 2017 his recommendations for resolving the remaining issues under negotiation.

If all delegations can accept those recommendations within two months, there will be no need for another round of negotiations, although a meeting of experts may be convened to conduct a legal and technical review of the draft Agreement.

If one or more delegations cannot accept those recommendations, it is anticipated that another round of negotiations will occur in the near future with a view to finalizing the text.

Delegations expressed their sincere gratitude to the Government and people of Iceland for their excellent work in hosting and organizing the meeting and for their warm hospitality.

Printed by Printforce, the Netherlands